Lecture Notes in Computer Science 4679

Commenced Publication in 1973
Founding and Former Series Editors:
Gerhard Goos, Juris Hartmanis, and Jan van Leeuwen

Editorial Board

David Hutchison
  Lancaster University, UK
Takeo Kanade
  Carnegie Mellon University, Pittsburgh, PA, USA
Josef Kittler
  University of Surrey, Guildford, UK
Jon M. Kleinberg
  Cornell University, Ithaca, NY, USA
Friedemann Mattern
  ETH Zurich, Switzerland
John C. Mitchell
  Stanford University, CA, USA
Moni Naor
  Weizmann Institute of Science, Rehovot, Israel
Oscar Nierstrasz
  University of Bern, Switzerland
C. Pandu Rangan
  Indian Institute of Technology, Madras, India
Bernhard Steffen
  University of Dortmund, Germany
Madhu Sudan
  Massachusetts Institute of Technology, MA, USA
Demetri Terzopoulos
  University of California, Los Angeles, CA, USA
Doug Tygar
  University of California, Berkeley, CA, USA
Moshe Y. Vardi
  Rice University, Houston, TX, USA
Gerhard Weikum
  Max-Planck Institute of Computer Science, Saarbruecken, Germany

T0189746

Alan L. Yuille   Song-Chun Zhu
Daniel Cremers   Yongtian Wang (Eds.)

Energy Minimization Methods in Computer Vision and Pattern Recognition

6th International Conference, EMMCVPR 2007
Ezhou, China, August 27-29, 2007
Proceedings

 Springer

Volume Editors

Alan L. Yuille
University of California, Los Angeles, USA
E-mail: yuille@stat.ucla.edu

Song-Chun Zhu
University of California, Los Angles, USA
E-mail: sczhu@stat.ucla.edu

Daniel Cremers
University of Bonn, Germany
E-mail: dcremers@cs.uni-bonn.de

Yongtian Wang
Beijing Insitute of Technology, China
E-mail: wyt@bit.edu.cn

Library of Congress Control Number: 2007932816

CR Subject Classification (1998): I.5, I.4, I.2.10, I.3.5, F.2.2, F.1.1

LNCS Sublibrary: SL 6 – Image Processing, Computer Vision, Pattern Recognition, and Graphics

ISSN   0302-9743
ISBN-10   3-540-74195-X Springer Berlin Heidelberg New York
ISBN-13   978-3-540-74195-4 Springer Berlin Heidelberg New York

This work is subject to copyright. All rights are reserved, whether the whole or part of the material is concerned, specifically the rights of translation, reprinting, re-use of illustrations, recitation, broadcasting, reproduction on microfilms or in any other way, and storage in data banks. Duplication of this publication or parts thereof is permitted only under the provisions of the German Copyright Law of September 9, 1965, in its current version, and permission for use must always be obtained from Springer. Violations are liable to prosecution under the German Copyright Law.

Springer is a part of Springer Science+Business Media

springer.com

© Springer-Verlag Berlin Heidelberg 2007
Printed in Germany

Typesetting: Camera-ready by author, data conversion by Scientific Publishing Services, Chennai, India
Printed on acid-free paper   SPIN: 12107827   06/3180   5 4 3 2 1 0

Preface

This volume contains the papers presented at the Sixth International Conference on Energy Minimization Methods on Computer Vision and Pattern Recognition (EMMCVPR 2007), held at the Lotus Hill Institute, Ezhou, Hubei, China, August 27–29, 2007. The motivation for this conference is the realization that many problems in computer vision and pattern recognition can be formulated in terms of probabilistic inference or optimization of energy functions. EMMCVPR 2007 addressed the critical issues of representation, learning, and inference. Important new themes include probabilistic grammars, image parsing, and the use of datasets with ground-truth to act as benchmarks for evaluating algorithms and as a way to train learning algorithms. Other themes include the development of efficient inference algorithms using advanced techniques from statistics, computer science, and applied mathematics.

We received 140 submissions for this workshop. Each paper was reviewed by three committee members. Based on these reviews we selected 22 papers for oral presentation and 15 papers for poster presentation. This book makes no distinction between oral and poster papers. We have organized these papers in seven sections on algorithms, applications, image parsing, image processing, motion, shape, and three-dimensional processing.

Finally, we thank those people who helped make this workshop happen. We acknowledge the Program Committee for their help in reviewing the papers. We are grateful to Kent Shi and Jose Hales-Garcia for their help with organizing Web pages and dealing with the mechanics of collecting reviews, organizing registration, and so no. We also thank Mrs. Wenhau Xia at Lotus Hill Institute for her assistance. We are very grateful to Alfred Hofmann at Springer for agreeing to publish this volume in the *Lecture Notes in Computer Science* series.

June 2007

Alan Yuille
Song-Chun Zhu
Daniel Cremers
Yongtian Wang

Organization

Co-chairs

Alan L. Yuille, University of California, Los Angeles, USA
Song-Chun Zhu, University of California, Los Angeles, USA; Lotus Hill Institute, China
Daniel Cremers, University of Bonn, Genmany
Yongtian Wang, Beijing Institute of Technology, China

Program Committee

Serge Belongie, University of California, San Diego, USA
Andrew Blake, Microsoft Research Cambridge, UK
Yuri Boykov, University of Western Ontario, USA
Thomas Brox, University of Bonn, Germany
Joachim Buhmann, Swiss Federal Institute of Technology, Switzerland
Xilin Chen, Institute of Computing Technology, CAS, China
Jim Clark, McGill University, Canada
Laurent Cohen, Universite Paris IX Dauphine, France
Tim Cootes, University of Manchester, UK
Jason Corso, University of California, Los Angeles, USA
James Coughlan, Smith-Kettlewell Eye Research Institute, USA
Frank Dellaert, Georgia Institute of Technology, USA
Sven Dickinson, University of Toronto, Canada
Paolo Favaro, Heriot-Watt University, UK
Mário Figueiredo, Instituto Superior Técnico, Portugal
Daniel Freedman, Rensselaer Polytechnic Institute, USA
Wen Gao,Peking University, China
Feng Han, Sarnoff Corporation, USA
Edwin Hancock, University of York, USA
Martial Hebert, Carnegie Mellon University, USA
Zhiguo Jiang ,Bei-Hang University ,China
Dan Kersten, University of Minnesota, USA
Iasonas Kokkinos, National Tech University of Athens, Greece
Vladimir Kolmogorov, University College London, UK
Fei Fei Li, Princeton University, USA
Stan Li, Institute of Automation, CAS, China
Xiuwen Liu, Florida State University, USA
Yanxi Liu, Carnegie Mellon University, USA
HongJing Lu, University of Hong Kong, China
Petros Maragos, National Technical University of Athens, Greece
Stephen Maybank, Birkbeck College, University of London, UK

Dimitris Metaxas, State University of New Jersey, USA
Marcello Pelillo, Università Ca' Foscari di Venezia, Italy
Anand Rangarajan, University of Florida, USA
Nong Sang, HuaZhong University of Science and Technology, China
Guillermo Sapiro, University of Minnesota, USA
Christoph Schnoerr, University of Mannheim, Germany
Bernard Schoelkopf, MPI for Biological Cybernetics, Germany
Eitan Sharon, University of California, Los Angeles, USA
Jianbo Shi, University of Pennsylvania, USA
Anuj Srivastava, Florida State University, USA
Larry Staib, Yale University, USA
Xiaoou Tang, Microsoft Research Asia,China
Philip Torr, Oxford Brookes University, UK
Alain Trouve, Ecole Normale Supérieure de Cachan, France
Zhuowen Tu, University of California, Los Angeles, USA
Baba Vemuri, University of Florida, USA
Alessandro Verri, Università degli Studi di Genova, Italy
Luminita Vese, University of California, Los Angeles, USA
Yizhou Wang, Palo Alto Research Center, USA
Ying Wu, Northwestern University, USA
Laurent Younes, Johns Hopkins University, USA
Changshui Zhang, Tsinghua University,China
Tianxu Zhang, HuaZhong University of Science and Technology, China

Sponsors

The Lotus Hill Institute
The International Association for Pattern Recognition

Table of Contents

Algorithms

Applications to Faces and Text

Image Parsing

Image Processing

Motion Analysis

Shape Analysis

Three-Dimensional Processing

An Effective Multi-level Algorithm Based on Simulated Annealing for Bisecting Graph

Lingyu Sun[1] and Ming Leng[2]

[1] Department of Computer Science,
Jinggangshan College, Ji'an, PR China 343009
[2] School of Computer Engineering and Science,
Shanghai University, Shanghai, PR China 200072
sunlingyu@jgsu.edu.cn,lengming@shu.edu.cn

Abstract. Partitioning is a fundamental problem in diverse fields of study such as knowledge discovery, data mining, image segmentation and grouping. The min-cut bipartitioning problem is a fundamental graph partitioning problem and is NP-Complete. In this paper, we present an effective multi-level algorithm based on simulated annealing for bisecting graph. The success of our algorithm relies on exploiting both the simulated annealing procedure and the concept of the graph core. Our experimental evaluations on 18 different graphs show that our algorithm produces encouraging solutions compared with those produced by MeTiS that is a state-of-the-art partitioner in the literature.

1 Introduction

Partitioning is a fundamental problem with extensive applications to many areas using a graph model, including VLSI design [1], knowledge discovery [2], data mining [3],[4], image segmentation and grouping [5],[6]. For example, inspired by spectral graph theory, Shi and Malik [6] formulate visual grouping as a graph partitioning problem. The nodes of the graph are image pixels. The edges between two nodes correspond to the strength with which these two nodes belong to one group. In image segmentation, the weights on the edges of the graph corresponds to how much two pixels agree in brightness, color, etc. Intuitively, the criterion for partitioning the graph will be to minimize the sum of weights of connections across the groups and maximize the sum of weights of connections within the groups. The *min-cut bipartitioning problem* is a fundamental partitioning problem and is NP-Complete [7]. The survey by Alpert and Kahng [1] provides a detailed description and comparison of various such schemes which can be classified as *move-based* approaches, *geometric representations*, *combinatorial* formulations, and *clustering* approaches.

Most existing partitioning algorithms are heuristics in nature and they seek to obtain reasonably good solutions in a reasonable amount of time. Kernighan and Lin (KL) [8] proposed a heuristic algorithm for partitioning graphs. The KL algorithm is an iterative improvement algorithm that consists of making several improvement passes. It starts with an initial bipartitioning and tries to

A.L. Yuille et al. (Eds.): EMMCVPR 2007, LNCS 4679, pp. 1–12, 2007.
© Springer-Verlag Berlin Heidelberg 2007

improve it by every pass. A pass consists of the identification of two subsets of vertices, one from each part such that can lead to an improved partitioning if the vertices in the two subsets switch sides. Fiduccia and Mattheyses (FM) [9] proposed a fast heuristic algorithm for bisecting a weighted graph by introducing the concept of cell *gain* into the KL algorithm. These algorithms belong to the class of *move-based* approaches in which the solution is built iteratively from an initial solution by applying a move or transformation to the current solution. Move-based approaches are the most frequently combined with stochastic hill-descending algorithms such as those based on Tabu Search[10],[11], Genetic Algorithms [12], Neural Networks [13], Ant Colony Optimization[14], Particle Swarm Optimization[15], Swarm Intelligence[16] etc., which allow movements towards solutions worse than the current one in order to escape from local minima.

As the problem sizes reach new levels of complexity, a new class of graph partitioning algorithms have been developed that are based on the multi-level paradigm. The multi-level graph partitioning schemes consist of three phases [17],[18],[19]. The *coarsening phase* is to reduce the size of the graph by collapsing vertex and edge until its size is smaller than a given threshold. The *initial partitioning phase* is to compute initial partition of the coarsest graph. The *uncoarsening phase* is to project successively the partition of the smaller graph back to the next level finer graph while applying an iterative refinement algorithm.

In this paper, we present a multi-level algorithm which integrates a new simulated annealing-based refinement approach and an effective matching-based coarsening scheme. Our work is motivated by the multi-level refined mixed simulated annealing and tabu search algorithm(MLrMSATS) of Gil which can be considered as a hybrid heuristic with additional elements of a tabu search in a simulated annealing algorithm for refining the partitioning in [20] and Karypis who introduces the concept of the graph *core* for coarsening the graph in [19] and supplies **MeTiS** [17], distributed as open source software package for partitioning unstructured graphs. We test our algorithm on 18 graphs that are converted from the hypergraphs of the ISPD98 benchmark suite [21]. Our comparative experiments show that our algorithm produces excellent partitions that are better than those produced by **MeTiS** in a reasonable time.

The rest of the paper is organized as follows. Section 2 provides some definitions and describes the notation used throughout the paper. Section 3 describes the motivation behind our algorithm. Section 4 presents an effective multi-level simulated annealing refinement algorithm. Section 5 experimentally evaluates our algorithm and compares it with **MeTiS**. Finally, Section 6 provides some concluding remarks and indicates the directions for further research.

2   Mathematical Description

A graph $G=(V,E)$ consists of a set of vertices V and a set of edges E such that each edge is a subset of two vertices in V. Throughout this paper, n and m denote the number of vertices and edges respectively. The vertices are numbered from 1 to n and each vertex $v \in V$ has an integer weight $S(v)$. The edges are numbered from *1* to m and each edge $e \in E$ has an integer weight $W(e)$. A decomposition of a graph V into two disjoint subsets V^1 and V^2, such that $V^1 \cup V^2 = V$ and $V^1 \cap V^2 = \varnothing$, is called a *bipartitioning* of V. Let $S(A) = \sum_{v \in A} S(v)$ denotes the size of a subset $A \subseteq V$. Let ID_v be denoted as v's *internal degree* and is equal to the sum of the edge-weights of the adjacent vertices of v that are in the same side of the partitioning as v, and v's *external degree* denoted by ED_v is equal to the sum of edge-weights of the adjacent vertices of v that are in different sides. The *cut* of a *bipartitioning* $P = \{V^1, V^2\}$ is the sum of weights of edges which contain two vertices in V^1 and V^2 respectively. Naturally, vertex v belongs at the boundary if and only if $ED_v > 0$ and the *cut* of P is also equal to $0.5 \sum_{v \in V} ED_v$.

Given a balance constraint b, the *min-cut bipartitioning problem* seeks a solution $P = \{V^1, V^2\}$ that minimizes *cut(P)* subject to $(1-b)S(V)/2 \leq S(V^1), S(V^2) \leq (1+b)S(V)/2$. A *bipartitioning* is *bisection* if b is as small as possible. The task of minimizing *cut(P)* can be considered as the *objective* and the requirement that solution P will be of the same size can be considered as the *constraint*.

3 Motivation

Simulated annealing belongs to the probabilistic and iterative class of algorithms. It is a combinatorial optimization technique that is analogous to the annealing process used for metals [22]. The metal is heated to a very high temperature, so the atoms gain enough energy to break chemical bonds and become free to move. The metal is then carefully cooled down so that its atoms crystallize into high ordered state. In simulated annealing, the combinatorial optimization cost function is analogous to the energy $E(s)$ of a system in state s which must be minimized to achieve a stable system.

The main idea of simulated annealing is as follows: Starting from an initial configuration, different configurations of the system states are generated at random. A *perturbation* of a system state consists of reconfiguring the system from its current state to a next state within a neighborhood of the solution space. The change in energy cost between the two configurations is determined and used to compute the *probability* p of the system moving from the present state to the next. The *probability* p is given by $\exp(-\frac{\triangle E}{T})$, where $\triangle E$ is the increase in the energy cost and T is the temperature of the system. If $\triangle E$ is negative, then the change in state is always accepted. If not, then a random number r between 0 and 1 is generated and the new state of the system is accepted if $r \leq p$, else the system is returned to its original state. Initially, the temperature is high meaning that a large number of *perturbations* are accepted. The temperature is reduced gradually according to a *cooling schedule*, while allowing the system to reach equilibrium at each temperature through the cooling process.

In [23], Gil proposed the refinement of mixed simulated annealing and tabu search algorithm(RMSATS) that allows the search process to escape from local minima by the simulated annealing procedure, while simultaneously the occurrence of cycles is prevented by a simple tabu search strategy. At each iteration of

RMSATS, the hybrid heuristic strategy is used to obtain a new partitioning $\bar{s}$ in the neighbourhood, $N(s)$, of the current partitioning s through moving vertex v to the other side of the partitioning s. Every feasible partitioning, $\bar{s} \in N(s)$, is evaluated according to the cost function $c(\bar{s})$ to be optimized, thus determining a change in the value of the cost function, $c(\bar{s}) - c(s)$. The problem with local search techniques and hill climbing is that the searching may stop at local optimum. In order to overcome this drawback and reach the global optimum, RMSATS must sometimes accept the worse partitioning to jump out from a local optimum. Therefore, admissible moves are applied to the current partitioning allowing transitions that increase the cost function as in simulated annealing. When a move increasing the cost function is accepted, the reverse move should be forbidden during some iterations in order to avoid cycling, as in tabu search. In [20], Gil presents the MLrMSATS approach that is enhancement of the RMSATS algorithm with the multi-level paradigm and uses the RMSATS algorithm during the *uncoarsening and refinement phase* to improve the quality of the finer graph $G_l(V_l, E_l)$ partitioning $P_{G_l} = \{V_l^1, V_l^2\}$ which is projected from the partitioning $P_{G_{l+1}} = \{V_{l+1}^1, V_{l+1}^2\}$ of the coarser graph $G_{l+1}(V_{l+1}, E_{l+1})$.

In this paper, we present a new multi-level simulated annealing refinement algorithm(MLSAR) that combines the simulated annealing procedure with a boundary refinement policy. It has distinguishing features which are different from the MLrMSATS algorithm. First, MLSAR introduces the conception of *move-direction* to maintain the balance constraint of a new partitioning $\bar{s}$. Second, MLSAR defines $c(\bar{s}) = cut(\bar{s})$ and exploits the concept of *gain* to fast the computation of $c(\bar{s}) - c(s)$ that is computed by $ED(v)$-$ID(v)$, where the vertex v is chosen to move to the other side of the partitioning s. MLSAR also uses two buckets with the last-in first-out (LIFO) scheme to fast storage and update the gains of boundary vertices of two sides and facilitate retrieval the highest-gain vertex. Finally, MLSAR doesn't select vertex v to move at random in boundary vertices as in MLrMSATS, but always chooses to move a highest-gain vertex v from the larger side of the partitioning. It is important for simulated annealing to strengthen its effectiveness and achieve significant speedups for high quality solutions with well-designed heuristics and properly move generation strategy.

In [17], Karypis presents the sorted heavy-edge matching (SHEM) algorithm that identifies and collapses together groups of vertices that are highly connected. Firstly, SHEM sorts the vertices of the graph ascendingly based on the *degree* of the vertices. Next, the vertices are visited in this order and SHEM matches the vertex v with unmatched vertex u such that the weight of the edge $W(v, u)$ is maximum over all incident edges. In [19], Amine and Karypis introduce the concept of the graph *core* for coarsening the *power-law* graphs. In [11], Leng and Yu present the core-sorted heavy-edge matching (CSHEM) algorithm that combines the concept of the graph *core* with the SHEM scheme. Firstly, CSHEM sorts the vertices of the graph descendingly based on the *core* number of the vertices by the algorithm in [24]. Next, the vertices are visited in this order and CSHEM matches the vertex v with its unmatched neighboring vertex whose edge-weight is maximum.

In our multi-level algorithm, we adopt the MLSAR algorithm during the *refinement phase* and an effective matching-based coarsening scheme during the *coarsening phase* that uses the CSHEM algorithm on the original graph and the SHEM algorithm on the coarser graphs. The pseudocode of our multi-level algorithm is shown in Algorithm 1.

Algorithm 1 (our multi-level algorithm)

> INPUT: original graph $G(V,E)$
> OUTPUT: the partitioning P_G of graph G
> /*$coarsening\ phase$*/
> $l = 0$
> $G_l(V_l,E_l){=}G(V,E)$
> $G_{l+1}(V_{l+1},E_{l+1}){=}\mathrm{CSHEM}(G_l(V_l,E_l))$
> While ($|V_{l+1}| > 20$) do
> $l = l + 1$
> $G_{l+1}(V_{l+1},E_{l+1}){=}\mathrm{SHEM}(G_l(V_l,E_l))$
> End While
> /*$initial\ partitioning\ phase$*/
> $P_{G_l}{=}\mathrm{GGGP}(G_l)$
> /*$refinement\ phase$*/
> While ($l \geq 1$) do
> $P'_{G_l}{=}\mathrm{MLSAR}(G_l,P_{G_l})$
>   Project P'_{G_l} to $P_{G_{l-1}}$;
> $l = l - 1$
> End While
> $P_G{=}\mathrm{MLSAR}(G_l,P_{G_l})$
> Return P_G

4   An Effective Multi-level Simulated Annealing Refinement Algorithm

Informally, the MLSAR algorithm works as follows: At cycle zero, an initialization phase takes place during which the initial partitioning Q is projected from the partitioning $P_{G_{l+1}}$ of the coarser graph G_{l+1}, the Markov chain length L is set to be the number of vertices of the current level graph G_l, the internal and external degrees of all vertices are computed and etc. The main structure of MLSAR consists of a nested loop. The outer loop detects the frozen condition by an appropriate termination criterion whether the current temperature T_k is less than final temperature; the inner loop determines whether a thermal equilibrium at temperature T_k is reached by using the following criterions: The number of attempted moves exceeds L, or the bucket of the start side of the *move-direction* is empty. In the inner loop of the MLSAR algorithm, a neighbor of the current partitioning P is generated by selecting the vertex v with the highest gain from the larger side of the partitioning P and performing the move according to the following rule: The move is certainly accepted if it improves $cut(P)$, or probabilistically accepted according to a random number uniformly distributed on the interval [0,1]. In the latter case, if the acceptance test is negative then no move is performed, and the current partitioning P is left unchanged. The pseudocode of MLSAR is shown in Algorithm 2. The cycles counter is denoted by k and L represents the Markov chain length. Let $Best$ be the best partitioning seen so far and P be the current partitioning. At cycle k, T_k represents the current temperature and the counter of neighbors sampled is denoted by L_k.

Algorithm 2 (MLSAR)

INPUT: initial bipartitioning Q,balance constraint b,attenuation rate α
initial temperature T_i,final temperature T_f
OUTPUT: the best partitioning $Best$, cut of the best partitioning $cut(Best)$
MLSAR(
/*Initialization*/
$k = 0$
$T_k = T_i$
Set current parition $P = Q$;
Set the best parition $Best = Q$;
Set Markov chain length $L=|V|$;
For every vertex v in $G = (V, E)$ do

$$ID_v = \sum_{(v,u)\in E \wedge P[v]=P[u]} W(v,u)$$

$$ED_v = \sum_{(v,u)\in E \wedge P[v] \neq P[u]} W(v,u)$$

Store v in $boundary\ hash\text{-}table$ if and only if $ED_v > 0$;
End For
/*Main loop*/
While $T_k \geq T_f$ do
$L_k=1$
Compute the gains of boundary vertices of two sides;
Insert the gains of boundary vertices of two sides in buckets respectively;
While $L_k \leq L$ do
Decide the *move-direction* of the current move;
If (the bucket of the start side of the *move-direction* is empty) then
  Break;
Else
  Select the vertex v with the highest gain in the bucket;
  Designate the vertex v as tabu status by inserting v in tabu list;
  If $(random(0,1) \leq min(1, exp(\frac{(ED_v-ID_v)\times|boundary\ hash\text{-}table|}{2\times cut(Q)\times T_k})))$ then
  $L_k=L_k+1$
  Update P by moving the vertex v to the other side;
  original *cut* Minus its original *gain* as the *cut* of new partition P;
  Update the *internal* and *external degrees* of its neighboring vertices;
  Update the *gains* of its neighboring vertices in two buckets;
  Update *boundary status* of its neighboring vertices in *boundary hash-table*;

If (the *cut* is minimum and satisfies balance constraint *b*) then
    Best=P
    Record roll back point;
    Record new *cut* minumum;
  End If /* *cut is minimum**/
  End If /* $r \le p^*$/
 End If /* *the bucket is empty**/
End While /* *thermal equilibrium* $L_k \le L^*$/
Roll back to minumum *cut* point by undoing all moves and updating the
           internal and *external degrees* and *boundary hash-table*;
Empty the tabu list and two buckets;
$T_{(k+1)} = \alpha \times T_k$
$k = k + 1$
End While /* *frozen criterion* $T_k \ge T_f$*/
Return *Best* and *cut*(*Best*)

The MLSAR algorithm uses a tabu list, which is a short-term memory of
moves that are forbidden to execute, to avoid cycling near local optimum and
to enable moves towards worse solutions, as in the MLrMSATS algorithm. In
the terminology of tabu search [25], the MLSAR strategy is a simple form of
tabu restriction without aspiration criterion whose prohibition period is fixed
at $|V_l|$. Because the MLSAR algorithm aggressively selects the best admissible
vertex based on the tabu restriction, it must examine and compare a number of
boundary vertices by the bucket that allows to storage, retrieval and update the
gains of vertices very quickly. It is important to obtain the efficiency of MLSAR
by using the bucket with the LIFO scheme, as tabu search memory structure.
The *internal* and *external degrees* of all vertices, as complementary tabu search
memory structures, help MLSAR to facilitate computation of vertex *gain* and
judgement of boundary vertex. We also use a *boundary hash-table*, as another
complementary tabu search memory structure, to store the boundary vertices
whose *external degree* is greater than zero.

During each iteration of MLSAR, the *internal* and *external degrees* and *gains*
of all vertices are kept consistent with respect to the current partitioning *P*.
This can be done by updating the *degrees* and *gains* of the vertex *v*'s neighboring vertices. Of course, the *boundary hash-table* might change as the current
partitioning *P* changes. For example, due to a move in an other boundary vertex,
a boundary vertex would no longer be such a boundary vertex and should be
removed from the *boundary hash-table*. Furthermore, a no-boundary vertex can
become such a vertex if it is connected to a boundary vertex which is moved to
the other side and should be inserted in the *boundary hash-table*.

5   Experimental Results

We use the 18 graphs in our experiments that are converted from the hypergraphs
of the ISPD98 benchmark suite [21] and range from 12,752 to 210,613 vertices.

Table 1. The characteristics of 18 graphs to evaluate our algorithm

benchmark	vertices	hyperedges	edges
ibm01	12752	14111	109183
ibm02	19601	19584	343409
ibm03	23136	27401	206069
ibm04	27507	31970	220423
ibm05	29347	28446	349676
ibm06	32498	34826	321308
ibm07	45926	48117	373328
ibm08	51309	50513	732550
ibm09	53395	60902	478777
ibm10	69429	75196	707969
ibm11	70558	81454	508442
ibm12	71076	77240	748371
ibm13	84199	99666	744500
ibm14	147605	152772	1125147
ibm15	161570	186608	1751474
ibm16	183484	190048	1923995
ibm17	185495	189581	2235716
ibm18	210613	201920	2221860

Each benchmark comes with 3 files, a .net file, a .are file and a .netD file. Each hyperedge is a subset of two or more vertices in hypergraph and is stored in .net file. We convert hyperedges into edges by the rule that every subset of two vertices in hyperedge can be seemed as edge. We create the edge with unit weight if the edge that connects two vertices doesn't exist, else add unit weight to the weight of the edge. Next, we get the weights of vertices from .are file. Finally, we store 18 edge-weighted and vertex-weighted graphs in format of **MeTiS** [17]. The characteristics of these graphs are shown in Table 1.

We implement the MLSAR algorithm in ANSI C and integrate it with the leading edge partitioner **MeTiS**. In the evaluation of our multi-level algorithm, we must make sure that the results produced by our algorithm can be easily compared against those produced by **MeTiS**. We use the same balance constraint b and random seed in every comparison. In the scheme choices of three phases offered by **MeTiS**, we use the SHEM algorithm during the *coarsening phase*, the greedy graph growing partition algorithm during the *initial partitioning phase* that consistently finds smaller edge-cuts than other algorithms, the boundary KL (BKL) refinement algorithm during the *uncoarsening and refinement phase* because BKL can produce smaller edge-cuts when coupled with the SHEM algorithm. These measures are sufficient to guarantee that our experimental evaluations are not biased in any way.

The quality of partitions is evaluated by looking at two different quality measures, which are the minimum *cut* (MinCut) and the average *cut* (AveCut). To ensure the statistical significance of our experimental results, two measures are obtained in twenty runs whose random seed is different to each other. For all

Table 2. Min-cut bipartitioning results with up to 2% deviation from exact bisection

benchmark	vertices	edges	Metis(α)		our algorithm(β)		ratio(β:α)	
			MinCut	AveCut	MinCut	AveCut	MinCut	AveCut
ibm01	12752	109183	517	1091	354	575	0.685	0.527
ibm02	19601	343409	4268	11076	4208	6858	0.986	0.619
ibm03	23136	206069	10190	12353	6941	8650	0.681	0.700
ibm04	27507	220423	2273	5716	2075	3542	0.913	0.620
ibm05	29347	349676	12093	15058	8300	10222	0.686	0.679
ibm06	32498	321308	7408	13586	3525	8667	**0.476**	0.638
ibm07	45926	373328	3219	4140	2599	3403	0.807	0.822
ibm08	51309	732550	11980	38180	11226	16788	0.937	**0.440**
ibm09	53395	478777	2888	4772	2890	3375	1.001	0.707
ibm10	69429	707969	10066	17747	5717	8917	0.568	0.502
ibm11	70558	508442	2452	5095	2376	3446	0.969	0.676
ibm12	71076	748371	12911	27691	11638	16132	0.901	0.583
ibm13	84199	744500	6395	13469	4768	7670	0.746	0.569
ibm14	147605	1125147	8142	12903	8203	9950	**1.007**	0.771
ibm15	161570	1751474	22525	46187	14505	32700	0.644	0.708
ibm16	183484	1923995	11534	22156	9939	17172	0.862	0.775
ibm17	185495	2235716	16146	26202	14251	17126	0.883	0.654
ibm18	210613	2221860	15470	20018	15430	18248	0.997	**0.912**
average							**0.819**	**0.661**

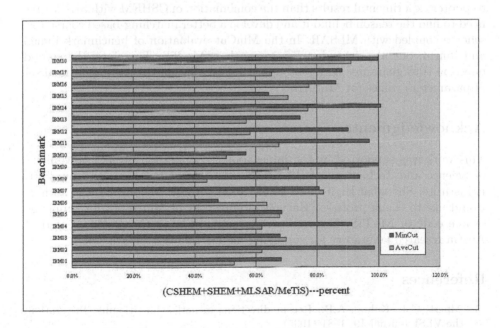

Fig. 1. The MinCut and AveCut comparisons of two algorithms on 18 graphs

experiments, we use a 49-51 *bipartitioning* balance constraint by setting b to 0.02. Furthermore, we adopt the experimentally determined optimal set of parameters values for MLSAR, $\alpha=0.9$, $T_i=10.0$, $T_f=0.01$.

Table 2 presents *min-cut bipartitioning* results allowing up to 2% deviation from exact bisection and Fig. 1 illustrates the MinCut and AveCut comparisons of two algorithms on 18 graphs. As expected, our algorithm reduces the AveCut by 8.8% to 56.0% and reaches 33.9% average AveCut improvement. Although our algorithm produces partitioning whose MinCut is up to 0.7% worse than that of **MeTiS** on two benchmarks, we still obtain 18.1% average MinCut improvement and between -0.7% and 52.4% improvement in MinCut. All evaluations that twenty runs of two algorithms on 18 graphs are run on an 1800MHz AMD Athlon2200 with 512M memory and can be done in four hours.

6   Conclusions

In this paper, we have presented an effective multi-level algorithm based on simulated annealing. The success of our algorithm relies on exploiting both the simulated annealing procedure and the concept of the graph core. We obtain excellent *bipartitioning* results compared with those produced by **MeTiS**. Although it has the ability to find cuts that are lower than the result of **MeTiS** in a reasonable time, there are several ways in which this algorithm can be improved. For example, we note that adopting the CSHEM algorithm alone leads to poorer experimental results than the combination of CSHEM with SHEM. We need to find the reason behind it and develop a better matching-based coarsening scheme coupled with MLSAR. In the MinCut evaluation of benchmark ibm09 and ibm14, our algorithm is 0.7% worse than **MeTiS**. Therefore, the second question is to guarantee find good approximate solutions by setting optimal set of parameters values for MLSAR.

Acknowledgments

This work was supported by the international cooperation project of Ministry of Science and Technology of PR China, grant No. CB 7-2-01, and by "SEC E-Institute: Shanghai High Institutions Grid" project. Meanwhile, the authors would like to thank professor Karypis of university of Minnesota for supplying source code of **MeTiS**. The authors also would like to thank Alpert of IBM Austin research laboratory for supplying the ISPD98 benchmark suite.

References

1. Alpert, C.J., Kahng, A.B.: Recent directions in netlist partitioning. Integration, the VLSI Journal 19, 1–81 (1995)
2. Hsu, W.H., Anvil, L.S.: Self-organizing systems for knowledge discovery in large databases. In: International Joint Conference on Neural Networks, pp. 2480–2485 (1999)

3. Zha, H., He, X., Ding, C., Simon, H., Gu, M.: Bipartite graph partitioning and data clustering. In: Proc. ACM Conf. Information and Knowledge Management, pp. 25–32 (2001)
4. Ding, C., He, X., Zha, H., Gu, M., Simon, H.: A Min-Max cut algorithm for graph partitioning and data clustering. In: Proc. IEEE Conf. Data Mining, pp. 107–114 (2001)
5. Shi, J., Malik, J.: Normalized cuts and image segmentation. In: Proc. IEEE Conf. Computer Vision and Pattern Recognition, pp. 731–737 (1997)
6. Shi, J., Malik, J.: Normalized cuts and image segmentation. IEEE Trans. Pattern Anal. Mach. Intell, 888–905 (2000)
7. Garey, M.R., Johnson, D.S.: Computers and intractability: A guide to the theory of NP-completeness. WH Freeman, New York (1979)
8. Kernighan, B.W., Lin, S.: An efficient heuristic procedure for partitioning graphs. Bell System Technical Journal 49, 291–307 (1970)
9. Fiduccia, C., Mattheyses, R.: A linear-time heuristics for improving network partitions. In: Proc. 19th Design Automation Conf. pp. 175–181 (1982)
10. Leng, M., Yu, S., Chen, Y.: An effective refinement algorithm based on multi-level paradigm for graph bipartitioning. In: The IFIP TC5 International Conference on Knowledge Enterprise. IFIP Series, pp. 294–303. Springer (2006)
11. Leng, M., Yu, S.: An effective multi-level algorithm for bisecting graph. In: The 2nd International Conference on Advanced Data Mining and Applications. LNCS/LNAI, pp. 493–500. Springer, Heidelberg (2006)
12. Żola, J., Wyrzykowski, R.: Application of genetic algorithm for mesh partitioning. In: Proc. Workshop on Parallel Numerics, pp. 209–217 (2000)
13. Bahreininejad, A., Topping, B.H.V., Khan, A.I.: Finite element mesh partitioning using neural networks. Advances in Engineering Software, 103–115 (1996)
14. Leng, M., Yu, S.: An effective multi-level algorithm based on ant colony optimization for bisecting graph. In: The 11th PacificAsia Conference on Knowledge Discovery and Data Mining. LNCS/LNAI, pp. 138–149. Springer, Heidelberg (2007)
15. Sun, L., Leng, M., Yu, S.: A new multi-level algorithm based on particle swarm optimization for bisecting graph. In: The 3rd International Conference on Advanced Data Mining and Applications. LNCS/LNAI, Springer, Heidelberg (2007)
16. Sun, L., Leng, M.: An effective refinement algorithm based on swarm intelligence for graph bipartitioning. In: The International symposium on combinatorics, algorithms, probabilistic and experimental methodologies. LNCS/LNAI, Springer, Heidelberg (2007)
17. Karypis, G., Kumar, V.: MeTiS 4.0: Unstructured graphs partitioning and sparse matrix ordering system. Technical Report, Department of Computer Science, University of Minnesota (1998)
18. Selvakkumaran, N., Karypis, G.: Multi-objective hypergraph partitioning algorithms for cut and maximum subdomain degree minimization. IEEE Trans. Computer Aided Design 25, 504–517 (2006)
19. Amine, A.B., Karypis, G.: Multi-level algorithms for partitioning power-law Graphs. Technical Report, Department of Computer Science, University of Minnesota (2005), Available on the WWW at URL http://www.cs.umn.edu/~metis
20. Gil, C., Ortega, J., Montoya, M.G.: Parallel heuristic search in multilevel graph partitioning. In: Proc. 12th Euromicro Conference on Parallel, Distributed and Network-Based Processing, pp. 88–95 (2004)

21. Alpert, C.J.: The ISPD98 circuit benchmark suite. In: Proc. Intel Symposium of Physical Design, pp. 80–85 (1998)
22. Aarts, E., Korst, J.: Simulated annealing and boltzmann machines. A Stochastic Approach to Combinatorial Optimization and Neural Computing. John Wiley and Sons, New York (1990)
23. Gil, C., Ortega, J., Montoya, M.G., Basnos, R.: A mixed heuristic for circuit partitioning. Computational Optimization and Applications 23, 321–340 (2002)
24. Batagelj, V., Zaversnik, M.: Generalized cores. Journal of the ACM, 1–8 (2002)
25. Glover, F., Manuel, L.: Tabu search: Modern heuristic techniques for combinatorial problems, pp. 70–150. Blackwell Scientific Publications, Oxford (1993)

Szemerédi's Regularity Lemma and Its Applications to Pairwise Clustering and Segmentation

Anna Sperotto[1] and Marcello Pelillo[2]

[1] Department of Electrical Engineering, Mathematics, and Computer Science
University of Twente, The Netherlands
a.sperotto@utwente.nl
[2] Dipartimento di Informatica
Università Ca' Foscari di Venezia, Italy
pelillo@dsi.unive.it

Abstract. Szemerédi's regularity lemma is a deep result from extremal graph theory which states that every graph can be well-approximated by the union of a constant number of random-like bipartite graphs, called regular pairs. Although the original proof was non-constructive, efficient (i.e., polynomial-time) algorithms have been developed to determine regular partitions for arbitrary graphs. This paper reports a first attempt at applying Szemerédi's result to computer vision and pattern recognition problems. Motivated by a powerful auxiliary result which, given a partitioned graph, allows one to construct a small reduced graph which inherits many properties of the original one, we develop a two-step pairwise clustering strategy in an attempt to reduce computational costs while preserving satisfactory classification accuracy. Specifically, Szemerédi's partitioning process is used as a preclustering step to substantially reduce the size of the input graph in a way which takes advantage of the strong notion of edge-density regularity. Clustering is then performed on the reduced graph using standard algorithms and the solutions obtained are then mapped back into the original graph to create the final groups. Experimental results conducted on standard benchmark datasets from the UCI machine learning repository as well as on image segmentation tasks confirm the effectiveness of the proposed approach.

1 Introduction

Graph-theoretic representations and algorithms have long been an important tool in computer vision and pattern recognition, especially because of their representational power and flexibility. However, there is now a renewed and growing interest toward explicitly formulating computer vision problems within a graph-theoretic setting. This is in fact particularly advantageous because it allows vision problems to be cast in a pure, abstract setting with solid theoretical underpinnings and also permits access to the full arsenal of graph algorithms developed in computer science and operations research. Graph-theoretic problems which have proven to be relevant to computer vision include maximum

A.L. Yuille et al. (Eds.): EMMCVPR 2007, LNCS 4679, pp. 13–27, 2007.
© Springer-Verlag Berlin Heidelberg 2007

flow, minimum spanning tree, maximum clique, shortest path, maximal common subtree/subgraph, etc. In addition, a number of fundamental techniques that were designed in the graph algorithms community have recently been applied to computer vision problems. Examples include spectral methods and fractional rounding.

In 1941, the Hungarian mathematician P. Turán provided an answer to the following innocent-looking question. What is the maximal number of edges in a graph with n vertices not containing a complete subgraph of order k, for a given k? This graph is now known as a Turán graph and contains no more than $n^2(k-2)/2(k-1)$ edges. Later, in another classical paper, T. S. Motzkin and E. G. Straus [12] provided a novel proof of Turán's theorem using a continuous characterization of the clique number of a graph. Thanks to the contributions of P. Erdös, B. Bollobás, M. Simonovits, E. Szemerédi, and others, Turán study developed soon into one of the richest branches of 20th-century graph theory, known as *extremal graph theory*, which has intriguing connections with Ramsey theory, random graph theory, algebraic constructions, etc. Very roughly, extremal graph theory studies how the intrinsic structure of graphs ensures certain types of properties (e.g., cliques, colorings and spanning subgraphs) under appropriate conditions (e.g., edge density and minimum degree) (see, e.g., [3]).

Among the many achievements of extremal graph theory, Szemerédi's *regularity lemma* is certainly one of the best known [6]. It states, essentially, that every graph can be partitioned into a small number of random-like bipartite graphs, called regular pairs, and a few leftover edges. Szemerédi's result was introduced in the mid-seventies as a tool for his celebrated proof of the Erdös-Turán conjecture on arithmetic progressions in dense sets of integers. Since then, the lemma has emerged as a fundamental tool not only in extremal graph theory but also in theoretical computer science, combinatorial number theory, etc. We refer to [11,10] for a survey of the (theoretical) applications of Szemerédi's lemma and its generalizations.

The regularity lemma is basically an existence predicate. In its original proof, Szemerédi demonstrated the existence of a regular partition under the most general conditions, but he did not provide a constructive proof to obtain such a partition. However, in 1992, Alon et al. [1] succeeded in developing the first algorithm to create a regular partition on arbitrary graphs, and showed that it has polynomial computational complexity. Other polynomial-time algorithms can be found in [8,5,9].

This paper reports perhaps the first practical application of the regularity lemma and related algorithms. Our original motivation was to study how to take advantage of the information provided by Szemerédi's regular partitions in a pairwise clustering context. Indeed, pairwise (or similarity-based) data clustering techniques are gaining increasing popularity over traditional feature-based grouping algorithms. In many application domains, in fact, the objects to be clustered are not naturally representable in terms of a vector of features. On the other hand, quite often it is possible to obtain a measure of the similarity/dissimilarity between objects. Hence, it is natural to map (possibly implicitly) the data to be clustered to the nodes of a weighted graph, with edge-weights representing similarity or dissimilarity relations. However, a typical problem associated to pairwise grouping algorithms is the scaling behavior with the number of data. On a dataset containing N examples, the number of potential comparisons scales with $O(N^2)$, thereby hindering their applicability to problems involving very large data sets, such as high-resolution imagery and spatio-temporal data. Recent sophisticated attempts to deal with this problem use optimal embeddings [15], the Nystrom method [2,7], and out-of-sample dominant sets [14].

The solution outlined in this paper combines the notion of regularity introduced by Szemerédi with that of graph-based clustering. In this context, the regularity lemma is used as a preclustering strategy, in an attempt to work on a more compact, yet informative, structure. Indeed, this structure is well-known in extremal graph theory and is commonly referred to as the *reduced graph*. An important auxiliary result, the so-called Key Lemma, reveals that this graph does inherit many of the essential structural properties of the original graph. In summary, our approach consists basically in a two-phase procedure. In the first phase, the input graph is decomposed into small pieces using Szemerédi's partitioning process and the corresponding (weighted) reduced graph is constructed, the weights of which reflect edge-densities between class pairs of the original partition. Next, a standard graph-based clustering procedure is run on the reduced graph and the solution found is mapped back into original graph to obtain the final groups. In our simulations we used dominant-set algorithms to perform the clustering on the reduced graph, but any other pairwise algorithm such as Normalized Cut would work equally well. Note that our approach differs from other attempts aimed at reducing the complexity of pairwise grouping processes, such as [2,7,14], as we perform no sampling of the original data but work instead on a derived structure which does retain the important features of the original one.

We performed some experiments both on standard datasets from the UCI machine learning repository and on image segmentation tasks, and the preliminary results obtained confirm the effectiveness of the proposed approach.

2 Szemerédi's Regularity Lemma

Let $G = (V, E)$ be an undirected graph with no self-loops, where V is the set of vertices and E is the set of edges, and let $X, Y \subseteq V$ be two disjoint subsets of vertices of G. We define the *edge density* of the pair (X, Y) as:

$$d(X, Y) = \frac{e(X, Y)}{|X||Y|} \tag{1}$$

where $e(X, Y)$ denotes the number of edges of G with an endpoint in X and an endpoint in Y, and $|\cdot|$ denotes the cardinality of a set. Note that edge densities are real numbers between 0 and 1.

Given a positive constant $\varepsilon > 0$, we say that the pair (A, B) of disjoint vertex sets $A, B \subseteq V$ is ε-*regular* if for every $X \subseteq A$ and $Y \subseteq B$ satisfying

$$|X| > \varepsilon|A| \quad \text{and} \quad |Y| > \varepsilon|B| \tag{2}$$

we have

$$|d(X, Y) - d(A, B)| < \varepsilon. \tag{3}$$

Thus, in an ε-regular pair the edges are distributed fairly uniformly.

A partition of V into pairwise disjoint classes $C_0, C_1, \ldots, C_k$ is said *equitable* if all the classes C_i ($1 \leq i \leq k$) have the same cardinality. The *exceptional set* C_0 (which may be empty) has only a technical purpose: it makes it possible that all other classes have exactly the same number of vertices. An equitable partition $C_0, C_1, \ldots, C_k$, with C_0 being the exceptional set, is called ε-*regular* if $|C_0| < \varepsilon|V|$ and all but at most εk^2 of the pairs (C_i, C_j) are ε-regular ($1 \leq i < j \leq k$).

Theorem 1 (Szemerédi's regularity lemma [16]). *For every positive real ε and for every positive integer m, there are positive integers $N = N(\varepsilon, m)$ and $M = M(\varepsilon, m)$ with the following property: for every graph G with $|V| \geq N$ there is an ε-regular partition of G into $k + 1$ classes such that $m \leq k \leq M$.*

Given an $r \times r$ symmetric matrix (p_{ij}) with $0 \leq p_{ij} \leq 1$, and positive integers $n_1, n_2, \ldots, n_r$, a *generalized random graph* R_n for $n = n_1 + n_2 + \ldots + n_r$ is obtained by partitioning n vertices into classes C_i of size n_i and joining the vertices $x \in V_i$, $y \in V_j$ with probability p_{ij}, independently for all pairs $\{x, y\}$. Now, as pointed out by Komlós and Simonovits [11], the regularity lemma asserts basically that every graph can be approximated by generalized random graphs. Note that, for the lemma to be useful, the graph has to to be dense. Indeed, for sparse graphs it becomes trivial as all densities of pairs tend to zero [6].

The lemma allows us to specify a lower bound m on the number of classes. A large value of m ensures that the partition classes C_i are sufficiently small, thereby increasing the proportion of (inter-class) edges subject to the regularity condition and reducing the intra-class ones. The upper bound M on the number of partitions guarantees that for large graphs the partition sets are large too. Finally, it should be noted that a singleton partition is ε-regular for every value of ε and m.

The regularity lemma permits εk^2 pairs to be irregular. Following a path pointed out by Szemerédi [16], many researchers studied if the result could be strengthened, avoiding the presence of such pairs. However, it turned out that forcing the lemma in that way would affect its generality [1].

The problem of checking if a given partition is ε-regular is a quite surprising one from a computational complexity point of view. In fact, as proven in [1], it turns out that constructing an ε-regular partition is easier than checking if a given one responds to the ε-regularity criterion.

Theorem 2. *The following decision problem is co-NP-complete. Given a graph G, an integer $k \geq 1$ and a parameter $\varepsilon > 0$, and a partition of the set of vertex of G into $k + 1$ parts. Decide if the given partition is ε-regular.*

Recall that the complexity class *co-NP-complete* collects all problems which complement are NP-complete [4]. In other words, a *co-NP-complete* problem is one for which there are efficiently verifiable proofs of its no-instances, i.e., its counterexamples.

3 Finding Regular Partitions in Polynomial Time

The original proof of Szemerédi's lemma [16] is not constructive, yet this has not narrowed the range of its applications in the fields of extremal graph theory, number theory and combinatorics. In the early 1990's, Alon et al. [1] proposed a new formulation of the lemma which emphasizes the algorithmic nature of the result.

Theorem 3 (A constructive version of the regularity lemma [1]). *For every $\varepsilon > 0$ and every positive integer t there is an integer $Q = Q(\varepsilon, t)$ such that every graph with $n > Q$ vertices has an ε-regular partition into $k + 1$ classes, where $t \leq k \leq Q$. For every fixed $\varepsilon > 0$ and $t \geq 1$ such a partition con be found in $O(M(n))$ sequential time, where $M(n) = O(n^{2.376})$ is the time for multiplying two $n \times n$ matrices with $0, 1$ entries over the integers. The partition can be found in time $O(\log n)$ on a EREW PRAM with a polynomial number of parallel processor.*

In the remaining of this section we shall derive the algorithm alluded to in the previous theorem. We refer the reader to the original paper [1] for more technical details and proofs.

Let H be a bipartite graph with equal color classes $|A| = |B| = n$. We define the *average degree* of H as

$$d = \frac{1}{2n} \sum_{i \in A \cup B} deg(i) \tag{4}$$

where, $deg(i)$ is the degree of the vertex i.

Let y_1 and y_2 be two distinct vertices, such that $y_1, y_2 \in B$. Alon et al. [1] define the *neighborhood deviation* of y_1 and y_2 by

$$\sigma(y_1, y_2) = |N(y_1) \cap N(y_2)| - \frac{d^2}{n} \tag{5}$$

where $N(x)$ is the set of neighbors of the vertex x. For a subset $Y \subseteq B$, the *deviation* of Y is defined as:

$$\sigma(Y) = \frac{\sum_{y_1, y_2 \in Y} \sigma(y_1, y_2)}{|Y|^2} \tag{6}$$

Now, let $0 < \varepsilon < \frac{1}{16}$. It can be shown that if there exists $Y \subseteq B$, $|Y| > \varepsilon n$ such that $\sigma(Y) \geq \frac{\varepsilon^3}{2} n$ then at least one of the following cases occurs [1]:

1. $d < \varepsilon^3 n$ (which amounts to saying that H is ε-regular);
2. there exists in B a set of more than $\frac{1}{8}\varepsilon^4 n$ vertices whose degrees deviate from d by at least $\varepsilon^4 n$;
3. there are subsets $A' \subseteq A$, $B' \subseteq B$, $|A'| \geq \frac{\varepsilon^4}{4} n$, $|B'| \geq \frac{\varepsilon^4}{4} n$ and $|d(A', B') - d(A, B)| \geq \varepsilon^4$.

Note that one can easily check if 1 holds in time $O(n^2)$. Similarly, it is trivial to check if 2 holds in $O(n^2)$ time, and in case it holds to exhibit the required subset of B establishing this fact. If both cases above fail we can proceed as follows. For each $y_0 \in B$ with $|deg(y_0) - d| < \varepsilon^4 n$ we find the set of vertices $B_{y_0} = \{y \in B : \sigma(y_0, y) \geq 2\varepsilon^4 n\}$. It can be shown that there exists at least one such y_0 for which $|B_{y_0}| \geq \frac{\varepsilon^4}{4}n$. The subsets $B' = B_{y_0}$ and $A' = N(y_0)$ are the required ones. Since the computation of the quantities $\sigma(y, y')$, for $y, y' \in B$, can be done by squaring the adjacency matrix of H, the overall complexity of this algorithms is $O(M(n)) = O(n^{2.376})$.

In order to understand the final partitioning algorithm we need the following two lemmas.

Lemma 1 (Alon et al. [1]). *Let H be a bipartite graph with equal classes $|A| = |B| = n$. Let $2n^{-1/4} < \varepsilon < \frac{1}{16}$. There is an $O(n^{2.376})$ algorithm that verifies that H is ε-regular or finds two subsets $A' \subseteq A$, $B' \subseteq B$, $|A'| \geq \frac{\varepsilon^4}{4}n$, $|B'| \geq \frac{\varepsilon^4}{4}n$ and $|d(A', B') - d(A, B)| \geq \varepsilon^4$.*

It is quite easy to check that the regularity condition can be rephrased in terms of the average degree of H. Indeed, it can be seen that if $d < \varepsilon^3 n$, then H is ε-regular, and this can be tested in $O(n^2)$ time. Next, it is necessary to count the number of vertices in B whose degrees deviate from d by at least $\varepsilon^4 n$. Again, this operation takes $O(n^2)$ time. If the number of deviating vertices is more than $\frac{\varepsilon^4}{8}n$, then the degrees of at least half of them deviate in the same direction and if we let B' be such a set of vertices and $A' = A$ we are done. Otherwise, it can be shown that there must exist $Y \subseteq B$ such that $|Y| \geq \varepsilon n$ and $\sigma(Y) \geq \frac{\varepsilon^3}{2}n$. Hence, our previous discussion shows that the required subsets A' and B' can be found in $O(n^{2.376})$ time.

Given an equitable partition P of a graph $G = (V, E)$ into classes $C_0, C_1 \ldots C_k$, Szemerédi [16] defines a measure called *index of partition*:

$$ind(P) = \frac{1}{k^2} \sum_{s=1}^{k} \sum_{t=s+1}^{k} d(C_s, C_t)^2. \tag{7}$$

Since $0 \leq d(C_s, C_t) \leq 1$, $1 \leq s, t, \leq k$, it can be seen that $ind(P) \leq \frac{1}{2}$.

The following lemma is the core of Szemerédi's original proof.

Lemma 2 (Szemerédi [16]). *Let $G = (V, E)$ be a graph with n vertices. Let P be an equitable partition of V into classes $C_0, C_1 \ldots C_k$, with C_0 being the exceptional class. Let $\gamma > 0$. Let k be the least positive integer such that $4^k > 600\gamma^{-5}$. If more than γk^2 pairs (C_s, C_t) $(1 \leq s < t \leq k)$ are γ-irregular, then there is an equitable partition Q of V into $1 + k4^k$ classes, the cardinality of the exceptional class being at most*

$$|C_0| + \frac{n}{4^k} \tag{8}$$

and such that

$$ind(Q) > ind(P) + \frac{\gamma^5}{20}. \tag{9}$$

The idea formalized in the previous lemma is that, if a partition violates the regularity condition, then it can be refined by a new partition and, in this case, the $ind(P)$ measure can be improved. On the other hand, the new partition adds only "few" elements to the current exceptional set so that, in the end, its cardinality will respect the definition of equitable partition.

We are now in a position to sketch the complete partitioning algorithm. The procedure is divided into two main steps: in the first step all the constants needed during the next computation are set; in the second one, the partition is iteratively created. An iteration is called *refinement step*, because, at each iteration, the current partition is closer to a regular one.

Given any $\varepsilon > 0$ and a positive integer t, the constants $N = N(\varepsilon, t)$ and $T = T(\varepsilon, t)$ are defined as follows. Let b be the least positive integer such that

$$4^b > 600\left(\frac{\varepsilon^4}{16}\right)^{-5}, \quad b \geq t. \tag{10}$$

Let f be the integer-valued function defined as:

$$f(0) = b, \quad f(i+1) = f(i)4^{f(i)}. \tag{11}$$

Put $T = f\left(\lceil 10(\frac{\varepsilon^4}{16})^{-5}\rceil\right)$ and $N = \max\{T4^{2T}, \frac{32T}{\varepsilon^5}\}$. Given a graph $G = (V, E)$ with $n \geq N$ vertices, an ε-regular partition of G into $k+1$ classes, where $t \leq k \leq T$, can be constructed using the following $O(M(n)) = O(n^{2.376})$ algorithm.[1]

1. **Create the initial partition:** Arbitrarily divide the set V into an equitable partition P_1 with classes $C_0, C_1, \ldots, C_b$ where $|C_i| = \lfloor n/b \rfloor$, $i = 1 \ldots b$ and $|C_0| < b$. Let $k_1 = b$.
2. **Check regularity:** For every pair C_r, C_s of P_i verify if it is ε-regular or find $X \subseteq C_r$, $Y \subseteq C_s$, $|X| \geq \frac{\varepsilon^4}{16}|C_r|$, $|Y| \geq \frac{\varepsilon^4}{16}|C_s|$ such that

$$|d(X, Y) - d(C_s, C_r)| \geq \varepsilon^4. \tag{12}$$

3. **Count regular pairs:** If there are at most $\varepsilon\binom{k_i}{2}$ pairs that are not verified as ε-regular, then halt. P_i is an ε-regular partition.
4. **Refine:** Apply Lemma 2 where $P = P_i$, $k = k_i$, $\gamma = \frac{\varepsilon^4}{16}$ and obtain a partition P' with $1 + k_i 4^{k_i}$ classes.
5. Let $k_{i+1} = k_i 4^{k_i}$, $P_{i+1} = P'$, $i = i + 1$ and go to step 2.

Before concluding this section we mention that after Alon et al.'s contribution, other algorithms have been proposed for finding Szemerédi's partitions. In particular, we mention Frieze and Kannan's approach [8], which is based on an intriguing relation between the regularity conditions and the singular values of matrices, and Czygrinow and Rödl's [5], who proposed a new algorithmic version of Szemerédi's lemma for hypergraphs.

---

[1] Note that Alon et al. [1] proved Theorem 3 with $Q = N$ $(\geq T)$.

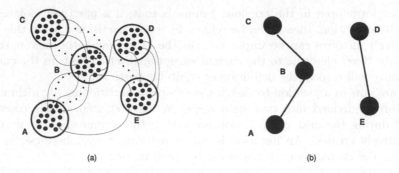

(a)

(b)

Fig. 1. Example of reduced graph creation. Left: original regular partition. Here, only the pairs (A, B), (B, C) and (D, E) are regular. Right: the graph obtained after the reduction process.

4   The Regularity Lemma and Pairwise Clustering

The regularity lemma postulates the existence of a partition of any graph into disjoint subsets satisfying a uniformity criterion, and the algorithm described in the previous section is able to find such a partition in polynomial time. On the other hand, pairwise (or graph-based) clustering refers to the process of partitioning an edge-weighted similarity graph in such a way as to produce maximally homogeneous classes. It is far too easy to see that, although both processes aim at producing a partition of a given graph, the way in which they attempt to do it is substantially different. Indeed, Szemerédi's partitioning process is very strict as it imposes the creation of same-size classes (this condition being relaxed only for the exceptional set), and this would clearly be too restrictive within a clustering context. Also, the notion of a regular partition, while emphasizing the role of inter-class relations, pays no explicit attention to the relations among vertices in the same class. Indeed, the very nature of the regularity concept indicates that the less intra-class edges, the "better" the partition.

A regular partition reveals the existence of a hidden structure in a graph. Hence, despite their dissimilarities, it could be interesting to try to combine the two partitioning approaches in order to obtain a novel and efficient clustering strategy. In fact, extremal graph theory provides us with some interesting results and abstract structures that can be conveniently employed for our purpose: these are the notion of the *reduced graph* and the so-called *Key Lemma* [11].

Given a graph $G = (V, E)$, a partition P of the vertex-set V into the sets $C_1, C_2, \ldots C_k$ and two parameters ε and d, the reduced graph R is defined as follows. The vertices of R are the clusters $C_1, C_2, \ldots C_k$, and C_i is adjacent to C_j if (C_i, C_j) is ε-regular with density more than d. Figure 1 shows an example of transformation from a partitioned graph to its reduced graph.

Consider now a graph R and an integer t. Let $R(t)$ be the graph obtained form R by replacing each vertex $x \in V(R)$ by a set V_x of t independent vertices, and joining $u \in V_x$ to $v \in V_y$ if and only if (x, y) is an edge in R. $R(t)$ is a graph in which every edge of R is replaced by a copy of the complete bipartite graph K_{tt}.

The following Lemma shows how to use the reduced graph R and its modification $R(t)$ to infer properties of a more complex graph.

Theorem 4 (Key Lemma [11]). *Given $d > \varepsilon > 0$, a graph R and a positive integer m, construct a graph G following these steps:*

1. *replace every vertex of R by m vertices*
2. *replace the edges of R with regular pairs of density at least d.*

Let H be a subgraph of $R(t)$ with h vertices and maximum degree $\Delta > 0$, and let $\delta = d - \varepsilon$ and $\varepsilon_0 = \delta^\Delta/(2 + \Delta)$. If $\varepsilon \le \varepsilon_0$ and $t - 1 \le \varepsilon_0 m$, then H is embeddable into G (i.e., G contains a subgraph isomorphic to H). In fact, we have

$$\|H \to G\| > (\varepsilon_0 m)^h \tag{13}$$

where $\|H \to G\|$ denotes the number of labeled copies of H in G.

Given a graph R, the Key Lemma furnishes rules to expand R to a more complex partitioned graph G which respects edge-density bounds. On the other hand, we have another expanded graph, $R(t)$. Because of their construction, $R(t)$ and G are very similar, but they can have a different vertex cardinality. In addition, note that the only densities allowed between vertex subsets in $R(t)$ are 0 and 1. The Key Lemma establishes constraints to the edge density d and the subset size t in order to assure the existence of a fixed graph H embeddable into $R(t)$, which is also a subgraph of G. Let H be a subgraph of $R(t)$. If t is sufficiently small with respect to m, and d is sufficiently high with respect to ε and Δ, it is possible to find small subsets of vertices such that they are connected with a sufficiently high number of edges. The copies of H are constructed vertex by vertex by picking up elements from the previous identified subsets.

As described in [11], a common and helpful combined use of the reduced graph and Key Lemma is as follows (see Figure 2):

- Start with a graph $G = (V, E)$ and apply the regularity lemma, finding a regular partition P
- Construct the reduced graph R of G, w.r.t. the partition P.
- Analyze the properties of R, in particular its subgraphs.
- As it is assured by Theorem 4, every small subgraph of R is also a subgraph of G.

In summary, a direct consequence of the Key Lemma is that it is possible to search for significant substructures in a reduced graph R in order to find common subgraphs of R and the original graph.

Now, let us put ourselves in a pairwise clustering context. We are given a large weighted (similarity) graph $G = (V, E, \omega)$ where the vertices correspond to data points, edges represent neighborhood relationships, and edge-weights reflect similarity between pairs of linked vertices. Motivated by the previous discussion,

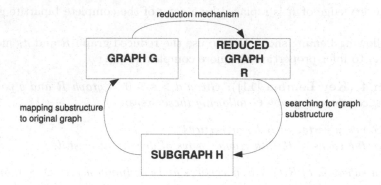

Fig. 2. Reduction strategy to find significant substructures in a graph

here we propose a *two-phase clustering strategy*. In the first phase, we apply the regularity lemma in order to build a compact, yet informative representation of the original data set, in the form of a reduced graph. In the second phase, a pairwise clustering method is used to find meaningful structures in the reduced graph. In the experiments described in the next section, for example, we used the dominant-set approach described in [13], but other pairwise algorithms could be used equally well.

Note that an algorithmic formulation of the regularity lemma for weighted graphs was explicitly introduced in [5], together with an extension of the result concerning hypergraphs. Nevertheless, simple verifications on the previous unweighted definitions show that the algorithms are not influenced by edge-weights. In this case, the *density* between the sets of a pair (A, B) in a becomes:

$$d = d_\omega(A, B) = \frac{\sum_{i=1}^{|A|} \sum_{j=1}^{|B|} \omega(a_i, b_j)}{|A||B|} \tag{14}$$

which is the weighted counterpart of (1).

Since the notion of regularity emphasizes only inter-class relations, there must be a criterion for taking into account intra-class similarities as well. Our approach consists of introducing in the partioning algorithm a rule to select elements to be dispatched to new subsets. Specifically, the criterion used is the average weighted degree

$$\text{awdeg}_S(i) = \frac{1}{|S|} \sum_{j \in S} \omega(i, j) \quad S \subseteq V. \tag{15}$$

All elements in the current subset S are listed in decreasing order by average weighted degree. In so doing, the partition of S takes place simply by subdividing the ordered sequence of elements into the desired number of subsets. The decreasing order has been preferred because of the presence of the exceptional set: in this way we assume that only the less connected vertices join the exceptional set. Hence, the obtained regular partition contains classes the elements of which can already be considered similar to each other.

The vertex set cardinality in the reduced graph is a fundamental quantity during the clustering process because it affects the precision of the solution found. In particular, a large cardinality implies a greater precision, as the clusters have to be found in a more articulated context. A low cardinality, instead, makes the clustering phase less accurate because the inclusion or the exclusion of a vertex in a cluster may change dramatically the solution at the finer level of the original graph. Our experience shows that working with medium-size subsets produces the best results in terms of accuracy and speed.

5   Experimental Results

We performed some preliminary experiments aimed at assessing the potential of the approach described in the previous sections. Before presenting the experimental setting and the results obtained, a few technical observations concerning our implementation of Alon et al's algorithm are in order. First, note that the original algorithm stops when a regular partition has been found. In practice, the number of iterations and the vertex set cardinality required is simply too big to be considered. Hence, in our experiments we decide to stop the process either when a regular partition has been found or when the subset size becomes smaller than a predetermined threshold. Further, the next-iteration number of subsets of the original procedure is also intractable, and we therefore decided to split every subset, from an iteration to the next one, using a (typically small) user-defined parameter. Finally, note that in the original formulation of the regularity lemma, the exceptional set is only a technicality to ensure that all other classes have the same cardinality. In a clustering context, the exceptional set is a nuisance as it cannot be reasonably considered as a coherent class. It is therefore necessary to somehow assign its elements to the groups found by the clustering procedure. A naive solution, adopted in our experiments, is to assign them to the closest cluster according to a predefined distance measure.

We conducted a first series of experiments on standard datasets from the UCI machine learning repository.[2] Specifically, we selected the following datasets: the Johns Hopkins University Ionosphere database (352 elements, 34 attributes, two classes), the Haberman's Survival database (306 elements, 3 attributes, two classes), and the Pima Indians Diabetes database (768 elements, 8 attributes, two classes). The similarity between data items was computed as a decreasing function of the Euclidean distance between corresponding attribute vectors, i.e., $w(i,j) = \exp(-\|\mathbf{v}_i - \mathbf{v}_j\|^2/\sigma^2)$, where $\mathbf{v}_i$ is the i-th vector of the dataset and σ is a positive real number which affects the decreasing rate of w.

Table 1 summarizes the result obtained on these data by showing the classification accuracy obtained by our two-phase strategy for each database considered. Recall that in the second phase we used the dominant-set algorithm [13] for clustering the reduced graph. Further, for the sake of comparison, we present the results produced by a direct application of the dominant-set algorithm to the original similarity graph without any pre-clustering step. As can be seen,

---

[2] http://www.ics.uci.edu/~mlearn/MLRepository.html

Table 1. Results obtained on the UCI benchmark datasets. Each row represents a dataset, while the columns represent (left to right): the number of elements in the corresponding dataset, the classification accuracy obtained by the proposed two-phase strategy and that obtained using the plain dominant-set algorithm, respectively, the speedup achieved using our approach w.r.t. plain dominant-set, the size of the reduced graphs, and the compression rate.

Dataset	Size	Classif. Accuracy		Speedup	R.G. size	Compression rate
		Two-phase	Plain DS			
Ionosphere	351	72%	67%	1.13	4	98.9%
Haberman	306	74%	73%	2.16	128	58.1%
Pima	768	65%	65%	2.45	256	66.7%

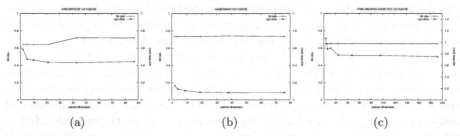

(a)   (b)   (c)

Fig. 3. Behavior of classification accuracy (left y-axis) and CPU time (right y-axis) as a function of the size of regularity classes for the three UCI datasets: Ionosphere (a), Haberman (b), and Pima (c)

our combined approach substantially outperforms the plain algorithm. Table 1 shows also the speedup achieved using our approach with respect to a direct application of the dominant-set algorithm. Note that, while the construction of a regular partition depends only on the number of vertices in the original graph and the subset dimension, the dominant set detection is affected by the edge weights. So it can converge faster or slower depending on the dataset we are analyzing. As can be seen, with our pre-clustering strategy we are able to achieve a speedup up to 2.45 on these data.

As observed at the end of the previous section, the cardinality of the reduced graph (or, alternatively, the size of the regularity subsets which in our implementation is a user-defined parameter) is of crucial importance as introduces a trade-off between the second-phase precision and the overall performance. The last two columns of Table 1 show the size of the reduced graphs and the corresponding compression rates used in the experiments. These values were chosen manually so as to optimize both accuracy and speed. Note how, using the regularity partitioning process, we were able to achieve compression rates from 66.7% to 98.9% while improving classification accuracy. It is of interest to analyze the behavior of our approach as this parameter is varied. Figure 3 plots the classification accuracy and the CPU time as a function of the size of the regularity classes (which is inversely related to the cardinality of the reduced graph) for

Table 2. The table summarizes information concerning the image segmentation experiments. Each row represents an image, while the columns represent (left to right): the number of pixels in the original image, the number of vertices in the reduced graph (with the subset dimension in parenthesis), the compression rate, and the speedup achieved using our approach w.r.t. plain dominant-set.

Image	Pixels	R.G. size	Compression rate	Speedup
Airplane	9600	128(75)	98.7%	4.23
Elephants	9600	32(300)	99.7%	17.79
Lake	9600	128(75)	98.7%	20.94
Tree	9600	64(150)	99.3%	16.63

Fig. 4. Segmentation results for 120×80 grayscale images. Left column: original images. Central column: results with our two-phase approach. Right column: results with plain dominant sets.

all datasets considered. The CPU time curve highlights how a small subset dimension requires a higher computational time compared to larger sizes. At the same time, note how the classification accuracy is not substantially affected by the class size: this means that the preclustering method is resistant to parameter variation and we are therefore allowed to use bigger subset dimensions in order to achieve better performance.

Our two-phase strategy was also applied to the problem of segmenting brightness images. Here, the image is represented as an edge-weighted graph where vertices represent pixels and edge-weights reflect the similarity between pixels. The measure adopted here to assign edge-weights is based on brightness proximity, as in [13]. Specifically, similarity between pixels i and j was measured by $w(i,j) = \exp(-(I(i) - I(j))^2/\sigma^2)$, where $I(i)$ is the normalized intensity value at node i.

Table 2 summarizes the main technical details of our experiments, in particular the compression rate and the speedup, while Figure 4 shows the segmentation results. Overall, both algorithms produced comparable segmentation results, despite the fact the with our approach we are looking for dominant sets in graphs which are at least 98% smaller than the original graph images and with a speedup of up to 20.94.

6   Conclusions

With this paper we have tried to import into the computer vision and pattern recognition fields a profound result from extremal graph theory which asserts essentially that all graphs can be decomposed in such a way as to satisfy a certain uniformity criterion. Since its introduction in the mid-seventies Szemerédi's regularity lemma has emerged as an invaluable tool not only in graph theory but also in theoretical computer science, combinatorial number theory, etc. [10]. Here, we have proposed to take advantage of the properties of regular partitions in a pairwise clustering context. Specifically, in our approach, Szemerédi's decomposition process is used as a preclustering step to substantially reduce the size of the input similarity graph. Clustering is in fact performed on a more compact derived graph using standard algorithms and the solutions obtained are then mapped back into the original graph to create the final groups. Experimental results conducted on standard benchmark datasets from the UCI machine learning repository as well as on image segmentation tasks confirm the effectiveness of the proposed approach. The power of the regularity lemma and the preliminary results obtained on our application make us confident that Szemerédi's result could be potentially used in a variety of other computer vision and pattern recognition problems, whenever large and dense graphs arise.

References

1. Alon, N., Duke, R.A., Lefmann, H., Rödl, V., Yuster, R.: The algorithmic aspects of the regularity lemma. J. of Algorithms, vol. 16, pp. 80–109 (1994) Also, Proc. 33rd IEEE FOCS, Pittsburgh, IEEE, 473–481 (1992)
2. Bengio, Y., Paiement, J., Vincent, P., Delalleau, O., Roux, N.L., Ouimet, M.: Out-of-sample extensions for lle, isomap, mds, eigenmaps, and spectral clustering. In: Thrun, S., Saul, L., Schölkopf, B. (eds.) Advances in Neural Information Processing Systems 16, MIT Press, Cambridge, MA (2004)

3. Bollobás, B.: Extremal Graph Theory. Academic Press, London (1978)
4. Cormen, T.H., Leiserson, C.E., Rivest, R.L., Stein, C.: Introduction to Algorithms, 2nd edn. MIT Press, Cambridge, MA (2001)
5. Czygrinow, A., Rödl, V.: An algorithmic regularity lemma for hypergraphs. SIAM J. Comput. 30(4), 1041–1066 (2000)
6. Diestel, R.: Graph Theory, 3rd edn. Springer, New York (2005)
7. Fowlkes, C., Belongie, S., Chung, F.R.K., Malik, J.: Spectral grouping using the Nyström method. IEEE Trans. Pattern Anal. Mach. Intell. 26, 214–225 (2004)
8. Frieze, A.M., Kannan, R.: A simple algorithm for constructing Szemerédi's regularity partition. Electron. J. Comb. 6 (1999)
9. Kohayakawa, Y., Rödl, V., Thoma, L.: An optimal algorithm for checking regularity. SIAM J. Comput. 32(5), 1210–1235 (2003)
10. Komlós, J., Shokoufandeh, A., Simonovits, M., Szemerédi, E.: The regularity lemma and its applications in graph theory. In: Khosrovshahi, G.B., Shokoufandeh, A., Shokrollahi, A. (eds.) Theoretical Aspects of Computer Science: Advanced Lectures, pp. 84–112. Springer, New York (2002)
11. Komlós, J., Simonovits, M.: Szemerédi's regularity lemma and its applications in graph theory. In: Miklós, D., Szonyi, T., Sós, V.T. (eds.) Combinatorics, Paul Erdös is Eighty (Vol. 2), pp. 295–352. Bolyai Society Mathematical Studies 2, Budapest (1996)
12. Motzkin, T.S., Straus, E.G.: Maxima for graphs and a new proof of a theorem of Turán. Canad. J. Math. 17, 533–540 (1965)
13. Pavan, M., Pelillo, M.: Dominant sets and pairwise clustering. IEEE Trans. Pattern Anal. Mach. Intell. 29, 167–172 (2007)
14. Pavan, M., Pelillo, M.: Efficient out-of-sample extension of dominant-set clusters. In: Saul, L.K., Weiss, Y., Bottou, L. (eds.) Advances in Neural Information Processing Systems 17, pp. 1057–1064. MIT Press, Cambridge, MA (2005)
15. Roth, V., Laub, J., Kawanabe, M., Buhmann, J.M.: Optimal cluster preserving embedding of nonmetric proximity data. IEEE Trans. Pattern Anal. Mach. Intell. 25, 1540–1551 (2003)
16. Szemerédi, E.: Regular partitions of graphs. In: Colloques Internationaux CNRS 260—Problèmes Combinatoires et Théorie des Graphes, Orsay, pp. 399–401 (1976)

Exact Solution of Permuted Submodular MinSum Problems

Dmitrij Schlesinger

Dresden University of Technology

Abstract. In this work we show, that for each permuted submodular MinSum problem (Energy Minimization Task) the corresponding submodular MinSum problem can be found in polynomial time. It follows, that permuted submodular MinSum problems are exactly solvable by transforming them into corresponding submodular tasks followed by applying standart approaches (e.g. using MinCut-MaxFlow based techniques).

1 Introduction

Labeling problems and especially MinSum problems (Energy Minimization tasks) have attracted attention of many researchers in the last years. They are used for many computer vision tasks like image restoration [1,5,8], stereo reconstruction [2,10,15,18], different kinds of segmentation [4,7,8,12] and others. Very promising results were obtained especially for submodular MinSum problems [6,11,16,17,18]. In this work we study *permuted submodular* MinSum problems. We show, that for each permuted submodular MinSum problem the corresponding submodular MinSum problem can be found in polynomial time. It follows, that permuted submodular MinSum problems are exactly solvable by transforming them into corresponding submodular tasks followed by applying standart approaches (e.g. using MinCut-MaxFlow based techniques).

2 Notations and Definitions

We begin with the definition of a MinSum problem followed by considerations about submodularity and permuted submodularity. A MinSum problem is defined by means of the following components. Let

$V = (R, E)$	be an unoriented graph with the set R of nodes and the set E of edges. Nodes and edges are denoted by $r \in R$ and $e = (r, r') \in E$ respectively;
K	be a finite set of states. Its elements (called states or labels) are denoted by k;
$q_r : K \to \mathbb{R}$	be a function for each node, that assigns a real value to each state in that node;
$g_{rr'} : K \times K \to \mathbb{R}$	be a function for each edge, that assigns a real value to each state pair on that edge.

A.L. Yuille et al. (Eds.): EMMCVPR 2007, LNCS 4679, pp. 28–38, 2007.
© Springer-Verlag Berlin Heidelberg 2007

A labeling f is a mapping $f : R \to K$, that maps the set of nodes into the set of states, where $f(r) \in K$ denotes the state chosen in the node r. Let $\mathcal{F} = K^R$ denote the set of all labelings. The task is to calculate

$$(\text{arg}) \min_{f \in \mathcal{F}} \left[\sum_{r \in R} q_r\big(f(r)\big) + \sum_{(rr') \in E} g_{rr'}\big(f(r), f(r')\big) \right]. \tag{1}$$

We use the tuple $\mathcal{A} = (V, K, q, g)$ of task components to refer a given task[1].

We would like to point out here, that MinSum problems considered above are a special case of more general labeling problems [13]. The generalization consists in considering a general semiring $(W, \oplus, \otimes)$ instead of the MinSum case $(\mathbb{R}, \min, +)$. Besides of already considered MinSum tasks, we will use also the notation of OrAnd problems, i.e. labeling problems, where the semiring is $(\{0, 1\}, \vee, \wedge)$. In this case the task is to calculate

$$\bigvee_{f \in \mathcal{F}} \left[\bigwedge_{r \in R} q_r\big(f(r)\big) \wedge \bigwedge_{(rr') \in E} g_{rr'}\big(f(r), f(r')\big) \right], \tag{2}$$

where the functions $q_r : K \to \{0, 1\}$ and $g_{rr'} : K \times K \to \{0, 1\}$ are boolean constraints. Such tasks are mainly known in the literature as Constraint Satisfaction problems.

Now we recall the definition of submodularity. In order to do that, it is necessary to introduce an additional notation – the order of states in each node. Let the set K of states be completely ordered in each node r of the graph (in general differently for each node). It means, that a reflexive, transitive and asymmetric relation $\succ$ is defined for pairs of states k_1 and k_2 in each node r (we will call it "above/below"), i.e. $k_2 \succ k_1$ means "the state k_2 is located above the state k_1". We denote an order, chosen in a node r, as π_r and the set of all possible orders of the state set K as Π. The tuple $\pi = (\pi_r \mid r \in R)$ denotes all ordering relations of the task and will be called ordering of the task. The set of all orderings π of the task is denoted by $\mathcal{P} = \Pi^R$.

A task, where the state set is ordered in each node, will be called ordered task and referred by the tuple $\tilde{\mathcal{A}} = (V, K, q, g, \pi)$ of its components, where π is the ordering of the task. According to this the set of all unordered tasks $\mathcal{A} = (V, K, q, g)$ can be seen as a partition of the set of all ordered tasks:

$$\mathcal{A} = (V, K, q, g) = \{ \tilde{\mathcal{A}} = (V, K, q, g, \pi) \mid \pi \in \mathcal{P} \}. \tag{3}$$

Therefore we use also notations $\tilde{\mathcal{A}} \in \mathcal{A}$, having in mind, that the (V, K, q, g) - components of both tasks $\mathcal{A}$ and $\tilde{\mathcal{A}}$ are the same.

A function $g_{rr'}$ is called submodular with respect to given orders π_r and $\pi_{r'}$ iff

$$g_{rr'}(k_1, k_1') + g_{rr'}(k_2, k_2') \leq g_{rr'}(k_1, k_2') + g_{rr'}(k_2, k_1') \tag{4}$$

---

[1] For simplicity we consider in detail only MinSum problems of second order, where the state set K is the same for all nodes. A generalization will be given later.

holds for each four-tuple with $k_2 \succ k_1$ according to π_r and $k_2' \succ k_1'$ according to $\pi_{r'}$.[2] An ordered MinSum task $\tilde{\mathcal{A}} = (V, K, q, g, \pi)$ is called submodular if all functions $g_{rr'}$ are submodular with respect to the corresponding orders π_r and $\pi_{r'}$. Functions q_r can be thereby arbitrary.

To be able to "manipulate" with requirements (4) and to simplify considerations, let us introduce an auxiliary notation – the numbers

$$\alpha_{rr'}(k_1, k_2, k_1', k_2') = g_{rr'}(k_1, k_1') + g_{rr'}(k_2, k_2') - g_{rr'}(k_1, k_2') - g_{rr'}(k_2, k_1'). \quad (5)$$

We would like to note, that these numbers can be defined without any respect to the order of states. Besides of this, these numbers have the following properties, which we will need later:

$$\alpha_{rr'}(k_1, k_2, k_1', k_2') = -\alpha_{rr'}(k_2, k_1, k_1', k_2') \quad (6)$$

and

$$\alpha_{rr'}(k_1, k_2, k_1', k_2') + \alpha_{rr'}(k_2, k_3, k_1', k_2') = \alpha_{rr'}(k_1, k_3, k_1', k_2'). \quad (7)$$

Using this notation, the submodularity conditions (4) can be rewritten in the following form. A function $g_{rr'}$ is called submodular with respect to the introduced orders π_r and $\pi_{r'}$ iff

$$\alpha_{rr'}(k_1, k_2, k_1', k_2') \leq 0 \quad (8)$$

holds for each four-tuple with $k_2 \succ k_1$ and $k_2' \succ k_1'$.

As we can see, the notation of submodularity relates explicitly to the orders, introduced for the state set in each node. A MinSum task, that is given in the conventional manner (1) (unordered), which is submodular with respect to a particular ordering π, is in general not submodular with respect to another one. Consequently it is not possible to speak about submodularity itself as about a property of a MinSum task, i.e. submodularity can not be defined for a given MinSum task, but only for a given *ordered* MinSum task.

A natural generalization of submodularity is the notation of *permuted* submodularity. An unordered MinSum task $\mathcal{A} = (V, K, q, g)$ is called permuted submodular if *there exists* an ordering π, so that the ordered task $\tilde{\mathcal{A}} = (V, K, q, g, \pi) \in \mathcal{A}$ is submodular. The aim of this work is to build a method, which recognizes for a general unordered task $\mathcal{A} = (V, K, q, g)$, whether it is permuted submodular or not. In addition, if the task is permuted submodular it is necessary to find an ordering π, such that the ordered task $\tilde{\mathcal{A}} = (V, K, q, g, \pi) \in \mathcal{A}$ is submodular. In the next section we show, that both problems can be solved simultaneously in polynomial time.

3   Transforming a Permuted Submodular Problem into a Submodular One

We split the consideration in two stages. First, we introduce an auxiliary binary OrAnd task (an OrAnd task with only two states), which represents necessary conditions for the initial MinSum task to be permuted submodular. Second, we prove, that the considered necessary conditions are sufficient as well. At the same time we give a scheme to

---

[2] If $k_1 = k_2$ or $k_1' = k_2'$ the condition (4) is always satisfied.

construct the necessary ordering π of the task from the solution of the auxiliary OrAnd task (if it exists).

The auxiliary OrAnd task will be given by means of its components, i.e. $\mathcal{B} = (V_b = (R_b, E_b), K = \{0, 1\}, q_b, g_b)$. The task is built as follows:

- All possible orders in each node of the initial task are "encoded" in a certain suitable way using binary variables. Each such variable is represented by a node in the constructed OrAnd task, its states are possible boolean values, i.e. 0 or 1.
- The submodularity requirements (8) are expressed in form of boolean constraints, which are defined for certain pairs of the introduced binary variables.

Let us consider these steps in detail.

To build the set of nodes R_b for the constructed OrAnd task, let us consider the set of states K for a node r of the initial MinSum task. We introduce in the new task one node (denoted by $s \in R_b$) for each pair of states (k_1, k_2) of the initial set of states K. The states in the introduced nodes s correspond to the possible order relations for the corresponding state pairs (k_1, k_2). There are only two possibilities: either the second state is located above the first one – i.e. $k_1 \prec k_2$ (state 0 in the corresponding node s) or vice versa – i.e. $k_1 \succ k_2$ (state 1). The constructed set of nodes is illustrated by an example in Fig. 1. In this figure the nodes are shown by squares (in both, the initial MinSum task and the constructed OrAnd task), states are depicted by circles. The left part of the figure shows a node of the MinSum task with three states k_1, k_2 and k_3. In the right side the corresponding part of the constructed OrAnd problem is shown. All orders π_r are encoded using three binary variables – s_1 (corresponds to the pair (k_1, k_2) of the initial MinSum task), s_2 (corresponds to the pair (k_2, k_3)) and s_3 (corresponds to the pair (k_1, k_3)). The state 0 in the node s_1 means "the state k_1 is located below the state k_2" etc. The labeling $f(s_1) = 0$, $f(s_2) = 1$ and $f(s_3) = 0$ corresponds e.g. to the

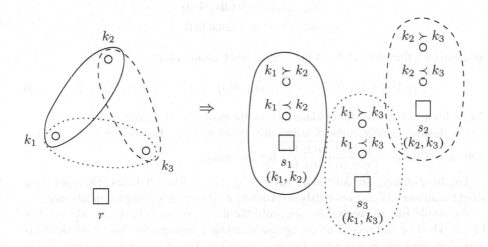

Fig. 1. Constructed set of nodes

order $k_1 \prec k_3 \prec k_2$ in the initial MinSum task. Summarized, the node set R_b of the constructed OrAnd task is

$$R_b = \{(r, k_1, k_2) \mid r \in R; \; k_1, k_2 \in K; \; k_1 \neq k_2\},$$
$$\text{so that} \quad (r, k_1, k_2) \in R_b \Leftrightarrow (r, k_2, k_1) \notin R_b. \tag{9}$$

The state set is $K_b = \{0, 1\}$, labelings are mappings $f : R_b \to \{0, 1\}$, the set of all labelings is denoted by $\mathcal{F}_b = \{0, 1\}^{R_b}$.

It can be easily seen, that not all labelings f in the constructed OrAnd task correspond to an ordering π of the initial MinSum task, because an arbitrary chosen labeling corresponds in general to a relation for the set of states, which is not necessarily transitive[3]. Therefore the set of labelings $\mathcal{F}_b$ in the constructed OrAnd task is an overset of the set of orderings $\mathcal{P}$ of the initial MinSum problem – i.e. $\mathcal{F}_b \supset \mathcal{P}$.

To express the submodularity conditions (8) let us consider a state pair (k_1, k_2) for a node r (a node s in the constructed OrAnd task) and a state pair (k_1', k_2') for another node r' (node s'). Let us assume, that $\alpha_{rr'}(k_1, k_2, k_1', k_2') < 0$ holds (other cases $\alpha_{rr'}(k_1, k_2, k_1', k_2') > 0$ and $\alpha_{rr'}(k_1, k_2, k_1', k_2') = 0$ can be considered analogously). It follows from the requirements (8) and property (6), that only two combinations of relations for these state pairs are possible in those orderings, which transform the MinSum task into a submodular one: $k_1 \prec k_2, \; k_1' \prec k_2'$ or $k_2 \prec k_1, \; k_2' \prec k_1'$ (these two state pairs should be ordered "coherently"). To represent this condition we link the nodes s and s' by an edge and define the function $g_{bss'} : \{0, 1\} \times \{0, 1\} \to \{0, 1\}$ on this edge as follows. We disable those combinations of orders, which violate submodularity. For example in case $\alpha < 0$

$$g_{bss'}(0, 0) = 1 \text{ (enabled)}$$
$$g_{bss'}(0, 1) = 0 \text{ (disabled)}$$
$$g_{bss'}(1, 0) = 0 \text{ (disabled)}$$
$$g_{bss'}(1, 1) = 1 \text{ (enabled).}$$

Summarized, the edge set E_b of the constructed OrAnd task is

$$E_b = \{(s, s') = \big((r, k_1, k_2), (r', k_1', k_2')\big) \mid s, s' \in R_b; \; (r, r') \in E\}. \tag{10}$$

The rules to express the submodularity conditions (8) are summarized in Fig. 2. In this figure allowed combinations of orders are shown by lines. Binary functions $g_{bss'}$ are written in form $g = \dfrac{g(1,0) \; \mid \; g(1,1)}{g(0,0) \; \mid \; g(0,1)}$ for shortness.

The functions q_{bs} are defined as $q_{bs}(0) = q_{bs}(1) = 1$ for all nodes s, because for a single state pair (k_1, k_2) obviously no ordering relation can be disabled in advance.

We would like to emphasize especially the third case $\alpha_{rr'}(k_1, k_2, k_1', k_2') = 0$ in Fig. 2. The first and second cases represent certain constrains for combinations of orders. The third one represents in fact no constraints. This case can be interpreted as if such edges would not exist in the constructed OrAnd task at all. For instance, if for a

---

[3] Reflexivity and asymmetry are obviously preserved by construction.

Case	Constraints	$g_{bss'}$
$\alpha_{rr'}(k_1, k_2, k_1', k_2') < 0$		$\begin{array}{\|c\|c\|}\hline 0 & 1 \\\hline 1 & 0 \\\hline\end{array}$
$\alpha_{rr'}(k_1, k_2, k_1', k_2') > 0$		$\begin{array}{\|c\|c\|}\hline 1 & 0 \\\hline 0 & 1 \\\hline\end{array}$
$\alpha_{rr'}(k_1, k_2, k_1', k_2') = 0$		$\begin{array}{\|c\|c\|}\hline 1 & 1 \\\hline 1 & 1 \\\hline\end{array}$

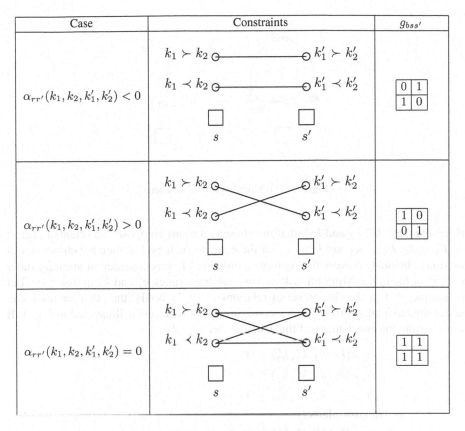

Fig. 2. Construction of functions $g_{bss'}$ for the auxiliary OrAnd task

node s all edges (s, s') are of the third type, then an arbitrary state $f(s)$ can be chosen – i.e. the consistency of a labeling does not depend on the state, chosen in that node.

Since the constructed OrAnd task is binary and only of second order, it is exactly solvable (see [3,17] for details). It is easy to see, that in our case it can be solved even more efficiently as in the general case due to the specific type of the functions g_b.

Obviously, the initial MinSum task is not permuted submodular if there is no consistent labeling in the constructed OrAnd problem. Let us consider the opposite case – there exists a consistent labeling, i.e. a tuple π of relations, which satisfy submodularity conditions (8). As already said, the relations chosen in such a way do not necessarily represent complete orderings of the state set in each node, because this relations may be not transitive. Now we show, that the OrAnd task described above, represents not only the necessary conditions, but sufficient conditions as well. We prove this by showing the following. If there exists at least one consistent labeling in the constructed OrAnd task, than there exists at least one labeling, which corresponds to a complete ordering π of the initial MinSum task.

Let us consider a *consistent* labeling, which *does not* correspond to a complete ordering. It means, that there exist a node r of the initial task and such three states (let

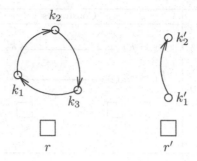

Fig. 3. An "oriented cycle" of relations

us denote them k_1, k_2 and k_3), that the chosen relations compose "an oriented cycle", i.e. $k_1 \prec k_2$, $k_2 \prec k_3$ and $k_3 \prec k_1$ (if there are no such cycles, then the chosen set of relations obviously represents a complete ordering). Let us consider an arbitrary other node r' of the initial MinSum task and two arbitrary states k_1' and k_2' in that node. Let us assume, that in the chosen set of relations $k_1' \prec k_2'$ holds (the other situation can be considered analogously). Such a configuration of relations is illustrated in Fig. 3. It follows from the consistency of this labeling (see Fig. 2):

$$\alpha_{rr'}(k_1, k_2, k_1', k_2') \leq 0$$
$$\alpha_{rr'}(k_2, k_3, k_1', k_2') \leq 0$$
$$\alpha_{rr'}(k_3, k_1, k_1', k_2') \leq 0$$

or equivalentely

$$\alpha_{rr'}(k_1, k_3, k_1', k_2') \geq 0 \quad \text{(due to property (6) of } \alpha\text{-s).}$$

On the other hand

$$\alpha_{rr'}(k_1, k_2, k_1', k_2') + \alpha_{rr'}(k_2, k_3, k_1', k_2') = \alpha_{rr'}(k_1, k_3, k_1', k_2')$$

holds (property (7) of α-s). This yields

$$\alpha_{rr'}(k_1, k_2, k_1', k_2') = \alpha_{rr'}(k_2, k_3, k_1', k_2') = \alpha_{rr'}(k_3, k_1, k_1', k_2') = 0.$$

Let us consider for instance the pair (k_1, k_2) in the node r, which corresponds to a node s in the constructed OrAnd task (the same holds for other pairs (k_2, k_3) and (k_1, k_3)). Since the node r' and the states k_1' and k_2' can be chosen arbitrary, it follows:

$$\alpha_{rr'}(k_1, k_2, k_1', k_2') = 0 \quad \text{for all} \quad (r, r') \in E; \ k_1', k_2' \in K$$

or equivalently (see Fig. 2)

$$g_{bss'} = \begin{array}{|c|c|} \hline 1 & 1 \\ \hline 1 & 1 \\ \hline \end{array} \quad \text{for all} \ (s, s') \in E_b.$$

It means, that there are no constraints for the state pair (k_1, k_2) in the initial MinSum task. It can be interpreted as if the corresponding node s in the constructed OrAnd task is not connected to any other node. Consequently the state in s (i.e. relation for k_1 and k_2) can be chosen arbitrary without affecting the consistency of the labeling. Therefore it is possible to avoid the situation given in Fig. 3 by choosing e.g. $k_1 \succ k_2$ keeping thereby the whole labeling consistent. In doing so submodularity conditions (8) remain satisfied.

Start with $K' = \{k \in K\}$ (an arbitrary state from K);
Repeat until $K' = K$:
  1 Take a state $k \in K/K'$, which is not in the ordered subset K' yet;
  2 If $k \prec \sup K'$ according to π_r, then set $k \prec k'$ for all $k' \in K'$, go to 5;
  3 Let $k^* = \inf\{k' \in K' \mid k' \prec k\}$;
  4 Set $k \succ k'$ for all $k' \in K'$, $k' \prec k^*$;
  5 Insert k into K'.

Fig. 4. Algorithm for ordering the state set

It is easy to see, that it is possible to avoid all such directed cycles and to obtain complete orders of the state set in all nodes r of the initial MinSum task. Obviously, it can be done for each node independently e.g. by the following procedure. Let us denote by $K' \subseteq K$ a completely ordered subset of K. Consider a node r and the relations π_r given by a solution of the auxiliary OrAnd task (π_r may be not transitive). The algorithm for ordering the state set in the node r is given in Fig. 4. Since the subset K' is completely ordered, the operations sup and inf are well defined for each non empty subset of K'. Hence it is possible to prove the condition $k \prec \sup K'$ in step 2 and take k^* in step 3. Step 2 is necessary only in order to avoid the situation, that the set $\{k' \in K' \mid k' \prec k\}$ in step 3 is empty. In steps 2 and 4 the relations given by the OrAnd task can be changed, but only for those state pairs, which compose an

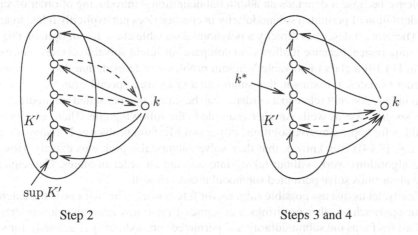

Step 2   Steps 3 and 4

Fig. 5. Algorithm for ordering the state set (an example)

oriented cycle. Therefore the submodularity conditions remain preserved (as explained before). After the reassignments in steps 2 or 4 there are no oriented cycles in the set $K' \cup \{k\}$, i.e. this set is completely ordered. Therefore the set K' remains completely ordered after step 5. The ordering procedure is illustrated in Fig. 5. Relations, given by the OrAnd task, which are reassigned by this procedure (an example), are shown by dashed arrows.

4   Conclusion

We have shown in this work, that for each *permuted submodular* MinSum problem the corresponding *submodular* MinSum problem can be found in polynomial time. It follows, that permuted submodular MinSum problems are exactly solvable by transforming them into corresponding submodular tasks followed by applying standart approaches.

At this point we would like to discuss possible generalizations of our method. It is easy to see, how to extend it for the case of different state sets for each node. It is also not hard to see, that the approach can be extended for permuted submodular tasks of order, higher than two. The main idea of this generalization is based on the fact, that the submodularity of a MinSum task (of arbitrary order) can be formulated using constraints, that are similar to (8) – i.e. they are always defined only for four-tuples of states, independent of the order of the task. Therefore it is again possible to formulate the "task of ordering estimation" as a binary OrAnd problem of only second order – i.e. a solvable OrAnd task. Furthermore, it can be easily seen, that the properties (6) and (7) of α-s also remain. Therefore it can be again stated, that the constructed OrAnd task represents both necessary and sufficient conditions for the initial MinSum problem to be permuted submodular.

We have already mentioned the disadvantages of the conventional definition of submodularity. This definition does not allow to characterize a class of solvable MinSum problems, because it demands an additional definition – introducing of order of states. The definition of permuted submodularity in contrast does not explicitly relate to an order. Therefore it describes directly a polynomial solvable class. In our opinion this fact is of importance, because it allows "to compare" different solvable classes. For example, in [14,19] a class of solvable MinSum problems is defined, that does not relate to an order of states. It is shown, that submodular tasks are a special case. Since the class given in [14] does not relate to an order, it can be stated, that permuted submodular tasks are a special case as well. Another example is the following one. There are many algorithms for approximative solutions of general MinSum problems. For some of them (see e.g. [9,14]) it is known, that they solve submodular problems exactly. However these algorithms work without taking into account an order of states. Consequently, these algorithms solve permuted submodular tasks as well.

Finally, let us discuss possible subjects for future work. The first one is the extention of our approach to other semirings. For some of them this generalization is straightforward (as far as the submodularity and permuted submodularity is defined), for some other cases not. The main difficulty is, that the numbers α (see (5)) can not be defined, if the considered semiring has no operation, inverse to $\otimes$ – for example the OrAnd semiring has no operation, inverse to $\wedge$. Even for MinSum tasks our approach does not work, if there are infinite qualities of $g_{rr'}$ in the task.

At the moment, we do not see clearly a practical application, where the optimization of a permuted submodular functional appears, because in practice functionals to be optimized (their terms of second or higher order) are as a rule known in advance – they often mirror prior knowledge about the particular application. The situation is different, if the parameters of the model (for example the functions $g_{rr'}$) are learned, i.e. not a-priori known. In this context it is necessary to approximate a particular MinSum task by a "nearest" permuted submodular one. Such type of tasks can be seen as a "fuzzy variant" of the auxiliary OrAnd task, described in this work.

Acknowledgment

This work was partially supported by the EU INTAS project PRINCESS 04-77-7347.

References

1. Besag, J.: On the statistical analysis of dirty pictures (with discussion). Journal of the Royal Statistical Society, Series B 48(3), 259–302 (1986)
2. Boykov, Y., Veksler, O., Zabih, R.: Fast approximate energy minimization via graph cuts, ICCV, pp. 377–384 (1999)
3. Flach, B.: Strukturelle Bilderkennung: Habilitationsschrift, Dresden University of Technology, in German (2003)
4. Flach, B., Schlesinger, D., Kask, E., Skulisch, A.: Unifying registration and segmentation for multi-sensor images. In: Van Gool, L. (ed.) Pattern Recognition. LNCS, vol. 2449, pp. 190–197. Springer, Heidelberg (2002)
5. Greig, D.M., Porteous, B.T., Seheult, A.H.: Exact maximum a posteriori estimation for binary images. J. R. Statist. Soc. 51(2), 271–279 (1989)
6. Ishikawa, H.: Exact optimization for markov random fields with convex priors. IEEE Transactions on Pattern Analysis and Machine Intelligence 25(10), 1333–1336 (2003)
7. Ishikawa, H., Geiger, D.: Segmentation by grouping junctions. In: IEEE Computer Society Conference on Computer Vision and Pattern Recognition (1998)
8. Keuchel, J., Schnörr, C., Schellewald, C., Cremers, D.: Binary partitioning, perceptual grouping, and restoration with semidefinite programming. IEEE Transactions on Pattern Analysis and Machine Intelligence 25(11), 1364–1379 (2003)
9. Kolmogorov, V.: Convergent tree-reweighted message passing for energy minimization. IEEE Transactions on Pattern Analysis and Machine Intelligence (PAMI) 28(10), 1568–1583 (2006)
10. Kolmogorov, V., Zabih, R.: Computing visual correspondence with occlusions via graph cuts. In: International Conference on Computer Vision, pp. 508–515 (2001)
11. Kolmogorov, V., Zabih, R.: What energy functions can be minimized via graph cuts? In: Heyden, A., Sparr, G., Nielsen, M., Johansen, P. (eds.) ECCV 2002. LNCS, vol. 2352, pp. 65–81. Springer, Heidelberg (2002)
12. Kovtun, I.: Texture segmentation of images on the basis of markov random fields, Tech. report, TUD-FI03 (May 2003)
13. Schlesinger, M.I., Hlaváč, V.: Ten lectures on statistical and structural pattern recognition. Kluwer Academic Publishers, Dordrecht (2002)

14. Schlesinger, M.I., Giginyak, V.V.: Solution to structural recognition (max,+)-problems by their equivalent transformations, Control Systems and Machines, Naukova Dumka, Kiev, no. 1,2, in Russian (2007)
15. Schlesinger, D.: Gibbs probability distributions for stereo reconstruction. In: Michaelis, B., Krell, G. (eds.) Pattern Recognition. LNCS, vol. 2781, pp. 394–401. Springer, Heidelberg (2003)
16. Schlesinger, D., Flach, B.: Transforming an arbitrary minsum problem into a binary one, Tech. report, Dresden University of Technology, TUD-FI06-01 (April 2005), http://www.bv.inf.tu-dresden.de/~ds24/tr_kto2.pdf
17. Schlesinger, M.I., Flach, B.: Some solvable subclasses of structural recognition problems. In: Czech Pattern Recognition Workshop 2000 (Svoboda, T. (ed.)), pp. 55–62 (2000)
18. Shlezinger, D.: Strukturelle Ansätze für die Stereorekonstruktion, Ph.D. thesis, Dresden University of Technology, in German (2005), http://nbn-resolving.de/urn:nbn:de:swb:14-1126171326473-57594
19. Werner, T.: A linear programming approach to max-sum problem: A review, Tech. Report CTU–CMP–2005–25, Center for Machine Perception, K13133 FEE Czech Technical University (December 2005)

Efficient Shape Matching Via Graph Cuts*

Frank R. Schmidt[1], Eno Töppe[1], Daniel Cremers[1], and Yuri Boykov[2]

[1] Department of Computer Science,
University of Bonn, Germany
{schmidtf,toeppe,dcremers}@iai.uni-bonn.de
[2] Department of Computer Science,
University of Western Ontario,
London ON, Canada
yuri@csd.uwo.ca

Abstract. Meaningful notions of distance between planar shapes typically involve the computation of a correspondence between points on one shape and points on the other. To determine an optimal correspondence is a computationally challenging combinatorial problem. Traditionally it has been formulated as a shortest path problem which can be solved efficiently by Dynamic Time Warping.

In this paper, we show that shape matching can be cast as a problem of finding a minimum cut through a graph which can be solved efficiently by computing the maximum network flow. In particular, we show the equivalence of the minimum cut formulation and the shortest path formulation, i.e. we show that there exists a one-to-one correspondence of a shortest path and a graph cut and that the length of the path is identical to the cost of the cut. In addition, we provide and analyze some examples for which the proposed algorithm is faster resp. slower than the shortest path method.

1 Introduction

1.1 Metrics for Shapes

The definition of metrics for different classes of objects is a fundamental challenge in Computer Vision. To quantify how similar two given objects are is of central importance for object recognition, clustering, classification, retrieval and statistical modeling. The definition of metrics for a given object class is not a trivial matter, it is typically coupled to the problem of determining a correspondence between parts of one object and parts of the other. The efficient computation of an optimal correspondence is generally a hard computational challenge.

In this paper, we are concerned with the definition and efficient computation of metrics for the class of planar shapes, i.e. closed curves embedded in $\mathbb{C}$. In order to abstract from location and rotation, in this paper the term *shape* refers to the equivalence class of a closed curve in $\mathbb{C}$ under the action of the Special

* This work was supported by the German Research Foundation, grant #CR-250/1-1.

A.L. Yuille et al. (Eds.): EMMCVPR 2007, LNCS 4679, pp. 39–54, 2007.
© Springer-Verlag Berlin Heidelberg 2007

Euclidean group $SE(2)$. Thus, two curves have the same shape if one can be transformed into the other by rotation and translation.

Yet, how do we quantify the distance between two shapes if they are not identical, i.e. if one curve cannot be obtained by rotation and translation of the other? To merely compute the L_2-distance between the two curves (up to rotation and translation) will generally not lead to a distance that is consistent with human notions of shape similarity. To mimic human notions of similarity, one needs to take into account that a shape consists of several parts which may be dislocated, articulated or missing from one shape to the other. The computation of the shape distance will then require to robustly match respective points of one curve to points of the other. To propose a novel framework for shape matching which allows to efficiently compute this correspondence is the key contribution of this paper.

1.2   Related Work and Contribution

The study of shape and shape similarity has a long tradition going back to works of Galilei [8] and Thompson [21]. From a wealth of literature on shape and shape metrics, we will merely discuss a few more closely related works. A review of the history of shape research can be found in [5]. A mathematical definition of shape as the equivalence class under a certain transformation group goes back to Kendall [13]. A shape metric based on the computation of elastic deformations between two shapes was developed in [3]. There exist numerous shape descriptors which capture the local shape by means of differential or integral invariants, the most commonly considered descriptor being curvature [17]. For a detailed discussion of different kinds of invariants we refer to [15].

In this work, we are not focused on the introduction of new invariants, but rather on the question of how to efficiently compute a matching given any local shape descriptor. Ideally the matching should aim at putting in correspondence points on each shape that have similar local descriptors, at the same time it should penalize local stretching or shrinking of one curve with respect to the other. Traditionally this shape matching has been cast as a shortest path problem through a two-dimensional planar graph, the edge weights of which incorporate the distance of the local descriptors and a penalty for local stretching [16,10,3,2,9,14,22,19,18]. While a shortest path through the respective graph can be computed efficiently using *Dynamic Time Warping* (DTW), one of the key drawbacks of this approach is that Dynamic Time Warping requires a corresponding point pair for initialization. The most current methods therefore apply DTW for all possible initial correspondences, and then select the minimum of all computed shortest paths as the distance between the two shapes. More efficient formulations were proposed in [19,1].

In this work, we will cast the shape matching problem as a problem of finding the minimum cut through a graph. In contrast to the shortest path formulation, the graph cut approach does not require a separate optimization over the initial correspondence. While the graph cut method has recently obtained considerable attention in the Computer Vision community, it has been mainly applied to the problems of image segmentation and stereo reconstruction [11,12]. To the best of our knowledge, the idea of shape matching using graph cuts is new.

In the next section, we will construct a graph such that the matching of two shapes is equivalent to a cut through the graph. As a consequence, an optimal matching can be computed by finding the minimal cut. In Section 3, we show the equivalence of the graph cut formulation with the traditional shortest path formulation. In Section 4, we present an integral descriptor to approximate the curvature. This descriptor provides a robust matching with respect to articulations and noise as we will show in Section 5. To analyze the runtime, we compare an example for which the proposed method outperforms the shortest path method with an example of the opposite property and in Section 6, we provide a conclusion.

2   Shape Matching Via Graph Cuts

In the following, we will cast the problem of matching two planar shapes as a problem of cutting an appropriate graph. First, we will present the connection between continuous matchings and cutting a cylinder into two different surfaces. Afterwards, we will formulate the cylinder as a graph and the matching problem as a graph cut problem.

2.1   Connection Between Matchings and Graph Cuts

It is well known that the curvature of a curve $c : \mathbb{S}^1 \to \mathbb{C}$ is invariant under rigid body motions. Thus, every *shape* $\mathcal{C}$ is uniquely represented by its curvature function $\kappa : \mathbb{S}^1 \to \mathbb{R}$. The set of all shapes will form the *shape space* $\mathcal{S}$ which can formally be defined as the orbit space of all uniform embeddings $\mathrm{Emb}_u(\mathbb{S}^1, \mathbb{C})$ under the left action of the Special Euclidian Group SE(2).

Besides rigid body motions, other *shape transformations* are possible. In most applications, we want to detect local stretching or contraction. Hence, we are looking for a direct correspondence mapping which maps the points of one shape to the correspondent points of the other shape. Since the points of a shape form an arbitrary subset of the plane $\mathbb{C}$, it is easier to find the correspondence directly on the parameterization domain $\mathbb{S}^1$ (cf. Figure 1).

To avoid self-occlusions during the matching process, a *matching* can be modeled via an orientation-preserving diffeomorphism $m : \mathbb{S}^1 \to \mathbb{S}^1$ that maps points of the first parameterization domain to the corresponding points of the second parameterization domain. On the space of these matchings, we will define a functional $E : \mathrm{Diff}^+(\mathbb{S}^1) \to \mathbb{R}^+$ that measures the goodness of a matching. The goal of a matching algorithm is to find the minum of E which will mainly measure the L^2-distance of the curvature functions. But since any matching m allows to stretch and to contract a given shape until it fits to another shape, it is possible that a part of one shape collides to an arbitrary small interval on the other

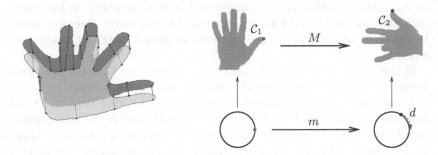

Fig. 1. Matching and disparity function. *Left hand side:* Matching two shapes amounts to computing a correspondence between pairs of points on both shapes. *Right hand side:* Instead of looking for a mapping $M : \mathcal{C}_1 \to \mathcal{C}_2$, a matching $m : \mathbb{S}^1 \to \mathbb{S}^1$ is defined on the parameterization domain. The distance between s and $m(s)$ defines the disparity function $d : \mathbb{S}^1 \to \mathbb{R}$.

shape. To avoid this side effect, the elastic variation of a matching is penalized. Thus, we are interested in the following energy functional

$$E(m) := \int_{\mathbb{S}^1} \left[\kappa_1(s) - \kappa_2(m(s)) \right]^2 \sqrt{1 + m'(s)^2} \, ds + \alpha \cdot \int_{\mathbb{S}^1} |1 - m'(s)| \, ds.$$

In most applications, we are interested in the disparity function $d : [0; 2\pi] \to \mathbb{R}$. This disparity $d(s) := s - m(s)$ can be used to denote the displacement of each point on one contour when mapped to the other. (cf. Figure 1). The energy functional becomes with respect to d the energy functional that was used in [15][1]:

$$E(d) := \int\limits_0^{2\pi} \left[\kappa_1(s) - \kappa_2(s - d(s)) \right]^2 \sqrt{1 + (1 - d')^2} \, ds + \alpha \int\limits_0^{2\pi} |d'(s)| \, ds. \tag{1}$$

In the following, the matching problem will be formulated as a disparity problem. Since a disparity d has the circle $\mathbb{S}^1$ as parameterization domain and the real numbers $\mathbb{R}$ as image set, the graph $\Gamma(d)$ of d is a closed curve on the cylinder $\mathbb{S}^1 \times \mathbb{R}$. We will call this cylinder the *disparity cylinder*.

On the other hand, the graph $\Gamma(m)$ of a matching mapping $m : \mathbb{S}^1 \to \mathbb{S}^1$ presents a loop on the torus $\mathbb{S}^1 \times \mathbb{S}^1$. Therefore, the question arises how the loop $\Gamma(m)$ on the torus relates to the loop $\Gamma(d)$ on the cylinder. To illustrate this connection, we present a geometrical approach to obtain the disparity cylinder. In Figure 2, the graph of the matching mapping id is represented by a *diagonal loop*. If we cut the given torus open along $\Gamma(\mathrm{id})$, we obtain a cylinder with $\Gamma(\mathrm{id})$ as left *and* right boundary. These boundaries represent the graph of the disparity

---

[1] While the authors of [15] implemented (1), they omitted the scaling factor $\sqrt{1 + (1 - d')^2}$ in their formulation. Moreover, they claimed to integrate the L^2-distance of d' instead of the L^1-distance.

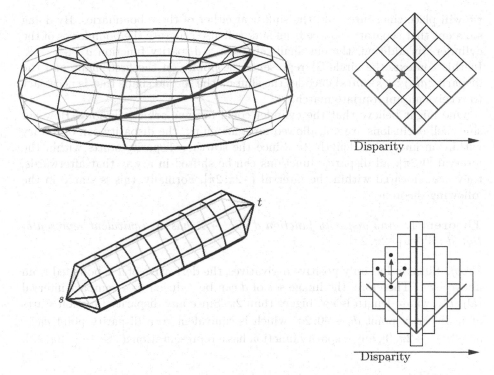

Fig. 2. Disparity cylinder. The two curves C_1 and C_2 are both parameterized over $\mathbb{S}^1$. The product space $\mathbb{S}^1 \times \mathbb{S}^1$ of all possible correspondences forms a torus (left hand side of the top row). If we cut this torus open along the diagonal, we receive a cylinder of which a small patch is shown in the top row on the right hand side. Here every vertical row shows matchings of constant disparity which can be obtained by an $s - t$-separating graph cut. To allow the dual edge set of a graph cut to pass through faces of constant disparity, we use a cylinder of which a small patch is shown in the bottom row on the right hand side.

function $d(s) = 0$ and $d(s) = 2\pi$ respectively. Since we do not like to restrict the values of any disparity function to the interval $[0; 2\pi]$, we glue different copies of this cylinder together to obtain a bigger cylinder. On this cylinder, the former torus loop $\Gamma(\mathrm{id})$ becomes the cylinder loop $\Gamma(0)$. Therefore, disparity loops and matching loops are directly coupled and any disparity loop provides a boundary separating cut.

In the next section, we will model the cylinder via an algebraic graph G. In this construction, we have to take into account that the dual edge set of any cut will be a sufficient representation of a disparity function and vice versa (cf. Theorem 3).

2.2   Graph Construction

As we have sketched above, we want to define a *cylindrical graph* in a way that the minimal graph cut will separate the two boundaries from one another. Therefore, we will place the source and the sink near either of these boundaries. By doing so, a cut that separates source from sink will also separate the boundaries of the cylinder. In addition, the cut shall represent a disparity function $d : \mathbb{S}^1 \to \mathbb{R}$. In this context, the circle $\mathbb{S}^1$ represents the points on the first shape and the disparity $d(s) := s - m(s)$ encodes the shift on the second shape that is necessary to receive an appropriate match.

One might believe that the cylinder could become arbitrarily long. However since self-occlusions are not allowed for a matching, the disparity can only vary within an interval of length 2π. Since the starting disparity starts within the interval $[0; 2\pi]$, all disparity functions can be shifted in a way that afterwards, they are relocated within the interval $[-2\pi; 2\pi]$. Formally, this is stated in the following theorem.

Theorem 1. *Any disparity function $d : \mathbb{S}^1 \to \mathbb{R}$ has an equivalent representation $\hat{d} : \mathbb{S}^1 \to [-2\pi; 2\pi]$.*

Proof. Since m has only positive derivatives, the derivative of d is bounded from above by 1. Therefore, the image set of d can be reduced to a compact interval $[D_1; D_2]$ whose length is not bigger than 2π. Since any disparity function starts at a disparity point $d_0 \in [0; 2\pi]$ which is equivalent to a disparity point $\hat{d}_0 = d_0 - 2\pi \in [-2\pi; 0]$, any disparity function has a representation $\hat{d} : \mathbb{S}^1 \to [-2\pi; 2\pi]$. $\qquad\square$

As illustrated at the right hand side in the top row of Figure 2, the canonical graph inhibits a path along vertices of constant disparity. Therefore, we choose a graph G with the property that the dual graph G^* allows three different transitions that are sketched on the right hand side in the bottom row of Figure 2. The explicit construction of this graph is the goal of this section.

According to Theorem 1, it suffices to model a compact cylinder instead of a cylinder whose boundaries are positioned at infinity. This cylinder shall be represented by the Cartesian product of an interval I and a circle S:

$$I := \{-(N + 0.5), \ldots, -1.5, -0.5, 0.5, 1.5, \ldots, N + 0.5\}$$
$$S := \{0, 0.5, 1, \ldots, N - 1, N - 0.5\}$$

Note that the circle is represented by a modulo space to assure that by increasing the points on the circle, we will eventually return to the starting point. Formally, $(N - 0.5) + 0.5 \equiv 0 \mod N$.

Now, we can construct the cylindrical graph $G = (V, E, s, t, c)$ with vertices V, edges $E \subset V \times V$, source $s \in V$, sink $t \in V$ and the capacity function $c : E \to [0; \infty]$. We define the set of vertices $V := Z \sqcup \{s, t\}$ as the disjoint union of the cylinder $Z := I \times S$ with the set of the source s and the sink t. Therefore, every vertex $v = (v_d, v_x) \in Z$ consists of a disparity value v_d and a point v_x on a circle. Here, v_x shall encode a point on the first shape $\mathcal{C}_1$ and $v_x - v_d$ a point on the second shape $\mathcal{C}_2$. Thus, v_d encodes the disparity $d(v_x)$. Note that if v_d is an integer, the pair (v_d, v_x) is not an element of Z. Some of these integer pairs shall in fact represent the faces of the cylinder. Every graph cut cuts the cylinder open along these faces and describes a closed path in the dual graph G^*. Therefore, the pairs (v_d, v_x) with an integer v_d are mainly reserved for a convenient representation of the graph cut. Let us return to the graph construction. The edges E shall connect the source s with the *left boundary* of the cylinder and the *right boundary* of the cylinder with the sink t (cf. Figure 2). Moreover, some direct neighbors on the cylinder are connected in a way that a structure like a brick wall emerges (cf. Figure 3):

$$
\begin{aligned}
E := \quad & \{s\} & \times & \quad (\{-(N+0.5)\} \times S) & \cup \\
& (\{N+0.5\} \times S) \times & & \quad \{t\} & \cup \\
& \left\{ (v^1, v^2) \in Z \times Z \,\middle|\, v^2 - v^1 = \pm\binom{0}{0.5} \right\} & & & \cup \\
& \left\{ (v^1, v^2) \in Z \times Z \,\middle|\, v^2 - v^1 = \binom{1}{0}, v^1 + \binom{0.5}{-0.5} \in (2\mathbb{Z}) \times \mathbb{Z} \right\} & & & \cup \\
& \left\{ (v^1, v^2) \in Z \times Z \,\middle|\, v^2 - v^1 = \binom{1}{0}, v^1 - \binom{0.5}{0} \in (2\mathbb{Z}) \times \mathbb{Z} \right\} & & & \cup \\
& \left\{ (v^1, v^2) \in Z \times Z \,\middle|\, v^1 - v^2 = \binom{1}{0}, v^2 + \binom{0.5}{-0.5} \in (2\mathbb{Z}) \times \mathbb{Z} \right\} & & & \cup \\
& \left\{ (v^1, v^2) \in Z \times Z \,\middle|\, v^1 - v^2 = \binom{1}{0}, v^2 - \binom{0.5}{0} \in (2\mathbb{Z}) \times \mathbb{Z} \right\} & & &
\end{aligned}
$$

With this construction every rectangular patch

$$
F_{d,x} := [d-0.5; d+0.5] \times \left[x - \frac{d+1}{2}; x - \frac{d-1}{2} \right]
$$

will carry the disparity information of an appropriate matching. In the next section, we show the equivalence of the graph cut approach and the usual shortest path method. Since the capacity c shall encode the functional (1), we are interested in the squared curvature difference as a measure of similarity. Therefore, we may define the *curvature similarity function* $(x, y) \mapsto (\kappa_1(x) - \kappa_2(y))^2$ between a point x on shape C_1 and a point $y := x - d$ on shape C_2. Now, this function can be discretized via a similarity matrix $M \in \mathbb{R}^{N \times N}$ for any given $N \in \mathbb{N}$. This matrix measures the similarity on the vertices of discretized shapes. Since the capacities of the graph G shall carry the similarity of shape edges, this is done by integrating the curvature similarity function. Together with the smoothness term α in (1), we introduce the capacity function:

$$
c(e) := \begin{cases}
\frac{m_{x, x-d} + m_{x+1, x+1-(d+1)}}{2} + \alpha & , \text{ if } e = \left(\binom{d+0.5}{x-\frac{d-1}{2}}, \binom{d+0.5}{x-\frac{d}{2}} \right) \\[2ex]
\frac{m_{x, x-d} + m_{x+1, x+1-d}}{\sqrt{2}} & , \text{ if } e = \left(\binom{d-0.5}{x-\frac{d-1}{2}}, \binom{d+0.5}{x-\frac{d-1}{2}} \right) \\[2ex]
\frac{m_{x, x-d} + m_{x, x-(d+1)}}{2} + \alpha & , \text{ if } e = \left(\binom{d-0.5}{x-\frac{d}{2}}, \binom{d-0.5}{x-\frac{d-1}{2}} \right) \\[2ex]
\infty & , \text{ else}
\end{cases}
\tag{2}
$$

In the next section, we will see that the explicit choice of the capacities will in fact lead to the equivalence of the graph cut approach and the shortest path

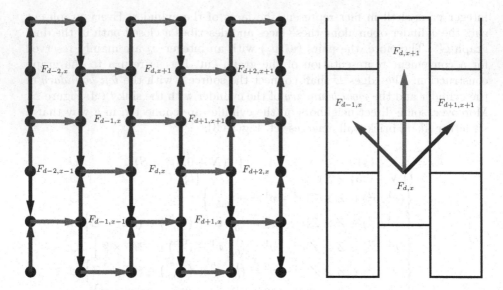

Fig. 3. Graph Construction used for Shape Matching. Every rectangle $F_{d,x}$ corresponds to a point x on shape one and its disparity d in respect to a point on shape two. A cut through this graph therefore assigns a disparity to each point on shape one. The edge weights need to be chosen in such a way that the cut edges measure the difference of curvatures of corresponding points on both shapes (cf. (2)). The three arrows on the right hand side indicate permissible transits: the left arrow amounts to a step along shape two, the right arrow corresponds to a step along shape one, while the vertical arrow indicates a step along both shapes (thus keeping the disparity constant).

method. To conclude the construction, we like to define the dissimilarity of two given shapes in the mean of minimal graph cuts:

Definition 1 (Shape Distance). *Given two shapes $\mathcal{C}_1$ and $\mathcal{C}_2$ with their discretized curvature dissimilarity matrix $M \in \mathbb{R}^{N \times N}$. Via this matrix the above graph $G = (V, E, s, t, c)$ is defined. We will call*

$$\mathrm{dist}(\mathcal{C}_1, \mathcal{C}_2) := \min_{\substack{V = S \sqcup T \\ s \in S, t \in T}} \sum_{e \in E \cap (S \times T)} c(e)$$

the distance between the shapes.

3   Equivalence to Shortest Path Formulation

In this section, we present the equivalence between the graph cut method and the shortest path method. To this end, we will show that every cut of the graph represents a disparity function and vice versa. In addition, we will show that a minimal graph cut of G represents a disparity function that minimizes the functional (1).

As we have pointed out, the graph G describes the surface of a closed cylinder and thus, induces a set $F = \{F_{d,x}\}$ of faces. Since every edge $e \in E$ separates a *right face* $f_r(e) \in F$ from a *left face* $f_l(e) \in F$, a weighted dual graph $G^* = (F, E^*, w)$ can be defined as follows:

$$e^* := (f_r(e), f_l(e)) \qquad\qquad w(e^*) := c(e).$$

In the following theorem, we will show that any graph cut of G will provide a cycle in the dual graph G^* which separates the two boundaries of the cylinder from one another.

Theorem 2. *Let $G = (V, E, s, t, c)$ the cylindrical graph introduced in the last section and $S, T \subset V$ a minimal cut with cut edges $X = E \cap (S \times T)$. Then, the dual set $X^* \subset E^*$ is a cycle in the dual graph G^*. Moreover, X^* separates the two boundaries of the cylinder from one another.*

Proof. Since G is a planar graph, every cut edge set and especially the minimal cut edge set X provides a cycle X^* in G^* [23]. Because the cut $(\widetilde{S}, \widetilde{T})$ with

$$\widetilde{S} := \{(v_x, v_d) \in Z | v_d < 0\} \sqcup \{s\} \qquad \text{and}$$
$$\widetilde{T} := \{(v_x, v_d) \in Z | v_d > 0\} \sqcup \{t\}$$

has finite cut edge costs, there is no edge of infinite capacity in the minimal cut edge set X. Therefore, the left boundary of the cylinder Z belongs to S, the right boundary of Z belongs to T and the path X^* separates the two boundaries from one another. $\qquad\qquad\square$

Since every disparity d separates the two boundaries of the cylinder from one another, d encodes a graph cut in G. According to the possible transits of the dual cut path X^* (right hand side of Figure 3), X^* describes the discretized graph of a function and not a relation which would imply a movement of a path to the right or to the left that do not have an upward component. Therefore, we have proven the following theorem

Theorem 3. *Given the graph G, the dual edge set of any cut is a representation of a disparity function and vice versa.* $\qquad\qquad\square$

Summarized, we have shown that the graph cut method covers the whole space of disparity functions. Now, we will approach the energy functional (1) itself and show that the cut edge costs of any cut X is equal to the cost of an equivalent path according to the shortest path method. To this purpose, we will revisit this method. It chooses an initial matching $(1, y) \in \mathbb{S}^1 \times \mathbb{S}^1$ and afterwards, cuts the image set $\mathbb{S}^1 \times \mathbb{S}^1$ open along the two curves $\{1\} \times \mathbb{S}^1$ and $\mathbb{S}^1 \times \{y\}$. By doing so, one receives a square and the graph of an arbitrary matching $m : \mathbb{S}^1 \to \mathbb{S}^1, m(1) = y$ becomes a path from $(0, 0)$ to $(2\pi, 2\pi)$. Discretizing this square and using the Dynamic Time Warping algorithm, the energy $E(m)$ is minimized over all mappings fulfilling $m(1) = y$. By varying now the fixed value

$y \in \mathbb{S}^1$, the minimum of $E(m)$ will eventually be found. Therefore the algorithm can mathematically be summarized as

$$\min_{m \in \mathrm{Diff}^+(\mathbb{S}^1)} E(m) = \min_{y \in \mathbb{S}^1} \min_{\substack{m \in \mathrm{Diff}^+(\mathbb{S}^1) \\ m(1)=y}} E(m).$$

The proposed graph cut approach on the other hand, calculates the minimum of E directly by exploiting the dual properties of planar graphs. We want to emphasize that the choice of an initial matching is a challenging task, since a continuous variation of a shape will lead to discrete jumps in the matching. By using graph cuts, we solve this problem *continously*, since the used graph cut algorithm [4] uses the max-flow minimal-cut theorem [6,7]. In [15] the functional (1) was represented by a graph with the following edge costs c_{SP}:

$$c_{SP}((x,y),(x+1,y)) = \frac{m_{x,y} + m_{x+1,y}}{2} + \alpha$$

$$c_{SP}((x,y),(x+1,y+1)) = \frac{m_{x,y} + m_{x,y+1}}{\sqrt{2}}$$

$$c_{SP}((x,y),(x,y+1)) = \frac{m_{x,y} + m_{x,y+1}}{2} + \alpha$$

In the proposed graph cut approach, we have to translate the notion (x,y) into the notion (d,x), whereas $d := x - y$. The graph cut costs c_{GC} become therefore respectively (cf. Figure 3 and (2)):

$$c_{GC}(F_{d,x}, F_{d+1,x+1}) = \frac{m_{x,x-d} + m_{x+1,x+1-(d+1)}}{2} + \alpha = c_{SP}((x,y),(x+1,y))$$

$$c_{GC}(F_{d,x}, F_{d,x+1}) = \frac{m_{x,x-d} + m_{x+1,x+1-d}}{\sqrt{2}} = c_{SP}((x,y),(x+1,y+1))$$

$$c_{GC}(F_{d,x}, F_{d-1,x}) = \frac{m_{x,x-d} + m_{x,x-(d-1)}}{2} + \alpha = c_{SP}((x,y),(x,y+1))$$

This proves the equivalence of the graph cut algorithm and the shortest path method to calculate an optimal matching of two given shapes.

4   Integral Invariants and Curvature

The shape matching via graph cuts as introduced above relies on local features such as curvature. In practice, these need to compute in a robust manner. To this end, [15] introduced features via integrals. Since these features were invariant under rigid body motions, they were called *integral invariants*. One of these integral invariants approximated the curvature by calculating the intersection of the shape's interior and a circle of fixed radius r (cf. Figure 4).

In contrast to [15], we perform a Taylor approximation of the invariant which is *exact* up to first order. Therefore, consider $c : \mathbb{S}^1 \to \mathbb{C}$ a closed curve with

Fig. 4. Curvature calculation. The curvature at any point along the curve can be estimated from the intersection A_r of a ball with radius r centered at the curve point with the interior of the shape. $R = \frac{1}{\kappa}$ is the radius of the osculating curve (cf. (3)).

its curvature function $\kappa : \mathbb{S}^1 \rightarrow \mathbb{R}$. Near the point $c(t)$, the curve c can be described via a circle with radius $R(t) := \frac{1}{\kappa(t)}$. This approximation is a second order approximation of c. For a radius r, let A_r be the area of the set $\{x \text{ inside } c| \, \|x - c(t)\|^2 \leq r^2\}$. Thus, we obtain

$$A_r \approx \int_{-a}^{a} \sqrt{R^2 - \tau^2} - \left[R - \sqrt{r^2 - \tau^2} \right] d\tau$$

$$= R^2 \sin^{-1}\left(\frac{a}{R}\right) + r^2 \sin^{-1}\left(\frac{a}{r}\right) - Ra$$

whereas $a = \sqrt{r^2 - \left(\frac{r^2}{2R}\right)^2}$. Introducing $\varphi := \sin^{-1}\left(\frac{r}{2R}\right)$, we receive

$$\frac{A_r}{r^2} \approx \frac{1}{2}\left(\frac{\varphi}{\sin(\varphi)^2} - \frac{\cos(\varphi)}{\sin(\varphi)} \right) + \frac{\pi}{2} - \varphi$$

The linear Taylor approximation of the right hand side leads to the expression $\frac{\pi}{2} - \frac{2}{3}\varphi$. Therefore, the curvature κ can be approximated via

$$\kappa \approx \frac{2}{r} \sin\left(\frac{3\pi}{4} - \frac{3A_r}{2r^2} \right) \tag{3}$$

Note that the quadratic approximation error can be reduced by decreasing the radius r. Moreover, $\kappa = \lim_{r \to 0} \frac{2}{r} \sin\left(\frac{3\pi}{4} - \frac{3A_r}{2r^2}\right)$. In our implementation, we used the right hand side of (3) to calculate the curvature function of a given curve.

5 Experimental Results

In this section, we will present the results of the presented matching method. by starting with some applications like articulations and noise. In the last subsection, we will analyze the runtime of the graph cut method and the shortest path method. Except for the noise examples, we always used shapes that were provided by the LEMS laboratory of the Brown University [20].

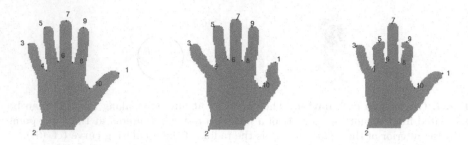

Fig. 5. Articulation. The matching is visualized via numbered shape points. *Left:* The original shape with 10 selected points. *Middle:* A shape with an articulated thumb. *Right:* A shape with two articulated fingers.

Fig. 6. Gaussian noise. The matching between an original hand and a hand added with Gaussian noise is visualized. From left to right the standard deviation is $\sigma = 0, 0.5, 1, 3, 4$. At $\sigma = 4$ a matching starts to collapse (cf. point 4).

5.1 Matching with Articulated Parts

In practice, a shape is the 2D-projection of a given 3D-object. To match two different images of the same object, we have to take the flexibility of the 3D-object into account. The simplest way in doing so, is to allow a certain bending of the given shape. In Figure 5, a matching is visualized by showing some correspondent points on three given shapes. As we can see, the graph cut approach handles the matching of articulated parts in both cases very well.

5.2 Robustness to Noise

Every task in Computer Vision has to deal with the uncertainty of the observed data. Thus, any matching method has to handle this task in the best way possible. Since we are using an integral description of the curvature (cf. Section 4), this task is handled until we reach a point where even a human has its problems to recognize the object. In Figure 6, we see a hand with increasing Gaussian noise which is added in normal direction of every shape point. Until a standard deviation of $\sigma = 4$ is reached, the matching method works accurately. Besides these generated examples, Figure 7 shows that the matching for real data works also very well.

Fig. 7. Matching examples. Horses, cows (incl. one donkey) and jets are matched accurately.

5.3   Comparison with Dynamic Time Warping

In comparison with the shortest path method using Dynamic Time Warping, it is not clear whether the Graph Cut method is faster or slower. In fact, there are examples for which the proposed Graph Cut method is faster and other examples for which the Dynamic Time Warping method outperforms the Graph Cut method. In this section, we will analyze two of these examples to make some performance assumptions.

In Figure 8, we see that for similar shapes the proposed method outperforms the DTW method. On the other hand, for different shapes the opposite is the case. Therefore, it looks like the Graph Cut method handles similar shapes quite easier than un-similar shapes. It is a known fact that the efficient calculation of a maximum flow within a network depends highly on the edges' capacities. Unfortunately, this disadvantage of graph cut applications can create the bottleneck of the shape matching method for some examples. If two shapes $C_1, C_2 \in S$ are similar to one another, $\text{dist}(C_1, C_2)$ and thus the maximum flow is quite small. In other words, the maximum flow is close to the initial flow which is zero. Therefore, the amount of augmented paths that has to be examined by

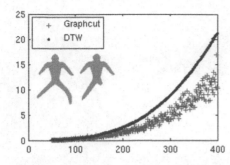 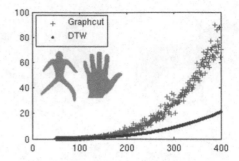

Fig. 8. Runtimes of shape matching. Here, the runtime of the proposes graph cut method and the commonly known shortest path method using DTW is plotted against the sampling rate of both given shapes. We can see that there are cases where the graph cut method outruns the DTW method. But that this is not always the case.

the Graph Cut algorithm is rather small and the proposed method works nearly instantly.

Note that the central advantage of the Graph Cut approach is the fact that every cut provides a *cycle* within the dual graph. Therefore the path X^* that has been induced by a minimal cut X fulfills always the constraint of being a *closed path*. Since this constraint has to be forced on the shortest path method, the DTW method always need $\mathcal{O}(N^3)$ calculation steps which is a disadvantage for similar shapes. For very different shapes, the DTW method does its job well and outperforms the more sophisticated Graph Cut method in finding the *minimal matching* whose semantic in the meaning of matching is of course quite questionable.

5.4 Graph Cut Algorithm

For the implementation of the shape matching, we used the implementation presented in [4] which works for segmentation problems in linear time. Unfortunately, this algorithm does not always provide a linear runtime in the size of the graph for the shape matching context. But as we have seen above, it outperforms the shortest path algorithm in some cases.

The minimum-cut maximum-flow algorithm of [4] was developed to handle energy minimization problems in Computer Vision. This algorithm looks for augmenting paths from source to sink and updates the flow accordingly. To this end, it constructs a search tree to decide which paths are good candidates of an augmenting path. Since the method depends highly on the amount of augmented paths that have to be considered, this method is fast for a shape matching scenario which starts with a flow that is close to the maximum flow. This is always the case for similar shapes, i.e. $\mathrm{dist}(\mathcal{C}_1, \mathcal{C}_2) \approx 0$. Because the method looks for the whole matching in a continuous manner, the difficult task of finding a starting match is done on the fly by the used method.

6   Conclusion and Future Work

In this paper, we proposed a polynomial-time algorithm for matching two planar shapes which is based on casting the matching problem as one of computing the minimal cut through a 2D graph embedded in $\mathbb{R}^3$. We proved that the graph cut problem is equivalent to the traditional shortest path formulation. However, in contrast to the previously proposed solution by Dynamic Time Warping (DTW), the graph cut formulation allows to circumvent the complete search over an initial correspondence. The minimum cut is computed by solving the dual maximum flow problem. The matching is by construction invariant to rigid body motions. In addition, experimental results show that shapes can be reliably matched despite articulation of parts and significant amounts of noise. Runtime comparisons between the proposed graph cut formulation and DTW indicate that the proposed method outruns the DTW method at least for similar shapes. Further effort is focused on obtaining additional speed-up, by considering a more suitable max-flow algorithm. There have been practical improvements of the DTW method [1,19]. At the same time, we expect that more adapted graph cut algorithms may lead to similar speedups.

References

1. Appleton, B.C.: Globally minimal contours and surfaces for image segmentation. PhD thesis, University of Queensland, Australia (December 2004)
2. Bakircioglu, M., Grenander, M., Khaneja, N., Miller, M.I.: Curve matching on brain surfaces using frenet distances. Human Brain Mapping, pp. 329–333 (1998)
3. Basri, R., Costa, L., Geiger, D., Jacobs, D.: Determining the similarity of deformable shapes. Vision Research 38, 2365–2385 (1998)
4. Boykov, Y., Kolmogorov, V.: An experimental comparison of min-cut/max-flow algorithms for energy minimization in vision. IEEE Trans. on Patt. Anal. and Mach. Intell. 26(9), 1124–1137 (2004)
5. Dryden, I.L., Mardia, K.V.: Statistical Shape Analysis. Wiley, Chichester (1998)
6. Elias, P., Feinstein, A., Shannon, C.E.: A note on the maximum flow through a network. IRE Transactions on Information Theory (later IEEE Transactions on Information Theory) IT-2, 117–199 (1956)
7. Ford Jr., L.R., Fulkerson, D.R.: Maximal flow through a network. Canadian Journal of Mathematics 8, 399–404 (1956)
8. Galilei, G.: Discorsi e dimostrazioni matematiche, informo a due nuoue scienze attenti alla mecanica i movimenti locali. appresso gli Elsevirii; Opere VIII (2) (1638)
9. Gdalyahu, Y., Weinshall, D.: Flexible syntactic matching of curves and its application to automatic hierarchical classication of silhouettes. IEEE Trans. on Patt. Anal. and Mach. Intell. 21(12), 1312–1328 (1999)
10. Geiger, D., Gupta, A., Costa, L.A., Vlontzos, J.: Dynamic programming for detecting, tracking and matching deformable contours. IEEE Trans. on Patt. Anal. and Mach. Intell. 17(3), 294–302 (1995)
11. Greig, D.M., Porteous, B.T., Seheult, A.H.: Exact maximum a posteriori estimation for binary images. J. Roy. Statist. Soc. Ser. B. 51(2), 271–279 (1989)

12. Ishikawa, H.: Exact optimization for Markov random fields with convex priors. IEEE Trans. on Patt. Anal. and Mach. Intell. 25(10), 1333–1336 (2003)
13. Kendall, D.G.: The diffusion of shape. Advances in Applied Probability 9, 428–430 (1977)
14. Latecki, L.J., Lakämper, R.: Shape similarity measure based on correspondence of visual parts. IEEE Trans. on Patt. Anal. and Mach. Intell. 22(10), 1185–1190 (2000)
15. Manay, S., Cremers, D., Hong, B.-W., Yezzi, A., Soatto, S.: Integral invariants for shape matching. IEEE Trans. on Patt. Anal. and Mach. Intell. 28(10), 1602–1618 (2006)
16. McConnell, R., Kwok, R., Curlander, J.C., Kober, W., Pang, S.S.: $\psi - s$ correlation and dynamic time warping: two methods for tracking ice floes in sar images. IEEE Trans. on Geosc. and Rem. Sens. 29, 1004–1012 (1991)
17. Mokhtarian, F., Mackworth, A.: A theory of multiscale, curvature-based shape representation for planar curves. IEEE Trans. on Patt. Anal. and Mach. Intell. 14, 789–805 (1992)
18. Pitiot, A., Delingette, H., Toga, A., Thompson, P.: Learning object correspondences with the observed transport shape measure. In: Information Processing in Medical Imaging, pp. 25–37 (July 2003)
19. Sebastian, T., Klein, P., Kimia, B.: On aligning curves. IEEE Trans. on Patt. Anal. and Mach. Intell. 25(1), 116–125 (2003)
20. Sharvit, D., Chan, J., Tek, H., Kimia, B.: Symmetry-based indexing of image databases (1998)
21. Thompson, D.W.: On Growth and Form. Cambridge University Press, Cambridge (1917)
22. Trouve, A., Younes, L.: Diffeomorphic matching problems in one dimension: Designing and minimizing matching functions. In: Europ. Conf. on Computer Vision, pp. 573–587 (2000)
23. Whitney, H.: Congruent graphs and the connectivity of graphs. Amer. J. Math. 54, 150–168 (1932)

Simulating Classic Mosaics with Graph Cuts

Yu Liu, Olga Veksler, and Olivier Juan

Department of Computer Science
University of Western Ontario
London, Ontario
Canada, N6A 5B7
Tel.: +1-519-661-2111; Ext.: 81417
yliu382@csd.uwo.ca, olga@csd.uwo.ca, juan@csd.uwo.ca

Abstract. Classic mosaic is one of the oldest and most durable art forms. There has been a growing interest in simulating classic mosaics from digital images recently. To be visually pleasing, a mosaic should satisfy the following constraints: tiles should be non-overlapping, tiles should align to the perceptually important edges in the underlying digital image, and orientation of the neighbouring tiles should vary smoothly across the mosaic. Most of the existing approaches operate in two steps: first they generate tile orientation field and then pack the tiles according to this field. However, previous methods perform these two steps based on heuristics or local optimisation which, in some cases, is not guaranteed to converge. Some other major disadvantages of previous approaches are: (i) either substantial user interaction or hard decision making such as edge detection is required before mosaicing starts (ii) the number of tiles per mosaic must be fixed beforehand, which may cause either undesired overlap or gap space between the tiles. In this work, we propose a novel approach by formulating the mosaic simulating problem in a global energy optimisation framework. Our algorithm also follows the two-step approach, but each step is performed with global optimisation. For the first step, we observe that the tile orientation constraints can be naturally formulated in an energy function that can be optimised with the α-expansion algorithm. For the second step of tightly packing the tiles, we develop a novel graph cuts based algorithm. Our approach does not require user interaction, explicit edge detection, or fixing the number of tiles, while producing results that are visually pleasing.

1   Introduction

Mosaic is one of the most ancient and durable art forms. Since ancient Greek and Roman times, people used beautiful, fascinating mosaics to decorate floor pavements, wall murals and ceilings. Classic mosaics are composed of a huge number of small tiles with regular shapes, such as rectangles and squares. Simulating classic mosaics automatically is one of the areas in non-photorealistic rendering that has been investigated by many researchers. In recent years, there has been a rapid growth in non-photorealistic rendering techniques, since such techniques can emphasise important aspects of a scene and create digital art.

A.L. Yuille et al. (Eds.): EMMCVPR 2007, LNCS 4679, pp. 55–70, 2007.
© Springer-Verlag Berlin Heidelberg 2007

There are two main challenges for mosaic simulating. First, tile orientations should emphasise the edges of perceptually important shapes in the image. This is achieved by placing tiles parallel to the edges to be emphasised. In addition, tiles must be packed tightly while preserving their completeness. Inspired by the artists' work, many approaches have been developed for simulating this process.

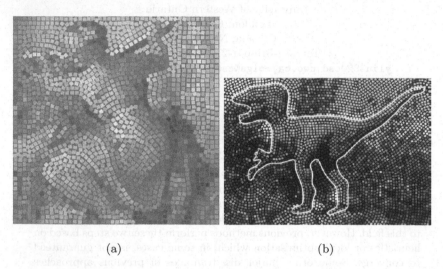

(a)   (b)

Fig. 1. Previous work: (a) Result from Hausner [1]. (b) Artifacts created by Elber and Wolberg [2]. The feature curve outlined by the user is in white. Notice that even the tiles that are far away in the background are still aligned with the outline of the dinosaur, even though perceptually significant edges are present in the background.

An early and successful work on rendering classic mosaics was introduced by Hausner [1], see Fig. 1(a). In Hausner, the user is required to identify the perceptually important edges. Then the distance transform to these edges is computed, and the gradients of this distance map are adopted as the tile orientations. Finally the Centroidal Voronoi Diagram (CVD) based on Manhattan distance is used to pack the tiles, assuming that the number of tiles is known.

Elber and Wolberg [2] also require a user to draw several closed curves around the edges to be emphasised. A set of offset curves parallel to the feature curves is computed to generate the orientation guide curves. These curves are trimmed to eliminate self-intersection. Tiles are packed along these orientation guide lines under the constraint that they should not touch each other and the gap space between them is small. See Fig. 1(b) for their result.

The above approaches can produce visually pleasing results. However both approaches are based on local heuristics which lead to certain problems. In Hausner, the two main problems are: (a) the convergence of CVD algorithm is not guaranteed and (b) the tile orientations change drastically at the discontinuities of the distance transform gradient map, but these discontinuities usually do not coincide with intensity edges. In Elber and Wolberg's approach, there is

big gap space where the curvature of the orientation guide curve is very large
and discontinuity in tile orientations at the skeletons of the closed feature curves
provided by the user. In addition, both of these approaches require large amount
of user interaction to draw the feature edges. Furthermore, in [1], the number of
tiles per mosaic image has to be fixed before the algorithm's start, which causes
either a lot of undesired overlapping or large gap space. An additional drawback
of Elber and Wolberg's approach is that the tiles in the background which are
far away from the foreground objects are also aligned with the feature curves
outlined by the user, which creates the artifacts shown in Fig. 1(b).

Other approaches for simulating classic mosaics include Di Blasi and Gallo [3],
Battiato et. al [4], Schlechtweg et al. [5]. All of these approaches are based on
local optimisation. Some of them replace the user drawn curves with the results
of an edge detector. However the state of the art in edge detection does not
allow yet to find perceptually important edge robustly. Some commercial image
processing software also tries to simulating the visual effects of mosaics, such
as Adobe Photo shop [6], but they actually perform simple resolution reducing
processing which does not create effects similar to classic mosaics.

The goal of our work is automatic classic mosaic simulation without user
interaction or explicit edge detection. We observe that we can formulate the
tile orientation estimation and tile packing steps in a global energy optimisation
framework, which helps to avoid the problems of local optimisation mentioned
above. Like the previous approaches, we wish to make tiles align to perceptually
important edges in the image while optimising the gap space between tiles. Unlike
the previous approaches, we completely prohibit overlapping tiles.

Our algorithm has three major steps. In the first step, we generate a tile orientation field. Each tile is encouraged to take orientation that aligns it to the nearby
strong intensity edges (if any) in the underlying image. In addition, tile orientations are encouraged to vary smoothly over the image, which forces the tiles to
create a pleasing visual effect and also helps to reduce gap space between tiles,
since there is less gap space between neighbouring tiles with similar orientations.
We can encode the strong edge alignment and smoothness constraints in a global
energy function which we then optimise with the α-expansion algorithm of [7].
The benefits of this approach to tile orientation generation is that smoothness
of tile orientations is enforced globally in the whole image (unlike the distance
transform approaches), and explicit edge detection (either by the user or an edge
detector) is eliminated. None of the existing methods addresses the smoothness
of tile orientations with global optimisation. We chose the α-expansion algorithm
of [7] because it has been successfully applied to many optimisation problems
in vision and graphics, such as texture synthesis [8], image stitching [9][10] and
stereo correspondence [7][11][12].

After we have generated a smooth orientation field for the tiles, we can begin
mosaic building. Unfortunately, if we formulate tile packing for the final mosaic
as a global optimisation in the straightforward manner, the resulting energy is
prohibitively expensive to optimise, since the tile packing problem is essentially
a bin-packing problem. Our solution is inspired by the texture synthesis [8] and

image stitching algorithms [9][10] . We generate several mosaic layers and stitch them together to form a final mosaic

Thus the second step of our algorithm is generating multiple candidate mosaic layers, using a reasonable heuristic. These layers are usually not good enough to be visually pleasing. There may be large gap space between tiles and many tiles might cross sharp intensity edges, which blurs the final mosaic since the tile colour is the average colour of the image pixels it covers. In the third and final step, we develop a novel layer stitching algorithm, that selects good parts from all the candidate layers and stitches them together to form the final mosaic. Here "a good part" means a region in a candidate mosaic layer where the tiles are packed tightly and do not cross strong intensity edges. This third step is also done in the energy optimisation framework. The constraints of gap space minimisation, edge avoidance, and prohibiting tile overlap are encoded in an energy function, which is then optimised with a novel graph cuts based optimisation algorithm.

2   Energy Optimisation with Graph Cuts

In this section, we briefly review graph cuts based optimisation. Many problems in vision and graphics can be naturally represented as labelling problems, such as image segmentation, stereo, and motion. Let $\mathcal{P}$ be the set of all image pixels and suppose for each $p \in \mathcal{P}$ we wish to assign some label $f_p \in \mathcal{L}$. Let $f = \{f_p | p \in \mathcal{P}\}$ be a labelling that assigns each pixel $p \in \mathcal{P}$ a label $f_p \in \mathcal{L}$. The following energy function is formulated to measure the quality of f:

$$E(f) = E_{smooth}(f) + E_{data}(f) \tag{1}$$

Here, E_{smooth}, which is often called the smoothness term, measures the extent to which f is not smooth. E_{data}, usually called the data term, measures how pixels in $\mathcal{P}$ like the labels that f assigns them. E_{data} is often formulated as

$$E_{data}(f) = \sum_{p \in \mathcal{P}} D_p(f_p) \tag{2}$$

where D_p is the penalty for assigning pixel p the label f_p. A typical choice for E_{smooth} is

$$E_{smooth} = \sum_{\{p,q\} \in \mathcal{N}} V_{pq}(f_p, f_q) \tag{3}$$

Usually, $\mathcal{N}$ consists of pairs of immediately adjacent pixels, that is the interactions are given by the standard 4-connected grid. Graph cut [7] has been proved to be very successful in optimising these types of energies [13]. We use max-flow implementation of [14] for computing minimum cut.

3   Simulating Classic Mosaics with Graph Cuts

In this section, we give a detailed description of our algorithm. In section 3.1, we give some necessary definitions, in section 3.2 we explain the first step of our

algorithm, which is generating smooth tile orientations, in section 3.3 we explain
the second step of our algorithm, which is generating multiple candidate mosaic
layers, and finally, in section 3.4 we explain our last step, which is stitching the
candidate mosaic layers together to obtain the final optimised mosaic.

3.1 Notation and Definitions

Let I be the rectangular image grid where we want to place the tiles and let $\mathcal{P}$
be the collection of all pixels inside I. We assume that all tiles are square with
the side $tSize$. A tile is denoted by $t = \{p_t, \varphi_t, \mathcal{T}_t\}$. Here p_t is a pixel inside the
image I such that p_t is at the centre of tile t. The angle $\varphi_t \in [0, \frac{\pi}{2})$ is the tile
orientation. Since our tiles are rotationally symmetric, we need angles only in
the range of $[0, \frac{\pi}{2})$. $\mathcal{T}_t$ is the set of pixels in the image that tile t covers. To find
the pixels in $\mathcal{T}_t$, we build a coordinate system with origin at p_t, and the horizontal and vertical axes parallel to that of the image plane I. If the orientation
of tile t is $\varphi_t \in [0, \frac{\pi}{2})$, then a pixel (x, y) is inside tile $\mathcal{T}_t$, if its coordinates (x, y)
satisfy:

$$|x \times \cos(\varphi_t) + y \times \sin(\varphi_t)| \leq \tfrac{tSize}{2}$$
$$|y \times \cos(\varphi_t) - x \times \sin(\varphi_t)| \leq \tfrac{tSize}{2} \quad (4)$$

Our discrete optimisation framework requires that the set of angles is finite.
Here we discretise the tile orientations into n angles. Let l_i be the ith angle, then
it is defined as:

$$l_i = \frac{\pi}{2n} \times (i - 1), \quad i = 1, 2, \ldots, n$$

A label set $\mathcal{L} = \{l_1, l_2, \ldots, l_n\}$ represent the orientations. If p_t is assigned l_i,
then tile $t = \{p_t, \varphi_t, \mathcal{T}_t\}$ has an orientation of $\varphi_t = l_i$. We set n to 16.

3.2 Generating Tile Orientations

In this section, we describe how we compute the tile orientation field. That is
for each pixel p, we compute the appropriate tile orientation, assuming that a
tile will be placed with its centre at pixel p in the final mosaic. Of course, in the
final mosaic only a fraction of all pixels will, in fact, become tile centres. Most
of the pixels will be covered by some tile, but will not be in the centre of the tile
that covers them, and some pixels will be in the gap space, that is they will not
be covered by any tile. Selecting pixels that will become tile centres is addressed
in the second and third step of the algorithm, described in sections 3.3 and 3.4.

Our energy function for smooth tile orientation field is given by equations (1),
(2), and (3), where f_p is the orientation label assigned to pixel p. This energy
encodes the constraints of edge alignment and strong edge avoidance in the data
term $D_p(f_p)$, and the smoothness of orientation between neighbouring pixels
in the smoothness term $V_{pq}(f_p, f_q)$. The α-expansion method based on graph
cuts [7] is applied to minimise this energy.

60   Y. Liu, O. Veksler, and O. Juan

Data Term. We first discuss our data term, which has the following form:

$$D_p(f_p) = E_p^{align}(f_p) + E_p^{avoid}(f_p), \tag{5}$$

where $E_p^{align}(f_p)$ encodes edge alignment constraints, $E_p^{avoid}(f_p)$ encodes edge avoidance constraints. Assuming that a pixel p will become a tile centre, and knowing the tile size, it is fairly easy to estimate which orientation is appropriate for a square tile of fixed size with the centre at pixel p, so that it aligns with a strong intensity edge in its neighbourhood, if any. It is defined as:

$$E_p^{align}(f_p) = w_e \times \max_{i=1...4} \|C_{r_i}(p) - C_{b_i}(p)\| \tag{6}$$

Let t be the tile $t = \{p, f_p, \mathcal{T}_t\}$. The darker yellow and green rectangles in Fig. 2 show how regions b_i are defined, and the light yellow and green rectangles in Fig. 2 illustrate the r_i regions. $C_{r_i}(p)$ is the average colour vector in region r_i and $C_{b_i}(p)$ is the average colour vector in region b_i. $\|C_{r_i}(p) - C_{b_i}(p)\|$ is the magnitude of the average colour difference between regions r_i and b_i.

This term encourages pixel p which is close to an edge to be assigned the orientation that aligns the tile centred at p with the edge. The weight of the colour difference term w_e is set to be negative. Thus when there is a high response on the colour difference between the pixels inside the tile and that outside the tile, the term $E_p^{align}(f_p)$ is negative, encouraging high contrast.

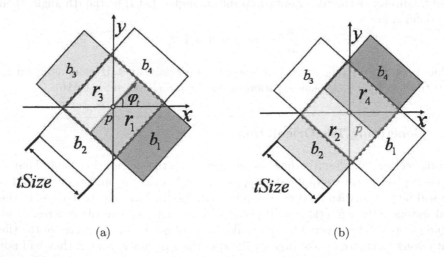

(a)   (b)

Fig. 2. Definition of $E_p^{align}(f_p)$. A tile $t = \{p_t, \varphi_t, \mathcal{T}_t\}$ is shown by the red dotted squares in (a) and (b). Region r_i is inside tile t and it shares one edge with tile t, for example, in (a), r_1 and r_3 are shown in lighter green and yellow rectangles inside the tile t. In (b), regions r_2 and r_4 are illustrated by the lighter yellow and lighter green rectangles. Region b_i is a rectangle on the border of tile t and b_i shares the same edges with tile t and region r_i. The darker green and darker yellow rectangles in (a) are regions b_1 and b_3 when (b) shows regions b_2 and b_4. The average colour differences between r_i and b_i is measured in E_p^{align}.

Strong edge avoidance is encoded in $E_p^{avoid}(f_p)$, which is defined as:

$$E_p^{avoid}(f_p) = w_v \times \sum_{q \in T_t} \|g(q)\| \tag{7}$$

Let $I(p)$ be the intensity at pixel p and $\|g(p)\|$ be the magnitude of the gradient at pixel p. Let $p = (x, y)$, then $g(p) = g(x, y) = \langle g_x(x, y), g_y(x, y) \rangle$. We approximate gradient by the standard Sobel operator. This term measures the intensity variance inside the tile, therefore we call it the variance term. If the label f_p makes the tile overlap a strong intensity edge, then the variance term will penalise the overlap between the edge and the tile. This term is particularly important for pixels close to the edges of an object. The weight of w_v is set to be positive, since we want high gradient to be penalised.

Notice that the $D_p(f_p)$ term involves summation over a potentially large group of pixels (if $tSize$ is large). To compute the data term $D_p(f_p)$ efficiently, the "integral image" [15] is used in our approach. The integral image approach allows computing $D_p(f_p)$ in constant time, independent of the tile size $tSize$.

Smoothness Term. We define the interaction term $V_{pq}(f_p, f_q)$ as:

$$V_{pq}(f_p, f_q) = w_s \times |f_p - f_q|_{\mathrm{mod}(\frac{\pi}{2})} \tag{8}$$

where

$$|f_p - f_q|_{\mathrm{mod}(\frac{\pi}{2})} = \begin{cases} |f_p - f_q| & \text{if } |f_p - f_q| \leq \frac{\pi}{4} \\ \frac{\pi}{2} - |f_p - f_q| & \text{otherwise} \end{cases}$$

This interaction reflects the fact that the angle of $\frac{\pi}{2}$ leads to the same tile placement as the angle of 0, due to the symmetry of the square shape. The smoothness term encourages orientations to propagate smoothly over the image.

3.3 Generating Mosaic Layers

In the second step of our algorithm, we generate a set of mosaic layers. Recall that in the first step of our algorithm, for each pixel p we compute an orientation f_p s.t. if a tile t is placed with its centre at pixel p, it will have orientation f_p. Thus for each pixel, we have a "candidate" tile t for possible inclusion in the final mosaic. We build multiple mosaic layers out of these candidate tiles. To give every candidate tile a chance to be included in the final mosaic while keeping the number of candidate layers to a minimum, we insure that the candidate mosaic layers have no tiles in common, and that every candidate tile is present in one of the candidate layers. We also insure that each candidate mosaic layer does not have overlapping tiles, this is crucial for our algorithm in step 3, see section 3.4.

Mosaic layers are generated iteratively, in a region growing manner. In each iteration we build one mosaic layer and make sure that the current mosaic layer does not contain any of the tiles which are already present in a previous mosaic layer. To build one mosaic layer, we starting at some random pixel s which is not included in any candidate mosaic layer yet. We greedily choose a nearby pixel

p in such a way that the candidate tiles centred at s and p do not overlap but the gap space between the tiles s and p is small. In addition, the candidate tile centred at p must not have been selected yet for any candidate layer.

3.4   Stitching Two Mosaic Layers

After generating a set of candidate mosaic layers, our last step is to stitch them together to form the final mosaic. The stitching should minimise the gap space and should not contain any "broken" tiles. In addition, tiles are encouraged to avoid crossing strong intensity edges. Therefore, we must take edge avoidance into account. In this section, we develop a novel graph cuts algorithm to solve this stitching problem in an energy optimisation framework.

Minimising gap spaces and prohibiting tile overlap. Suppose all the candidate tiles are indexed as $t_1, t_2, \ldots, t_n$, and $D = \{t_i : i = 1, \ldots, n\}$. A mosaic layer is just a subset of tiles in D. Let $M_1, M_2, \ldots, M_k$ be the mosaic layers. In the second step, we generated them in such a way that for any $i \neq j$, $M_i \cap M_j = \emptyset$, and $\bigcup_{i=1,\ldots,k} M_i = D$. Also, for all $i = 1, \ldots, k$, if $t \neq t' \in M_i$, then tiles t and t' do not overlap, that is for $t = \{p_t, \varphi_t, \mathcal{T}_t\}$ and $t' = \{p_{t'}, \varphi_{t'}, \mathcal{T}_{t'}\}$, $\mathcal{T}_t \cap \mathcal{T}_{t'} = \emptyset$.

Our stitching algorithm is iterative. We start from a random mosaic layer M_i, setting it to be our current solution M_c. In each iteration, we select a random layer $M = M_j$, and then find a tile preserving cut between the current solution M_c and the random layer M. Let R be the result of stitching layers $M_c \neq M$, that is $R \subset (M_c \cup M)$.

Table 1. Energy for stitching two layers M_c, M

f_s	$D_s(f_s)$	Layer		f_s	$D_s(f_s)$	Layer		(f_s, f_r)	$V_{sr}(f_s, f_r)$
$f_s = 0$	0	$s \in M_c$		$f_s = 0$	$w_g \times \|g_s\|$	$s \in M_c$		(0,0)	0
$f_s = 1$	A_s			$f_s = 1$	A_s			(0,1)	∞
$f_s = 1$	0	$s \in M$		$f_s = 1$	$w_g \times \|g_s\|$	$s \in M$		(1,0)	O_{sr}
$f_s = 0$	A_s			$f_s = 0$	A_s			(1,1)	0

(a): $D_s(f_s)$ without Edge Avoidance.	(b): $D_s(f_s)$ with Edge Avoidance.	(c): $V_{sr}(f_s, f_r)$

We formulate the problem of stitching two mosaic layers $M_c \neq M$ together as a labelling problem. Let tile set $S = M \cup M_c$. Our label set is $\{0, 1\}$. Labelling $f = \{f_s | s \in S\}$ assigns a label $f_s \in \{0, 1\}$ to every $s \in S$. Any labelling f corresponds to a stitching $R^f \subset S$ in the following way. For every $t \in M_c$, if $f_t = 0$, then tile $t \in R^f$ and if $f_t = 1$, tile $t \notin R^f$. For every $t \in M$, $f_t = 1$ means that tile $t \in R^f$ and if $f_t = 0$, then $t \notin R^f$. Notice that the meaning of labels $\{0, 1\}$ is reversed for tiles in layers M_c and M.

We define the following energy function:

$$E(f) = \sum_{s \in S} D_s(f_s) + \sum_{s, r \in \mathcal{N}} V_{sr}(f_s, f_r), \tag{9}$$

where the neighbourhood system $\mathcal{N}$ is:

$$\mathcal{N} = \{(s,r)|s \in M_c, r \in M \text{ and } T_s \cap T_r \neq \emptyset\}$$

The data term $D_s(f_s)$ is given in Table 1(a). In this table, A_s is the size of the region which is covered only by tile s, see Fig. 3.

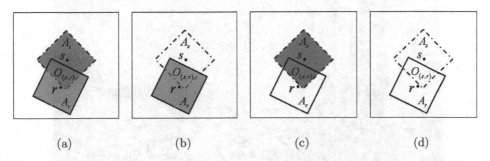

(a)	(b)	(c)	(d)

Fig. 3. A simple example to illustrate data and interaction terms. In (a), mosaic layer $M = \{r\}$ is shown by the solid square and $M_c = \{s\}$ is shown by the dotted square. A_r is the region covered only by tile r and A_s is the region covered only by tile s. O_{sr} is the size of overlapping region between s and r. Both tiles s and r presented in the result stitching, therefore, the energy is ∞. In (b), tile s is assigned label 0 and not present in the result stitching. The energy for this case is A_s. In (c), tile r is removed and the energy is A_r. In (d), both tiles are removed, then $V_{sr}(1,0) = O_{sr}$.

The interaction term $V_{sr}(f_s, f_r)$ is defined as table 1(c). Here O_{sr} denotes the size of the overlapping region between tiles $s \in M_c$ and $r \in M$, see Fig. 3. This interaction term insures that no tile overlap occurs in the stitching R which has finite energy $E(f^R)$, see Fig. 3. Notice that the gap space that mosaic layers M_c and M have in common can not be filled in by their stitching R. If M is a mosaic layer, let

$$G(M) = \{p \in \mathcal{P}|p \notin \bigcup_{s \in M} T_s\},$$

That is $G(M)$ is the gap space in mosaic layer M. It is easy to see that for any f^R such that $E(f^R) < \infty$,

$$E(f^R) = |G(R)| - |G(M) \cap G(M_c)| \tag{10}$$

That is the energy of f^R is the number of pixels in the gap space of R which were not in the gap space of either M or M_c. Since for any M_c and M, $|G(M) \cap G(M_c)|$ is a constant with respect to our optimisation problem, the optimal f^R will correspond to the stitching R which has $G(R)$, or the gap space, as small as possible. This is exactly the kind of stitching that we desire, minimising the gap space while prohibiting tile overlap.

(a) Input painting

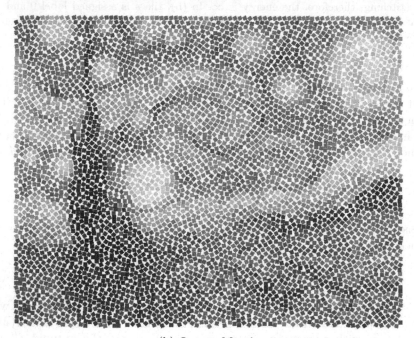

(b) Output Mosaic

Fig. 4. Starry Night

Fig. 5. Libyan Sibyl Mosaic. The input image is zoomed out and illustrated in the right bottom corner.

Edge avoidance. There is an additional constraint that we wish to add to our stitching algorithm. For a visually pleasing mosaic, tiles should be placed to avoid crossing strong intensity edges inside the underlying image, since otherwise the resulting mosaic is blurred (a colour of a tile is just the average colour of the underlying image pixels that the tile covers). Let g be the intensity gradients for the given image I, where the gradients are computed by standard Sobel operator. The sum of gradient values inside tile $s = \{p_s, \varphi_s, \mathcal{T}_s\}$ is denoted by $|g_s| = \sum_{p \in \mathcal{T}_s} |g(p)|$, where $|g(p)|$ is the magnitude of gradient at pixel p. To incorporate the intensity variance with our energy function defined in Eq. (9), we redefine the data term $D_s(f_s)$ as Table 1(b). The weight of g_s, denoted by w_g, is positive.

With the introduction of the edge avoidance constraint, the energy of stitching current layer changes from that in Eq. (10) to the one below:

$$E(f^R) = |G(R)| - |G(M) \cap G(M_c)| + w_g \times \sum_{t \in R} |g(t)|. \tag{11}$$

It can be seen that the energy in Eq. (11) is submodular [16], and therefore we can optimised it exactly by computing a minimum cut in a certain graph [16]. However, this energy is not quite what we need to measure the quality of stitching R. The $\widetilde{E}(R)$ below is what we really need, since it accounts for the total gap space:

$$\widetilde{E}(f^R) = |G(R)| + w_g \times \sum_{t \in R} |g(t)| = E(f^R) + |G(M) \cap G(M_c)| \tag{12}$$

$\widetilde{E}(f^R)$ simply adds the area of the gap space and the intensity variance inside the mosaic tiles. The problem is that even though for a given layer M, optimising the energy in Eq. (11) gives us the best stitching $R \subset (M_c \cup M)$ of current layer M_c and the new layer M, the common gap space $|G(M) \cap G(M_c)|$ that we cannot optimise for may be too large to give the actual decrease of the desired energy in Eq. (12). The way we solve the problem is as follows. After optimising the energy in Eq. (11), we only switch to the stitching R if $\widetilde{E}(f^R)$ goes down. The steps generating a new layer M, stitching it with the current layer M_c, and updating the current solution in case of decrease in $\widetilde{E}(f^R)$ are performed until the maximum number of iterations.

4   Experimental Results

In this section, we show some results generated by our algorithm. We used the following settings of parameters for all the experiments in this paper: the edge alignment $w_e = -50$, the edge avoidance $w_v = 20$, the smoothness $w_s = 20$, and, finally, the layer stitching $w_g = 0.015$. These values were determined experimentally to give good results for all the images.

In Figs. 5 and 4 we show the original images and our simulated mosaics for the Libyan Sibyl and the Starry Night paintings. The mosaics are visually pleasing, possessing the desired effects. The tile orientations emphasise the shapes of the objects and vary smoothly across the image. For the dinosaur image, see Fig. 6, the tiles in the mosaic are aligned with the edges of the dinosaurs inside the image, and the tiles in the background are aligned with significant background

(a) Input Image

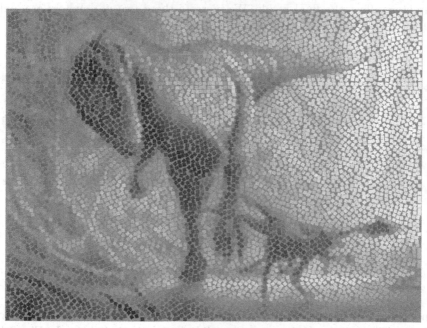

(b) Output Mosaic

Fig. 6. Dinosaur

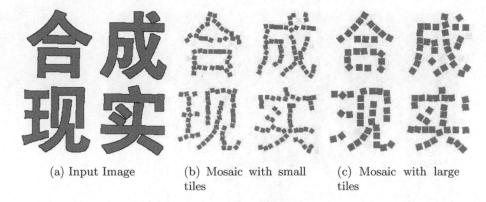

(a) Input Image   (b) Mosaic with small   (c) Mosaic with large
                           tiles   tiles

Fig. 7. Mosaic images for characters

shapes, unlike the results of Elber and Wolberg [2] in Fig. 1(b). This is because we do not make hard decisions about the boundaries to be emphasised, rather we let our algorithm to discover those boundaries automatically.

To make sure every candidate tile has a chance to appear in the final mosaic, the number of candidate mosaic layers and the number of iterations when stitching the layers together must be large enough. Figs. 8 and 9 show how these two parameters affect the final result. We increase these two parameters gradually until the final result is visually pleasing.

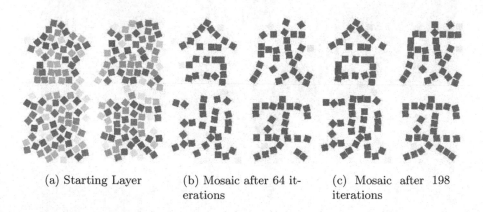

(a) Starting Layer   (b) Mosaic after 64 it-   (c) Mosaic after 198
                              erations   iterations

Fig. 8. Mosaics generated with different numbers of iterations

In our work, we set all the tiles to have the same size and same square shape. It is important to choose an appropriate *tSize* for synthetic images. In Fig. 7(b), the tile size is too small compared to the strokes of the characters. To keep the resulting mosaic from being blurred, the tiles located at the edges of the characters are removed, therefore, the characters in the mosaic look much thinner

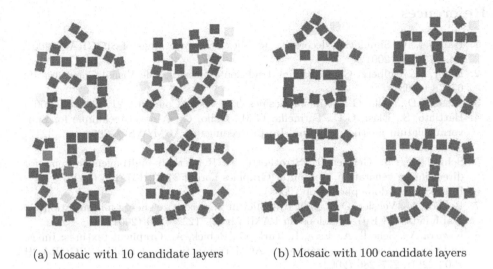

(a) Mosaic with 10 candidate layers   (b) Mosaic with 100 candidate layers

Fig. 9. Mosaics generated with different numbers of candidate layers

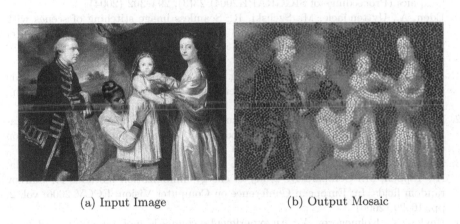

(a) Input Image   (b) Output Mosaic

Fig. 10. Family

than those in the input image. In Fig. 7(c), the previous problem is solved by increasing the tiles size, such that *tSize* is almost the thickness of the character strokes, and the result is much improved.

For some images, we need to vary the tile size for different regions of the image. Small tiles are needed in the image regions which have many fine details, and large tiles work well in areas with larger details. For example, in Fig. 10(b), the background is visually pleasing with the tile size of 10 × 10 pixels, however, the facial details of the people inside the image are blurred, because the tile size is too large to represent the fine facial details. We plan to incorporate different tile sizes and shapes to our future work.

References

1. Hausner, A.: Simulating decorative mosaics. In: Proceedings of SIGGRAPH2001, pp. 573–580 (2001)
2. Elber, G., Wolberg, G.: Rendering traditional mosaics. The Visual Computer 19, 67–78 (2003)
3. Blasi, G.D., Gallo, G.: Artificial mosaics. The Visual Computer 21, 373–383 (2005)
4. Battiato, S., Blasi, G.D., Farinella, G.M., Gallo, G.: A novel technique for opus vermiculatum mosaic rendering. In: proceedings of ACM/WSCG2006, pp. 3247–3259 (2004)
5. Schlechtweg, S., Germer, T., Strothotte, T.: Renderbots-multi-agent systems for direct image generation. Computer Graphics Forum 24(2), 137–148 (2005)
6. PhotoShop: Adobe photoshop (2006)
7. Boykov, Y., Veksler, O., Zabih, R.: Efficient approximate energy minimization via graph cuts. IEEE transactions on PAMI 21(12), 1222–1239 (2001)
8. Kwatra, V., Schodl, A., Essa, I., Turk, G., Bobick, A.: Graphcut textures: Image and video synthesis using graph cuts. ACM Transactions on Graphics, SIGGRAPH 2003 22(3), 277–286 (2003)
9. Agarwala, A., Dontcheva, M., Agrawala, M., Drucker, S., Colburn, A., Curless, B., Salesin, D., Cohen, M.: Interactive digital photomontage. ACM Transaction on Graphics (Proceedings of SIGGRAPH2004) 23(3), 294–302 (2004)
10. Eden, A., Uyttendaele, M., Szeliski, R.: Seamless image stitching of scenes with large motions and exposure differences. In: Proceedings of 2006 IEEE Computer Society Conference on Computer Vision and Pattern Recognition. vol. 2, pp. 2498–2505 (2006)
11. Kolmogorov, V., Zabih, R.: Computing visual correspondence with occlusion via graph cuts. In: Proceedings of IEEE International Conference on Computer Vision, pp. 508–515 (2001)
12. Kim, J., Kolmogorov, V., Zabih, R.: Visual correspondence using energy minimization and mutual information. In: Proceedings of IEEE International Conference on Computer Vision. vol. 2, pp. 1033–1040 (2003)
13. Szeliski, R., Zabih, R., Scharstein, D., Veksler, O., Kolmogorov, V., Agarwala, A., Tappen, M.: Comparative study of energy minimization methods for markov random fields. In: European Conference on Computer Vision, ECCV 2006, vol. 2, pp. 16–29 (2006)
14. Boykov, Y., Kolmogorov, V.: An experimental comparison of min-cut/max-flow algorithms for energy minimization in vision. IEEE Transactions on Pattern Analysis and Machine Intelligence(PAMI) 24, 137–148 (2004)
15. Viola, P., Jones, M.: Rapid object detection using a boosted cascade of simple features. In: Proceedings of the IEEE CVPR2001 1, 511–518 (2001)
16. Kolmogorov, V., Zabih, R.: What energy functions can be minimized via graph cuts? IEEE Transactions on Pattern Analysis and Machine Intelligence 26(2), 147–159 (2004)

An Energy Minimisation Approach to Attributed Graph Regularisation

Zhouyu Fu[1] and Antonio Robles-Kelly[1,2]

[1] Department of Information Engineering, ANU, Canberra, Australia
[2] NICTA* RSISE Bldg. 115, Australian National University, ACT 0200, Australia

Abstract. In this paper, we propose a novel approach to graph regularisation based on energy minimisation. Our method hinges in the use of a Ginzburg-Landau functional whose extremum is achieved efficiently by a gradient descend optimisation process. As a result of the treatment given in this paper to the regularisation problem, constraints can be enforced in a straightforward manner. This provides a means to solve a number of problems in computer vision and pattern recognition. To illustrate the general nature of our graph regularisation algorithm, we show results on two application vehicles, photometric stereo and image segmentation. Our experimental results demonstrate the efficacy of our method for both applications under study.

1 Introduction

Many computer vision problems involve the regularisation of structured data, such as vector fields and graphs. Regularisation methods have been reported in the contexts of optical flow computation [1], curvature-based surface shape representation [2] and the smoothing of stereo disparity fields [3]. Image correlation and regularisation have been used for recovering depth-maps from one or multiple views [4]. In the case of depth information recovery, the problem is often formulated in an energy minimisation setting solved using graph cuts [5,6,7] or belief propagation [8].

An important issue underpinning the smoothing process is that of preserving the underlying relational structure while satisfying constraints on data-closeness. In fact, data-closeness constraints can sometimes be re-cast as constraints on a set of attributes or components of a vector-field. For example, in shape-from-shading, when Lambertian reflectance is assumed, the component of the surface normal in the light source direction is constrained to be equal to the inverse cosine of the normalised image brightness [9]. For the purposes of optical flow or photometric stereo computation, brightness constraints are frequently used [10]. Furthermore, tasks such as semi-supervised image segmentation can be viewed as smoothing ones, where the cluster indexes are assigned so as to ensure smoothness while preserving data-closeness.

* National ICT Australia is funded by the Australian Governments Backing Australia's Ability initiative, in part through the Australian Research Council.

A.L. Yuille et al. (Eds.): EMMCVPR 2007, LNCS 4679, pp. 71–86, 2007.
© Springer-Verlag Berlin Heidelberg 2007

A number of methods have been proposed for graph regularisation [11,12,13]. The bulk of these methods are aimed at solving semi-supervised learning problems, which hinge in finding a smooth labeling over the graph consistent with hand-labeled ground truth. From an alternative viewpoint, Zhang and Hancock [14] have used the heat equation to regularise MRI data abstracted to a weighted graph. This treatment is related to regularisation processes on tensors, where data constrained on a manifold is regularised making use of partial differential equations [15,16].

In this paper, we present a novel energy minimisation method for attributed graph regularisation based upon an energy minimisation framework which is capable of accommodating constraints. This has a number of advantages. Firstly, it permits the inclusion of data-closeness terms into the regularisation process. Secondly, such a treatment lends itself to the use of an energy functional which can be extremised efficiently by making use of continuous optimisation techniques. Moreover, the formulation presented here is quite general and provides a link between optimisation on manifolds and regularisation of relational structures. The paper is organised as follows. In the following section, we introduce the preliminaries concerning the energy functional used here. Section 3 describes our graph regularisation approach in detail. In Section 4, we provide results on semi-supervised image segmentation and photometric stereo. Finally, we offer conclusions on the work in Section 5.

2   Graph Regularisation and Manifold-Constrained Energy Minimisation

In the mathematics literature, there is a considerable body of work aimed at understanding how graphs can be embedded on a manifold so as to minimize a measure of distortion. Broadly speaking, there are three ways in which the problem can be addressed. First, the graph can be interpolated by a surface whose genus is determined by the number of nodes, edges and faces of the graph. Second, the graph can be interpolated by a hyperbolic surface which has the same pattern of geodesic (internode) distances as the graph [17]. Third, a manifold can be constructed whose triangulation is the simplicial complex of the graph [18]. A review of methods for efficiently computing distance via embedding is presented in the recent paper of Hjaltason and Samet [19].

In the pattern analysis community, there has recently been renewed interest in the use of embedding methods motivated by graph theory. One of the best known of these is ISOMAP [20]. Here, a neighbourhood ball is used to convert data-points to a graph, and Dijkstra's algorithm is used to compute the shortest (geodesic) distances between nodes. By applying multidimensional scaling (MDS) to the matrix of geodesic distances, the manifold is reconstructed. The resulting algorithm has been demonstrated to locate well-formed manifolds for a number of complex data-sets. Related algorithms include locally linear embedding, which is a variant of PCA that restricts the complexity of the input data using a nearest neighbour graph [21], and the Laplacian eigenmap that constructs an adjacency weight matrix for the data-points and projects the data onto the principal eigenvectors of the associated Laplacian matrix (the degree matrix minus the weight matrix) [22]. Hein *et al.* [23] have established the degree of point-wise consistency for graph Laplacians with data-dependent weights to a weighted Laplace operator. Collectively, these methods are sometimes referred to as manifold learning theory.

Here, we tackle the problem of graph regularisation from a geometric analysis perspective which combines the strengths of tensor regularisation methods and manifold learning techniques. We incorporate constraints to the energy minimisation process by making use of an energy functional defined on a manifold. Thus, in this section, we present the theoretical foundation of our method, which hinges in the use of a Ginzburg-Landau functional on a Riemannian manifold [24].

2.1 Riemannian Manifolds

To commence, we characterise the edges of a graph using a geodesic on a Riemannian manifold. Analogously, the nodes of the graph can be viewed as points on the manifold, whereas the node attributes are given by a real-valued function at the point under consideration. Viewed in this way, graph regularisation can be cast as a manifold-constrained optimisation based upon an energy minimisation framework whose goal is to extremise a free-energy functional.

Let $G = (V, A, E, W)$ denote a weighted graph with index-set V, attribute-set $A : V \to \Re^n$, edge-set $E = \{(u, v)|(u, v) \in V \times V, u \neq v\}$ and the edge-weight function set $W : E \to [0, 1]$. View the nodes in the graph as points on a manifold M, then the weight $W_{u,v}$ associated with the edge connecting the pair of nodes u and v can be interpreted as the arc-length $\mathcal{E}_{p_u, p_v}$ over the geodesic connecting the pair of points p_u and p_v in M. Analogously, the attributes of a node v can be viewed as a vector field X at point p_v, i.e. $X \mid_{p_v}$.

As we are interested in modeling the edges in the graph as geodesics in a manifold, we also consider the parameterised curve $\gamma : t \in [\alpha, \beta] \mapsto M$. From Riemannian geometry [25], we know that for γ to be a geodesic, it must satisfy the condition $\nabla_{\gamma'}\gamma' = 0$ where ∇ is a Levi-Civita connection [26]. Further, Levi-Civita connections are metric preserving, unique and are guaranteed to exist.

To provide a characterisation invariant over isometric transformations, we use the curvature tensor $R(\gamma', Y)$ and the Jacobi field along γ, which is a differentiable vector field $Y \in M_p$ orthogonal to γ satisfying Jacobi's equation $\nabla_t^2 Y + R(\gamma', Y)\gamma' = 0$, where ∇ is a Levi-Civita connection. Hence, from Jacobi's equation follows that $\nabla_t^2 Y = -\mathcal{K}(\gamma', Y)Y$.

The fact above is important since it relates the curvature tensor with a vector field along a geodesic. Including the curvature term $\nabla_t^2 Y$ into a Lagrangian to be extremised as an aim of computation opens-up the possibility of constraining the topology of the manifold through its curvature, hence allowing us to formulate the graph regularisation problem as a manifold constrained one.

2.2   The Ginzburg-Landau Functional

With the notation introduced in the previous section, we can now turn our attention to the energy functional which we aim at minimising. To do this, we use a Lagrangian of the following form $\mathcal{L} = K(Y) - U(Y)$, where $K(Y)$ and $U(Y)$ are the kinetic and potential energies of the vector field $Y \in M$, respectively. Furthermore, let $K(Y) = \mid \nabla_t Y \mid^2 + \mid \nabla_t^2 Y \mid^2$. This yields the Ginzburg-Landau energy functional [27], which is given by

$$S = \int_M K(Y) - U(Y) = \int_M \gamma_1 \mid \nabla_t^2 Y \mid^2 + \gamma_2 \mid \nabla_t Y \mid^2 + \gamma_3 V(Y) \qquad (1)$$

which is the quantity we aim at minimising. Recall that we have required the vector field Y to be a Jacobi field. As a result, we can use the shorthand $\nabla_t Y = \mathcal{K}(\gamma', Y)Y$ and write the Lagrangian as follows

$$\mathcal{L} = \gamma_1 \mid \mathcal{K}(\gamma', Y)Y \mid^2 + \gamma_2 \mid \nabla_t Y \mid^2 + \gamma_3 V(Y) \qquad (2)$$

where γ_i is a real-valued constant.

Note that the Lagrangian above has two terms related to the kinetic energy $K(Y)$. The first of these is related to the curvature of the manifold upon which the field Y is constrained. The second one is governed by the connection of Y with respect to the parameter t. This is an important observation since it permits the introduction of both, a structural term and an manifold-constraining one. In the following section, we elaborate further on this and relate the above functional to the graphical model under consideration.

2.3   Energy Minimisation on Graphs

Having introduced the Ginzburg-Landau functional in the previous section, we now turn our attention to the combinatorial analogue of the Lagrangian above. We commence by noting that the term $\mid \nabla_t Y \mid^2$ in Equation 2 can be viewed as the gradient of Y with respect to t. Moreover, here we focus our attention in the case where the term $\frac{1}{2} \mid \nabla_Y \mid^2$, can be approximated making use of the first difference around a region $\mathcal{R}_u$ centered at point $p_u \in M$. We can write

$$\mid \nabla_t Y \mid_{p_u}^2 = \text{Vol}[\mathcal{R}(u)] \int_{p_v \in \mathcal{R}(u)} \beta(u,v) \left(\frac{Y \mid_{p_u}}{\sqrt{\text{Vol}[\mathcal{R}(u)]}} - \frac{Y \mid_{p_v}}{\sqrt{\text{Vol}[\mathcal{R}(u)]}} \right)^2 du \qquad (3)$$

where $\text{Vol}[\mathcal{R}(u)]$ is the volume operator for the region $R(u)$ around the point $p_u \in M$.

The use of the shorthand above ensures that, in the case the terms $\mid \mathcal{K}(\gamma', Y)Y \mid^2$ and $\gamma_3 V(Y)$ are null, i.e. $\mathcal{L} = \gamma_1 \mid \mathcal{K}(\gamma', Y)Y \mid^2 + \gamma_2 \mid \nabla_t Y \mid^2$, the Euler-Lagrange equation for the Lagrangian $\mid \nabla_t Y \mid^2$ is given by $\nabla_t Y = 0$. In other words, the choice of shorthand above exploits the relationship between the normalised Combinatorial Laplacian $\mathcal{L}$ and its diffusion processes, by viewing the Laplacian $\mathcal{L}$ as an operator in the field Y which satisfies

$$\mathcal{L}Y(v) = \frac{1}{\sqrt{deg(v)}} \sum_{u \in M} \left(\frac{Y(v)}{\sqrt{deg(v)}} - \frac{Y(u)}{\sqrt{deg(u)}} \right) W(v,u) \qquad (4)$$

It is worth noting in passing that Equation 4 is consistent with those expressions in [28] corresponding to binary graphs with non time-dependent spaces of functions over their edges. Furthermore, the use of the Laplacian generalises those results that were only known for k-regular graphs. Consequently, it provides a coherent framework for a more general treatment of the graph regularisation problem. The link weight matrix W is related to the normalised Laplacian through the equation $\mathcal{L} = D^{-\frac{1}{2}}(D - W)D^{-\frac{1}{2}} = D^{-\frac{1}{2}}LD^{-\frac{1}{2}}$, where D is a diagonal matrix such that $D = diag(deg(1),deg(2),\ldots,deg(|V|))$ and $deg(v)$ is the degree of the node indexed v, i.e. $deg(v) = \sum_{u\sim v;u\in V} W(v,u)$.

We will explore the use of the curvature term in the following section. Before proceeding, we examine the potential term $V(Y)$. In computer vision and pattern recognition, we often deal with settings such that we aim at regularising a graph based on node attributes that originate from observed data. Let these attributes be a vector field X in M. The regularised field Y can then be viewed as being proportional to X up to a linear operator $\mathcal{Q}$, i.e. $Y \propto \mathcal{Q}Y$. By making use of an L-2 norm, we can define the potential $V(Y)$ to be

$$V(Y) = ||X - \mathcal{Q}Y||^2 \tag{5}$$

where $\mathcal{Q}$ is a linear operator in Y and X is the "observed" vector field. As a result, the potential term is given by the squared difference between the node-attribute set and a linear transformation on the regularised graph attributes.

2.4   Introducing Constraints

So far, we have not imposed any explicit constraint on the vector field Y, but rather viewed the term $|\mathcal{K}(\gamma',Y)Y|^2$ as an implicit constraint on the intrinsic geometry of the manifold M. The geometric meaning of this treatment implies that, by setting $|\mathcal{K}(\gamma',Y)Y|^2$ to zero, we would, effectively, effect a regularisation process in which the vector field Y can assume values in the whole of the Euclidean space.

However, in practice, each attribute vector $y_u \in Y$ for every node indexed u in V can be associated with constraints determined by the aim of computation or application vehicle. For instance, for probabilistic estimation, we may impose non-negativity and sum-to-one constraints upon $y_v = [y_v(1), y_v(2), \ldots,$ $y_v(|y_v|)]^T$, i.e. $\sum_{i=1}^{|y_v|} y_v(i) = 1$ and $y_v(i) \geq 0 \forall i \in \{1,2,\ldots,|y_v|\}$. In other cases, we may require the node attributes to be normalised to unity, i.e. $|y_v|^2 = 1$.

As a result, by expressing these constraints as a multivariate function $c(y_v, t) = 0$ over the graph-node attribute set subject to the time parameter t, we can make use of the curvature term $|\mathcal{K}(\gamma',Y)Y|^2$ so as to obtain a general form of the energy functional based on a quadratic penalty for the kinetic energy terms in Equation 2 [29]. As a result, the combinatorial analogue of Equation 1 becomes

$$\mathcal{S} = \sum_{u\in V}\left\{\gamma_1 c(u,t)^2 + \gamma_2 \sum_{v\tilde{u}} W(u,v)\left(\frac{y_u}{\sqrt{\deg(u)}} - \frac{y_v}{\sqrt{\deg(v)}}\right)^2 + \gamma_3||X - \mathcal{Q}Y||^2\right\} \tag{6}$$

which is the energy functional we aim at extremising. In the next section, we introduce a simple method for recovering the vector field Y which extremises the equation above.

3 Implementation Issues

In this section, we present the optimisation procedure we employ so as to extremise the Lagrangian in Equation 6. To do this, we rewrite the Lagrangian $\mathcal{L}$ in matrix form. Recall that given an "observed" node-attribute set X, we aim at recovering the vector field Y which corresponds to the regularised node attributes subject to a linear operator $\mathcal{Q}$ and a edge-weight matrix W. This is, given an attribute set X and a linear operator $\mathcal{Q}$, we aim at recovering the regularised node-attribute set Y subject to connectivity constraints.

To this end, we view the vector fields X and Y as $V \times n$ matrices whose i^{th} row is given by the n attributes of the node indexed i in the graph under study. Similarly, the linear operator $\mathcal{Q}$ is, in practice, a $n \times V$ matrix which, in the simplest case is the identity matrix. With these ingredients, we make use of the Stone-Weierstrass theorem [30]. Recall that the Stone-Weierstrass theorem states that any continuous, real-valued function defined in a closed interval can be approximated as closely as desired by a polynomial function in two variables. Hence, we subsequently consider a product of polynomial functions such that $c(v, t) = f(t)g(y_v)$. As a result, we have

$$\mathcal{S}_t = \gamma_1 \sum_{v \in V} \left(f(t)g(y_v) \right)^2 + \gamma_2 \mathrm{tr}[Y^T \mathcal{L} Y] + \gamma_3 ||X - \mathcal{Q}Y||^2 \qquad (7)$$

where we have written $\mathcal{S}_t$ so as to emphasise the fact that, following the introduction of the time-dependent function $c(v, t)$, the functional $\mathcal{S}$ depends on the parameter t.

The advantage of the formulation above is that we can now treat the function $f : \Re^+ \mapsto \Re$ as a penalty parameter so as to govern the degree up to which the constraint function $g(y_v)$ is enforced. By considering a non-negative, monotonically increasing function $f(t)$, we can gradually enforce the constraints. We do this by following an iterative gradient-descend method for which the gradient of $\mathcal{S}$ is given by

$$\nabla_Y \mathcal{S}_t = 2\gamma_1 \sum_{v \in V} f(t)^2 g(y_v) \frac{\partial g(y_v)}{\partial y_v} + 2\gamma_2 \mathcal{L} Y + 2\gamma_3 \mathcal{Q}^T (\mathcal{Q}Y - X) \qquad (8)$$

where $f(t)$ gradually increases in value with respect to iteration number.

We can give a geometric interpretation to the treatment above. Consider the case where $f(0) = 0$. This corresponds to the unconstrained case at $t = 0$. As a result of the monotonicity condition on $f(t)$, as the time parameter increases in value, so does the function $f(t)$. Hence, as t increases, the the function $c(v, t) = f(t)g(y_v)$ becomes dominant in the optimisation process. This can be viewed as a simulated annealing procedure [31] in which the function $f(t)$ plays the role of the annealing schedule.

4   Applications

In this section, we illustrate the utility of our approach on two applications. These are calibrated photometric stereo and image segmentation. Despite the differences in terms of goal of computation between the two applications under study, both tasks can be recast into a graph regularisation setting. In the following section we show how can this be effected.

4.1   Robust Normal Estimation for Photometric Stereo

Photometric stereo aims at recovering the surface albedo and geometry given multiple images of the object in the same viewpoint illuminated by light sources from different directions[32]. In contrast with uncalibrated settings [33], here we assume known light source directions and focus on the calibrated setting [34]. Assuming distant point light sources and a matte material surface, the intensity $I_{v,j}$ of the vth pixel in the jth image is given by Lambert's Law, i.e.

$$I_{v,j} = \mathbf{L}_j^T \mathbf{b}_v = \rho_v \mathbf{L}_j^T \mathbf{n}_v \tag{9}$$

where $\mathbf{b}_v = \rho_v \mathbf{n}_v$, ρ_v and $\mathbf{n}_v$ are the surface albedo and the unit surface normal at the pixel indexed v and $\mathbf{L}_j$ is the vector corresponding to the jth light source, whose intensity and direction are given by the magnitude and orientation of $\mathbf{L}_j$, respectively.

The albedo term in Equation 9 can be eliminated by taking the ratio of intensity values in two arbitrary images i and j

$$R_u(i,j) = \frac{I_{u,i}}{I_{u,j}} = \frac{\mathbf{L}_i^T \mathbf{n}_u}{\mathbf{L}_j^T \mathbf{n}_u} \Rightarrow (R_u(i,j)\mathbf{L}_j - \mathbf{L}_i)^T \mathbf{n}_u = 0 \tag{10}$$

Equation 10 is a linear homogeneous equation in $\mathbf{n}_u$. For k-source photometric stereo, we can establish $k(k-1)/2$ similar equations for each pixel. The surface normal $\mathbf{n}_u$ at the pixel-site indexed u can then be solved in a least squares sense for $k \geq 3$ by applying singular value decomposition (SVD) to $(R_u(i,j)\mathbf{L}_j - \mathbf{L}_i)$ and taking the singular vector with the smallest singular value.

The main argument against this treatment relates to the ratio operation, which makes the method above prone to noise corruption. To overcome this problem, we commence by selecting a subset of pixels, which we denote Ω, whose intensity values are high and photometric variations are low across different views. This is, we only consider pixels whose intensities are greater than a threshold τ and which are linearly dependent between views, i.e. they differ by a scale factor following Lambert's Law.

Making use of the subset Ω, we then formulate the normal estimation problem for photometric stereo as that of recovering the surface normals $\mathbf{n}_u^{(t)}$ at iteration t which minimise

$$S_t = \gamma_1 f(t)^2 \sum_u (\|\mathbf{n}_u^{(t)}\| - 1)^2 + \gamma_2 tr(\mathbf{N}_t^T \mathcal{L} \mathbf{N}_t) + \gamma_3 \|\mathbf{N}_{t-1} - E\mathbf{N}_t\| \tag{11}$$

where $f(t)$ is a function of t as before, $\mathbf{N}_t = [\mathbf{n}_1^{(t)}, \mathbf{n}_2^{(t)}, \ldots, \mathbf{n}_{|V|}^{(t)}]^T$ and E is an indicator matrix which slots over the $|V|$ pixels in the image such that $E_{u,v} = 1$ if the vth pixel in Ω corresponds to the uth pixel in the image and $E_{u,v} = 0$ otherwise. For purposes of initialisation, we recover the initial values $\mathbf{n}_u^{(0)}$ of the surface normals making use of the SVD-based procedure described earlier.

As in the previous section, $\mathcal{L}$ is the normalised graph Laplacian computed making use of the weight-matrix whose entries are given by

$$
W(u,v) = \begin{cases} exp\Big(-\dfrac{||\mathbf{n}_u^{(t)} - \mathbf{n}_v^{(t)})||^2}{\sigma_n^2} - \dfrac{||p_u - p_v||^2}{\sigma_s^2} \Big) & \text{If } u \text{ and } v \text{ are adjacent} \\ 0 & \text{otherwise} \end{cases}
$$
(12)

where p_u and p_v denote the image coordinates of those pixels indexed u and v and σ_c and σ_s are two bandwidth parameters. The weight scheme, as defined above, takes into account both the spatial and feature consistencies of the label fields and is analogous to the underlying principle of bilateral filtering [35].

The gradient of Equation 11 w.r.t. $\mathbf{N}_t$ is then given by

$$
\nabla_{\mathbf{N}_t}\mathcal{S} = 2\gamma_1 f(t)^2 C + 2\gamma_2 \mathcal{L}\mathbf{N}_t + 2\gamma_3 E^T(E\mathbf{N}_t - \mathbf{N}_{t-1})
$$
(13)

where C is the gradient term calculated from the constraint, whose v^{th} row is given by $\mathbf{n}_v^{(t)} - \dfrac{\mathbf{n}_v^{(t)}}{||\mathbf{n}_v^{(t)}||}$ and $f(t)$ is a monotonically increasing function governed by iteration number. Thus, this procedure can can be viewed as a renormalisation on $\mathbf{N}_t$ so as to satisfy unit norm constraints in which, as the iteration number increases, the penalty imposed by the constraint term on $\mathcal{S}_t$ increases in dominance.

4.2   Semi-supervised Image Segmentation

Image segmentation is a classical problem in computer vision. While early methods are mainly focused on automatic unsupervised image segmentation, here we focus our attention in the problem of interactive, semi-supervised image segmentation which takes user-specified foreground/background pixels as ground truth at input and aims at segmenting the remaining pixels in the image accordingly. Semi-supervised image segmentation, posed as a problem of energy minimisation over binary labels, can be efficiently solved by graph cuts. In this section, we propose an alternative probabilistic approach to the problem by applying the energy minimisation method presented in the previous sections. Our treatment is defined over the continuous label fields and can handle binary as well as k-way segmentation.

Let $\Phi_j(j \in \{1, \ldots, k\})$ denote the set of pixels labeled by the user in region j and $\mathcal{U}$ denote the set of unlabeled pixels for the image under study. We can define the desired label field entry $\mathbf{x}_v$ for pixel v as follows

$$
\mathbf{x}_v = \begin{cases} \mathbf{e}_j & v \in \Phi_j \\ \mathbf{b}_v & v \in \mathcal{U} \end{cases}
$$
(14)

where $\mathbf{e}_j$ is the jth row taken from a $k \times k$ identity matrix which is an indicator of membership of pixel v to the class j, and $\mathbf{b}_v = [b_{v,1}, \ldots, b_{v,k}]$ is a k-vector with $b_{v,j}$ being the posterior probability of pixel v in class j. The posterior can be computed via the following Bayes rule

$$b_{v,j} = \frac{P(v|C_j)P(C_j)}{\sum_j P(v|C_j)P(C_j)} = \frac{P(v|C_j)}{\sum_j P(v|C_j)} \tag{15}$$

where $p(C_j)$ is the prior probability of class j and $P(v|C_j)$ the conditional probability of pixel v in class j. Here, we have assumed equal priors for every class. The class conditional probabilities are calculated from the normalised colour histogram of labeled pixels in the class. Similar to the section above, the entries of the weight matrix W are given by

$$W(i,j) = \begin{cases} exp(-\dfrac{||I(v) - I(u)||^2}{\sigma_c^2} - \dfrac{||p_v - p_u||^2}{\sigma_s^2}) & \text{if } u \text{ and } v \text{ are adjacent} \\ 0 & \text{otherwise} \end{cases} \tag{16}$$

where $I(v)$ and p_v denote the value and image coordinate of pixel v and σ_c and σ_s are two bandwidth parameters.

With these ingredients, we view the semi-supervised image segmentation problem as that of recovering a set of labels $\mathbf{Y}_t$ at iteration t which minimises the cost function

$$\mathcal{S}_t = \gamma_1 f(t)||\mathbf{Y}_t \mathbf{1}_k - \mathbf{1}_N||^2 + \gamma_2 tr(\mathbf{Y}_t^T \mathcal{L} \mathbf{Y}_t) + \gamma_3 ||\mathbf{Y}_{t-1} - \mathbf{Y}_t||^2 \tag{17}$$

where $\mathbf{1}_n$ is a $n \times 1$ vector of ones, $\mathbf{Y}_t$ is a $|V| \times k$ matrix of labels whose v^{th} row corresponds to the label vector for pixel indexed v and $\mathcal{L}$ is the normalised graph Laplacian. In the equation above, the constraint term enforces the sum-to-one constraint over each label field $\mathbf{y}_v^{(t)}$.

For gradient based optimisation, the initial value of label $\mathbf{y}_v^{(t)}$ for pixel v is set to $\mathbf{x}_v$. The gradient of the cost function in Equation 17 w.r.t. $\mathbf{Y}_t$ is given by

$$\nabla_{\mathbf{Y}_t} \mathcal{S} = 2\gamma_1 f(t)(\mathbf{Y}_t \mathbf{1}_k - \mathbf{1}_N)\mathbf{1}_k + 2\gamma_2)\mathcal{L}\mathbf{Y}_t + 2\gamma_3(\mathbf{Y}_t - \mathbf{Y}_{t-1}) \tag{18}$$

which is akin to the treatment given to our photometric stereo application vehicle in the previous section.

5   Experiments

In this section, we show results on photometric stereo and semi-supervised image segmentation. In all our experiments, we have set $\gamma_3 = 0.05$ and $\gamma_2 = 1 - \gamma_3 = 0.95$. For the constraint term, we have used $\gamma_1 f(t) = exp(t - 5)$, where $t = \{1, 2, 3, \ldots, 10\}$.

Fig. 1. Top row: Input images; Middle row, left-hand panel: Needlemap recovered by SVD and gaussian smoothing; Middle row, right-hand panel: Surface normals recovered by our method; Bottom row: Surfaces reconstructed from the corresponding normal fields in the Middle row

5.1 Photometric Stereo

In this section, we show results of our regularisation method for robust normal estimation in photometric stereo on two image data-sets depicting a porcelain Buddha and a wooden owl[1]. Both image-sets were acquired under controlled lighting conditions using distant point light sources whose directions and intensities were

[1] The datasets are available at http://www.cs.washington.edu/education/courses/
cse455/06wi/projects/project3/web/project3.htm

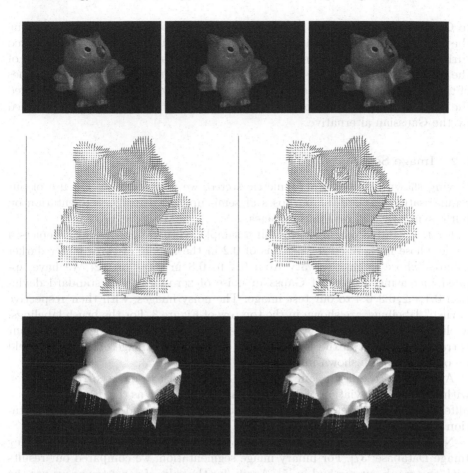

Fig. 2. Top row: Input images; Middle row, left-hand panel: Needlemap recovered by SVD and gaussian smoothing; Middle row, right-hand panel: Surface normals recovered by our method; Bottom row: Surfaces reconstructed from the corresponding normal fields in the middle row

recorded at acquisition time. Each data-set comprised 12 images corresponding to different illuminant directions. For our experiments, we used the last 3 images for each object under study.

The input images are shown in the top row of Figures 1 and 2, respectively. We compared our results to those yield by Gaussian smoothing on the surface normals recovered by SVD. In the middle row of Figures 1 and 2, we show the surface normals delivered by both, our method and the alternative. The bottom row shows the surface reconstruction results yielded by the Least Squares fitting method in [36] making use of the surface normals shown in the middle row of Figure 1. As we can see from the results on both image-sets, our method can reduce the noise corruption in the normal estimation process while still preserving edges and discontinuities in the object surface normals. In contrast to Gaussian smoothing, our method has removed noise without over-smoothing the surface normals. This, in turn, reduces the spurious artifacts in the recovered surfaces and prevents oversmoothing. This can be observed in the lap and face of the Buddha, where the Gaussian smoothing has lost the details on the eyelashes of the left eye and the lower end of the robe. Similarly, our method has preserved the discontinuities on the owl face and neck, which have been oversmoothed by the Gaussian alternative.

5.2   Image Segmentation

Having shown results on photometric stereo, we now illustrate the use of our regularisation method for purposes of semi-supervised image segmentation on both, synthetic and real-world images.

First, we tested our algorithm on a sample synthetic image which depicts a circle whose normalised intensity is of 0.2 in the middle of three larger darker squares whose intensities vary from 0.4 to 0.8 in steps of 0.2. We have applied increasing degrees of Gaussian noise of zero-mean and standard deviation 0.1, 0.2, 0.3, 0.4 to our test image. The noisy images, with their respective "brush" labelings, are shown in the top row of Figure 3. For the brush labelings, we have used a mask for the sake of consistency and, in the panels, each colour corresponds to one of the four different classes in the image. The results yield by our method are shown in the bottom row of Figure 3.

As shown in the figure, our method is quite robust to noise corruption. Even with a standard deviation of 0.4, which is twice as much as the distance between different regions in clean images, our method can still yield plausible segmentation results.

Next, we turn our attention to real-world images selected from the Berkeley Image Database [37]. For binary image segmentation, we compared our results with the graph cut method in [5]. Again, for the sake of consistency, we use the

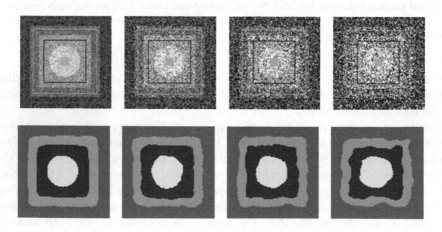

Fig. 3. Results on the synthetic image. Top row: input images with increasing levels of Gaussian noises. Bottom row: Results of our regularisation approach.

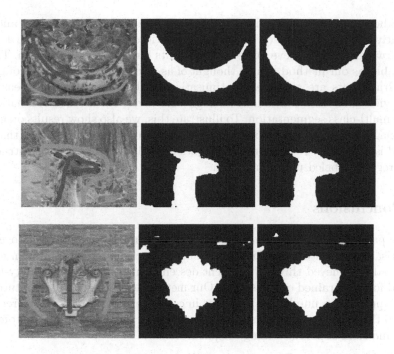

Fig. 4. Examples of binary foreground/background segmentation; Left-hand column: Input images with brush labelings; Middle-hand column: Results yield by the method in [5]. Right-hand column: Results yield by our approach.

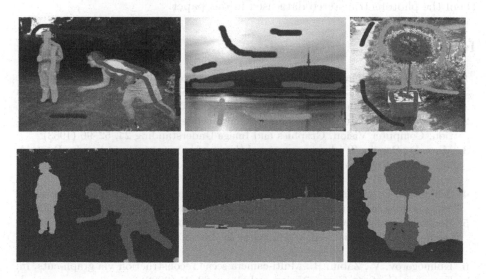

Fig. 5. Examples of multi-class image segmentation; Top row: Input images with brush labelings; Bottom row: Results recovered by our approach

same labeling mask and bandwidth parameters for both, our method and the alternative. Sample results on binary segmentation are shown in Figure 4.

The two methods achieve quite similar results for binary classification. This is reasonable, as our method can be thought of as a continuous relaxation of graph cuts. Graph cuts are a discrete optimisation method in nature, whereas ours can be viewed as their continuous analogue. However, our method can naturally handle multi-class segmentation. To illustrate this, we also show results on multi-class image segmentation in Figure 5. From the figure, we can conclude that the method is capable of recovering segmentation results that capture the structure of the regions labeled using the brush.

6   Conclusions

In this paper, we have presented an energy minimization approach to graph regularisation based on the Ginzburg-Landau functional. The minimum of the functional is achieved through gradient descent which employs a penalty-based method for constrained optimisation. Our method is quite general in nature and can be applied to a number of problems in computer vision and pattern recognition. To this end, we have illustrated the utility of our method on photometric stereo and semi-supervised image segmentation.

Acknowledgment

The authors would like to thank Dan Goldman and Steven Seitz for providing them the photometric stereo data used in this paper.

References

1. Nagel, H., Enkelmann, W.: An investigation of smoothness constraints for the estimation of displacement vector fields from image sequences. IEEE Trans. on Pattern Analysis and Machine Intelligence 8, 565–593 (1986)
2. Terzopoulos, D.: Multilevel computational processes for visual surface reconstruction. Computer Vision, Graphics and Image Understanding 24, 52–96 (1983)
3. Marr, D., Poggio, T.: A computational theory of human stereo vision. In: Proceedings of the Royal Society of London. Series B, Biological Sciences. vol. 204, pp. 301–328 (1979)
4. Scharstein, D., Szeliski, R.: A taxonomy and evaluation of dense two-frame stereo correspondence algorithms. Int. Journal of Computer Vision 47(13), 7–42 (2002)
5. Boykov, Y., Jolly, M.P.: Interactive graph cuts for optimal boundary & region segmentation of objects in n-d images. In: Intl. Conf. on Computer Vision, pp. 105–112 (2001)
6. Kolmogorov, V., Zabih, R.: Multi-camera scene reconstruction via graph-cuts. In: European Conf. on Comp. Vision. vol. 3, pp. 82–96 (2002)
7. Vogiatzis, G., Torr, P., Cipolla, R.: Multi-view stereo via volumetric graph-cuts. In: IEEE Conf. on Computer Vision and Pattern Recognition, vol. II, pp. 391–398 (2005)

8. Sun, J., Shum, H.Y., Zheng, N.N.: Stereo matching using belief propagation. In: European Conf. on Comp. Vision, pp. 510–524 (2002)
9. Worthington, P.L., Hancock, E.R.: New constraints on data-closeness and needle map consistency for shape-from-shading. IEEE Transactions on Pattern Analysis and Machine Intelligence 21(12), 1250–1267 (1999)
10. Barron, J.L., Fleet, D.J., Beauchemin, S.S.: Performance of optical flow techniques. Int. Journal of Computer Vision 12(1), 43–77 (1994)
11. Blum, A., Chawla, S.: Learning from labeled and unlabeld data using graph min-cuts. In: Proc. of Intl. Conf. on Machine Learning, pp. 19–26 (2001)
12. Zhu, X., Ghahramani, Z., Lafferty, J.: Semi-supervised learning using gaussian fields and harmonic functions. In: 20th Intl. Conf. on Machine Learning (2003)
13. Zhou, D., Bousquet, O., Lal, T., Weston, J., Schölkopf, B.: Learning with local and global consistency. In: Neural Information Processing Systems (2003)
14. Zhang, F., Hancock, E.R.: Tensor mri regularization via graph diffusion. In: British Machine Vision Conference, vol. II, pp. 589–598 (2006)
15. Chefd'hotel, C., Tschumperle, D., Deriche, R., Faugeras, O.D.: Constrained flows of matrix-valued functions: Application to diffusion tensor regularization. In: European Conf. on Comp. Vision, vol. I, pp. 251–265 (2002)
16. Tschumperle, D., Deriche, R.: Diffusion tensor regularization with constraints preservation. In: IEEE Conf. on Computer Vision and Pattern Recognition, vol. I, pp. 948–953 (2001)
17. Busemann, H.: The geometry of geodesics. Academic Press, London (1955)
18. Ranicki, A.: Algebraic l-theory and topological manifolds. Cambridge University Press, Cambridge (1955)
19. Hjaltason, G.R., Samet, H.: Properties of embedding methods for similarity searching in metric spaces. IEEE Trans. on Pattern Analysis and Machine Intelligence 25, 530–549 (2003)
20. Tenenbaum, J.B., de Silva, V., Langford, J.C.: A global geometric framework for nonlinear dimensionality reduction. Science 290(5500), 2319–2323 (2000)
21. Roweis, S.T., Saul, L.K.: Nonlinear dimensionality reduction by locally linear embedding. Science 290, 2323–2326 (2000)
22. Belkin, M., Niyogi, P.: Laplacian eigenmaps and spectral techniques for embedding and clustering. Neural Information Processing Systems 14, 634–640 (2002)
23. Hein, M., Audibert, J., von Luxburg, U.: From graphs to manifolds - weak and strong pointwise consistency of graph laplacians. In: Proceedings of the 18th Conference on Learning Theory (COLT), pp. 470–485 (2005)
24. Ginzburg, V., Landau, L.: On the theory of superconductivity. Zh. Eksp. Teor. Fiz. 20, 1064–1082 (1950)
25. Berger, M.: A Panoramic View of Riemannian Geometry. Springer, Heidelberg (2003)
26. Chavel, I.: Riemannian Geometry: A Modern Introduction. Cambridge University Press, Cambridge (1995)
27. Jost, J.: Riemannian Geometry and Geometric Analysis. Springer, Heidelberg (2002)
28. Chung, F.R.K.: Spectral Graph Theory. American Mathematical Society, Providence (1997)
29. Nocedal, J., Wright, S.: Numerical Optimization. Springer, Heidelberg (2000)
30. Stone, M.H.: The generalized weierstrass approximation theorem. Mathematics Magazine 21(4), 167–184 (1948)
31. Kirkpatrick, S., Gelatt, C.D., Vecchi, M.P.: Optimization by simulated annealing. Science 220(4598), 671–680 (1983)

32. Woodham, R.: Photometric methods for determining surface orientation from multiple images. Optical Engineering 19(1), 139–144 (1980)
33. Yuille, A., Coughlan, J.: Twenty questions, focus of attention, and a*: A theoretical comparison of optimization strategies. In: Energy Minimization Methods in Computer Vision and Pattern Recognition, pp. 197–212 (1999)
34. Hertzmann, A., Seitz, S.: Example-based photometric stereo: Shape reconstruction with general, varying brdfs. IEEE Trans. on Pattern Analysis and Machine Intelligence 27(8), 1254–1264 (2005)
35. Tomasi, C., Manduchi, R.: Bilateral filtering for gray and color images. In: Intl. Conf. on Computer Vision, pp. 839–846 (1998)
36. Basri, R., Jacobs, D.: Photometric stereo with general, unknown lighting. In: Proc. Computer Vision and Pattern Recognition, pp. 374–381 (2001)
37. Martin, D., Fowlkes, C., Tal, D., Malik, J.: A database of human segmented natural images and its application to evaluating segmentation algorithms and measuring ecological statistics. In: Int. Conf. on Computer Vision. vol. 2, pp. 416–423 (July 2001)

A Pupil Localization Algorithm Based on Adaptive Gabor Filtering and Negative Radial Symmetry

Fei Xiong[1], Yi Zhang[2], and Guilin Zhang[2]

[1] Institute for PRAI, Huazhong University of Science and Technology
Wuhan, China, 430074
shionefaye@gmail.com
[2] Institute for PRAI, Huazhong University of Science and Technology
Wuhan, China, 430074

Abstract. The pupil localization algorithm is very important for a face recognition system. Traditional pupil localization algorithms are easy to be affected by uneven illuminations and accessories. Aiming at those limitations, a novel pupil localization algorithm is proposed in this paper. The algorithm firstly implements face image tilt adjustment and extracts eye region through the horizontal intensity gradient integral projection and Gabor filtering. Then in order to increase the eye detection accuracy, PCA is applied to select Gabor filter, and the projection enhancement algorithm is presented. At last the Negative Radial Symmetry is presented to locate the pupil position precisely in eye windows. Experimental results show that the method can locate the pupil position accurately, and demonstrate robustness to uneven lighting, noise, accessories and pose variations.

1 Introduction

Face alignment is of great significance in Face Recognition System and directly affects the recognition rate. A crucial step in face alignment is to get accurate pupil position. Generally, two kinds of pupil localization method exist. One is based on appearance, such as the Template Matching[1], integral projection[1], Snake[2], Deformable Template[3], Hough Transform[4], and Active Appearance Models (AAMs)[5]. Since the image intensity is used as features, these algorithms are affected by image appearance variation. The other kind of methods, such as the GaborEye[6] model method, mainly treat the frequency response as the judgment. Since the Gabor-based algorithms extract Gabor feature to locate the eye-and-brow[6] region, they can resist the uneven illumination and noise effects. But filter direction selection is unavailable and it has poor robustness to the accessories appearing on the eye-and-brow region. In this paper, the face tilt angle detection and adjustment algorithm, which adjust the face image to level, simplifies the direction selection of Gabor wavelet filter. And the horizontal integral projection of x-directional intensity gradient and Gabor wavelet filter is applied to extract the eye windows. Then the PCA is applied to select the Gabor filter with best performance and the projection enhancement

A.L. Yuille et al. (Eds.): EMMCVPR 2007, LNCS 4679, pp. 87–96, 2007.
© Springer-Verlag Berlin Heidelberg 2007

algorithm is presented to increase the eye detection accuracy. At last, NRS (Negative Radial Symmetry) is proposed as the amended RS (Radial Symmetry) operator to locate the pupil position more accurately. The algorithm flow is shown as Fig. 1. Section 2 introduces the eye region localization algorithm, including the rotation adjustment; section 3 clarifies the amendment of the eye region localization algorithm; section 4 describes NRS operator.

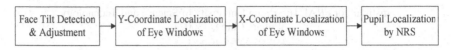

Fig. 1. The flow chart of pupil localization algorithm

2   Eye Detection

The contrast of gray intensity between eye and the other skin region is great, and there are a lot of edges with fierce gradient. It has been proved by GaborEye that the fierce response to specific Gabor filter of eye-and-brow region is valuable to extract eye windows in most cases. In this paper, the calculation of eye windows' x-coordinate and y-coordinate is respectively based on Gabor wavelet filtering and horizontal integral projection of x-directional gradient. Since not only Gabor filter direction should be consistent with the face tilt angle but also a level face image is the precondition of eyes' y-coordinate localization algorithm, we calculate the tilt angle and adjust face images to level firstly.

2.1   Face Tilt Detection and Adjustment

In the reality, face gesture and tilt angle are arbitrary. The face tilt angle is defined as the degrees from the vertical of the line connecting the two eyes. In order to calculate the tilt angle, we need to protrude the eyes from the whole image. Because of the low gray intensity of eyes, the bottom-hat operator[7] is applied to protrude the eyes:

$$V = I \bullet B - I \tag{1}$$

where B is the element structure, I is the original image, with size of $H \times W$, H denotes the height, and W represents the width. Due to the round intensity valley field on eye regions, a disk element structure is used in this morphological operation. The radius of B is decided by the original image size. In this paper, the radius is 7 while the image is 100×100 pixel size. Take I for example, the eyes and eyebrows in the image V, shown as Fig. 2, are with high intensity while the hairs and background in I are eliminated. While rotating the image from -45 to 45 degrees, we calculate the x-directional integral projection of V on each degree. The standard variation σ of the projection vector is the maximum when the rotation makes the two eyes are located on the same level. The object function is as follows:

$$\begin{cases} P_{i\alpha} = \dfrac{\sum\limits_{j=1}^{W} V_{i,j}}{W} \\ \sigma_{\alpha} = \dfrac{\sum\limits_{i}(P_{i\alpha} - \overline{P_{i\alpha}})^2}{H-1} \end{cases} ,i=1,...,H,\alpha \in [-45,45] \qquad (2)$$

$$\alpha^{*} = \arg \max_{\alpha}(\sigma_{\alpha}, \alpha \in [-45,45]) \qquad (3)$$

where α is the searching rotation angle, and α^{*} is the calculated tilt angle. According to equation (3) σ_{α} reaches the maximum when $\alpha^{*} = -8$. The adjusted image is shown as Fig. 2, and the face is level.

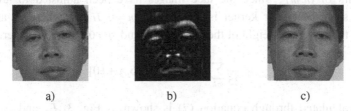

a)   b)   c)

Fig. 2. a) original image, b) valley field image, c) level image

2.2 Y-Coordinate Localization

Since the eye is composed of the pupil, white eyeball and the canthus, the horizontal intensity changed much more frequently and fiercely than the nose and mouth region. So the approximate y-coordinate can be calculated by the horizontal integral projection of x-directional gradient. In order to eliminate the edge of hairs and background, the gradient image S is the convolution of valley field image and the horizontal Sobel edge detector. Moreover, because the eyes are located in the upper part of the face images, this area is projected to calculate y^{*} according to the equation (4) and (5), which is the y-coordinate of the eyes.

$$Ps_i = \frac{\sum\limits_{j=1}^{W} S_{i,j}}{W}, i=1,\cdots,H \qquad (4)$$

$$y^{*} = \arg \max_{i}(Ps_i, i \in [1, H/2]) \qquad (5)$$

2.3 X-Coordinate Localization Based on Gabor Filtering

The vertical intensity gradient changes obviously after the face tilt is eliminated, i.e. there will be intense frequency response. Since the convolution with 2-D Gabor filter can provide the energy of any frequency and direction, therefore eye-and-brow region with special frequency and direction could be extracted by the much fiercer response than the other regions via a 2-D Gabor wavelet filter. The function of 2-D Gabor wavelet filter[8] is as follows:

$$G_{U,V}(z) = \frac{\left\| k_{U,V} \right\|^2}{\sigma^2} e^{(-\left\| k_{U,V} \right\|^2 \left\| z \right\|^2 / 2\sigma^2)} e^{-jk_{U,V}/2\sigma^2} \tag{6}$$

where $k_{U,V} = k_v e^{i\phi_n}$, $\sigma = 2\pi$, $z = (x,y)^T$ is the coordinate; $U = \dfrac{k_v \cos\phi_n}{2\pi}$,

$V = \dfrac{k_v \sin\phi_n}{2\pi}$, $k_v = \dfrac{k_{max}}{f^m}$ is the frequency, and normally $f = \sqrt{2}, m \in N$;

$\phi_n = \dfrac{n\pi}{8}, \phi_n \in [0,\pi]$ decides the direction. So various filters can be obtained with

different pairs of (n,m) . Since the face images have been adjusted to level, here $n = 0$. The scale of Gabor kernel is denoted as $m = \eta \cdot H$ [6], where m and H is respectively the scale and height of the face image, and $\eta = 0.03$ in this paper.

$$Project_{eyes} = \frac{1}{25} \sum_{i \in h} C_{i,j}, h \in (y-15, y+10] \tag{7}$$

$Project_{eyes}$ calculated through equation (7) is shown as Fig. 3. E_L and E_R are the maximums of $Project_{eyes}$ in left and right half respectively, the threshold is defined as $T = \alpha E$, where $\alpha = 0.75$. Hence the x-coordinate range of eye window is obtained and the eye windows can be extracted with the x and y coordinates.

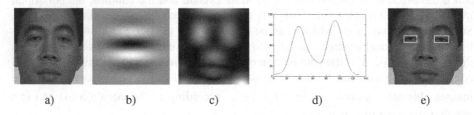

a) b) c) d) e)

Fig. 3. a) level face image, b) filter image, c) filtered image, d) vertical integral projection of eye region, e) extracted eye windows

3 Improvement on X-Coordinate Localization of Eye Windows

3.1 Enhancing the Double Peak Property of the Y-Directional Integral Projection

When wearing accessories, $Project_{eyes}$ would not appear as double peaks, and another noise peak would emerge in between. Because the middle peak location matches the single peak location of $Project_{mouth}$, equation (8) and (9) are proposed to weaken the middle peak in $Project_{eyes}$.

$$Project_{mouth} = \frac{5}{2H} \sum_{i \in h} C_{i,j}, h \in (H/2, 9H/10] \qquad (8)$$

$$Project = Project_{eyes} - \lambda \times Project_{mouth} \qquad (9)$$

where λ is a coefficient which is normally 0.5, $Project_{eyes}$ and $Project_{mouth}$ are respectively the vertical integral projection of eye and mouth regions. Shown as Fig. 4, the middle peak is eliminated and the double peak property is enhanced.

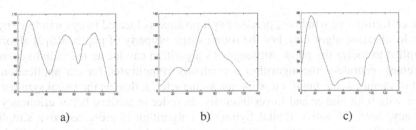

a)   b)   c)

Fig. 4. a) and b) are respectively the vertical integral projection of eye and mouth region, c) enhanced vertical integral projection of eye region

3.2 Automatic Gabor Filter Selection Based on PCA (Principle Component Analysis)

As 3-D flexible objects, the appearance of face images is changed with the various expressions, gestures, illumination and so on, which will affect their frequency distribution property. Mostly, the Gabor filter with the coefficient selected according to section 2.3 performs well. However the frequency component in the eye windows from different samples varies sometimes, and only one single filter can neither fully represent the way in which people recognize eyes, nor describe the exact difference between eyes and the other regions. In this paper, the PCA is applied to analyze the principle component and select the best $Project_n$ for the purpose of localizing the eyes' x-coordinate. Besides the filter with $m = 3$ mentioned above, another two filters with $m = 2, 4$ are introduced to acquire different frequency response M_n and $Project_n$, where $n = 1, 2, 3$.

$$V_n = U^T (P'_n - \bar{P}) \qquad (10)$$

$$\bar{P} = \frac{1}{n_P} \sum_P P'_i \qquad (11)$$

where U is constructed by the eigenvectors corresponding to large eigenvalues of the covariance matrix of the training projection data set $\mathbf{P} = \{P'_1, \ldots\ldots, P'_{n_p}\}$, which the correct eye localization can be made based on. And P' is standardized $Project$ whose mean is 0 and variation is 1. And the dimension of P' is unified to 50.

Then the best filter can be decided via equation (12)

$$n^* = \arg\max_n (\|V_n\|^2)$$ (12)

And the corresponding $Project_n$ can be chosen as the basis of calculating the x-coordinate of eyes instead of $Project_{eyes}$.

4   Pupil Localization

After extracting eye windows, precise pupil position is located in eye windows by the pupil localization algorithm. For the round shape property of pupils, RS[9] algorithm is applied to locate the pupil. Although RS algorithm can locate the radial symmetrical center accurately, the algorithm is relatively complicated for our application. Instead of aiming at the pupil's low intensity character, it detects the radial symmetrical center with both higher and lower intensity. In order to achieve better efficiency and accuracy, NRS (Negative Radial Symmetry) algorithm is presented as a simplified version of RS algorithm, which only detects the radial symmetrical center with lower intensity. Particularly, NRS carries out the 3×3 Sobel edge detector[10] to calculate the vertical and horizontal gradient vector G. And then negative point p_{-ve} of original pixel p is calculated via equation (13).

$$p_{-ve} = p - round(\frac{G(p)}{\|G(p)\|}n)$$ (13)

where $G(p)$ is a 2-D gradient vector, round means rounding each vector element to the nearest integer. And $n \in R$, where R is the range of possible pupil radius.

$$O_n(p_{-ve}(p)) = O_n(p_{-ve}(p)) - 1$$ (14)

$$M_n(p_{-ve}(p)) = M_n(p_{-ve}(p)) - \|G(p)\|$$ (15)

F_n is calculated by O_n and M_n via equation (16).

$$F_n = \frac{M_n}{k_n}(\frac{O_n}{k_n})^\alpha$$ (16)

where $\alpha = 2$, and $O_n(p) = \begin{cases} O_n(p), & if \quad O_n(p) < k_n \\ k_n, & \text{el se} \end{cases}$, $k_n = \begin{cases} 8, & if \quad n = 1 \\ 9, & else \end{cases}$

$$S_n = F_n * A_n$$ (17)

where A_n is a Gaussian widow with size of $n \times n$, and its standard deviation is $0.5n$. S is available via equation (18):

$$S = \sum_{N} S_n \tag{18}$$

The pupil center is positioned at the points with minimum S in the eye region. Computing time for NRS algorithm takes only 50% as much as RS, and further experiment proves that the localization accuracy is improved.

5 Experiment Result

926 pieces of images from FERET database are selected as the test set, the subjects of which are from different races, with various expressions and accessories, and under different illumination environment. The correct detection is defined as the eyes are detected and marked within a 15×29 rectangle, which is the common size of the eye window when the size of detected face images is 100×100, so is the pupil localization with the window size of 11×11, which means the error distance between the localized center and the real pupil center is less than 5 pixels.

A set of experiment is designed to prove the improvement of the eye detection and pupil localization by the Gabor filter selected via PCA. The eye detection rate and pupil localization rate is shown in Table 1 and Table 2.

Table 1. Eye detection rate for different filter selection

Filter Selection	Left eye detection rate	Right eye detection rate
n=2	79.2%	78.9%
n=3	95.8%	95.1%
n=4	88.4%	87.7%
the filter selected via PCA	98.3%	98.5%

Table 2. Pupil localization rate for different filter selection

Filter Selection	Left Pupil localization rate	Right Pupil localization rate
n=2	77.1%	76.4%
n=3	87.9%	89.9%
n=4	86.3%	87.0%
the filter selected via PCA	92.4%	93.2%

Shown as tables above, the performance based on the Gabor filter selected via PCA increases both the eye detection rate and pupil localization rate significantly. For the purpose of further evaluating the Gabor filter selection capability of PCA, the standard deviation of pupil localization error distance is calculated and shown as Table. 3.

Table 3. Pupil localization error distance standard deviation

Filter Selection	Left Pupil error distance standard deviation	Right Pupil error distance standard deviation
n=2	6.21	6.06
n=3	2.64	2.99
n=4	4.32	4.84
the filter selected via PCA	2.13	2.23

It can be concluded from the analysis of the two tables above that the application of the automatically selected Gabor filter via PCA performs better than any result based on single-band Gabor filter.

The better localization accuracy of NRS than RS is proven by the comparison between the two algorithms, which is presented in the followed table.

Table 4. Comparision of localization rate and error distance standard deviation between RS and NRS

	NRS	RS
Left Pupil localization rate	92.4%	90.1%
Right Pupil localization rate	93.2%	91.1%
Left Pupil error distance standard deviation	2.13	2.34
Right Pupil error distance standard deviation	2.28	2.42

NRS not only run faster than RS, but also shows better localization rate and smaller pupil localization error distance standard deviation. Hence the conclusion can be drawn that NRS performs better than RS.

In comparison with GaborEye model within the test set, the localization rate and pupil localization distance error standard deviation are shown as Table 5.

Table 5. Comparision of localization rate and error distance standard deviation between the adaptive Gabor filtering & NRS and GaborEye model & RS

	Adaptive Gabor filtering & NRS	GaborEye & RS
Left Pupil localization rate	92.4%	87.4%
Right Pupil localization rate	93.2%	88.2%
Left Pupil error distance standard deviation	2.13	3.07
Right Pupil error distance standard deviation	2.28	2.96

The data in Table 5 proves that the novel algorithm in this paper demonstrates a better localization rate and less localization error distance standard deviation. The result suggests that the algorithm based on multi-band Gabor wavelet filter fusion and NRS perform better than GaborEye model algorithm. The eye and pupil localization results are shown as Fig. 5.

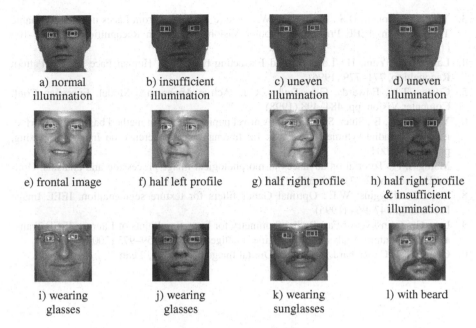

a) normal illumination b) insufficient illumination c) uneven illumination d) uneven illumination

e) frontal image f) half left profile g) half right profile h) half right profile & insufficient illumination

i) wearing glasses j) wearing glasses k) wearing sunglasses l) with beard

Fig. 5. Examples in FERET data base

6 Conclusion

The paper focuses on a novel pupil localization algorithm which aligns the face images under different illumination environment and with varieties of expressions and gesture. It extracts the eye region and then precisely locating the pupils. Advantages proved by the experiments based on the FERET database are as follows:

1. The tilt angle detection and adjustment problem is solved by calculating the x-directional integral projection vector's deviation. Moreover it simplifies the selection of the Gabor filter coefficients, with the improvement of the algorithm efficiency.
2. It locates the eyes' y-coordinate more precisely based on the horizontal gradient.
3. The PCA based Gabor filter selection algorithm is introduced to select the best filtered image for more accurate eye region extraction, and projection enhancement algorithm is proposed to resist the effect of accessories.
4. NRS operation is presented and applied with higher efficiency and accuracy.

The experiments prove that the new algorithm in this paper can correctly locate the pupil precisely and resist uneven illumination, various expression, and accessories.

References

1. Brunelli, R., Paggia, T.: Face recognition: features vs. templates. IEEE Trans. on PAMI 15(10), 1042–1052 (1993)
2. Kass, M., Witkin, A., Terzopoulos, D.: Snakes: Acfive contour models. Int. Journal of Computer Vision, 321–331 (1988)

3. Yuille, L., Cohen, D.S., Hallinan, P.W.: Feature Extraction from Faces using Deformable Templates. In: IEEE Proc. of Computer Vision and Pattern Recognition, pp. 104–109 (1989)
4. Lam, K.M., Yam, H.: Locating and Extracting the Eye in Human Face Images Pattern Recognition, 771–779 (1996)
5. Coates, T.F., Edwards, G.J., Taylor, C.J.: Active Appearance Model. European Conf. Computer Vision, pp. 484–498 (1998)
6. Yang, P., Du, B., Shan, S., Gao, W.: A novel pupil localization method based on GaborEye model and radial symmetry operator. In: International Conference on Image Processing, pp. 67–70 (2004)
7. Maragos, P.: Tutorial on advances in morphological image processing and analysis. Optical Engineering 26(7), 623–632 (1987)
8. Dum, D., Higgins, W.E.: Optimal Gabor filters for texture segmentation. IEEE Image Processing, 947–964 (1995)
9. Loy, G., Zetinsky, A.: Fast Radial Symmetry for Detecting Points of Interest. IEEE Transactions on Pattern Analysis and Machine Intelligence 25(8), 959–973 (2003)
10. Gonzalez, R.C., Richard, E.: Woods Digital Image Processing, 2 edn

Decomposing Document Images by Heuristic Search

Dashan Gao[1] and Yizhou Wang[2],*

[1] Dept. of Electrical and Computer Engineering
University of California, San Diego
9500 Gilman Drive
La Jolla, CA 92093-0409 USA
Tel.: 1-(858)-534-4538
dgao@ucsd.edu

[2] Palo Alto Research Center (PARC)
3333 Coyote Hill Rd.
Palo Alto, CA 94304-1314 USA
Tel.: 1-(650)-812-4772
Fax: 1-(650)-812-4334
Yizhou.Wang@parc.com

Abstract. Document decomposition is a basic but crucial step for many document related applications. This paper proposes a novel approach to decompose document images into zones. It first generates overlapping zone hypotheses based on generic visual features. Then, each candidate zone is evaluated quantitatively by a learned generative zone model. We infer the optimal set of non-overlapping zones that covers a given document image by a heuristic search algorithm. The experimental results demonstrate that the proposed method is very robust to document structure variation and noise.

1 Introduction

Document decomposition is a basic but crucial step for many document related tasks, such as document classification, recognition, and retrieval. For example, given a technical article, after it is decomposed into zones, the zones' properties can be used as indices for efficient document retrieval. Hence, an accurate, robust and efficient framework for document decomposition is a very important and demanding module of document analysis, which ensures success of subsequent tasks.

The goal of document image decomposition is to segment document images into zones. Each zone is a perceptually compact and consistent unit (at certain scale), e.g. a paragraph of text, a textural image patch. Methods for document image decomposition can be classified into three categories: bottom-up methods[5,9,15], top-down methods[1,7,8], and combination of the two[13]. Typical examples of bottom-up methods utilize detected connected components, and progressively aggregate them into higher level structures, e.g. words, text lines, and paragraphs (zones). Conversely, top-down methods decompose larger components into smaller ones. A typical top-down approach is the X-Y tree method [14], which splits a document image into rectangular

---

* The corresponding author.

A.L. Yuille et al. (Eds.): EMMCVPR 2007, LNCS 4679, pp. 97–111, 2007.
© Springer-Verlag Berlin Heidelberg 2007

areas (zones) recursively by alternating horizontal and vertical cuts along spaces. Usually, both approaches heavily depend on detecting connected components, separating graphics and white space, or certain document generating rules and heuristics. Both parameters for detecting document components and rules/heuristics used to segment documents are often manually tuned and defined by observing data from a development set. Thus, the adaptability and robustness of these methods are limited. It is hard for them to be generalized from document to document. When there exist ambiguities (e.g. text in document is noisy or has accidental proximity), neither type of method decomposes pages reliably. Methods based on statistical pattern analysis techniques [6,10] are generally more robust. However, current reported methods, so far, are still relatively naive. For example, features are ad hoc, and computation engines are greedy.

This paper proposes a novel approach to decomposing document images using machine learning and pattern recognition techniques. More specifically, given a document image, it first proposes over-complete overlapping zone hypotheses in a bottom-up way based on generic visual feature classifiers. Then, each candidate zone is evaluated and assigned a cost according to a learned generative probabilistic zone model. Finally, a zone inference module implemented as a heuristic search algorithm selects the optimal set of non-overlapping zones that covers the given document image corresponding to the global optimal page decomposition solution.

The most outstanding advantage of the proposed method is that it organically combines a convenient document representation, an elaborated computational data structure, and an efficient inference algorithm together to solve the page decomposition problem. In other words, it seamlessly incorporates data(documents), models(representation) and computing (data structure and algorithm) into an integrated framework. Thus, it makes the model effective for data representation and computation; and it also makes the computation efficient due to the convenient model and computational data structure. Moreover, this method is one of the very few methods providing globally optimal multi-column document decomposition solutions, besides the X-Y-tree-like context free grammar methods. Based on page decomposition results, further document analysis tasks, such as meta-data tagging, document recognition and retrieval, are expected to be more convenient.

We first introduce a document image representation in Section 2. We discuss data preparation, document models and learning in Section 3 and 4. In Section 5, we implement zone inference by a well informed heuristic search algorithm. Some results are shown in Section 6. Finally, we summarize the proposed method in Section 7.

2   Document Image Representation

We represent a document image by a 2-layer hierarchical model. The first layer is called the *primitive layer*. Given a document image I, we apply standard techniques, such as [17], to detect "words" as atomic primitives, and connect these words into a word-graph, denoted as G_w.

$$G_w = < V, E >,$$

where $V = \{v_i; i = 1, \ldots, N_w\}$, each "word" is a graph node v. N_w is the number of "words" in a document. The edge set, $E = \{(e = (i, j), w_{ij}) : v_i, v_j \in V, w_{ij} \in$

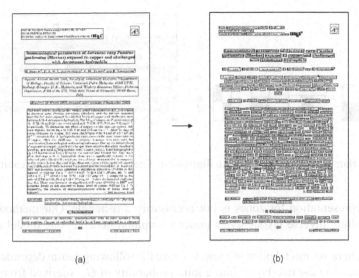

Fig. 1. Two layers of a document image model. a) Layout layer - segmented zones. b) Primitive layer - detected word bounding boxes.

$\mathbb{R}\}$, tells the neighborhood relation of pairs of "words." Each edge is associated with a weight, w_{ij}, representing bounding force between a pair of "words."

Note that these detected "words" need not be lexical words, a fraction of a word or an image patch is fine. The only purpose of this step is to reduce the image representation from pixels to a compact atomic "word" representation for the sake of computational efficiency.

The second layer is the *layout layer*, where the detected "words" are grouped into *zones* and form a zone-map, denoted as Z.

$$Z = (N_z, \{z_j : j = 1, \ldots, N_z\}), \tag{1}$$

where N_z is the number of zones. Each zone is defined as

$$z_j = (\{c_i^{(j)} : i = 1, \ldots, n_{cj}\}, \{v_k^{(j)} : k = 1, \ldots, n_{wj}\}), \tag{2}$$

which is a polygon representation; $c_i^{(j)}$ is a corner of a zone bounding polygon. n_{cj} is the number of vertices/corners of zone-j's bounding polygon. n_{wj} is the number of "words" comprising zone-j. Fig.1 shows the hierarchical representation of a document image.

Most conventional zoning algorithms heavily depend on connected component and white-space analysis, which involve *ad hoc* parameter tuning and rigid rule based reasoning. Consequently, the adaptability and robustness of the algorithms are limited. Our zone representation is based on corners, which is a well-known generic low-level robust visual feature, and it is independent of language. (Corners of the textural image bounding polygon is still obtained by connected component analysis.) Note that this

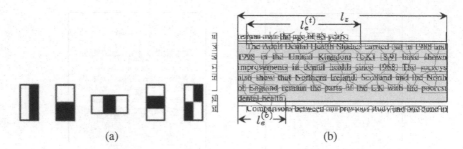

Fig. 2. (a) Harr-wavelet features. (b) An illustration of zone bounding box edge cutting spans.

polygon representation for zones is not necessarily a rectangle. Our method is capable of handling diverse layout styles under a common generic zone model as shown in Fig.6.

From generative model point of view, we have the following causal dependence $Z \rightarrow G_w$. We integrate the two layers into a joint probability of G_w (derived from an input document image I) and the hidden representation Z:

$$p(G_w, Z) = p(G_w|Z)p(Z),\qquad(3)$$

where $p(G_w|Z)$ is a generic zone likelihood model, and $p(Z)$ is a prior model for zone relations.

3 Data Preparation

3.1 Features

Generic visual features. In this project, we adopt 21 Harr-like filters to extract features from document images. These 21 filters are derived from 5 prototype Harr-filters (shown in Fig.2.(a), including a horizontal step edge, a vertical step edge, a horizontal bar(ridge), a vertical bar, and a diagonal blocks) by varying their size and scale. These features are generic and important visual features, and the filter responses can be computed in constant time at any scale and location using integral images[16].

"Word" related features. "Word" related features are very important and convenient features for document analysis. In this project, we identified six types of such feature on the word-graph g. We introduce the definition of each feature as follows.

1. "word" compactness in a zone: $f_w^{(0)}(g) = \frac{\sum_{i=1}^{k} A(g_i)}{A(z)}$ where g_i is the ith connected component of the word-graph within a candidate zone. Usually $k = 1$ in a zone, i.e. words are highly connected to one another within a zone. $A(\cdot)$ is the area of a connected component bounding box. This word-graph connected component is not the pixel-based connect component adopted by conventional document image analysis methods. $0 < f_w^{(0)}(g) \leq 1$.

2. "word" height(size) consistency in a zone: $f_w^{(1)}(g) = n_w^d(g)/n_w(g)$, where $n_w(g)$ is the number of "words" in a zone, and $n_w^d(g)$ is the number of "words" with dominant height in the zone. Usually, $0 < f_w^{(1)}(g) \le 1$. This feature tells the ratio of dominant sized "words" in a zone, which indicates the font size consistency of the zone.

3. zone bounding box top border edge-cutting span: $f_w^{(2)}(g) = l_e^{(t)}/l_z$, where l_z is the width of a zone, and $l_e^{(t)}$ is length of part of the zone bounding box top border that cuts word-graph edges. A graphical illustration of this feature is shown in Fig.2.(b). $0 \le f_w^{(2)}(g) \le 1$.

4. zone bounding box bottom border edge-cutting span: Similar to the above, $f_w^{(3)}(g) = l_e^{(b)}/l_z$, where $l_e^{(b)}$ is length of part of a zone bounding box bottom border that cuts word-graph edges as shown in Fig.2.(b). $0 \le f_w^{(3)}(g) \le 1$.

5. zone bounding box vertical border average edge-cutting weight: $f_w^{(4)}(g) = \frac{\sum_{i=1}^{n_e^{(v)}} w_e^{(i)}}{n_{tl}}$, where $n_e^{(v)}$ is number of edges cut by the two vertical borders of a zone bounding box. $w_e^{(i)}$ is the ith edge weight. n_{tl} is the number of text lines in the zone. This feature indicates the connection force of a proposed zone with its surroundings. The larger edge weight cut, less likely it is a valid zone.

6. text line alignment in a zone: $f_w^{(5)}(g) = \min(\text{var}(\mathbf{x}_l), \text{var}(\mathbf{x}_c), \text{var}(\mathbf{x}_r))$. It gives the minimum variance of the text lines' left, center and right x coordinates in a zone. The smaller the variance, the better the alignment.

These features are heuristic but independent of languages and layout style, as we try to avoid extracting syntax information from document to make our model more generalizable. These features are not necessarily independent, and they are utilized to evaluate the "goodness" of proposed zones.

3.2   Generating Word-Graph – The Primitive Layer

Given a document image I, we first compute a word-graph G_w using a neighbor finding algorithm based on the Voronoi tessellation algorithm[9] (Fig.3.(a)). Then, we compute edge weights, which tell how likely a pair of connected words are to belong to the same zone. The edge weights are posterior probabilities returned by an edge classifier discussed below. We adopt Support Vector Machines (SVM) to learn the binary edge classifier from word-graphs of training images as follows:

1. Data Preparation: We manually label zone bounding boxes on the training word-graphs. Positive edge samples are the edges within zone bounding boxes; negative samples are those cut by bounding box borders.

2. Feature Extraction: We extract a 22-dimensional feature vector including a feature accounting height difference of a pair of "words", and the 21 Harr-like filter responses (described in Section.3.1) from an image patch. The image patch is cut from I centering at the mid-point of an edge, and its area is four times large of the union of the two "words" bounding boxes.

3. SVM Training: We train a LibSVM[2] classifier on the extracted feature vectors.

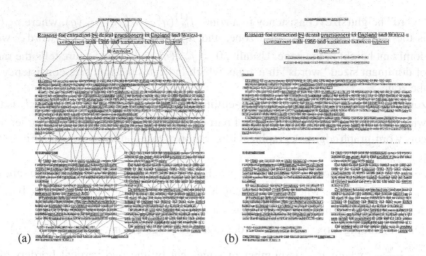

Fig. 3. (a) A Voronoi word-graph. (b) The same word-graph after prune edges whose weighs are less than 0.5 assigned by an edge classifier. Note that the edges between the paragraphs that are vertically adjacent to each other are not cut by the edge classifier.

Fig.3.(b) shows a word-graph after pruning edges whose weighs are less than 0.5 assigned by the SVM edge classifier. As the connection of a pair of "words" is computed based on generic features, the measure is more robust than pre-defined adhoc heuristic rules. Note that, in the figure, the edges between the paragraphs that are vertically adjacent to each other are not cut by the edge classifier.

3.3 Generating Zone Hypotheses

In Eqn.2, the zone representation is a generic polygon. In this paper, we demonstrate the power of the representation by simply using rectangles for zones without losing much generality due to the data set. Thus, Eqn.2 is reduced to $z_j = (c_{ul}, c_{lr}, \{v_k^{(j)} : k = 1, \ldots, n_{wj}\})$, where c_{ul} and c_{lr} are upper-left and lower-right corners of a zone bounding box.

In order to propose candidate zones efficiently, we train two classifiers, which detect upper-left and lower-right corners in document images, as follows:

1. Data Preparation: We obtain positive samples of zones' upper-left and lower-right corners using the labeled zones' corners in training word-graphs; Negative samples are collected by randomly selecting "word" bounding boxes' corners, which are not the corners of labeled zones.
2. Feature Extraction: We extract a 21-dimension generic visual feature vector (described in Section.3.1) from an image patch, which is cut from I centering at an upper-left or lower-right corner, and its size is 400×400 pixels.
3. SVM Training: We train LibSVM corner classifiers on the extracted feature vectors.

We augment the corner set by including bounding box corners of word-graph connected components in order not to miss any possible corners. Fig.4.(a) shows detected zone

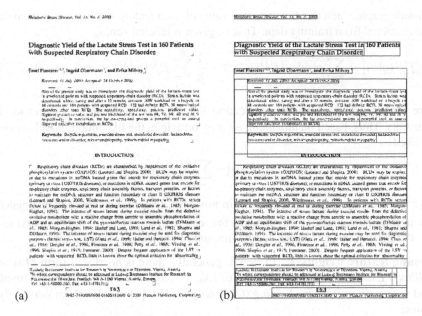

Fig. 4. (a) Detected corners. (b) Proposed top candidate zones.

corners. We propose all possible candidate zones by pairing all the detected upper-left with all lower-right corners. If heuristics are used in this process, candidate zones can be proposed more effectively by ruling out some bad configurations such as a candidate zone cannot cross line separators, etc. Fig.4.(b) shows the top 51 candidate zones with least costs proposed by this method. For the sake of computation efficiency, the rest of zones with higher costs are discarded. The zone costs are assigned by a learned generative zone model introduced below.

4   Models and Learning

4.1   A Likelihood Model for Zones

In Eqn.3, $p(G_w|Z)$ can be factorized into

$$p(G_w|Z) = p(g_{\bar{w}}) \prod_{i=1}^{N_z} p(g_i|z_i),$$

where $g_{\bar{w}}$ is sub-graphs of "words" not covered by any zone. $p(g_{\bar{w}}) = \exp(-|g_{\bar{w}}|)$, and $|\cdot|$ denotes the cardinality function. g_i is sub-word-graph(s) subsumed in zone-i, and $p(g_i|z_i)$ is a generative model for zones.

Intuitively, $p(g|z)$ governs how "words" are organized in zones in terms of the features $f_w^{(j)}(\cdot)$ described in Section.3.1. We want to construct a probabilistic model p on word-sub-graphs, such that the expected value of each feature is the same as its average value extracted from training data. That is, given n labeled zones,

$$E_j[f_w^{(j)}(g|z)] = \sum_{i=1}^{n} p(g_i|z_i) f_w^{(j)}(g_i|z_i) = \frac{1}{n} \sum_{i=1}^{n} f_w^{(j)}(g_i|z_i) = \mu_j, \quad j = 0, \ldots, 5, \tag{4}$$

where j indexes the zone features of Section.3.1. The observed feature statistics serve as constraints. Thus, based on *maximum entropy* principle, the likelihood model for zones is derived as

$$p(g|z) = c \exp\{-\sum_{j=0}^{5} \lambda_j f_w^{(j)}(g|z)\}, \tag{5}$$

where λ's are Lagrange multipliers or, in this case, feature weights to be estimated. c is the normalizing constant. Note that as the features $f_w^{(2)}, f_w^{(3)}, f_w^{(4)}$ are "context sensitive," the zone model encodes a certain amount of contextual information.

Learning feature weights λ_j. In Eqn.5, generally, there is no closed form *Maximum Likelihood Estimation* (MLE) solution for $(\lambda_0, \ldots, \lambda_5)$. We adopt a numerical method called *Generalized Iterative Scaling* (GIS) proposed by [4] to solve them iteratively as follows:

1. Given n labeled zones, compute each feature of each zone: $f_w^{(j)}(g_i|z_i)$, $(j = 0, \ldots, 5, i = 1, \ldots, n)$.
2. Compute the average of each feature extracted from the training data,

$$\mu_j = \frac{1}{n} \sum_{i=1}^{n} f_w^{(j)}(g_i|z_i), \quad j = 0, \ldots, 5.$$

3. Start iteration of GIS with $\lambda_j^{(0)} = 1$, $j = 0, \ldots, 5$.
4. At iteration t, with current parameter $\lambda_j^{(t)}$, use Eqn.5 to compute

$$E_j^{(t)}[f_w^{(j)}(g|z)] = \sum_{i=1}^{n} p^{(t)}(g_i|z_i) f_w^{(j)}(g_i|z_i), \quad j = 0, \ldots, 5$$

for each feature.
5. Update parameters

$$\lambda_j^{(t+1)} = \lambda_j^{(t)} + \frac{1}{C} \log \frac{\mu_j}{E_j^{(t)}}, \quad j = 0, \ldots, 5,$$

where C is the correction constant chosen large enough to cover an additional dummy feature[4]. ($C = 8$ in this project.)
6. Continue iteration from Step.4 until convergence.

4.2   A Prior Model for Zone-Map

The prior model of zone-maps governs not only each zone's shape, but also spatial distribution of zones in a page, e.g. similarity, proximity, symmetry. It is characterized by a statistical ensemble called *Gestalt ensemble* for various Gestalt patterns[18]. The model makes zone evaluation context sensitive. However, learning such a prior model is very expensive. In this project, as the documents in the public data set only contains rectangular zones, we take advantage of this specificity of the document set by simply enforcing that each zone is a rectangle and there is no overlap between any two zones, such that $p(\{z_1, \ldots, z_{N_z}\}) = \prod_{i \neq j} \delta(z_i \cap z_j)$, where $\delta(\cdot)$ is the Dirac delta function. Thus,

$$p(Z) = p(N_z) \prod_{i \neq j} \delta(z_i \cap z_j), \tag{6}$$

where $p(N_z)$ is prior knowledge on zone cardinality, which we assume to be a uniform distribution.

In summary, the joint probability of a word-graph G_w and zone partition Z is

$$p(G_w, Z) = p(G_w|Z)p(Z)$$
$$= p(g_{\bar{w}})\{\prod_{i=1}^{N_z} p(g_i|z_i)\} \cdot p(N_z) \prod_{i \neq j} \delta(z_i \cap z_j) \tag{7}$$

5   Zone Inference by Heuristic Search

Document image decomposition can be formulated into a Maximum A Posteriori (MAP) zone inference problem $(p(Z|G_w))$. However, to find the global optimal solution in this high dimensional space can be very expensive. In this paper, we propose a novel approach, which converts this challenging statistical inference problem into an optimal (covering) set selection problem by turning learned data statistics into costs and constraints. We design a well informed heuristic search algorithm, i.e. A^* search, to seek the global optimal page decomposition solution.

5.1   Generating Costs and Constraints from Learned Statistics

Instead of assigning costs and defining constraints in an *ad hoc* way, we derive them based on learned probabilistic models. In the page decomposition problem, we learn the following probabilistic models in Eqn.7: 1) a generative zone model, $p(g|z)$, and 2) a prior model about pairwise zone relation, $p(\{z_1, \ldots, z_{N_z}\})$.

We convert a probability $0 < P(\cdot) < 1$ to a cost as

$$c(\cdot) = \rho(-\log P(\cdot)), \tag{8}$$

where $\rho(x)$ is a robust function cutting off extreme values. When $P(\cdot) = 0$ or $P(\cdot) = 1$, there generates a binary constraint for that event. As a result, in this project, we have the following costs and constraints generated from the learned models: 1) individual cost for each zone, 2) a binary constraint that selected zones cover all "words" in a page, and 3) a binary constraint of no overlap between any pair of zones.

5.2 The A^* Algorithm

A^* algorithm is a best-first graph search algorithm, which finds a path from an initial node to a goal node. It maintains a set of partial solutions, i.e. paths through the graph starting at the start node, stored in a priority queue. The priority assigned to a path passing node x is determined by the function,

$$f(x) = g(x) + h(x), \tag{9}$$

where $g(x)$ is a *cost function*, which measures the cost it incurred from the initial node to the current node x, and $h(x)$ a *heuristic function* estimating the cost to the goal node from x. To ensure the search algorithm find the optimal solution, $h(x)$ must be admissible.

In this project, after candidate zones are proposed, page decomposition can be formulated as a weighted polygon partitioning problem in computational geometry: given a polygon (document page) and a set of candidate sub-polygons (zones), each with a weight(cost), the goal is to partition the polygon into a subset of *disjoint* sub-polygons in the candidate set so as to cover every "word" in a document image with minimum cost. This problem can be solved by an A^* search algorithm, which exploit heuristics from data to improve search performance.

As A^* search algorithm is a standard algorithm in the search literature, here we only introduce each term in the algorithm in the context of document decomposition.

State Variable x: Suppose that there are n candidate zones, we introduced a binary state vector $\mathbf{x} = (x_1, \ldots, x_n)$, where $x_i = 1$ means zone-i is selected; 0, otherwise. Any specific choice of 0's or 1's for the components of $\mathbf{x}$ corresponds to selecting a particular subset of the candidate zones.

The Goal State is every "word" in a given document is covered by only one zone.

The Cost Function $g(x)$**:** The cost of each path to $\mathbf{x}$ is defined as,

$$g(\mathbf{x}) = \mathbf{c}_z^T \mathbf{x}, \tag{10}$$

where $\mathbf{c}_z = (c_{z1}, \ldots, c_{zn})^T$ is the vector of individual zone costs, which was computed by Eqn.5 & Eqn.8 immediately after candidate zones are proposed.

The Heuristic Function $h(x)$**:** To insure the A^* algorithm admissible (or optimal), $h(\mathbf{x})$ must never overestimate the actual cost of reaching the goal. To achieve this and given the fact that both the document and the zones are represented by rectangles, the h-value of a path, from $\mathbf{x}$ to the goal state, is estimated by finding the minimum number of non-overlapping rectangles to partition the rest of the document page that has not been covered by the selected zones, $n_z(\mathbf{x})$,

$$h(\mathbf{x}) = n_z(\mathbf{x}) * c_{min}, \tag{11}$$

where c_{min} is the minimum zone cost learned from the training data. The estimate of $n_z(\mathbf{x})$ involves partitioning the complementary polygon of a state $\mathbf{x}$, which is created by removing the selected zones (rectangles) from the document page (a rectangle), into minimum number of non-overlapping rectangles. Partitioning arbitrary polygons

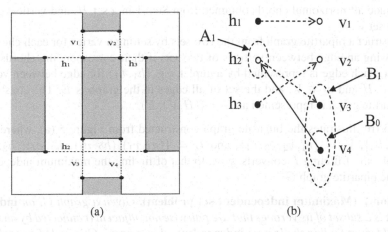

(a)   (b)

Fig. 5. An illustration of estimating a minimum rectangular partition of an document image. (a) A document image (the outer rectangle) with selected zones removed (shading rectangles). The black solid dots are reflex vertices. The vertical and horizontal chords between reflex vertices are labeled, each represented by a vertex of the bipartite graph in (b). (b) The bipartite graph constructed from the example in (a) and the derivation of a maximum independent set from a maximum matching (dashed edges are in the matching): vertices in the independent set shown as solid dots.

is NP-complete, but becomes tractable if it is restricted to only *orthogonal* polygons (whose edges are either horizontal or vertical), and can be estimated by the following theorem [11].

Theorem 1. *An orthogonal polygon can be minimally partitioned into $N - L - H + 1$ rectangles, where N is the number of reflex vertices[1] H is the number of holes and L is the maximum number of non-intersecting chords that can be drawn either horizontally or vertically between reflex vertices.*

One key computation in the theorem is that of finding L, the maximum number of non-intersecting chords that can be drawn either horizontally or vertically between reflex vertices. In this section we will show that this is equivalent to the problem of finding the *maximum number of independent vertices in the intersection (bipartite) graph of the vertical or horizontal chords between reflex vertices*, and derive a solution from a maximum matching of the bipartite graph [11,12].

As illustrated in Figure 5(a), the complementary of a document image (the outer rectangle), to be partitioned, is generated first by removing the selected zones (the shading rectangles) in a path **x**. To estimate L for this orthogonal polygon, a bipartite graph is first constructed in the following steps:

1. find all possible horizontal and vertical chords that can be drawn between all reflex vertices of the polygon (the black dots shown in Figure 5 (a));

---

[1] A reflex vertex is a vertex with interior angle greater than $180°$.

2. include all horizontal chords obtained from Step.1 in a set H, and vertical chords in a set V;
3. construct a bipartite graph from the two sets by setting a vertex for each chord and drawing an edge between vertices of two sets if their corresponding chords intercept. Each edge is represented by a duplet, e.g. (h_1, v_1) for edge between vertices $h_1 \in H$ and $v_1 \in V$, and the set of all edges in the graph is E. The constructed bipartite graph is represented as $G = ((H, V), E)$.

Figure 5 (b) illustrates the bipartite graph constructed from Figure 5 (a), where $H = \{h_1, h_2, h_3\}$, $V = \{v_1, v_2, v_3, v_4\}$, and $E = \{(h_1, v_1), (h_2, v_3), (h_2, v_4), (h_3, v_2)\}$. The problem of finding L converts, now, to that of finding the maximum independent set of the bipartite graph G.

Definition 1 (Maximum independent set problem). *Given a graph G, an* **independent set** *is a subset of its vertices that are pairwise not adjacent (connected by an edge). The problem of finding the largest independent set in a graph G is called a* **maximum independent set problem**.

The independent set problem of an arbitrary graph is known to be NP-complete, but a polynomial solution exists when the graph is a bipartite graph. This can be done by solving a *maximum bipartite matching* problem [11,12].

Definition 2 (Maximum bipartite matching problem). *Let $G = ((A, B), E)$ be an undirected bipartite graph, where A and B are sets of vertices, and E the set of edges of the form (a, b) with $a \in A$ and $b \in B$. A subset $M \subseteq E$ is a* **matching** *if no two edges in M are incident to the same vertex, that is no two edges share a common vertex. A vertex is called* **matched** *if it is incident to an edge in the match, and* **unmatched** *otherwise.* **Maximum bipartite matching** *is to find a matching M that contains the maximum number of edges of all possible matchings.*

The maximum bipartite matching problem is a well studied topic in graph theory, and often appears as a algorithm textbooks (e.g. [3]). As it is a standard algorithm, we omit the details for the sake of space limit.

In this project, we adopt graph-cut algorithm to solve the *maximum bipartite matching* for $G = ((H, V), E)$, so as to find the *maximum independent set* to solve L. Consequently, according to Theorem 1, the heuristic function $h(\mathbf{x})$ can be computed by Eq. 11. As the heuristic estimation is grounded on a solid theoretical foundation and generally gives a tight upper bound, the proposed algorithm becomes very efficient and a well-informed search strategy, which is also verified by the experiments.

Convenient Data Structures for Search. To enforce zones' non-overlapping constraint, we create a matrix called *candidate zone overlapping matrix O* to encode candidate zones' overlap. It is very easily generated as follows: Say, there are n candidate zones. O is initialized as a $n \times n$ matrix, each entry is set to 0. Then, for each pair of overlapping zones, say zone-i and zone-j, we set $O(i, j) = 1$ and $O(j, i) = 1$. During the search, whenever a candidate zone, say zone-i, is selected as a part of the solution, we check the i-th row of O, and exclude every candidate zone-j where $O(i, j) = 1$ from future selection. In this way, it is guaranteed that no overlapping zones can possibly appear in the solution.

Fig. 6. Page decomposition results. Note that the 1st result shows that the method is robust to noise and free from connected component restriction.

Another factor making the search efficient is due to a data structure called *word-to-candidate-zone index* $I_{w \to z}$. After n candidate zones are proposed, for each word in the document, there are pointers to the candidate zones that cover it. Note that each word can be covered by a number of overlapping candidate zones. But in the final solution, only one of them is selected. At each search state, say $\mathbf{x}$, we check each word which has not been covered by selected candidate zones so far to see any available candidate zones covers it. The number of available candidate zones is decreasing when more and more candidate zones as selected because the more candidate zone that are selected, the fewer available candidates; and the more overlapping candidate zones are excluded from the candidate list. $I_{w \to z}$ needs to be updated dynamically at each search step. If there exists an uncovered word neither covered by any available not-selected-yet candidate zones, $h(\mathbf{x}) = \infty$. It means that this search path is terminated earlier, even though there are possible additional non-overlapping rectangles to be partitioned solely based on the current shape of polygon (current configuration of decomposed document). This step makes the search much more efficient by pruning spurious search paths at an earlier stage.

6   Data Set and Experimental Results

We train and test our model on the first page of articles from NLM's MEDLINE database. We randomly select a set of first pages for training and a different set for testing. Some inference results are shown in Fig.6. We can see that our results are very accurate and robust to document layout variations and noise, and it is also free from connect component restriction. Moreover, we claim that due to the convenient document representation and computational data structure, together with the well informed heuristic search strategy, the algorithm is capable of efficiently solving most page decompositions within a second. This efficiency is also because we limit the number of candidate zones by only considering the top 20 to 50 proposals from the generative zone model. Otherwise, the search space grows exponentially with the number of candidates.

7   Conclusions

We have proposed a novel, generic and efficient learning and inference framework to solve a fundamental challenging problem of document analysis. It organically integrates data, models and computing to search for globally optimized multi-column document decomposition solutions. The learning part learns robust probabilistic models based on generic features. The inference module casts an expensive statistical inference to a well informed heuristic search problem. As a result, the proposed framework is very general, and it can be extended to a lot of machine learning applications.

Acknowledgements

We'd like to thank Dr. Eric Saund for his numerous discussions and insightful suggestions to this project.

References

1. Baird, H.S.: Background Structure In Document Images, Document Image Analysis. World Scientific, Singapore (1994)
2. Chang, C.-C., Lin, C.-J.: LIBSVM: a library for support vector machines, Software (2001), available at http://www.csie.ntu.edu.tw/~cjlin/libsvm
3. Cormen, T., Leiserson, C., Rivest, R., Stein, C.: Introduction to Algorithms, 2nd edn. MIT Press and McGraw-Hill, pp. 664–669 (2001)
4. Darroch, J.N., Ratcli, D.: Generalized Iterative Scaling for Log-Linear Models, The Annals of Mathematical Statistics, 43 (1972)
5. Drivas, D., Amin, A.: Page Segmentation and Classification Utilizing Bottom-Up Approach, ICDAR (1995)
6. Esposito, F., Malerba, D., Lisi, F.A.: Machine Learning for Intelligent Processing of Printed Documents. J. Intell. Inf. Syst. 14(2-3), 175–198 (2000)
7. Fujisawa, H., Nakano, Y.: A Top-Down Approach for the Analysis of Documents, ICPR (1990)
8. Ingold, R., Armangil, D.: A Top-Down Document Analysis Method for Logical Structure Recognition, ICDAR (1991)
9. Kise, K., Sato, A., Iwata, M.: Segmentation of page images using the area Voronoi diagram, CVIU (1998)
10. Laven, K., Leishman, S., Roweis, S.: A Statistical Learning Approach to Document Image Analysis, ICDAR (2005)
11. Lipski, W., Lodi, E., Luccio, F., Mugnai, C., Pagli, L.: On two dimensional data organization II. Fundamental Informaticae 2, 227–243 (1977)
12. Lipski Jr., W., Preparate, F.: Efficient algorithms for finding maximum matchings in convex bipartite graphs and related problems. Acta Informatica 15, 329–346 (1981)
13. Liu, J., Tang, Y., He, Q., Suen, C.: Adaptive Document Segmentation and Geometric Relation Labeling: Algorithms and Experimental Results, ICPR (1996)
14. Nagy, G., Seth, S., Viswanathan, M.: A Prototype Document Image Analysis System for Technical Journals, Computer, 25(7) (1992)
15. O'Gorman, L.: The Document Spectrum for Page Layout Analysis, PAMI, 15 (1993)
16. Viola, P., Jones, M.: Rapid Object Detection using a Boosted Cascade of Simple Features, CVPR (2001)
17. Wang, Y., Phillips, I.T., Haralick, R.: Statistical-based Approach to Word Segmentation, ICIP (2000)
18. Guo, C., Zhu, S.C., Wu, Y.: Modeling Visual Patterns by Integrating Descriptive and Generative Methods, IJCV (2003)

CIDER: Corrected Inverse-Denoising Filter for Image Restoration

You-Wei Wen[1], Michael Ng[2], and Wai-ki Ching[3]

[1] College of Science, South China Agricultural University, Guangzhou, P.R. China
wenyouwei@graduate.hku.hk
[2] Department of Mathematics, Hong Kong Baptist University, Kowloon Tong,
Hong Kong, P.R. China
mng@hkbu.edu.hk
[3] Department of Mathematics, The University of Hong Kong, Pokfulam Road,
Hong Kong, P.R. China
wkc@maths.hku.hk

Abstract. In this paper we propose and develop a new algorithm, Corrected Inverse-Denoising filtER (CIDER) to restore blurred and noisy images. The approach is motivated by a recent algorithm ForWaRD, which uses a regularized inverse filter followed by a wavelet denoising scheme. In ForWaRD, the restored image obtained by the regularized inverse filter is a biased estimate of the original image. In CIDER, the correction term is added to this restored image such that the resulting one is an unbiased estimator. Similarly, the wavelet denoising scheme can be applied to suppress the residual noise. Experimental results show that the performance of CIDER is better than other existing methods in our comparison study.

Keywords: Restoration, Wavelet Denoising, Inverse Filter, Regularization.

1 Introduction

Image restoration is an important problem in image processing and many real world applications. An observed image usually results from some blurring operation performed on the original image, and then corrupted by additive noises. The blurring of images often occurs when there is a relative motion between the camera and the original scene, from defocusing of the lens system, or from atmospheric turbulence.

In digital image processing, the discrete imaging model of the degradation process can be represented by using vectors and matrices. With the lexicographical ordering of the original image $\mathbf{x}$ with size M and the observed image $\mathbf{y}$, their relationship can be expressed as follows:

$$\mathbf{y} = \mathbf{H}\,\mathbf{x} + \mathbf{n}. \tag{1}$$

Here $\mathbf{H}$ is the blurring matrix and $\mathbf{n}$ is a vector of zero-mean Gaussian white noise with variance σ^2. The objective of image restoration is to recover the original image $\mathbf{x}$ from the observed image $\mathbf{y}$.

A.L. Yuille et al. (Eds.): EMMCVPR 2007, LNCS 4679, pp. 112–126, 2007.
© Springer-Verlag Berlin Heidelberg 2007

The problem of image restoration can be viewed as to design an operator $\mathbf{G}$ on image $\mathbf{y}$, such that the restoration image can be expressed as

$$\widehat{\mathbf{x}} = \mathbf{Gy} = \mathbf{GHx} + \mathbf{Gn}. \tag{2}$$

The Wiener filter and Constrained Least Squares (CLS) [9] are two common image restoration methods. The role of the Wiener filter is reduce the expected sum of squares of errors $\mathbf{E}\{(\widehat{\mathbf{x}} - \mathbf{x})^2\}$. The CLS method is to minimize the energy function, which is the weighted sum of the data fitting term $\|\mathbf{y} - \mathbf{Hx}\|_2^2$ and another term containing some prior information about the original image $\mathbf{x}$ to alleviate the problem of ill-conditioning characteristics.

The use of wavelet in image restoration is a relatively prevalent concept [2,8,16]. Its success is due to the fact that signals and images usually have sparse wavelet representations. Therefore a few but significant wavelet coefficients can be used in the representation. The most important advance in image restoration is that Neelamani *et al.* [15] designed a framework: ForWaRD (Fourier-Wavelet Regularized Deconvolution). ForWaRD is a two-stage method, the first stage is to take the regularized inverse of the convolution kernel in the Fourier domain and the second stage is to remove the residual noise in wavelet domain from the estimated image in the previous stage. Both theoretical analysis of ForWaRD and simulation results show that this two-stage method performs very well for image restoration. Recently, we developed a similar hybrid algorithm for spatial and wavelet domains image restoration [17]. The main idea is to use constrained least squares methods with wavelet denoising (CLS-W). The experimental results show the two-stage method is indeed effective for image restoration.

In the two-stage approach, the regularized inversion $\mathbf{G}_\alpha$ (α refers to the regularization parameter) is first employed to the observed image $\mathbf{y}$, we derive an estimated image $\widehat{\mathbf{x}}_\alpha = \mathbf{G}_\alpha \mathbf{y}$. Here we denote $\mathbf{x}_\alpha = \mathbf{G}_\alpha \mathbf{Hx}$ as a regularized version of the original image $\mathbf{x}$ and $\mathbf{n}_\alpha = \mathbf{G}_\alpha \mathbf{n}$ as the leaked noise (the residual noise). We know that $\widehat{\mathbf{x}}_\alpha$ is equal to a sum of the regularized image and the leaked noise, i.e.,

$$\widehat{\mathbf{x}}_\alpha = \mathbf{x}_\alpha + \mathbf{n}_\alpha,$$

In the second stage, a Wavelet domain Wiener Filter (WWF) [10] is performed on the current estimated image $\widehat{\mathbf{x}}_\alpha$ to remove the leaked noise $\mathbf{n}_\alpha$, the obtained image $\widehat{\mathbf{x}}$ is the restoration of the original image.

We remark that the final estimated image $\widehat{\mathbf{x}}$ is an estimate of the regularized image $\mathbf{x}_\alpha$ rather than the original image $\mathbf{x}$. This design is ignored in the two-stage method and the regularized image $\mathbf{x}_\alpha$ is used to approximate $\mathbf{x}$. The error between the original image $\mathbf{x}$ and the regularized image $\mathbf{x}_\alpha$ is then given by

$$\triangle \mathbf{x} = \mathbf{x} - \mathbf{x}_\alpha = (\mathbf{I} - \mathbf{G}_\alpha \mathbf{H})\mathbf{x}.$$

The error $\triangle \mathbf{x}$ can spoil a lot of the non-stationary image features such as edges and ridges. The error $\triangle \mathbf{x}$ can be controlled by choosing a very small regularization parameter. However, a small regularization parameter will amplify the noise. Thus it will be difficult to remove the residual noise $\mathbf{n}_\alpha$.

Our goal here is to design an algorithm in which the error $\triangle \mathbf{x}$ is added to the current estimated image $\widehat{\mathbf{x}}_\alpha$, and a distorted-free but noisy image $\mathbf{z} = \widehat{\mathbf{x}}_\alpha + \triangle \mathbf{x}$ is obtained. We then apply the wavelet denoising scheme to suppress the noise $\mathbf{n}_\alpha$ from the new observed image $\mathbf{z}$. This step gives a correction to the previous estimate after the new estimated image is obtained. Following the process of updating the previous estimation, the denoising scheme is employed. Thus, our proposed method is a three-stage method as follow:

- **Inversion:** Employing regularized inversion to obtain an estimate by using $\widehat{\mathbf{x}}_\alpha = \mathbf{G}_\alpha \mathbf{y}$.
- **Correction:** Updating the previous estimate, the error $\triangle \mathbf{x}$ is added to $\widehat{\mathbf{x}}_\alpha$ and a distorted-free but noisy image $\mathbf{z} = \widehat{\mathbf{x}}_\alpha + \triangle \mathbf{x}$ is obtained.
- **Denoising:** Using WWF to obtain a final estimated image $\widehat{\mathbf{x}}$ from $\mathbf{z}$.

We call our method CIDER (**C**orrected **I**nverse-**D**enoising Filt**ER**). The CIDER approach can be formulated as the following minimization problem:

$$\mathbf{z} = \operatorname{argmin}_{\mathbf{z}} \ \|\mathbf{Hz} - \mathbf{y}\|_2^2 + \alpha \|\mathbf{R}(\mathbf{z} - \mathbf{x}_p)\|_2^2; \tag{3}$$

$$\widehat{\mathbf{x}} = \operatorname{argmin}_{\mathbf{x}} \ \|\mathbf{z} - \mathbf{x}\|_2^2 + \|\mathbf{Wx}\|_{\mathbf{D}}^2. \tag{4}$$

Here $\mathbf{W}$ is the wavelet transform matrix and $\mathbf{D}$ is the weighted diagonal matrix.

In order to calculate the distortion error, we assume that the desired image $\mathbf{x}$ is known. It is an Oracle CIDER method, we show the process in Fig.1(a). One particular challenge is that the desired image $\mathbf{x}$ is not available. In the correction process, we propose an indirect approach based on the two-stage method. A pilot image estimate $\mathbf{x}_p$ is produced by the two-stage method, we then use this estimate to calculate the distortion error

$$\triangle \mathbf{x}_p = (\mathbf{I} - \mathbf{G}_\alpha \mathbf{H}) \mathbf{x}_p.$$

A new observed image $\mathbf{z}$ is obtained through adding the pilot error to the previous estimate. Wavelet denoising scheme is employed to remove the residual noise from this new observed image. It is an empirical approach. The overall process of CIDER is shown in Fig.1.

The outline of this paper is as follows. In Section 2, we first give a brief review on Wiener filter and wavelet domain Wiener filter. We then illustrate and analyze the CIDER method in Section 3. In Section 4, experimental results are given to illustrate the effectiveness of CIDER. Finally, concluding remarks are given in Section 5.

2   Wiener Filter and Wavelet-Domain Wiener Filter

2.1   Wiener Filter

Under the assumption of periodic boundary condition, the blurring model in (1) (see Andrew and Hunt [1]) can be expressed in frequency domain as

$$Y_k = H_k X_k + N_k,$$

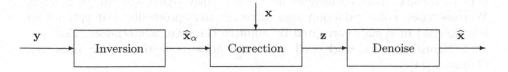

(a) Oracle CIDER. The true image is served as a pilot image.

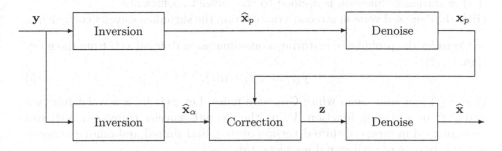

(b) Empirical CIDER. In the upper path, the two-stage approach is used to produce the pilot image estimate $\mathbf{x}_p$.

Fig. 1. CIDER: Corrected Inverse-Denoise FiltER

where X_k, Y_k, H_k and N_k are the discrete Fourier transforms of $\mathbf{x}, \mathbf{y}$, $\mathbf{H}$ and $\mathbf{n}$ respectively. A regularized inverse filter has been proposed by Nowak and Thul [16] when they studied the linear shift invariance inverse problem arising from photon-limited image. The method is further developed and studied in ForWaRD by Neelamani *et al.* [15], the regularized Wiener filter is defined as

$$G_{\alpha,k} = \left(\frac{1}{H_k} \right) \left(\frac{|H_k|^2 |X_k|^2}{|H_k|^2 |X_k|^2 + \alpha\sigma^2} \right).$$

When a flat signal spectrum X_k was assumed, it becomes the Tikhonv regularized method and

$$G_{\alpha,k} = \frac{H_k^*}{|H_k|^2 + \alpha},$$

where "$*$" denotes the conjugate transpose. The current estimated image is obtained through $\widehat{X}_{\alpha,k} = G_{\alpha,k} Y_k$.

2.2   Wavelet-Domain Denoising

When the blurring operator is scale homogeneous, Donoho [6] proposed a very efficient shrinkage procedure for denoising which is called the wavelet vaguelette denoising. If an image is corrupted by noises, then the noisy pixel values will be converted to noisy wavelet coefficients. The idea of this method is based on that large wavelet coefficients carry significant information and should be kept or shrunk; small coefficients are mostly noisy signal and can be ignored. Wavelet-based noise reduction algorithms are asymptotically near optimal for a wide class of signals corrupted by additive Gaussian white noises. However, such algorithms also work well when the noise is neither white noise nor Gaussian [11].

Wavelet denoising is done by a three-step procedure:

(i) the noisy signal is transformed to the wavelet coefficients;
(ii) a shrinkage function is applied to the wavelet coefficients;
(iii) the denoised version is reconstructed from the shrunken wavelet coefficients.

Consider the problem of restoring a one-dimensional signal $\mathbf{x}(t)$ from the noisy signal $\mathbf{y}(t)$

$$\mathbf{y}(t) = \mathbf{x}(t) + \mathbf{n}(t). \tag{5}$$

Here $\mathbf{n}(t)$ is a zero-mean white Gaussian noise. Let $\psi(t)$ be a wavelet function and $\varphi(t)$ be a scaling function. Using the discrete wavelet transform, $\mathbf{y}(t)$ can be expressed in terms of shifted version of $\varphi(t)$, and shifted and dilated versions of $\psi(t)$. In case of orthogonal wavelets, this gives

$$\mathbf{y}(t) = \sum_l \langle \mathbf{y}, \varphi_{J,l} \rangle \varphi_{J,l}(t) + \sum_{j,l} \langle \mathbf{y}, \psi_{j,l} \rangle \psi_{j,l}(t)$$

where

$$\varphi_{j,l}(t) = \sqrt{2^j}\varphi(2^j t - l) \quad \text{and} \quad \psi_{j,l}(t) = \sqrt{2^j}\psi(2^j t - 1).$$

The parameters j and l correspond to the scale and the location respectively, and the parameter J controls the resolution of the wavelet reconstruction.

A wavelet shrinkage estimator has been proposed to restore the original signal $\mathbf{x}(t)$, which attenuates independently each noisy coefficient $\langle \mathbf{y}, \psi_{j,l} \rangle$ with some non-linear function $s_{\lambda_{j,l}}(t)$, the restoration signal is given by Donoho and Johnstone [7]:

$$\widehat{\mathbf{x}}(t) = \sum_l \langle \mathbf{y}, \varphi_{J,l} \rangle \varphi_{J,l}(t) + \sum_{j,l} s_{\lambda_{j,l}}(\langle \mathbf{y}, \psi_{j,l} \rangle))\psi_{j,l}(t).$$

A number of different shrinkage functions have been considered in the literature. One of the most popular shrinkage functions is the hard threshold shrinkage function, which is defined as

$$s_{\lambda_{j,l}}(\langle \mathbf{y}, \psi_{j,l} \rangle) = \begin{cases} \langle \mathbf{y}, \psi_{j,l} \rangle, & |\langle \mathbf{y}, \psi_{j,l} \rangle| > \lambda_{j,l}, \\ 0, & |\langle \mathbf{y}, \psi_{j,l} \rangle| \leq \lambda_{j,l}. \end{cases} \tag{6}$$

Donoho and Johnstone [7] proposed a possible choice of parameter

$$\lambda_{j,l} = \sigma_j \sqrt{2\ln M}$$

with σ_j being the variance at wavelet scale j and M being the length of the signal.

Since the wavelet transform approximates the Karhunen-Loeve (KL) transform for a broad class of signals, Ghael *et al.* [10] proposed a Wavelet-domain Wiener Filter (WWF) as a denoising scheme which can improve the Mean Square Error (MSE) performance of the hard shrinkage method. The wavelet coefficients of the restored image is defined by employing the Wiener filtering on each wavelet coefficient, thus the function $s_{\lambda_{j,l}}(t)$ is expressed as follows:

$$s_{\lambda_{j,l}}(\langle \mathbf{y}, \psi_{j,l} \rangle) = \frac{\langle \mathbf{x}, \psi_{j,l} \rangle^2}{\langle \mathbf{x}, \psi_{j,l} \rangle^2 + \sigma_j^2} \langle \mathbf{y}, \psi_{j,l} \rangle. \tag{7}$$

But the coefficients $\langle \mathbf{x}, \psi_{j,l} \rangle$ are required to construct the unknowns $s_{\lambda_{j,l}}(\langle \mathbf{y}, \psi_{j,l} \rangle)$. Hence, we first estimate the coefficient $\langle \mathbf{x}, \psi_{j,l} \rangle$ by using the hard shrinkage method. The resulting minimum MSE of WWF is

$$\mathbf{E}\{\|\mathbf{x} - \widehat{\mathbf{x}}\|_2^2\} = \sum_{j,l} \frac{\langle \mathbf{x}, \psi_{j,l} \rangle^2 \sigma_j^2}{\langle \mathbf{x}, \psi_{j,l} \rangle^2 + \sigma_j^2}. \tag{8}$$

We remark that when $\langle \mathbf{x}, \psi_k \rangle \neq 0$, $s_{\lambda_{j,l}}(\langle \mathbf{y}, \psi_{j,l} \rangle)$ is the minimizer of the following problem

$$s_{\lambda_{j,l}}(\langle \mathbf{y}, \psi_{j,l} \rangle) = \mathrm{argmin}_z \, (z - \langle \mathbf{y}, \psi_k \rangle)^2 + \frac{\sigma_j^2}{\langle \mathbf{x}, \psi_k \rangle^2} z^2.$$

Let $\mathbf{W}$ be the wavelet transform matrix and $\mathbf{D}$ be the diagonal matrix with entries $\frac{\sigma_j^2}{\langle \mathbf{x}, \psi_k \rangle^2}$. By using the unitary invariance property of the 2-norm, the WWF approach can be formulated as

$$\widehat{\mathbf{x}} = \mathrm{argmin}_{\mathbf{x}} \|\mathbf{y} - \mathbf{x}\|_2^2 + \|\mathbf{x}\|_{\mathbf{D}}^2.$$

3   Analysis of CIDER

In this section, we analyze the CIDER method. We summarize the algorithm as follows:

The CIDER method:

(Stage 1) Deblurring by using the CLS method
  Estimate $\widehat{\mathbf{x}}_\alpha = (\mathbf{H}^*\mathbf{H} + \alpha\mathbf{R}^*\mathbf{R})^{-1}\mathbf{H}^*\mathbf{y}$;
(Stage 2) Correcting the distortion error
  (a) Use the ForWaRD method or the CLS-W method to estimate $\mathbf{x}_p$;
  (b) Compute the distortion error $\triangle\mathbf{x}_p = \alpha(\mathbf{H}^*\mathbf{H} + \alpha\mathbf{R}^*\mathbf{R})^{-1}\mathbf{R}^*\mathbf{R}\mathbf{x}_p$;
  (c) Update the previous estimate by using $\mathbf{z} = \widehat{\mathbf{x}}_\alpha + \triangle\mathbf{x}_p$;
(Stage 3) Denoising in a wavelet domain
  (a) Compute the wavelet coefficients $\langle \mathbf{z}, \psi_{j,l} \rangle$ of $\mathbf{z}$;
  (b) Compute the wavelet coefficients $\langle \mathbf{x}_p, \psi_{j,l} \rangle$;
  (c) Perform WWF by using $\langle \widehat{\mathbf{x}}_p, \psi_{j,l} \rangle = \frac{\langle \mathbf{x}_p, \psi_{j,l} \rangle^2}{\langle \mathbf{x}_p, \psi_{j,l} \rangle^2 + \sigma_j^2} \langle \mathbf{z}, \psi_{j,l} \rangle$;
  (d) Compute the inverse wavelet transform with $\langle \widehat{\mathbf{x}}, \psi_{j,l} \rangle$ to obtain the estimate $\widehat{\mathbf{x}}$.

Here we focus our discussion on the CLS-W method. We remark that one can perform a similar analysis for the ForWaRD method.

3.1   Mean Squares Error

In this subsection, we focus our discussion on Oracle CIDER method and assume that the desired image $\mathbf{x}$ is known and employ it as the pilot image, i.e., $\mathbf{x}_p = \mathbf{x}$. Let $\mathbf{e}$ be the difference between the original image $\mathbf{x}$ and the restored image $\widehat{\mathbf{x}}$, i.e., $\mathbf{e} = \mathbf{x} - \widehat{\mathbf{x}}$. The mean square error is then given by

$$J(\alpha) = \mathbf{E}\{\|\mathbf{e}\|_2^2\} = \sum_{j,l} \mathbf{E}\{\langle \mathbf{e}, \psi_{j,l}\rangle^2\}.$$

We note that $\widehat{\mathbf{x}}$ is obtained from $\mathbf{z}$ by using the WWF method, therefore we have

$$\langle \widehat{\mathbf{x}}, \psi_{j,l}\rangle = s_{j,l}\langle \mathbf{z}, \psi_{j,l}\rangle, \quad \text{where} \quad s_{j,l} = \frac{\langle \mathbf{x}, \psi_{j,l}\rangle^2}{\langle \mathbf{x}, \psi_{j,l}\rangle^2 + \sigma_j^2},$$

where σ_j^2 denotes the variance of $\mathbf{n}_\alpha$ at wavelet scale j. Since the leaked noise $\mathbf{n}_\alpha = \mathbf{G}_\alpha \mathbf{n}$ in $\widehat{\mathbf{x}}_\alpha$ is Gaussian but is not a white noise, we conclude that the expected value of $\mathbf{n}_\alpha$ is zero, and its covariance matrix is given by $\sigma^2 \mathbf{G}_\alpha \mathbf{G}_\alpha^*$. Let $\Psi_{j,l,k}$ be the discrete Fourier transform of $\psi_{j,l}$. Then we have

$$\sigma_j^2 = \mathbf{E}\{|\langle \mathbf{n}_\alpha, \psi_{j,l}\rangle|^2\} = \sum_k \frac{\sigma^2 |H_k|^2 |\Psi_{j,l,k}|^2}{(|H_k|^2 + \alpha |R_k|^2)^2}. \tag{9}$$

We note that $\langle \mathbf{e}, \psi_{j,l}\rangle = \langle \mathbf{x} - s_{j,l}\mathbf{z}, \psi_{j,l}\rangle$, we obtain $\mathbf{E}(\mathbf{z}) = \mathbf{x}$ from the assumption that $\mathbf{x}_p = \mathbf{x}$. It is clear that

$$\mathbf{E}(\langle \mathbf{e}, \psi_{j,l}\rangle) = (1 - s_{j,l})\langle \mathbf{x}, \psi_{j,l}\rangle,$$

and the variance is given by $\mathrm{Var}(\langle \mathbf{e}, \psi_{j,l}\rangle) = s_{j,l}^2 \sigma_j^2$.

As the MSE is the sum of a bias term and a variance term, we have

$$
\begin{aligned}
J(\alpha) &= \sum_{j,l} \{\mathbf{E}(\langle \mathbf{e}, \psi_{j,l}\rangle)\}^2 + \sum_{j,l} \mathrm{Var}(\langle \mathbf{e}, \psi_{j,l}\rangle) \\
&= \sum_{j,l} (1 - s_{j,l})^2 \langle \mathbf{x}, \psi_{j,l}\rangle^2 + \sum_{j,l} s_{j,l}^2 \sigma_j^2 \\
&= \mathrm{Bias}(\mathbf{e}) + \mathrm{Var}(\mathbf{e}).
\end{aligned}
$$

Summing up the bias term and the variance term, we obtain

$$J(\alpha) = \sum_{j,l} \frac{\langle \mathbf{x}, \psi_{j,l}\rangle^2 \sigma_j^2}{\langle \mathbf{x}, \psi_{j,l}\rangle^2 + \sigma_j^2}.$$

Next we present the following result to show that the method CIDER is more efficient than the CLS-W method.

Proposition 1. *Denote $\widetilde{\mathbf{x}}$ and $\widehat{\mathbf{x}}$ be the restored image by using CLS-W and Oracle CIDER respectively, and $\widetilde{\mathbf{e}} = \mathbf{x} - \widetilde{\mathbf{x}}, \mathbf{e} = \mathbf{x} - \widehat{\mathbf{x}}$. When we use the same regularization parameter α, we have*

$$\mathrm{Bias}(\widetilde{\mathbf{e}}) > \mathrm{Bias}(\mathbf{e}) \quad \text{and} \quad \mathrm{Var}(\widetilde{\mathbf{e}}) = \mathrm{Var}(\mathbf{e}),$$

and hence we conclude that the MSE of CLS-W is larger than that of CIDER.

Proof: We note that the restored image $\tilde{\mathbf{x}}$ is obtained by removing the residual noise from $\hat{\mathbf{x}}_\alpha$. The denoising scheme uses the ideal WWF, and the shrinkage coefficient is also given by $s_{j,l} = \frac{\langle \mathbf{x}, \psi_{j,l} \rangle^2}{\langle \mathbf{x}, \psi_{j,l} \rangle^2 + \sigma_j^2}$. Therefore, we have

$$\langle \tilde{\mathbf{e}}, \psi_{j,l} \rangle = \langle \mathbf{x} - s_{j,l} \hat{\mathbf{x}}_\alpha, \psi_{j,l} \rangle$$
$$= \langle \mathbf{x} - s_{j,l} \mathbf{x}_\alpha, \psi_{j,l} \rangle - s_{j,l} \langle \mathbf{n}_\alpha, \psi_{j,l} \rangle$$
$$= \langle (\mathbf{I} - s_{j,l} \mathbf{G}_\alpha \mathbf{H}) \mathbf{x}, \psi_{j,l} \rangle - s_{j,l} \langle \mathbf{n}_\alpha, \psi_{j,l} \rangle$$

Hence we obtain $\mathrm{Var}(\tilde{\mathbf{e}}) = s_{j,l}^2 \sigma_j^2$ and

$$\mathbf{E}(\langle \tilde{\mathbf{e}}, \psi_{j,l} \rangle) = \langle (\mathbf{I} - s_{j,l} \mathbf{G}_\alpha \mathbf{H}) \mathbf{x}, \psi_{j,l} \rangle$$
$$= \sum_k \left(1 - \frac{s_{j,l} |H_k|^2}{|H_k|^2 + \alpha |R_k|^2} \right) X_k \Psi_{j,l,k}.$$

We obtain

$$\{ \mathbf{E}(\langle \tilde{\mathbf{e}}, \psi_{j,l} \rangle) \}^2 \geq \sum_k (1 - s_{j,l})^2 |X_k|^2 |\Psi_{j,l,k}|^2$$
$$= \{ \mathbf{E}(\langle \mathbf{e}, \psi_{j,l} \rangle) \}^2$$

As the MSE of CLS-W is given by

$$\tilde{J}(\alpha) = \sum_{j,l} \{ \mathbf{E}(\langle \tilde{\mathbf{e}}, \psi_{j,l} \rangle) \}^2 + \sum_{j,l} \mathrm{Var}(\langle \tilde{\mathbf{c}}, \psi_{j,l} \rangle)$$
$$= \mathrm{Bias}(\tilde{\mathbf{e}}) + \mathrm{Var}(\tilde{\mathbf{e}}),$$

we conclude that

$$\mathrm{Bias}(\tilde{\mathbf{e}}) \geq \mathrm{Bias}(\mathbf{e}) \quad \text{and} \quad \mathrm{Var}(\tilde{\mathbf{e}}) = \mathrm{Var}(\mathbf{e}).$$

Therefore we have $\tilde{J}(\alpha) \geq J(\alpha)$. □

In the correction step, the distortion error

$$\triangle \mathbf{x} = \alpha (\mathbf{H}^* \mathbf{H} + \alpha \mathbf{R}^* \mathbf{R})^{-1} \mathbf{R}^* \mathbf{R} \mathbf{x}$$

is added to the current estimate. But the desired image $\mathbf{x}$ is not available. In practice, a pilot image $\mathbf{x}_p$ is used to approximate the desired image. We propose a cost function $J_p(\alpha)$ to approximate the final MSE, which is defined as

$$J_p(\alpha) = J_{p,1}(\alpha) + J_{p,2}(\alpha). \tag{10}$$

Here

$$J_{p,1}(\alpha) = \mathbf{E}(\| \triangle \mathbf{x} - \triangle \mathbf{x}_p \|_2^2) \quad \text{and} \quad J_{p,2} = \sum_{j,l} \frac{|\langle \mathbf{x}, \psi_{j,l} \rangle|^2 \sigma_j^2}{|\langle \mathbf{x}, \psi_{j,l} \rangle|^2 + \sigma_j^2}.$$

Denote $X_{p,k}$ to be the DFT of $\mathbf{x}_p$, we have

$$J_{p,1}(\alpha) = \mathbf{E}(\| \alpha (\mathbf{H}^* \mathbf{H} + \alpha \mathbf{R}^* \mathbf{R})^{-1} \mathbf{R}^* \mathbf{R} (\mathbf{x} - \mathbf{x}_p) \|_2^2)$$
$$= \sum_k \frac{\alpha^2 |R_k|^4 |X_k - X_{p,k}|^2}{(|H_k|^2 + \alpha |R_k|^2)^2}.$$

3.2   Regularization Parameter

The regularization parameter α controls the trade-off between fidelity to the observed image $\mathbf{y}$ and smoothness of the estimate $\hat{\mathbf{x}}$. Several methods are available for choosing the value of regularization parameter α [9,13]. In general, the regularization parameter is chosen to minimize the MSE between the estimated image and the original image. In the CIDER method, the regularization is chosen to minimize the cost function $J(\alpha)$.

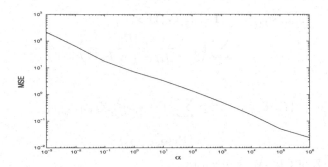

Fig. 2. Demonstration of the relationship between the regularization parameter α and the MSE in an Oracle CIDER method. A 256×256 Boat image with 9×9 box-blur and BSNR=30 dB is used. α is shown at x-axis and the MSE is shown on y-axis.

As $J(\alpha)$ is the mean squares error, we know that it is a strictly positive function in α. We then consider the derivative of $J(\alpha)$,

$$\frac{\partial J(\alpha)}{\partial \alpha} = \frac{\langle \mathbf{x}, \psi_{j,l} \rangle^4}{(\langle \mathbf{x}, \psi_{j,l} \rangle^2 + \sigma_j^2)^2} \frac{\partial \sigma_j^2}{\partial \alpha}.$$

From (9), we have

$$\frac{\partial \sigma_j^2}{\partial \alpha} = \sum_k (-2) \frac{\sigma^2 |H_k|^2 |R_k|^2 |\Psi_{j,l,k}|^2}{(|H_k|^2 + \alpha |R_k|^2)^3}.$$

Since $\frac{\partial \sigma_j^2}{\partial \alpha} < 0$, we know that $\frac{\partial J(\alpha)}{\partial \alpha} < 0$. This means that $J(\alpha)$ is a monotonic decreasing function in α. We conclude that when $\alpha \to \infty$, $\sigma_j^2 \to 0$ and $J(\alpha) \to 0$. Thus in the ideal case, CIDER can restore the original image $\mathbf{x}$ exactly. Fig.2 demonstrates the relationship between the regularization parameter and the MSE. We note that the MSE is decreasing when the regularization parameter α increases.

In the empirical CIDER, the regularization parameter α is also chosen to minimize the cost function $J_p(\alpha)$. We have

$$\frac{\partial J_p(\alpha)}{\partial \alpha} = \frac{\partial J_{p,1}(\alpha)}{\partial \alpha} + \frac{\partial J_{p,2}(\alpha)}{\partial \alpha} = 0.$$

Here

$$\frac{\partial J_{p,1}(\alpha)}{\partial \alpha} = \sum_k \frac{2\alpha |H_k|^2 |R_k|^4 |X_k - X_{p,k}|^2}{(|H_k|^2 + \alpha |R_k|^2)^3},$$

and

$$\frac{\partial J_{p,2}(\alpha)}{\partial \alpha} = (-2) \sum_{j,l,k} \frac{|\langle \mathbf{x}, \psi_{j,l} \rangle|^4}{(|\langle \mathbf{x}, \psi_{j,l} \rangle|^2 + \sigma_j^2)^2} \frac{\sigma^2 |H_k|^2 |R_k|^2 |\Psi_{j,l,k}|^2}{(|H_k|^2 + \alpha |R_k|^2)^3}.$$

We conclude that $\frac{\partial J_{p,1}(\alpha)}{\partial \alpha} > 0$ and $\frac{\partial J_{p,2}(\alpha)}{\partial \alpha} < 0$, it is easy to show that $J_{p,1}(\alpha)$ is a strictly positive, monotonic increasing function in α and $J_{p,2}(\alpha)$ is a strictly positive, monotonic decreasing function in α for $\alpha > 0$.

Denote

$$\eta_k = \sum_{j,l} \frac{\sigma^2 |\langle \mathbf{x}, \psi_{j,l} \rangle|^4 |\Psi_{j,l,k}|^2}{(|\langle \mathbf{x}, \psi_{j,l} \rangle|^2 + \sigma_j^2)^2} / (|R_k|^2 |X_k - X_{p,k}|^2),$$

we note that $\frac{\partial J_p(\alpha)}{\partial \alpha} < 0$ for $0 < \alpha < \min \eta_k$ and $\frac{\partial J_p(\alpha)}{\partial \alpha} > 0$ for $\alpha > \max \eta_k$. As the cost function $J_p(\alpha)$ is a sum of a monotonic increasing and a monotonic decreasing function in α, it has a unique minimum for $\alpha \geq 0$, which lies in the interval $[\min \eta_k, \max \eta_k]$.

4   Simulation Results

In this section, we demonstrate the performance of CIDER in image restoration. Our codes are written in Matlab. We use the second-order difference matrix as the regularization matrix $\mathbf{R}$.

We give an example to illustrate the non-stationary image features can be kept when our method is applied to restore the blurred and noisy images. The original image is blurred by a separable filter with weights [1 2 2 2 1]/8 and is corrupted by the Gaussian noise with $\sigma^2 = 9$. In Fig.3, we show the original image (left) and the blurred and noisy image (right). We use CIDER method to restore the image and the result is competitive to ForWaRD method. We also show the difference between the restored image and the original image, see Fig.4. We choose the regularization parameter $\alpha = 0.0027$ in ForWaRD method, and $\alpha = 0.0088$ in both Oracle CIDER method and empirical CIDER method. The regularization parameter α is chosen to be the optimal one in ForWaRD method and empirical CIDER method. We recall that the optimal regularization parameter for Oracle CIDER method is $\alpha = \infty$, we use the same regularization parameter as in the empirical CIDER method, to show that good results can be also obtained. As the denoising scheme is directly applied after the regularization inverse filter in ForWaRD, some image features were spoiled, see Fig.4 (Upper). When the correction error is added to the regularized image, the visualization of the difference obtained by empirical CIDER method is improved. As the pilot image is obtained by ForWaRD in empirical CIDER method, the visualization of the difference obtained by empirical CIDER method is worse than that by Oracle CIDER method, in which the true image is served as the pilot image.

Original Image    Blurred and Noisy Image

Fig. 3. The original image (left), and the blurred and noisy image (right) which is blurred by a separable filter with weights $[1\ 2\ 2\ 2\ 1]/8$ and is corrupted the noise with $\sigma^2 = 9$

Next, we demonstrate that our method can improve the Peak-Signal-to-Noise-Ratio (PSNR), which is defined as follows:

$$\text{PSNR} = 10\log_{10}\frac{||\mathbf{x}||^2}{||\mathbf{x}-\widehat{\mathbf{x}}||^2},$$

where $\mathbf{x}$ and $\widehat{\mathbf{x}}$ denote the original image and the restored image respectively, and the target image is of size n-by-m.

We compare empirical CIDER method with other four methods: the Wiener filter, the CLS method, the ForWaRD method and the CLS-W method. These four methods are to design an operator $\mathbf{G}$ for an estimated image $\widehat{\mathbf{x}} = \mathbf{Gy}$. We summarize their corresponding operators $\mathbf{G}$ in Table 1. The computational results by the Wiener filter and the ForWaRD method are generated by using the software "*ForWaRD*". For pilot images, we use the restored images obtained by the ForWaRD method and the CLS-W method. They refer to CIDER(F) and CIDER(C) respectively in the tables.

Table 1. Different inverse filters $\mathbf{G}$. Here $[\mathbf{D}_F]_{k,k} = \frac{H_k^* S_{xx,k}}{|H_k|^2 S_{xx,k}+\alpha S_{nn,k}}$, $[\widetilde{\mathbf{D}}_F]_{k,k} = \frac{H_k^*}{|H_k|^2+\alpha}$ and $[\mathbf{D}_W]_{k,k} = \frac{\langle \mathbf{x},\psi_{j,l}\rangle^2}{\langle \mathbf{x}_\alpha,\psi_{j,l}\rangle^2+\sigma_j^2}$, where $S_{xx,k}$ and $S_{nn,k}$ denote the k-th entries of the power spretra density (in a vector form) of $\mathbf{x}$ and $\mathbf{n}$ respectively, and $\mathbf{F}$ and $\mathbf{W}$ denote the discrete Fourier transform and wavelet transform matrices respectively.

	the inverse filter $\mathbf{G}$
Wiener	$\mathbf{F}^*\,\mathbf{D}_F\,\mathbf{F}$
CLS	$(\mathbf{H}^*\mathbf{H}+\alpha\mathbf{R}^*\mathbf{R})^{-1}\mathbf{H}^*$
ForWaRD	$(\mathbf{W}^*\mathbf{D}_W\mathbf{W})\,(\mathbf{F}^*\widetilde{\mathbf{D}}_F\mathbf{F})$
CLS-W	$(\mathbf{W}^*\mathbf{D}_W\mathbf{W})(\mathbf{H}^*\mathbf{H}+\alpha\mathbf{R}^*\mathbf{R})^{-1}\mathbf{H}^*$

Table 2. The PSNRs of the restored images for different algorithms. A 9×9 box-blur is used.

	Cameraman			Lenna			Theater			Boat		
	30 dB	40 dB	50 dB	30 dB	40 dB	50 dB	30 dB	40 dB	50 dB	30 dB	40 dB	50 dB
Wiener	19.01	20.79	22.94	19.25	20.99	23.09	20.31	22.37	24.42	22.23	23.28	24.86
CLS	19.24	21.32	23.87	19.61	21.74	24.30	20.79	23.84	26.80	23.00	25.81	28.80
ForWaRD	20.28	22.55	25.37	20.31	22.54	25.30	21.17	23.87	27.13	23.67	26.21	29.39
CLS-W	20.47	22.69	25.47	20.44	22.90	25.68	21.66	24.74	27.76	24.02	26.89	29.94
CIDER(F)	20.83	23.27	26.25	20.73	23.14	26.12	21.81	24.55	27.77	24.28	27.05	30.10
CIDER(C)	20.86	23.24	26.15	20.67	23.21	26.19	21.87	24.95	28.06	24.28	27.22	30.28

Table 3. The PSNRs of the restored images for different algorithms, the blur was tested in [8]

	Cameraman			Lenna			Theater			Boat		
	30 dB	40 dB	50 dB	30 dB	40 dB	50 dB	30 dB	40 dB	50 dB	30 dB	40 dB	50 dB
Wiener	13.73	21.24	28.22	18.31	21.71	28.53	15.60	22.67	30.00	16.05	22.87	30.21
CLS	21.53	24.88	30.02	22.52	25.82	30.53	23.60	27.61	32.88	25.37	29.00	34.36
ForWaRD	22.69	26.64	32.19	23.31	27.01	31.89	23.82	27.72	33.07	26.02	29.85	35.18
CLS-W	23.21	27.08	32.41	23.90	27.55	32.12	24.73	28.81	33.55	26.81	30.91	35.78
CIDER(F)	23.23	27.20	32.74	23.79	27.49	32.34	24.37	28.26	33.64	26.53	30.38	35.58
CIDER(C)	23.42	27.41	32.85	24.05	27.78	32.46	24.88	28.99	33.91	26.99	31.11	36.04

Table 4. The PSNRs of the restored images for different algorithms, Gaussian blur is used

	Cameraman			Lenna			Theater			Boat		
	30 dB	40 dB	50 dB	30 dB	40 dB	50 dB	30 dB	40 dB	50 dB	30 dB	40 dB	50 dB
Wiener	22.30	25.08	28.23	22.93	25.81	28.84	23.68	26.61	30.52	24.26	26.54	29.96
CLS	23.44	25.86	28.90	24.55	26.97	29.74	25.86	28.84	32.29	27.52	30.59	34.42
ForWaRD	24.59	27.17	30.06	25.41	27.80	30.16	26.20	28.80	32.21	28.37	31.10	34.75
CLS-W	24.72	27.30	30.14	25.59	28.09	30.55	26.69	29.62	32.63	28.75	31.95	35.53
CIDER(F)	24.90	27.75	30.83	25.68	28.29	31.07	26.58	29.44	33.13	28.70	31.75	35.68
CIDER(C)	24.93	27.75	30.93	25.70	28.32	31.02	26.81	29.84	33.19	28.88	32.19	36.06

Table 5. The PSNRs of the restored images for different algorithms, the blur was tested in [14]

	Cameraman			Lenna			Theater			Boat		
	30 dB	40 dB	50 dB	30 dB	40 dB	50 dB	30 dB	40 dB	50 dB	30 dB	40 dB	50 dB
Wiener	22.65	24.31	25.89	23.69	25.55	27.33	24.44	26.36	28.52	25.43	27.12	29.47
CLS	22.87	24.55	26.02	24.03	25.94	27.56	24.99	27.21	28.96	26.80	28.74	30.76
ForWaRD	23.62	25.28	26.70	24.79	26.58	28.09	24.85	27.42	29.19	27.05	29.44	31.47
CLS-W	23.85	25.33	26.73	24.86	26.64	28.14	25.67	27.70	29.33	27.68	29.65	31.67
CIDER(F)	23.97	25.52	27.07	24.96	26.82	28.50	25.23	27.68	29.69	27.49	29.75	32.06
CIDER(C)	24.01	25.54	27.04	24.95	26.78	28.49	25.75	27.83	29.59	27.78	29.82	32.15

We consider the following blurring functions:

- A 9×9-point box-car blur tested in [2] and [15], which is defined as:

$$h(i,j) = \begin{cases} \frac{1}{81}, & 0 \leq i,j \leq 8 \\ 0, & \text{otherwise.} \end{cases}$$

Fig. 4. The restored image and the difference image between the restored image and the original image. Upper: the restored image by ForWaRD method ($\alpha = 0.0027$, PSNR $= 22.07$) and the difference image. Middle: the restored image by empirical CIDER ($\alpha = 0.0088$, PSNR $= 22.50$) and the difference image. Bottom: the restored image by Oracle CIDER ($\alpha = 0.0088$, PSNR $= 32.87$) and the difference image.

- A blurring function is considered in Figueiredo and Nowak [8], and is given by

$$h(i,j) = \begin{cases} \frac{1}{1+i^2+j^2}, & -7 \le i, j \le 7 \\ 0, & \text{otherwise.} \end{cases}$$

- A Gaussian blur given by

$$h(i,j) = \begin{cases} \frac{1}{\sqrt{2\pi}\sigma} e^{-\frac{i^2+j^2}{2\sigma^2}}, & -7 \le i, j \le 7 \\ 0, & \text{otherwise.} \end{cases}$$

Here, we set $\sigma = 1$.
- A 5×5 separable filter with weights $[1,4,6,4,1]/16$ which was tested in Liu and Moulin [14].

The additive noise variances σ^2 are set such that the blurred SNRs (BSNRs) are 30dB, 40dB and 50dB respectively. Tables 2, 3, 4 and 5 show the PSNRs of the restored images obtained by different methods on a 256×256 "Cameraman", "Lenna", "Theater" and "Boat" images. According to the tables, the PSNRs of the restored images by CIDER are the largest than among those obtained by the other methods in general. Thus we conclude that CIDER outperforms the other methods.

5 Conclusion

In this paper, we presented a new method for image restoration. After the regularized inversion, the estimated image is a sum of a shrinkage image and a noise. We find that two-stage methods only consider removing the residual noise but ignore the fact that the shrinkage image is a distortion version of the original image. We correct the estimated image by an unbiased estimator and then the wavelet denoising method is applied to obtain the final restored image. Experimental results show that the performance of the proposed method is the best among all tested methods in the comparison study.

References

1. Andrew, H., Hunt, B.: Digital Image Restoration. Prentice-Hall, Englewood Cliffs, NJ (1977)
2. Banham, M.R., Katsaggelos, A.K.: Spatially Adaptive Wavelet-based Multiscale Image Restoration. IEEE Transactions on Image Processing 5, 619–634 (1996)
3. Biemond, J., Lagendijk, R.: Regularized Iterative Image Restoration in a Weighted Hilbert Space, Acoustics, Speech, and Signal Processing. In: IEEE International Conference on ICASSP '86., vol. 11, pp. 1485–1488 (1986)
4. Daubechies, I.: Orthonormal Bases of Compactly Supported Wavelets. Communications on Pure and Applied Mathematics 41, 909–996 (1988)
5. Donoho, D.L.: Denoising by Soft-thresholding. IEEE Trans. Inform. Theory 41, 613–627 (1995)

6. Donoho, D.L.: Nonlinear Solution of Linear Inverse Problems by Wavelet-vaguelette Decompositions. J. Appl. Comput. Harmon. Anal. 1, 100–115 (1995)
7. Donoho, D.L., Johnstone, I.M.: Ideal Spatial Adaptation by Wavelet Shrinkage. Biometrika 81, 425–455 (1994)
8. Figueiredo, M.A., Nowak, R.D.: An EM Algorithm for Wavelet-based Image Restoration. IEEE Transactions on Image Processing 8, 906–916 (2003)
9. Galatsanos, N.P., Katsaggelos, A.K.: Methods for Choosing the Regularization Parameter and Estimating the Noise Variance in Image Restoration and their Relation. IEEE Trans. Image Proc. 1, 322–336 (1992)
10. Ghael, S., Sayeed, A., Baraniuk, R.: Improved Wavelet Denoising via Empirical Wiener Filtering. Proc. SPIE. Wavelet Applications in Signal and Image Processing V 3169, 389–399 (1997)
11. Gopinath, R., Lang, M., Guo, H., Odegard, J.: Enhancement of Decompressed Images at Low Bit Rates. SPIE Math. Imagaging: Wavelet Applications in Signal and Image Processing 2303, 366–377 (1994)
12. Hillery, A., Chin, R.: Iterative Wiener Filters for Image Restoration. IEEE Transaction on Signal Processing 39, 1892–1899 (1991)
13. Kang, M.G., Katsaggelos, A.K.: General Choice of the Regularization Functional in Regularized Image Restoration. IEEE Transactions on Image Processing 4, 594–602 (1995)
14. Liu, J., Moulin, P.: Complexity-regularized image restoration. In: Proc. IEEE Int. Conf. on Image Processing-ICIP'98, Chicago, IL, vol. 1, pp. 555–559 (1998)
15. Neelamani, R., Choi, H., Baraniuk, R.: ForWaRD: Fourier-Wavelet Regularized Deconvolution for Ill-Conditioned Systems. IEEE Transactions on Signal Processing 52, 418–433 (2004)
16. Nowak, R.D., Thul, M.J.: Wavelet-vaguelette Restoration in Photon-limited Imaging. In: Proceedings of the 1998 IEEE International Conference on Acoustics, Speech, and Signal Processing, vol. 5, pp. 2869–2872 (1998)
17. Wen, Y., Ching, W., Ng, M.: A Hybrid Algorithm for Spatial and Wavelet Domains Image Restoration, Visual Communications and Image Processing 2005, Proceeding of the Society of Photo-optical Instrumentation Engineers, pp. 2004–2011

Skew Detection Algorithm for Form Document Based on Elongate Feature

Feng-ying Xie[1], Zhi-guo Jiang[1], and Lei Wang[2]

[1] Image Processing Center, BeiHang University, Beijing 100083,
China
[2] University Duisburg-Essen Duisburg International Studies in Engineering, Germany
47057
xfy_73@buaa.edu.cn

Abstract. One new and efficient skew detection algorithm is proposed for form documents according to the feature that the horizontal line has the same skew with the form. This algorithm includes the following steps: Firstly, all horizontal connected regions, including horizontal straight-lines, are extracted from the form document by directional region growing method presented in this paper; Secondly, the optimal line is selected from all horizontal connected regions based on the elongate of connected region; Thirdly, all the pixels belonging to the optimal line are considered to calculate the line parameters with linear least-square theory. The skew angle of the optimal line is just the form skew angle. One elongate function is defined in this paper which described the elongate feature correctly for the band-like connected region. The experiment results show that the form skew angle can be detected accurately, and this skew detection algorithm is fast and robust.

Keywords: form document, skew detection, directional region growing, connected region, elongate function.

1 Introduction

The skew angle estimation and correction of a document page is an important task for document analysis and optical character recognition (OCR) applications. In this paper, skew detection algorithm against form documents is researched. For form document, the skew angle can be estimated through computing the reference line parameter, and many skew estimating methods based on line detection have been developed during last several decades. Most of them can be classified into three general categories: Hough transform [1, 4-6], projection profile[7-11] and Run-lengths [3, 12-14].

The Hough transform (HT) has been recognized as one of the most popular and general methods for the straight-line detection. Briefly, the HT projects each point in the Cartesian image space (x, y) into the Hough parameter space (ρ, θ) using the

A.L. Yuille et al. (Eds.): EMMCVPR 2007, LNCS 4679, pp. 127–136, 2007.
© Springer-Verlag Berlin Heidelberg 2007

parametric representation: $\rho = x\cos\theta + y\sin\theta$. If the points are co-linear, the sinusoids intersect at a point (ρ, θ) corresponding to the parameter of the line. It can detect dashed and mildly broken lines. However it is very time consuming. To reduce the computational cost, many fast HT methods were proposed [5-6]. Unfortunately, the speed is still unsatisfied when the image data is large and/or the resolution of detection is high.

Projection profile method [7-11] is another kind of popular techniques for skew detection and works well for text-only documents. It is through the way firstly creating a histogram at each possible angle. Then a cost function is applied to these histograms. The skew angle is the angle at which this cost function is optimized. The method will fail if the projection of a line does not form a peak on the profile when it is mixed with text, and the error of estimated skew angle will be not acceptable when the lines are too short or severely broken. Chen and Lee proposed the strip projection method to alleviate this problem based on the fact that lines are more likely to form peaks on the projection profile in a small region [8]. For example, the horizontal line detection, they first divide an image into several vertical strips of equal width, and then perform horizontal projection on each strip. The detected collinear line segments in each strip are linked to form a line.

Run-lengths [3, 12-14] is often used as an image component in lines detecting. Yu and Jain proposed a data structure, called Block Adjacency Graph (BAG), to represent an image [13]. BAG is defined as $G(N, E)$, of which N is a set of block nodes and E is a set of edges indicating the connections between two nodes. Each node is a block that contains one or several horizontal run-lengths adjacently connected in the vertical direction and aligned on both left and right sides within a given tolerance. A line is detected by searching a connected sub-graph in the BAG with large aspect ratio. For detecting the form skew angle, every line parameter can be calculated by line fitting and the average skew angle of all lines is the form skew angle [14]. Because the final result is calculated with all horizontal lines not the optimal horizontal line, the skew angle is often not precise enough.

In this paper, a novel skew detection algorithm for form document is described. This algorithm treats the lines as connected components and the optimal horizontal line is extracted easily through calculating the values of elongate function for connected components defined in this paper. The form document skew angle can be obtained by fitting this optimal line based on linear least-square theory [15]. Through a large quantity of experiments, this proposed algorithm was proved to be very efficient for form documents.

2   Skew Detection for Form Document Based on Elongate Feature

2.1   Extracting Connected Components with DRG

Directional region growing algorithm (DRG) is a kind of region growing method which grows only in some specified direction, which can be illustrated in Fig.1. In Fig.1, the black dot denotes an object pixel, and the numbers from 1 to 8 are directions related to the object pixel. If the growth is along the direction 1 (the shadow region), only two neighbor pixels related to this direction are considered, they are right and upper-right. The growth will be continued if anyone of the two pixels related to this direction belongs to the object pixel, and otherwise stopped. The same reason, only upper-right and upper neighbor pixels are considered if growth is along direction 2.

Different with traditional region growing method, the connected component extracted by DRG will extend in specific direction and decline or even stop in other directions. Therefore the DRG is suitable for extracting linear connected regions that is of specific direction, which is illustrated in Fig.2.

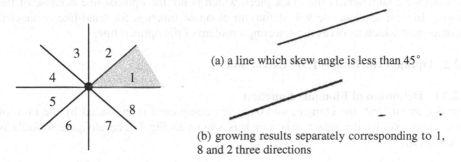

(a) a line which skew angle is less than 45°

(b) growing results separately corresponding to 1, 8 and 2 three directions

Fig. 1. A pixel and its 8 directions   **Fig. 2.** An example for directional region growing

In this paper only horizontal connected components need to be extracted for obtaining horizontal lines. Therefore the DRG are performed in 1 and 8 two directions for every object pixel, and so the lines skewed in the angle between ±45° will be extracted, at the same time those characters or vertical lines will be divided into many small and short connected components to be extracted too. Two DRG examples are

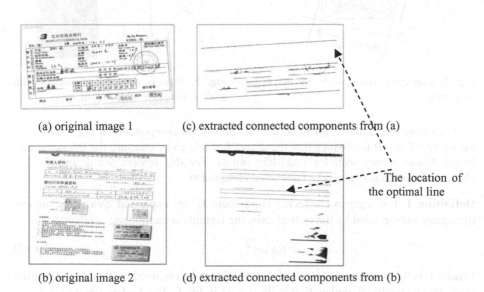

(a) original image 1   (c) extracted connected components from (a)

The location of the optimal line

(b) original image 2   (d) extracted connected components from (b)

Fig. 3. Horizontal connected components extracted by DRG algorithm

shown in Fig.3, and the short components in the form are removed for showing clearly. From the Fig.3, line objects are extracted as connected component, and there are burr noises at the cross between lines and characters.

After extracting horizontal connected components, the form skew angle can be calculated with one of horizontal lines. However, now the problem is that which kind of connected component is the line object and how an optimal line can be selected for calculating. One simple method is to select the longest connected component as the optimal line. For Fig.3-a, the method is feasible. But for Fig.3-b, the longest connected component is the black piece which is not the optimal line because of the blurr. In next section, we will define an elongate function for band-like connected component which resolves the selecting problem of the optimal line.

2.2 Optimal Horizontal Line Extraction

2.2.1 Definition of Elongate Function

In document [16], the elongate of connected component is measured by the ratio of the short axis by the major axis which is shown in Fig.4. The elongate formula is defined as:

$$e = \alpha / \beta .$$ (1)

Where α is the short axis length, β is the major axis length for the connected component.

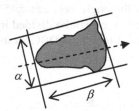
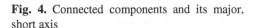

Fig. 4. Connected components and its major, short axis

Fig. 5. Two connected components with similar elongated value in [16]

According to the above formula, the two connected components in Fig.5 will have the close values. It is obvious that the above formula can't measure the thickness and length features very well for band-like object. For this reason, we define another elongate function for band-like connected component.

Definition 1. For a given connected component C, its expanded area is the area of the square outspreaded by the central axis, the formula is as follows:

$$EA = l^2 .$$ (2)

Where l is the central axis length of C, which can be obtained by thinning algorithm [17]. The geometrical explanation is illustrated in Fig.6. The broken line in Fig.6-a is the central axis, the area of the square in Fig.6-c is just the expanded area of Fig.6-a.

From Fig.6, for a connected component, the longer the central axis is, the bigger its expanded area is.

Definition 2. For a given connected component C, elongate function E is the ratio of its expanded area to its real area, the formula is as follows:

$$E = EA/area = l^2/area.$$ (3)

Where $area$ is the real area of C.

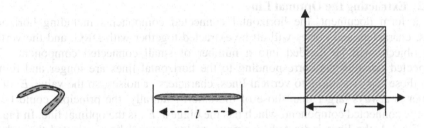

(a) connected component and   (b) outspreading (a) out   (c) square outspreaded from (a)
its central axis

Fig. 6. The illustration of elongated function definition in this paper

From the definition 2, the elongate function E measures the size of expanded area in unit area for connected component, it is related to two variables l and $area$. For the formula 3, we discuss as following:

1) Supposing l is unchanged, with the declining of $area$, the connected component will be thinner. In this case, its elongate becomes more and more strong, and E is increasing.
2) Supposing $area$ is unchanged, with the increasing of l, the connected component will be longer. In this case, its elongate becomes more and more strong, and E is increasing.

According to the above analysis we take the conclusion that the thinner and longer a band-like object is, the bigger the value of the E is. Fig.7 and table 1 is several connected components and their elongate function values.

Table 1. The values of E corresponding to connected components in Fig.7

No.	area (pixels)	l	E
1	200	20	2.0
2	99	33	11.0
3	396	66	11.0
4	198	66	22.0
5	655	144	31.7

Fig. 7. Several connected components, the label is ①, ②, ③, ④ and ⑤ from left to right

2.2.2 Extracting the Optimal Line

For a form document, the horizontal connected components including horizontal lines, characters and noises will all be extracted together with DRG, and the vertical line objects will be divided into a number of small connected components. The connected components corresponding to the horizontal lines are longer and thinner than those corresponding to vertical lines, characters or noises, so the value E of the former is always larger than those of the latter. Naturally, the principle could be set that the connected component, which has the biggest E, is the optimal line. In Fig.3-c and Fig.3-d, the lines indicated by arrow are the optimal lines selected through the feature function E for Fig.3-a and Fig.3-b.

2.3 Skew Estimation

Letting $(x_i, y_i), i = 1, 2, \cdots, N$ are the pixels in the optimal line obtained from above section, if the optimal line equation corresponding to this optimal line is:

$$y = ax + b. \tag{4}$$

According to the unitary linear regression equation, a and b will satisfy:

$$
\begin{cases}
a = \dfrac{N \sum\limits_{i=1}^{N} y_i x_i - \sum\limits_{i=1}^{N} x_i \sum\limits_{i=1}^{N} y_i}{N \sum\limits_{i=1}^{N} x_i^2 - \left(\sum\limits_{i=1}^{N} x_i \right)^2} \\[4ex]
b = \dfrac{\sum\limits_{i=1}^{N} y_i \sum\limits_{i=1}^{N} x_i^2 - \sum\limits_{i=1}^{N} x_i \sum\limits_{i=1}^{N} y_i x_i}{N \sum\limits_{i=1}^{N} x_i^2 - \left(\sum\limits_{i=1}^{N} x_i \right)^2}
\end{cases}
\tag{5}
$$

Where a is just the form's skew ratio.

The knowledge about line fitting can be referenced in the document [15]. Fig.8 shows the corrected result for Fig.3-b according to the skew angle estimated by the formula (5).

Fig. 8. The correction result for fig.3-b

3 Experiment Result Analysis

To analyze the effect of the method stated above, many comparative experiments were done between our algorithm and other skew detection algorithms in Pentium 2.6G, 512RAM under Windows XP. The other algorithms are coarse-to-fine HT(Method 1) and coarse-to-fine projection profile(Method 2).

Experiments 1: Data Analysis for Ideal Form Documents
Ideal form documents are generated through rotating the electronics form documents with different angle on computer. There are 9 types of and 72 tested form documents in all, and the skew angle range is between ±15° (The larger the skew angle is, the more the time consuming of method 1 or 2 is. But our method does not consume more time with the skew angle increasing.). Table 2 is the statistics for the tree methods.

Table 2. The statistics for ideal form images

	Mean processing time (s)	Mean absolute error
Method 1	0.293	0.0209
Method 2	0.172	0.0022
Our method	0.023	0.0001

Where mean absolute error is the error statistic for the skew rate.

Experiment 2: Data Analysis for Practical Scanning Form Documents
There are 20 types of and 358 practical form images input by scanner which are all filled in and stamped. And the skew angles range is between ±10°. Table 3 is the statistics.

Table 3. The statistics for practical form images

	Mean processing time(s)	Correctness ratio (%)
Method-1	0.730	90.2%
Method-2	0.493	95.8%
Our method	0.068	99.7%

Where, the correctness ratio is of the number of images, which are acceptable by vision from 10 people after corrected with the estimated skew angle by the algorithm described in this paper, over the number of all tested images.

Experiment 3: Data Analysis for degraded Form Documents
There are 4 types of and 14 ideal form images from experiment 1 to be blurred and noised at three levels from low to high. And three groups of degraded images are achieved. Fig.9 illustrates the effect of different level degraded image. The three levels of degraded images are separately treated as group 1 to 3. The mean absolute errors are calculated for the three groups of images with the three methods. Table 4 is the statistics.

(a) original image   (b) degraded image at   (c) degraded image at   (d) degraded image at
                      low level   middling level   high level

Fig. 9. The effect of degraded image

Table 4. The statistics of mean absolute error for three levels of degraded form images

	Group 1	Group 2	Group 3
Method-1	0.0143	0.0338	0.0344
Method-2	0.0018	0.0013	0.0049
Our method	0.00016	0.0003	0.00058

From above experiments, the conclusions for our method are as follows:

1) The detecting precision with our method is at a quantitative level above the other methods referred in this paper;
2) The detecting speed with our method is at a quantitative level above the other methods referred in this paper;
3) This method has a strong robustness and a wide applicability for blurred and noised form images.

4 Conclusion

In this paper, the form documents skew angle are detected through fitting the optimal horizontal line parameters according to the fact that the horizontal line has a same skew angle with the form. In the course of skew detecting, the objects, including horizontal lines, characters and vertical lines, are treated as connected components, and all the horizontal connected components are extracted, before skew detection, by the DRG algorithm proposed in this paper. To select the optimal line from these connected components correctly, an elongate function is proposed in this paper, which measures correctly the elongate degree for band-like objects and is a powerful tool for the optimal line selecting. At last, the skew angle is successfully obtained though line fitting based on optimization theory. From a large quantity of experiments, this method has fast speed and high precision, and the performance is robust.

References

1. Hinds, S.C., Fisher, J.L., D'Amato, D.P.: A document skew detection method using run length encoding and the Hough transform. In: Proc. 10th Int. Conf. Pattern Recognition, Atlantic City, pp. 464–468 (1990)
2. Xiping, H., Yunfeng, I.., Qingsheng, Z.: An Efficient Algorithm for Automatic Skew Correction of Color Document Image[J]. Journal of Image and Graphics 11(3), 367–371 (2006)
3. Yefeng, Z., Changsong, L., Xiaoqing, D., et al.: A Form Frame-Line Detection Algorithm Based on Directional Single-Connected Chain[J]. Journal of Software 22, 790–796 (2002)
4. Kwag, H.K., Kim, S.H., Jeong, S.H., Lee, G.S.: Efficient skew estimation and correction algorithm for document images. Image and Vision Computing, 2002 20, 25–35 (2002)
5. Shivakumar, P., Kumar, G.H., Guru, D.S., et al.: A new boundary growing and Hough transform based approach for accurate skew detection in binary document images. Proceedings of ICISIP 11, 140–146 (2005)
6. Kultanen, P., Xu, L., Oja, E.: Randomized Hough TransForm[J]. Proc. of ICPR'1990 1, 631–635 (1990)
7. Liu, J., Ding, X., Wu, Y.: Description and Recognition of Form and Automated Form Data Entry[C]. In: Proc. Int'l. Conf. Document Analysis and Recognition, pp. 579–582 (1995)
8. Chen, J.-L., Lee, H.-J.: An Efficient Algorithm for Form Structure Extraction Using Strip Projection[J]. Pattern Recognition 31(9), 1353–1368 (1998)
9. Qiu, Z., Lizhuang, M., Yan, G., et al.: Skew detection method for bill image based on directional projection[J]. Computer Applications 24(9), 50–51 (2004)
10. Shaoguang, Z., Li, X., Linya, T.: Detecting Line Segments by Using the Structural Features of Them in Digital Image[J]. Computer Engineering & Applications 22(2), 71–74 (2004)
11. Kao, C.-H., Don, H.-S.: Skew Detection of Document Images Using Line Structural Information[C]. In: Proceedings of the Third International Conference on Information Technology and Applications, vol. 1(7), pp. 704–709
12. Jibin, G., Delie, M.: Fast detecting and rectifying slanting image in table-form documents based on run-length features[J]. Journal Huazhong Univ. of Sci. & Tech. 33(8), 69–71 (2005)

13. Yu, B., Jain, A.K.: A Generic System for Form Dropout. IEEE Trans. Pattern Analysis and Machine Intelligence 18(11), 1127–1131 (1996)
14. Chhabra, A.K., Misra, V., Arias, J.: Detection of Horizontal Lines in Noisy Run Length Encoded Images: The FAST Method. In: Proc. IAPR Int'l. Workshop Graphics Recognition, pp. 35–48 (1995)
15. Baolin, C.: Optimization Theory & Algorithm[M], Beijing: Tsinghua University Press (1989) (in china)
16. Lee, H.Y., Park, W., et al.: Towards knowledge-based extraction of roads from 1 m-resolution satellite images[C], Image Analysis and Interpretation. In: Proc. 4th IEEE Southwest Symposium, pp. 171–176 (2000)
17. Guixiong, L., Baihua, S., Yunqing, F.: Thinning of binary image based on stroke trend analysis[J]. Optics and Precision Engineering 11(5), 527–530 (2003)
18. Yajun, L., Jirong, C., Xiaoliang, L.: Content Based Document Image Skew Adjusting[J]. Computer Simulation 23(12), 192–196 (2006)

Active Appearance Models Fitting with Occlusion

Xin Yu, Jinwen Tian, and Jian Liu

Huazhong University of Science and Technology,
State Key Laboratory for Multi-spectral Information Processing Technologies,
Wuhan 430074, P.R. China
uxinhenry@gmail.com

Abstract. In this paper, we propose an Active Appearance Models (AAMs) fitting algorithm, adaptive fitting algorithm, to localize an object in an image containing occlusion. The adaptive fitting algorithm conducts the fitting problem of AAMs containing object occlusion in a statistical framework. We assume that the residual errors can be treated as mixture statistical model of Gaussian and uniform model. We then reformulated the basic fitting algorithm and maximum a-posteriori (MAP) estimation algorithm of model parameter for AAMs to make the adaptive fitting algorithm. Extensive experiments are provided to demonstrate our algorithm.

1 Introduction

Accurate extraction and alignment of target objects from images are important for success of many computer vision and pattern recognition applications. Model-based methods offer potential solutions to this problem. Statistical shape models such as Active Shape Models (ASMs) and Active Appearance Models (AAMs) proposed by Cootes et al [1][2] have shown great potential in object localization and segmentation tasks. ASMs and AAMs are normally constructed by applying *Procrustes* analysis [6] followed by Principal Components Analysis (PCA) [10] to a collection of training images of objects with a mesh of canonical feature points on them.

Inherent to the AAMs algorithm described above is the assumption that each feature point in the source image has a corresponding match in the target image. This need may not always be the case. The occlusion may occur in the training data used to construct the AAMs, and/or in the input images to which the AAMs is fitted. Perhaps the most common cause of occlusion is 3D pose variation, which often causes self-occlusion. Other causes of occlusion include any objects placed in front of the interesting objects. Under such situations, the AAMs fitting algorithm typically fails. Then it is important to be able to: (1) construct AAMs from occluded training images, and (2) efficiently fit AAMs to novel images containing occlusion.

Previously, Gross et al [7] had proposed algorithms to construct AAMs from occluded training images and to track faces efficiently in videos containing

A.L. Yuille et al. (Eds.): EMMCVPR 2007, LNCS 4679, pp. 137–144, 2007.
© Springer-Verlag Berlin Heidelberg 2007

occlusion. In their approach, the mean shape, statistical model of shape and appearance in AAMs were reconstructed considering occlusion. Furthermore, robust and efficient AAMs fitting algorithms were proposed to track face with occlusion. We will use their reconstruction method for AAMs with occlusion as the prior step for AAMs fitting with occlusion in our approach framework.

In this paper, we conduct the fitting problem of AAMs containing object occlusion in a statistical framework. Instead of treating the residual texture error as a simple uniform Gaussian, we assume the residual error with occlusion come from two parts: non-occlusion part and occlusion part. Based on this assumption, we consider an appropriate distribution for the residual error. Then this distribution has been used to reformulate the basic AMMs searching approach. Furthermore, our statistical model can also be easily integrated into a maximum a-posteriori (MAP) formulation framework [1] and MAP estimation of model parameter has been reformulated considering occlusion.

This paper is organized as follows. In Section 2, we briefly review Gross's AAMs reconstruction method. The details of our AAMs fitting algorithm with occlusion are discussed in Section 3. Experiments are presented in Section 4. We conclude this paper in Section 5.

2   AAMs Construction with Occlusion

Assume a training set of shape-texture pairs to be $\Omega = \{(S_i, T_i^s)\}_{i=1}^N$, where N is the number of the shape-texture pairs. The Shape S_i of an AAM is defined by a triangulated mesh and in particular the vertex location of the mesh. The Texture T_i^s is the image patch enclosed by S_i.

The mean shape $\bar{S}$ is constructed using the *Procrustes* algorithm [7]. In case of occlusion the situation is complicated by the fact that not all of the mesh vertices are marked in every training image. The outline of the *Procrustes* algorithm stays the same, however only vertices visible in a given training image are used.

The shape parameters b_s are computed by firstly aligning shape vector S_i with the mean shape $\bar{S}$ using a similarity transform. Principal Components Analysis [3] is then performed on the aligned shape vector subtracted from the mean shape. In the presence of occlusion only the visible vertices are aligned to the mean shape. Principal Components Analysis with missing data [10][11] is then performed on the aligned shape vectors.

After shape warping, the texture T_i^s is warped correspondingly to $T_i \in R^L$, where L is the number of pixels enclosed by the mean shape $\bar{S}$. The warped textures T_i are also aligned to the tangent space of the mean texture $\bar{T}$ in the same approach as computing $\bar{S}$. Principal Components Analysis is then applied to the aligned texture vector. In the presence of occlusion, the shape normalized textures are incomplete. If any of the vertices of a triangle are not visible in the training image, that triangle will be missing in the training image. Again, we use Principal Components Analysis with missing data [10][11]. The details of complete construction algorithm are given in [7].

3   AAMs Fitting with Occlusion

In many scenarios there is the opportunity for occlusion. Since occlusion is so common, it is important to be able to efficiently fit AAMs to novel target images containing occlusion. We now describe how to find the parameters of AAMs which generate a synthetic image as close as possible to a particular occluded target image, assuming a reasonable starting approximation. We first describe previously proposed basic AAMs fitting algorithm [1] and show why it may fail in fitting to a novel target images containing occlusion. We then propose a different robust fitting algorithm, the Adaptive Fitting algorithm, which can be successful to fit the AAMs to a novel target images containing occlusion.

3.1   Basic AAMs Fitting

Fitting an AAM is usually be treated as an optimization process [1][8]. Given the current estimates of the shape parameters, it is possible to warp the input image onto the model coordinate frame and then compute an error image between the current model instance and the image that the AAM is been fitted to. A error vector can be defined:

$$\delta \mathbf{g} = \mathbf{g}_i - \mathbf{g}_m \tag{1}$$

where $\mathbf{g}_i$ is the vector of grey-level values in the target image, and $\mathbf{g}_m$ is the vector of grey-level values for the current model instance. In fitting procedure, it is to minimize the magnitude of the different vector $\triangle \mathbf{g} = |\delta \mathbf{g}|^2$ by varying the model parameters $\mathbf{c}$.

In basic AAMs fitting algorithms, the simple model chosen for the relationship between this error vector, $\delta \mathbf{g}$, and the additive incremental updated to the parameters, $\delta \mathbf{c}$, is constant linear:

$$\delta \mathbf{c} = \mathbf{R} \delta \mathbf{g} \tag{2}$$

The constant coefficients $\mathbf{R}$ in this linear relationship can be then found either by linear regression [4][5] or by other numerical methods [1][8].

In presence of occlusion, the above error function within the least-squares optimization framework always has bad performance. In order to deal with occluded pixels in the target image in a least-squares optimization framework, a robust error function should be carefully chosen.

3.2   Adaptive Fitting Algorithm

In the presence of occlusion, one way to contend with occlusion is to employ a preprocessing segmentation step. We propose, however, a more unified approach in which the fitting and segmentation are performed simultaneously.

In case of occlusion, instead of treating the residual texture error as a simple Gaussian variable, we assume the residual error come from two parts: non-occlusion part and occlusion part. It is equal to assume that each pixel in the error image between current model instance and target image can either be explained by a simple Gaussian error model M_g, or cannot be explained by this simple Gaussian model and therefore belongs to "outlier" model M_o. Pixels belonging to the outlier model are those that are occluded in the target images. By assuming that the pixels in error image are spatially independent and identically distributed, the conditional probability of observing the error image given the current appearance model is given by:

$$p(\delta \mathbf{g}|M) = p(\delta \mathbf{g}|M_g)P(M_g) + p(\delta \mathbf{g}|M_o)P(M_o) \tag{3}$$

Assuming that the prior probability of both model are equal, so the log-likelihood function of the conditional probability of observing the error image given the current appearance model is simply formulated:

$$p(\delta \mathbf{g}|M) = p(\delta \mathbf{g}|M_g) + p(\delta \mathbf{g}|M_o) \tag{4}$$

As described above, the conditional probability function of pixel without occlusion, $p(\delta \mathbf{g}|M_g)$, is assumed to be a simple Gaussian model (with variance σ^2). Meanwhile, the conditional probability function of pixel with occlusion, $p(\delta \mathbf{g}|M_o)$, is assumed to be a uniform function within a finite range. By special assumption on conditional probability function, we then get the robust error function which has the following form:

$$\delta \mathbf{g}' = \mathbf{W}\delta \mathbf{g}$$
$$= \mathbf{W}(\mathbf{g}_i - \mathbf{g}_m) \tag{5}$$

where $\mathbf{W}$ is a $L * L$ diagonal matrix. Each diagonal element in $\mathbf{W}$ is formulated as:

$$w_{kk} = \frac{e^{\frac{-(\mathbf{g}_i^k - \mathbf{g}_m^k)^2}{\sigma^2}}}{e^{\frac{-(\mathbf{g}_i^k - \mathbf{g}_m^k)^2}{\sigma^2}} + e^{-c^2}} \tag{6}$$

where $\mathbf{g}_i^k$ and $\mathbf{g}_m^k$ are the kth element of vector; σ^2 is variance of Gaussian model; c is constant scale value.

Comparing Eq. (1) and Eq. (5), it is obvious to find that the latter is actually the weighted version of the first one. We then can reformulate the basic AAM fitting algorithm very easily by just replacing the $\delta \mathbf{g}$ with $\delta \mathbf{g}'$. The simple model chosen for the relationship between this error vector and the additive incremental updated to the parameters is reformulate as:

$$\delta \mathbf{c} = -\mathbf{R}'\delta \mathbf{g}' \tag{7}$$

where

$$\mathbf{R}' = \left(\frac{\partial(\mathbf{W}\delta \mathbf{g})}{\partial \mathbf{c}}^T \frac{\partial(\mathbf{W}\delta \mathbf{g})}{\partial \mathbf{c}} \right)^{-1} \frac{\partial(\mathbf{W}\delta \mathbf{g})}{\partial \mathbf{c}}^T \tag{8}$$

Furthermore, our statistical model can also be easily integrated into a maximum a-posteriority (MAP) formulation [1] and MAP estimation of model parameter has been reformulated considering occlusion. In a maximum a-posterior formulation we seek to maximize

$$P(\text{model}|\text{data}) = P(\text{data}|\text{model})P(\text{model}) \tag{9}$$

By assuming that the residual errors can be treated as mixture of an Gaussian model and an uniform model, and that the model parameters are Gaussian with diagonal covariance $\mathbf{S}_c^2$, then maximizing eq. (9) is equivalent to minimizing

$$E(\mathbf{c}) = \sigma^{-2}(\mathbf{W}\delta\mathbf{g})^T(\mathbf{W}\delta\mathbf{g}) + \mathbf{c}^T(\mathbf{S}_c^{-1})\mathbf{c} \tag{10}$$

By forming the combined vector

$$\mathbf{y}(\mathbf{c}) = \begin{pmatrix} \sigma^{-1}(\mathbf{W}\delta\mathbf{g}) \\ \mathbf{S}_c^{-0.5}\mathbf{c} \end{pmatrix} \tag{11}$$

then $E = \mathbf{y}^T\mathbf{y}$. By applying a first order Taylor expansion to $\mathbf{y}$, we shown that $\delta\mathbf{c}$ must satisfy

$$\begin{pmatrix} \sigma^{-1}\frac{\partial(\mathbf{W}\delta\mathbf{g})}{\partial\mathbf{c}} \\ \mathbf{S}_c^{-0.5} \end{pmatrix} \delta\mathbf{c} = -\begin{pmatrix} \sigma^{-1}(\mathbf{W}\delta\mathbf{g}) \\ \mathbf{S}_c^{-0.5}\mathbf{c} \end{pmatrix} \tag{12}$$

In this case the optimal update step is shown to have the form

$$\delta\mathbf{c} = -(\mathbf{A}(\mathbf{W}\delta\mathbf{g})) + \mathbf{Bc} \tag{13}$$

where $\mathbf{A}$ and $\mathbf{B}$ have the form

$$\mathbf{A} = \sigma^{-1}\mathbf{C}\frac{\partial(\mathbf{W}\delta\mathbf{g})}{\partial\mathbf{c}} \tag{14}$$

$$\mathbf{B} = \mathbf{C}\mathbf{S}_c^{-0.5} \tag{15}$$

where $\mathbf{C}$ have the form

$$\mathbf{C} = \left(\sigma^{-2}\frac{\partial(\mathbf{W}\delta\mathbf{g})}{\partial\mathbf{c}}^T\frac{\partial(\mathbf{W}\delta\mathbf{g})}{\partial\mathbf{c}} + (\mathbf{S}_c^{-0.5})^T\mathbf{S}_c^{-0.5} \right)^{-1} \tag{16}$$

To find $\mathbf{R}'$ in Eq. (7), $\mathbf{A}$ in Eq. (14) and $\mathbf{B}$ in Eq. (15) respectively, multiple multivariate regression is performed on a sample of known model displacement and the corresponding different image. We apply Gaussian Processes as regression model in our approach. The Gaussian Processes model provides a kernel machine framework and has the state of the art of performance for regression and classification. For more details on Gaussian Processes regression, see our previous work in [12].

4 Experiments

To test the effects of occlusion to different AAMs fitting algorithm, we performed a series of experiments on images from the IMM face database [9]. 58 landmark points are manually labeled on the face. In the training stage, we start with fully labeled image sequences in which randomly selected regions are artificially occluded. Using artificially occluded data in this way enable us to systematic training AAMs containing occlusion with different degrees and types. In total 148 training images are used. By means of PCA with 95% total variances retained, the dimension of the shape parameter vector reduces to 66 and the dimension of the texture parameter vector reduces to 460. See Fig. 1 for examples.

Then testing images are input into AAMs constructed with occlusion. The AAMs fitting in the testing image (also the novel target image) is running to evaluate the robustness of different AAMs fitting algorithm in case of occlusion. Occlusions in testing images are from artificially occlusion which is generated in a similar way to that for training images. In total 74 testing images are used.

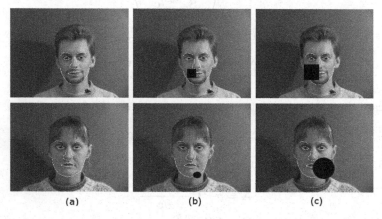

(a)   (b)   (c)

Fig. 1. Artificially occluded data. (a) Original images without occlusion. (b) Images with 10% of the face region occlude. (c) Images with 40% of the face region occlude.

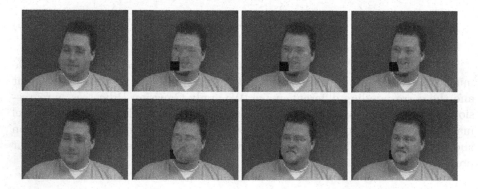

Fig. 2. Comparison of basic fitting algorithm (top row) and adaptive algorithm (bottom row) on one same testing images with artificial occlusion. The same initializing condition is imposed the two different algorithm. Bad performance in basic fitting algorithm, whereas excellent performance in adaptive algorithm.

4.1 Fitting Example

AAMs fitting is running by using the AAMs constructed with occlusion combined with different fitting algorithm. Here, we compare two fitting algorithms: basic fitting algorithm and our proposed adaptive fitting algorithm. In this subsection, we show several fitting example in all 74 fitting procedure. The key frames in optimization process are shown. Occlusions are introduced by artificial selecting random region in original test images. See Fig. 2 for examples. It is obvious to discover that adaptive fitting algorithm still success in presence of occlusion within testing image, whereas basic fitting algorithm failed under the same situation.

4.2 Fitting Evaluation

The average point-point distances between the searched shape and the manually labeled shape are compared. The searched shape is calculated by basic fitting algorithm and adaptive fitting algorithm. In total 74 testing images are considered in this experiment. In Fig. 3, the vertical axis represents the percentage of the situation in which the average point-point distances is smaller than the corresponding horizontal values. The artificial occlusion with 10% and 40% degree are imposed to the testing image.

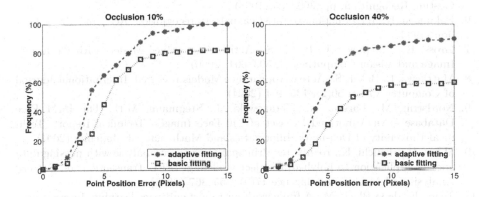

Fig. 3. Comparison of basic fitting algorithm and adaptive algorithm on one same testing images with artificial occlusion with 10% (left) and 40% (right) degree. The same initializing condition is imposed the two different algorithm. Bad performance in basic fitting algorithm, whereas excellent performance in adaptive algorithm.

5 Discussion

In this paper we proposed algorithms to and robustly fit AAMs with occlusion. We have reformulated the AAM fitting algorithm in a statistical framework, allowing it to be applicable in presence of occlusion. This is very useful for real applications where previous basic AAM maybe fail to obtain robust fitting in case of occlusion. In case of occlusion, instead of treating the residual texture error as a simple uniform Gaussian, we assume the residual error come from two parts: non-occlusion part and occlusion part and give a novel residual error function derivated in a statistical framework. We show that this new function can be seen as the weighed version of least-square error. We then give the weighed version of least-square optimization formulation, both in basic fitting and in MAP fitting. Experiments show that our proposed can still success in presence of occlusion in target image, whereas basic fitting algorithm has decreased performance.

References

1. Cootes, T.F., Taylor, C.J.: Statistical models of appearance for computer vision (2004), http://www.isbe.man.ac.uk/~bim/refs.html
2. Cootes, T.F., Edwards, G., Taylor, C.J.: Active appearance models. IEEE Transactions on Pattern Analysis and Machine Intelligence 23(6), 681–685 (2001)
3. Dryden, I.L., Mardia, K.V.: Statistical Shape Analysis. Wiley & Sons, Chichester (1998)
4. Edwards, G.J.: Learning to Identify Faces in Images and Video Sequences. PhD thesis, University of Manchester, Division of Imaging Science and Biomedical Engineering (1999)
5. Edwards, G.J., Taylor, C.J., Cootes, T.F.: Interpreting face images using active appearance models. In: Proc. International Conference on Automatic Face and Gesture Recognition, pp. 300–305 (1998)
6. Fukunaga, K.: Introduction to statistical pattern recognition. Academic Press, London (1990)
7. Gross, R., Matthews, I., Baker, S.: Active Appearance Models with Occlusion. Image and Vision Computing 24, 593–604 (2006)
8. Matthews, I., Baker, S.: Active Appearance Models revisited. International Journal of Computer Vision 60(2), 135–164 (2004)
9. Nordstrm, M., Larsen, M., Sierakowski, J., Stegmann, M.B.: The IMM Face Database - An Annotated Dataset of 240 Face Images. Technical Report, Technical University of Den-mark, Informatics and Mathematical Modeling (2004)
10. Shum, H., Ikeuchi, K., Reddy, R.: Principal component analysis with missing data and its application to polyhedral object modeling. IEEE Transactions on Pattern Analysis and Machine Intelligence 17(9), 855–867 (1995)
11. Torre, F., de la Black, M.: A framework for robust subspace learning. International Journal of Computer Vision 54(1), 117–142 (2003)
12. Yu, X., Liu, J., Tian, J.W.: A point matching approach based on Gaussian Processes. Journal of Huazhong University of Science and Technology: Nature Science 35(12) (2007)

Combining Left and Right Irises for Personal Authentication

Xiangqian Wu[1], Kuanquan Wang[1], David Zhang[2], and Ning Qi[1]

[1] School of Computer Science and Technology,
Harbin Institute of Technology (HIT), Harbin 150001, China
{xqwu, wangkq}@hit.edu.cn
http://biometrics.hit.edu.cn
[2] Biometric Research Centre, Department of Computing,
Hong Kong Polytechnic University, Kowloon, Hong Kong
csdzhang@comp.polyu.edu.hk

Abstract. Traditional personal authentication methods have many instinctive defects. Biometrics is an effective technology to overcome these defects. Among the available biometric approaches, iris recognition is one of the most accurate techniques. Combining the left and the right irises of same persons can improve the authentication accuracy and reduce the spoof attack risks. Furthermore, the fusion need not add any other hardware to the existing iris recognition systems. This paper investigates the feasibility of fusing both irises for personal authentication and the performance of some very simple fusion strategies. The experimental results show that the difference between the left and the right irises of the same persons is close to the difference between the irises captured from different persons. And combining the information of both irises can dramatically improve the authentication accuracy even when the quality of the iris images are not good enough. The results also show that the Minimum and the Product strategies can obtain the perfect performance, i.e. both FARs and FRRs of these two strategies can be reduce to 0%.

1 Introduction

Security is becoming increasingly important in the information based society. Personal authentication is one of the most important ways to enhance the security. However, the traditional personal authentication methods, including token-based ones (such as keys and cards, etc.) and knowledge-based ones (such as PINs and passwords, etc.), suffer from some instinctive defects: the token can be stolen or lost and the knowledge can be cracked or forgotten. Biometrics, which automatically uses the physiological or behavioral characteristic of people to recognize their identities, is one of the effective techniques to overcome these problems [1, 2, 3].

The iris recognition is one of the most accuracy methods among the current available biometric techniques and many effective algorithms have been developed. Daugman [4, 5] encoded the iris into a 256-bytes IrisCode by using

A.L. Yuille et al. (Eds.): EMMCVPR 2007, LNCS 4679, pp. 145–152, 2007.
© Springer-Verlag Berlin Heidelberg 2007

two-dimensional Gabor filters, and took the Hamming distance to match the codes. Wildes [6] matched the iris using Laplacian pyramid multi-resolution algorithms and a Fisher classifier. Boles et al. [7] extract iris features using a one-dimensional wavelet transform, but this method has been tested only on a small database. Ma et al. [8] construct a bank of spatial filters whose kernels are suitable for use in iris recognition. They have also developed a preliminary Gaussian-Hermite moments-based method which uses local intensity variations of the iris [9]. Later, they proposed an improved method based on characterizing key local variations [10]. All of these algorithms only use sole (left or right) iris of each person for authorization. The accuracy of these algorithms are heavily affected by the quality of iris images, so it is very difficult for a traditional iris recognition system to obtain a high accuracy if the system cannot get high quality images. Combining both irises can overcome this problem. That is, even if the qualities of the images obtained from both irises are not good enough, the system combining the both irises still can get a high accuracy rate. Besides, the system fusing both irises can reduce the spoof attack risk since faking one iris is much easier than faking two. The users should provide both (Left and Right) irises when they use this system combining both irises. Moreover, we can implement the fusion system without adding any other hardware to the existing iris recognition system. This paper investigates the feasibility of fusing both irises for personal authentication and the performance of some simple fusion strategies.

The rest of this paper is organized as follows. Section 2 reviews the IrisCode extraction and matching. Section 3 presents several fusion strategies. Section 4 contains some experimental results and analysis. And Section 5 provides some conclusions.

2   Feature Extraction and Matching

This section reviews the IrisCode method devised by Daugman [4,5,1].

2.1   Iris Preprocessing

As the iris images used in this paper contain some reflection (see Fig. 1), we first use a threshold to remove these reflection points from the image. Then the two circular boundaries of iris are searched by using the integrodifferential operators. Finally, the disk-like iris area is unwrapped to a rectangular region by using doubly dimensionless projection.

2.2   IrisCode Extraction

Iris code is produced by 2-D complex Gabor filters. Let I denote a preprocessed iris image. The IrisCode C contains two parts, i.e. the real part C_r and the imaginary part C_i, which are computed as following:

$$I_G = I * G. \tag{1}$$

$$C_r(i,j) = \begin{cases} 1, \text{ if } \mathbf{Re}[I_G(i,j)] \geq 0; \\ 0, \text{ otherwise.} \end{cases} \qquad (2)$$

$$C_i(i,j) = \begin{cases} 1, \text{ if } \mathbf{Im}[I_G(i,j)] \geq 0; \\ 0, \text{ otherwise.} \end{cases} \qquad (3)$$

where G is a complex 2-D Gabor filter and $*$ is the convolution operation.

2.3   IrisCode Comparison

IrisCode comparison is based on Hamming distance. The Boolean operator (XOR) is used to compare each pair of IrisCodes C_1 and C_2 bit by bit. The Boolean operator equals 1 if and only if the two bits are different; otherwise, it equals to 0. The normalized Hamming distance equation is given below,

$$HD = \frac{1}{N} \sum_{j=0}^{N-1} [C_1(j) \otimes C_2(j)] \qquad (4)$$

where $\otimes$ is the Boolean operator **XOR** and N is the number of the total points in a IrisCode (including both the real and imaginary parts).

3   Score Fusion

Denote x_1 and x_2 as the scores obtained from the left irises matching and right irises matching between two persons, respectively. To obtain the final matching score x, we fuse these two scores by following simple strategies, which need not any prior knowledge or training.

S_1: *Maximum Strategy:*
$$x = \max(x_1, x_2) \qquad (5)$$

S_2: *Minimum Strategy:*
$$x = \min(x_1, x_2) \qquad (6)$$

S_3: *Product Strategy:*
$$x = \sqrt{x_1 x_2} \qquad (7)$$

S_4: *Sum Strategy:*
$$x = \frac{x_1 + x_2}{2} \qquad (8)$$

4   Experimental Results and Analysis

4.1   Database

We have collected $21,240$ iris images from the left and right eyes of 2124 people by CCD-based devices. Each eye provides 5 images. The size of the images in the database is 768 × 568. Using only one iris (Left or Right), most people can be effectively recognized while there are still some people cannot be recognized. We choose 120 persons' iris images, using which the traditional iris recognition system cannot get a very high accuracy, to build a database to investigate the proposed approach. Since each person has five 5 irises and 5 left irises, we can get 5 pairs (left and right) of irises for each person. That is, total 120 × 5 = 600 pairs of irises can be obtained from this database. Fig. 1 shows the left and the right irises of some persons in the database.

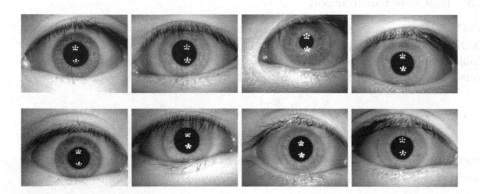

Fig. 1. The Left and the Right Irises of Some Persons in the Database. Top Row: Left Irises; Botton Row: Right Irises.

4.2 Difference Between Left and Right Irises

To fuse the features of both iris, we should investigate the difference between them. If the iris images from the right and left eyes of same persons are very similar, the fusion cannot improve the performance much. To investigate this, two types of matching are conducted on the database: 1) Each iris image is matched against all of the other images from different persons; 2) Each left iris image is matched against the right ones of the same person. The matching distance distributions of these two type matchings are plotted in Fig. 2. This figure shows that these two distributions are similar. That is, the difference between the right and left irises of the same persons is close to the difference of the irises from different persons. Hence, the left and the right irises of the same person are independent. So they can be fused to improve the authentication accuracy.

4.3 Matching Tests

To test the performance of the different fusion strategies, each pair of irises in the database is matched against the other pairs using different fusion strategies. The matching score distributions of each strategy are showed in Fig. 3. This figure shows that the distribution curves of each strategy have two peaks, which

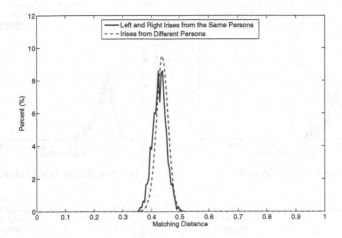

Fig. 2. The matching distance distributions of irises from different persons and the right and left irises from the same persons

respectively correspond the genuine and impostor matching distances. The two peaks of all strategies are widely separated. However, the genuine matching curve and the impostor matching curve of different strategies overlapped with different degree. The less the overlapping area is, the better the matching performance is. This overlapping area is defined as minimum total error (MTR), and the MTR of each strategy is listed in Table 1. From this table, the Maximum strategy improve the matching performance little, while the Sum, Minimum and Product strategies can improve the performance dramatically, especially, the MTRs of the Minimum and Product strategies have decreased to 0%, which is the ideal matching performance.

Table 1. MTR of Different Fusion Strategies

Strategy	Sole Left Iris	Sole Right Iris	Sum	Maximum	Minimum	Product
MTR (%)	1.6260	1.1972	0.0024	1.0987	0	0

4.4   Accuracy Tests

To evaluate the accuracies of the different fusion strategies, each pair of irises in the database is again matched with the other pairs. For each strategy, the false

Table 2. EER of Different Strategies

Strategy	Sole Left Iris	Sole Right Iris	Sum	Maximum	Minimum	Product
EER (%)	1.137	0.599	0.001	0.587	0	0

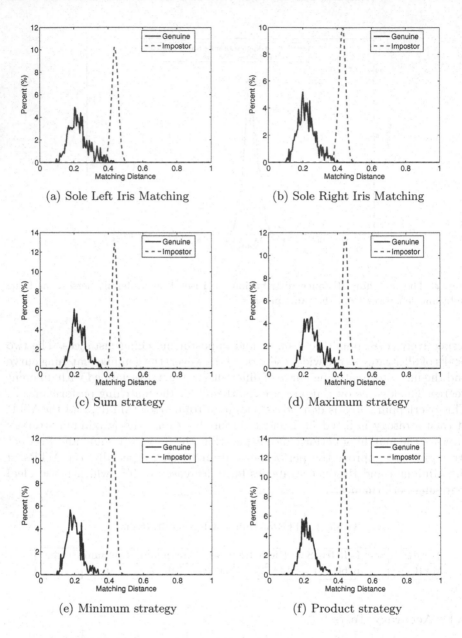

Fig. 3. Matching Score Distributions of Different Fusion strategies

accept rate (FAR) and false reject rate (FRR) at different distance thresholds is plotted in Fig. 4 and the equal error rate (EER) of them are listed in Table 2. This figure and table also demonstrate that the accuracy of the Maximum strategy is close to the one of the sole right iris matching, while the Sum, the Minimum and the Product strategies can improve the accuracy dramatically, especially the Minimum and the Product strategies can get the perfect accuracy, i.e. their EERs can be reduce to 0%.

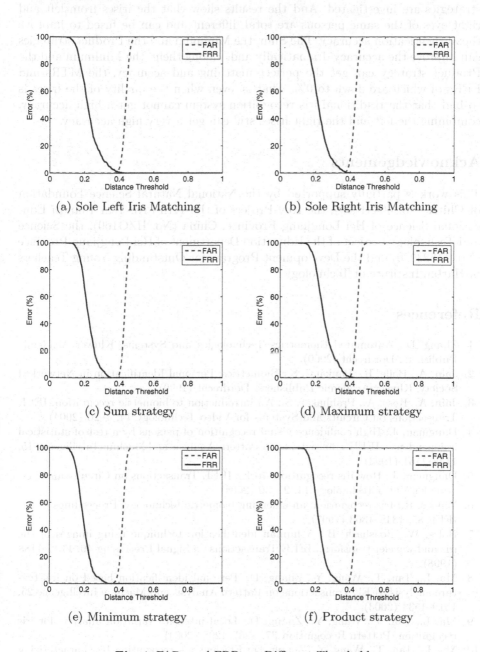

(a) Sole Left Iris Matching   (b) Sole Right Iris Matching

(c) Sum strategy   (d) Maximum strategy

(e) Minimum strategy   (f) Product strategy

Fig. 4. FARs and FRRs at Different Thresholds

5   Conclusions

This paper used both irises of a person for authentication. Several simple fusion strategies are investigated. And the results show that the irises from left and right eyes of the same persons are total different and can be fused to improve the authentication accuracy. The Sum, the Minimum and the Product strategies can improve the accuracy dramatically and, among them, the Minimum and the Product strategy can get the perfect matching and accuracy, the MTRs and EERs of which are down to 0%. That is, even when the quality of the irises is so bad that the traditional iris recognition system cannot get a high accuracy, combining the left and the right irises still can get a very high accuracy.

Acknowledgements

This work is partially supported by the National Natural Science Foundation of China (No. 60441005), the Key-Project of the 11th-Five-Year Plan of Educational Science of Hei Longjiang Province, China (No. HZG160), the Science and Technology Project of the Education Department of Hei Longjiang Province (No. 11523026) and the Development Program for Outstanding Young Teachers in Harbin Institute of Technology.

References

1. Zhang, D.: Automated Biometrics–Technologies and Systems. Kluwer Academic Publishers, Dordrecht (2000)
2. Jain, A., Bolle, R., Pankanti, S.: Biometrics: Personal Identification in Networked Society. Kluwer Academic Publishers, Dordrecht (1999)
3. Jain, A., Ross, A., Prabhakar, S.: An introduction to biometric recognition. IEEE Transactions on Circuits and Systems for Video Technology 14, 4–20 (2004)
4. Daugman, J.: High confidence visual recognition of persons by a test of statistical independence. IEEE Transactions on Pattern Analysis and Machine Intelligence 15, 1148–1161 (1993)
5. Daugman, J.: How iris recognition works. IEEE Transactions on Circuits and Systems for Video Technology 14, 21–30 (2004)
6. Wildes, R.: Iris recognition: an emerging biometric technology. Proceedings of the IEEE 85, 1348–1363 (1997)
7. Boles, W., Boashash, B.: A human identification technique using images of the iris and wavelet transform. IEEE Transactions on Signal Processing 46, 1185–1188 (1998)
8. Ma, L., Tan, T., Wang, Y., Zhang, D.: Personal identification based on iris texture analysis. IEEE Transactions on Pattern Analysis and Machine Intelligence 25, 1519–1533 (2003)
9. Ma, L., Tan, T., Wang, Y., Zhang, D.: Local intensity variation analysis for iris recognition. Pattern Recognition 37, 1287–1298 (2004)
10. Ma, L., Tan, T., Wang, Y., Zhang, D.: Efficient iris recognition by characterizing key local variations. IEEE Transactions on Image Processing 13, 739–749 (2004)

Bottom-Up Recognition and Parsing of the Human Body

Praveen Srinivasan and Jianbo Shi

GRASP Lab, University of Pennsylvania, 3330 Walnut Street Philadelphia, PA
19104, Tel.: 650-906-7334; Fax.: 215-573-2048
psrin@seas.upenn.edu, jshi@cis.upenn.edu

Abstract. Recognizing humans, estimating their pose and segmenting
their body parts are key to high-level image understanding. Because
humans are highly articulated, the range of deformations they undergo
makes this task extremely challenging. Previous methods have focused
largely on heuristics or pairwise part models in approaching this problem. We propose a bottom-up *growing*, similar to parsing, of increasingly
more complete partial body masks guided by a set of parse rules. At
each level of the growing process, we evaluate the partial body masks
directly via shape matching with exemplars (and also image features),
without regard to how the hypotheses are formed. The body is evaluated as a whole, not the sum of its parts, unlike previous approaches.
Multiple image segmentations are included at each of the levels of the
growing/parsing, to augment existing hypotheses or to introduce ones.
Our method yields both a pose estimate as well as a segmentation of
the human. We demonstrate competitive results on this challenging task
with relatively few training examples on a dataset of baseball players
with wide pose variation. Our method is comparatively simple and could
be easily extended to other objects. We also give a learning framework for
parse ranking that allows us to keep fewer parses for similar performance.

1 Introduction

Recognition, pose estimation and segmentation of humans and their body parts
remain important unsolved problems in high-level vision. Action understanding
and image search and retrieval are just a few of the areas that would benefit
enormously from this task. There has been good previous work on this topic,
but significant challenges remain ahead. We divide the previous literature on
this topic into three main areas:

Top-down approaches: [4] developed the well-known pictorial structures
(PS) method and applied it to human pose estimation. In the original formulation, PS does probablistic inference in a tree-structured graphical model usually
with the torso as the root. PS recovers locations, scales and orientations of rigid
rectangular part templates that represent an object. Pairwise potentials were
limited to simple geometric relations (relative position and angle), while unary
potentials were based on image gradients or edge detection. The tree structure is

A.L. Yuille et al. (Eds.): EMMCVPR 2007, LNCS 4679, pp. 153–168, 2007.
© Springer-Verlag Berlin Heidelberg 2007

a limitation since many cues (e.g., symmetry of appearance of right and left legs) cannot be encoded. [12] extended the original model to encode the fact that symmetric limb pairs have similar color, and that parts have consistent color or colors in general, but how to incorporate more general cues seems unclear. [13] track people by repeatedly detecting them with a top-down PS method. [16] introduced a non-parametric belief propagation (NBP) method with occlusion reasoning to determine the pose. All these approaches estimate pose, and do not provide an underlying segmentation of the image. Their ability to utilize more sophisticated cues beyond pixel-level cues and geometric constraints between parts is limited.

Search approaches: [11] utilized heuristic-guided search, starting from limbs detected as segments from Normalized Cuts (NCut) ([3]), and extending the limbs into a full-body pose and segmentation estimate. A follow up to this, [10], introduced an Markov-Chain Monte Carlo (MCMC) method for recovering pose and segmentation. [6] developed an MCMC technique for inferring 3-D body pose from 2-D images, but used skin and face detection as extra cues. [17] utilized a combination of top-down, MCMC and local search to infer 2-D pose.

Bottom-up/Top-down approaches [14] used bottom-up detection of parallel lines in the image as part hypotheses, and then combined these hypotheses into a full-body configuration via an integer quadratic program. [17] also fit into this category, as they use bottom-up cues such as skin pixel detection. Similarly, [5] integrated bottom-up skin color cues with a top-down, NBP process. [10] use superpixels to guide their search. While [2] estimate only segmentation and not pose for horses and humans in upright, running poses, they best utilize shape and segmentation information in their framework. [15] use bottom-up part detectors to detect part hypotheses, and then piece these hypotheses together using a simple dynamic programming (DP) procedure in much the same way as [4]. Lastly, our approach is similar to that described in [18], where bottom-up proposals are grouped hierarchically to produce a desired shape; unlike, [18] we use more sophisticated shape modelling than small groups of edges.

2   Overview of Our Parsing Method

Our goal is to combine a subset of salient shapes S (in our case, represented as binary masks, and provided by segmenting the image via NCut) detected in an image into a shape that is similar to that of a human body. Because the body has a very distinctive shape, we expect that it is very unlikely for this to occur by chance alone, and therefore should correspond to the actual human in the scene.

We formulate this as a parsing problem, where we provide a set of parsing rules that lead to a parse (also represented by a binary mask) for the body, as see in Figures 1 and 2. A subset of the initial shapes S are then parsed into a body. The rules are unary or binary, and hence a non-terminal can create a parse by *composing* the parses of one or two children nodes (via the pixel-wise OR operator). In addition the parses for a node can be formed directly from a shape from S, in addition to being formed from a child/children. Traditional parsing methods (DP methods) that exploit a subtree independence (SI) property in their scoring of a parse can search over an exponential number of parses in polynomial time.

We can define a traditional context-free grammar as a tuple

$$\langle V, T, A, R, S \rangle \tag{1}$$

V are parse non-terminals and T are the terminals, where A is the root non-terminal,

$$R = \{A_i \rightarrow B_i, C_i\} \tag{2}$$

is a set of production rules with $A_i \in V$ and $B_i, C_i \in V \cup T$ (we restrict ourselves to binary rules, and unary rules by making C_i degenerate), and S_i is a score for using rule R_i. Further, for each image, a terminal $T_i \in T$ will have potentially multiple instantiations $t_i^j, j = 1, ..., n_i$ each with its own score u_i^j for using $T_i \rightarrow t_i^j$ in a parse. Each terminal instantiation $t_i^j \in S$, corresponds to an initial shape S drawn from NCut segmentation. If the root is $A \in V$, then we can compute the score of the best parse (and therefore the best parse itself) recursively as

$$P(A) = \max_{r_i | r_i = (A \rightarrow B_i, C_i)} (S_i + P(B_i) + P(C_i)) \tag{3}$$

However, this subtree independence property greatly restricts the type of parse scoring function (PSF) that can be used.

By contrast, our approach seeks to maximize a shape scoring function F_A for A that takes as input two specific child parses b_i^j and c_i^k (or one, as we allow unary rules) corresponding to rule $A \rightarrow B_i, C_i$:

$$P(A) = \max_{r_i = (A \rightarrow B_i, C_i)} \max_{j,k} (F_A(b_i^j, c_i^k)) \tag{4}$$

Recall that we represent a parse b_i^j or t_i^j as a binary mask, not as the parse rules and terminals that form it. Note that the exact solution requires all parses for the children as opposed to just the best, since the scoring function F_A does not depend on the scores of the child parses. Because the exact solution is intractable, we instead solve this approximately by greedily pruning parses to a constant number. However, we use a richer PSF that has no subtree independence property. We can view the differences between the two methods along two dimensions: **proposal** and **evaluation**.

Proposal: DP methods explore all possible parses, and therefore have a trivial proposal step. Our method recursively groups bottom-up body part parses into increasingly larger parts of the body until an entire body parse is formed. For example, a lower body could be formed by grouping two Legs, or a Thigh+Lower leg and a Lower leg, or taken directly from S. In the worst case, creating parses from two children with n parses each could create n^2 new parses. Therefore, pruning occurs at each node to ensure that the number of parses does not grow exponentially further up the tree. To prune, we eliminate redundant or low scoring parses. Because there is pruning, our method does not evaluate all possible parses. However, we are still able to produce high quality parses due to a superior evaluation function.

Evaluation: On the evaluation side, DP employs evaluation functions with special structure, limiting the types of evaluation functions that can be used. Usually, this takes the form of evaluating a parse according to the parse rule used (chosen from a very limited set of choices) and the scores of the subparses that compose it, as in Equation (3). However, this does not allow scoring of the parse in a holistic fashion. Figure 3 gives an example; two shapes that on their own are not clearly parts of a disk, but when combined together, very clearly form a disk. Therefore, we associate with each node i a scoring function F_i (as in Equation (4)) that scores parses not based on the scores of their constituent parses or the parse rule, but simply based on their shape (we also extend this to include other features, such as probability of boundary and SIFT). The scoring function also allows for pruning of low-scoring and redundant parses. Note that our choice of F_i does not exhibit an SI property. Because of this, we are primarily interested in the actual result of the parse, a binary mask, as opposed to how it was generated from child parses or from S. In contrast to DP methods, a parse is evaluated irrespective of how it was generated. Our parse rules guide a search strategy, where evaluation is independent of the parsing rules used.

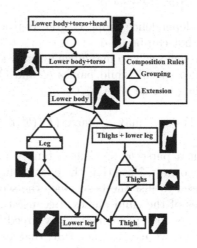

Fig. 1. Our body parse tree, shown with an exemplar shape from our training set for each node; the exemplars are used for shape scoring. Shape parsing begins at the leaf nodes of thigh and lower leg and proceeds upwards. Note that in addition to composing parses from children nodes, parses can always come from the initial shapes S.

2.1   Multiple Segmentations

To initialize our bottom-up parsing process, we need a set of intial shapes S. [11] noted that human limbs tend to be salient regions that NCut segmentation often isolate as a single segment. To make this initial shape hypothesis generation method more robust, we consider not one segmentation as in [11], but 12 different segmentations provided by NCut. We vary the number of segments from 5 to

- {Lower leg, Thigh} → Leg
- {Thigh, Thigh} → Thighs
- {Thighs, Lower leg} → Thighs+Lower leg
- {Thighs+Lower leg, Lower leg} → Lower body
- {Leg, Leg} → Lower body
- {Lower body} → Lower body+torso
- {Lower body+torso} → Lower body+torso+head

Fig. 2. Our parse rules. We write them in reverse format to emphasize the bottom-up nature of the parsing.

Fig. 3. The two shapes on the left bear little resemblance to a disk in isolation. However, when combined, the disk is clear.

60 in steps of 5, giving a total of 390 initial shapes per image. This allows us to segment out large parts of the body that are themselves salient, e.g. the lower body may appear as a single segment, as well as smaller parts like individual limbs or the head. Figure 4 shows for an image 2 of the 12 segmenations with overlaid boundaries. Segments from different segmentations can overlap, or be contained within another. In our system, these segments are all treated equally. These initial shapes could be generated by other methods besides segmentation, but we found segmentation to be very effective.

Fig. 4. Two segmentations of an image, 10 and 40 segments. Red lines indicate segment boundaries for 10 segments, green lines indicate boundaries for 40 segments, and yellow indicates boundaries common to both segmentations (best viewed in color).

2.2   Shape Comparison

For each node i, we have an associated shape comparison function F_i. For the root node, this ranks the final hypotheses for us. For all other nodes, F_i ranks hypotheses so that they can be pruned. All the shape comparison functions operate the same way: we match the boundary contour of a mask against boundary contours of a set of exemplar shapes using the inner-distance shape context (IDSC) of [7].

The IDSC is an extension of the original shape context proposed in [1]. In the original shape context formulation, given a contour of n points $x_1, ..., x_n$, a shape context was computed for point x_i by the histogram

$$\#(x_j, j \neq i : x_j - x_i \in bin(k)) \tag{5}$$

Ordinarily, the inclusion function $x_j - x_i \in bin(k)$ is based on the Euclidean distance $d = \| x_j - x_i \|_2$ and the angle $acos((x_j - x_i)/d)$. However, these measures are very sensitive to articulation. The IDSC replaces these with an *inner-distance* and an *inner-angle*.

The inner-distance between x_i and x_j is the shortest path between the two points traveling through the interior of the mask. This distance is less sensitive to articulation. The inner-angle between x_i and x_j is the angle between the contour tangent at the point x_i and tangent at x_i of the shortest path leading from x_i to x_j. Figure 5 shows the interior shortest path and contour tangent.

The inner-distances are normalized by the mean inner-distance between all pairs $\{(x_i, x_j)\}$, $i \neq j$ of points. This makes the IDSC scale invariant, since angles are also scale-invariant. The inner-angles and normalized log inner-distances are binned to form a histogram, the IDSC descriptor. For two shapes with points $x_1, ..., x_n$ and $y_1, .., y_n$, IDCSs are computed at all points on both contours. For every pair of points x_i, y_j, a matching score between the two associated IDCSs is found using the Chi-Square score ([1]). This forms an n-by-n cost matrix, for standard string matching via DP, establishing correspondence between the two contours. The algorithm also permits gaps in the matching with a user-specified penalty. We try the alignment at several different, equally spaced starting points on the exemplar mask to handle the cyclic nature of the closed contours, and keep the best scoring alignment (and the score). Because the DP algorithm minimizes a cost (smaller is better), we negate the score since our desire is to maximize F and all F_i. The complexity of the IDSC computation and matching is dominated by the matching; with n contour points and s different starting points, the complexity is $O(sn^2)$.

If s is chosen as n, then the complexity is $O(n^3)$. However, we instead use a coarse-to-fine strategy for aligning two shapes; based on the alignment of a subsampling of the points, we can narrow the range of possible alignments for successively larger number of points, thereby greatly reducing the complexity. We use a series of 25, 50 and 100 points to do the alignment.

2.3   Parse Rule Application Procedure

Our parse process consists of five basic steps that can be used to generate the hypotheses for each node. For a particular node A, given all the hypotheses for all children nodes, we perform the following steps:

Fig. 5. IDSC Computation. **Left:** We show: shortest interior path (green) from start (blue dot) to end (blue cross); boundary contour points (red); contour tangent at start (magenta). The length of interior path is the inner-distance; the angle between contour tangent and the start of the interior path is the inner-angle. **Center:** Lower body mask hypothesis; colored points indicate correspondence established by IDSC matching with exemplar on **right**.

Segment inclusion: Applies to all nodes. We include by default all the masks in S as hypotheses for A. This allows us to cope with an input image that is itself a silhouette, which would not necessarily be broken into different limbs, for example. A leg will often appear as a single segment, not as separate segments for the thigh and lower leg; it is easier to detect this as a single segment, rather than trying to split segments into two or more pieces, and then recognize them separately. This is the only source of masks for leaf nodes in the parse tree.

Grouping: $\{B, C\} \to A$ For binary rules, we can compose hypotheses from two children such as grouping two legs into a lower body, e.g. $\{$Leg, Leg$\} \to$ Lower body. For each child, based on the alignment of the best matching exemplar to the child, we can predict which part of the segment boundary is likely to be adjacent to another part.

A pair of masks, b from B and c from C, are taken if the two masks are within 30 pixels of each other (approximately 1/10th of the image size in our images), and combined with the pixel-wise OR operator. Because we need a single connected shape for shape comparison, if the two masks are not directly adjacent we search for a mask from the segmentations that is adjacent to both, and choose the smallest such mask m. m is then combined with b and c into a single mask. If no such mask m exists, we keep the larger of a and b. Figure 6 provides an example of the parse rule, $\{$Leg,Leg$\} \to$ LowerBody.

Extension: $\{B\} \to A$ For unary rules we generate hypotheses by projecting an expected location for an additional part based on correspondence with exemplars. This is useful when bottom-up detection of a part by shape, such as the torso or head, is difficult due to wide variation of shape, or lack of distinctive shape. Once we have a large piece of the body (at least the lower body), it is more reliable to directly project a position for hypotheses. Given a hypothesis of the lower body and its correspondence to a lower body exemplar shape, we can project the exemplar's quadrilateral (quad) representing the torso on to the hypothesis (we estimate a transform with translation, rotation and scale based on the correspondence of two contour points closest to the two bottom vertices of the torso quad).

Algorithm 1. $P_A = \text{Parse}(A, S)$: for a particular image, given initial segments S and part name A, produce ranked and pruned parses for A.

Input: Part name A and initial shapes S
Output: P_A: set of ranked and pruned parses for A
$P_A = S$;   // Initialize parses for A to intial segments S
foreach *rule* $\{B_i, C_i\} \to A$ *(or $B_i \to A$)* **do**
    $P_{B_i} = \text{Parse}(B_i, S)$;   // Recurse
    $P_{C_i} = \text{Parse}(C_i, S)$;   // If binary rule, recurse
    $P_A = P_A \cup \text{Group}(P_{B_i}, P_{C_i})$ (or $\text{Extend}(P_{B_i})$);   // Add to parses of A
end
$P_A = \text{RankByShape}(P_A)$ or $\text{RankByImageFeatures}(\text{GetFeatures}(F_A), w_A)$;
// GetFeature gets parse features, w_A is classifier
$P_A = \text{Prune}(P_A)$;   // Prune redundant/low scoring parses
return P_A;   // Return parses

Similarly, given a mask for the lower body and torso, and its correspondence to exemplars, we can project quads for the head. With these quads, we look for all masks in S which have at least half their area contained within the quad, and combine them with the existing mask to give a new hypothesis. For each hypothesis/exemplar pair, we compose a new hypothesis. We sumamrize the general parsing process with the recursive algorithm presented in Algorithm 1.

Scoring. Once hypotheses have been composed, they are scored by matching to the nearest exemplar with IDSCs and DP. Correspondence is also established with the exemplar, providing an estimate of pose.

Pruning. Many hypotheses are either low-scoring, redundant or both. We prune away these hypotheses with a simple greedy technique: we order the hypotheses by their shape score, from highest to lowest (best to worst). We add the best hypothesis to a representative set, and eliminate all other hypotheses which are similar to the just added hypothesis. We then recurse on the remaining hypotheses until the representative set reaches a fixed size. For mask similarity we use a simple mask overlap score O between masks a and b: $O(a, b) = \dfrac{area(a \bigcap b)}{area(a \bigcup b)}$, where $\bigcap$ performs pixel-wise AND, and the area is the number of "1" pixels. If $O(a, b)$ is greater than a threshold, a and b are considered to be similar. After this step, we have a set of hypotheses that can be passed higher in the tree, or to evaluate in the end if the node A is the root. Figure 6 illustrates the stages of the parsing process for generating the hypotheses for a single node. Also included are examples of grouping, extension, and shape matching/scoring.

3   Learning with Other Features

Besides shape, we also investigated using SIFT [8] and Probability of Boundary (PB) [9] features for ranking of parses. Traditional maximum-likelihood

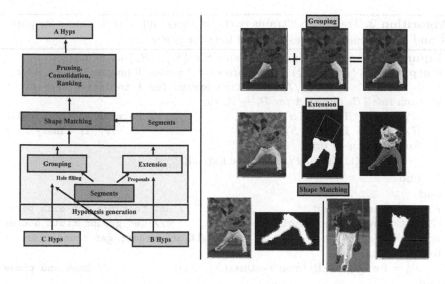

Fig. 6. Left: parse rule application. For binary rules, all pairs of child hypotheses within 30 pixels are grouped, with hole filling provided by segments if needed. For unary rules, hypotheses undergo extension using projected quads and segment proposals. Shape matching is performed on both the original segments as well as the composed/extended hypotheses. For leaf nodes, shape matching is performed only on the segments. After shape matching, the hypotheses are consolidated, pruned and ranked. **Right:** Grouping: two legs, on the left, are grouped into a lower body parse, on the right. *While recognizing legs alone from shape is not that robust, when they are combined into one shape, the distinctive shape of the lower body becomes apparent: the whole is not the sum of the parts.* Extension: the leftmost image shows a lower body parse with multiple torso quads projected from exemplars on to the image using the correspondence between the lower body hypothesis and the lower body exemplars; the center image shows the exemplar with its torso quad that yielded the best torso hypothesis, seen in the right image. Shape matching: two examples of shape matching. The lower body on the right was detected directly from the segments S, underscoring the importance of injecting the shapes from S into all levels of the parse tree.

learning (similar to that for AND-OR graphs) would require an explicit specification of the best parse for each rule. Unfortunately, this is difficult to specify a-priori due to the exponential number of different possible parses (as combinations of the initial input shapes). Instead, for each part we train a weighted logistic regression (WLR) classifier to rerank the parses according to a variety of features. These parses are then pruned and passed up to other rules. We summarize our training procedure in Algorithm 2. It is essentially the same as the original parsing procedure in Algorithm 1, but after generating the set of parses for a part, a WLR is trained to re-rank the parses. During testing, instead of ranking parses by shape score, they are ranked by the learnt WLR.

Algorithm 2. Train(A, S): trains part classifiers and returns parses for part A and all descendants; uses input of initial segments S.

Input: Part name A and initial segments $S = \{S_1, ..., S_n\}$ for all images
Output: $P_A = \{P_A^1, ..., P_A^n\}$: set of parses for A across all images; w_A: WLR for A
$P_A = S$;   // Initialize parses for A to intial segments S
foreach *rule* $\{B_i, C_i\} \to A$ *(or* $B_i \to A$*)* **do**
    $[P_{B_i}, w_{B_i}] = \text{Train}(B_i, S)$;   // Recurse to train child
    $[P_{C_i}, w_{C_i}] = \text{Train}(C_i, S)$;   // If binary rule
    foreach *image j* **do**
        $P_A^j = P_A^j \cup \text{Group}(P_{B_i}^j, P_{C_i}^j)$ (or $\text{Extend}(P_{B_i}^j)$);
    end
end
$F_A = \text{GetFeatures}(P_A)$;   // Get features for each parse
$G_A = \text{GetGroundTruthOverlapScores}(P_A)$;   // Get ground truth scores
$w_A = \text{WLR}(F_A, G_A)$;   // Do WLR on parses; get classifier w_A
foreach *image j* **do**
    $P_A^j = \text{Prune}(\text{RankByImageFeatures}(F_A^j, w_A))$;   // Rank and prune
end
return $[P_A, w_A]$;   // Return parses and classifer

3.1   Features

For PB, we simply computed the average PB value along the boundary of the mask. For SIFT features, we created a SIFT codebook and computed codebook histograms for each mask as additional features. We extracted SIFT features from a dataset of 450 additional baseball images, using the code from http://vision.ucla.edu/~vedaldi/code/sift/sift.html with the default settings. These features were clustered via k-means into a 200 center codebook. For a test image and a given mask, SIFT features are extracted, associated with the nearest center in the codebook, and then a histogram over the frequency of each center in the mask is computed. This gives a 200 dimension vector of codebook center frequencies.

3.2   Learning the Classifiers

The result is a 202 dimensional feature vector (texture: 200, shape: 1, PB: 1). Given these features, for each part we can learn a scoring function that best ranks the masks for that part. We also need ground truth scores for each part; from the ground truth segmentation of the parts in the image, we compute the overlap score between the mask and the ground truth labeling, giving a score in $[0, 1]$.

We optimize a WLR energy function that more heavily weights examples that have a high ground-truth score; a separate classifier is learnt for each part. f_i^j is the feature vector associated with the ith mask of the jth image, s_i^j is its ground truth score, and w parametrizes the energy function we wish to learn. For m images with n_j parses each, we have

$$E(w) = (\prod_{j=1}^{m} \prod_{i=1}^{n_j} (\frac{\exp(w^\mathsf{T} f_i^j)}{\sum_{i=1}^{n_j} \exp(w^\mathsf{T} f_i^j)})^{s_i^j})(\exp(\frac{-w^\mathsf{T} w}{\sigma^2})) \qquad (6)$$

$$\log E(w) = \sum_{j=1}^{m} (\sum_{i=1}^{n_j} s_i^j (w^\mathsf{T} f_i^j)) - (\sum_{i=1}^{n_j} s_i^j)(\log(\sum_{i=1}^{n_j} \exp(w^\mathsf{T} f_i^j))) - \frac{w^\mathsf{T} w}{\sigma^2} \qquad (7)$$

where we have added a regularization term $\exp(\frac{-w^\mathsf{T} w}{\sigma^2})$ that is a zero-mean isotropic Gaussian with variance σ^2. If, for each image j, exactly one of the ground truth scores s_i^j were 1 and the rest 0, this would be exactly mutli-class logistic regression (LR). But since there may be several good shapes, we modify the LR.

We can maximize this convex function via a BFGS quasi-newtonian solver using the function value above and the gradient below:

$$\partial_w \log E(w) = (\sum_{j=1}^{m} ((\sum_{i=1}^{n_j} s_i^j f_i^j) - (\sum_{i=1}^{n_j} s_i^j)\frac{\sum_{i=1}^{n_j} f_i^j \exp(w^\mathsf{T} f_i^j)}{\sum_{i=1}^{n_j} \exp(w^\mathsf{T} f_i^j)})) - 2\frac{w}{\sigma^2} \qquad (8)$$

Given a classifier w_p for each part p, we can rank a set of part hypotheses by simply evaluating $w_p^\mathsf{T} f_i^j$ and ranking the scores in descending order. We use the same parsing process, but with a different ranking function (a learned one), as opposed to using just shape. Algorithm 2 provides a pseudocode summary of this training procedure.

3.3 Results Using Shape Only

We present results on the baseball dataset used in [11] and [10]. This dataset contains challenging variations in pose and appearance. We used 15 images to construct shape exemplars, and tested on $|I| = 39$ images. To generate the IDSC descriptors, we used the code provided by the authors of [7]. Boundary contours of masks were computed and resampled to have 100 evenly-spaced points. The IDSC histograms had 5 distance and 12 angle bins (in $[0, 2\pi]$). The occlusion penalty for DP matching of contours was 0.6 * (average match score). For pruning, we used a threshold of 0.95 for the overlap score to decide if two masks were similar (a, b are similar $\iff O(a, b) \geq 0.95$) for the lower body+torso and lower body + torso + head, and 0.75 for all other pruning. In all cases, we pruned to 50 hypotheses.

Because we limit ourselves to shape cues, the best mask (in terms of segmentation and pose estimate) found by the parsing process is not always ranked first; although shape is a very strong cue, it alone is not quite enough to always yield a good parse. Our main purpose was to investigate the use of global shape features over large portions of the body via parsing. We evaluate our results in two different ways: segmentation score and projected joint position error. To the best of our knowledge, we are the first to present both segmentation and pose estimation results on this task.

3.4 Segmentation Scoring

We present our results in terms of an overlap score for a mask with a ground truth labeling. Our parsing procedure results in 50 final masks per image, ranked by

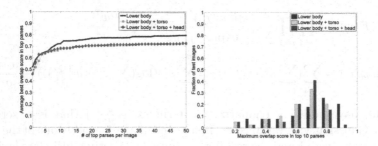

Fig. 7. Left: We plot the average (across all images) of the maximum overlap score as a function of the top k parses. **Right:** We focus on the top 10 parses, and histogram the best overlap score out of the top 10 for each image and region.

their shape score. We compute the overlap score $O(m, g)$ between each mask m and ground truth mask g. We then compute the cumulative maximum overlap score through the 50 masks. For an image i with ranked parses $p_1^i, ...p_n^i$, we compute overlap scores $o_1^i, ..., o_n^i$. From these scores, we compute the cumulative maximum $C^i(k) = \max(o_1^i, ..., o_k^i)$. The cumulative maximum gives us the best mask score we can hope to get by taking the top k parses.

To understand the behavior of the cumulative maximum over the entire dataset, we compute $M(k) = \frac{1}{|I|}\sum_{i=1}^{|I|} C^i(k)$, or the average of the cumulative maximum over all the test images for each $k = 1, ..., n$ ($n = 50$ in our case). This is the average of the best overlap score we could expect out of the top k parses for each image. We consider this a measure of both precision and recall; if our parsing procedure is good, it will have high scoring masks (recall) when k is small (precision). On left in Figure 7, we plot $M(k)$ against k for three different types of masks composed during our parsing process: lower body, lower body+torso, and lower body + head + torso. We can see that in the top 10 masks, we can expect to find a mask that is similar to the ground truth mask desired, with similarity 0.7 on average. This indicates that our parsing process does a good job of both generating hypotheses as well as ranking them.

While the above plot is informative, we can obtain greater insight into the overlap scores by examining all $C^i(k)$, $i = 1, ..., |I|$ for a fixed $k = 10$. We histogram the values of $C^i(10)$ on the right in Figure 7. We can see that most of the values are in fact well over 0.5, clustered mostly around 0.7. This confirms our belief that the parsing process effectively ranks and recalls hypotheses, and that shape is a useful cue for segmenting human shape.

3.5 Joint Position Scoring

We also examinine the error in joint positions predicted by the correspondence of a hypothesis to the nearest exemplar. We take 5 joints: head-torso, torso-left thigh, torso-right thigh, left thigh-left lower leg, right thigh-right lower leg. The positions of these joints are marked in the exemplars, and are mapped to a body hypothesis based on the correspondence between the two shapes. For a joint

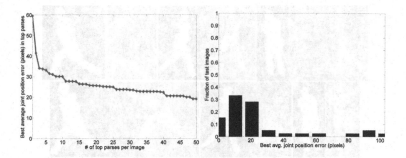

Fig. 8. Left: We plot the average (across all images) of the minimum average joint error in the top k parses as a function of k. **Right:** Taking the top 10 parses per image, we histogram the minimum average joint errors across all the images. We can see that the vast majority of average errors are roughly 20 pixels or less.

with position j in the exemplar, we locate the two closest boundary contour points p, q in the exemplar that have corresponding points p', q' in the shape mask. We compute a rotation, scaling and translation that transforms p, q to p', q', and apply these to j to obtain a joint estimate j' for the hypothesis mask. We compare j' with the ground truth joint position via Euclidean distance. For each mask, we compute the average error over the 5 joints. Given these scores, we can compute statistics in the same way as the overlap score for segmentation. On the left in Figure 8 we plot the average cumulative *minimum* $M(k)$, which gives the average best-case average joint error achieveable by keeping the top k masks. We see again that in the top 10 masks, there is a good chance of finding a mask with relatively low average joint error. On the right in Figure 8, we again histogram the data when $k = 10$.

Lastly, we show several example segmentations/registrations of images in Figure 9. Note that with the exception of the arms, our results are comparable to those of [10] (some of the images are the same), and in some cases our segmentation is better. As noted in [10], although quantitative measures may seem poor (e.g., average joint position error), qualitatively the results seem good.

4   Results with Learning

Figure 10 shows plots of comparisons of shape only and shape, SIFT, and PB. For training, we used an additional 16 baseball images to train the WLR classifiers since training directly on the same images from which the shape exemplars were taken would likely have emphasized shape too much in the learning. 10 fold cross validation was performed with a range of different regularization values σ^2 to avoid overfitting. We tested on 26 baseball images.

We use the same type of plot as in the segmentation scoring previously described, and plot the results over all parts used in the parsing process. We can see that the use of additional features, particularly for the smaller parts, results

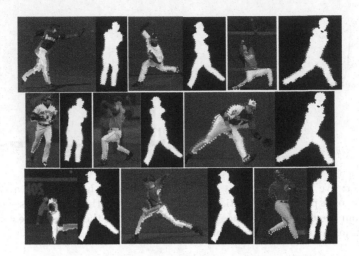

Fig. 9. Body detection results. Ssegmentation has been highlighted and correspondence to the best matching exemplar indicated by colored dots. All parses were the top scoring parses for that image (images are ordered row-major), with the exception of images 4 (2nd best), 8 (3rd best), 6 (3rd best). Some images were cropped and scaled for display purposes only. Full body overlap scores for each image (images are ordered row-major): 0.83, 0.66, 0.72, 0.74, 0.76, 0.70, 0.44, 0.57 and 0.84. Average joint position errors for each image: 12.28, 28, 27.76, 10.20, 18.87, 17.59, 37.96, 18.15, and 27.79.

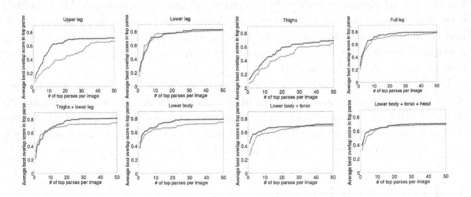

Fig. 10. Average cumulative maximum overlap scores for parsing with and without learning, across all parts. Red curves indicate performance with learning, green represents without learning. Plot methodology is same as that used for left plot in Figure 7.

in substantially better ranking of hypotheses. Even for larger regions, such as the lower body, the additional features have impact; the average cumulative maximum overlap score while keeping the top 10 parses with learning is approximately equal to the score when using only shape and keeping the top 20 parses, implying that we could keep half as many parses to obtain the same quality of result. For the largest regions, shape is clearly the most important cue, since performance is very similar. This also validates our choice of shape as a primary cue.

5   Conclusion

In summary, we present a shape parsing method that constructs and verifies shapes in a bottom-up fashion. In contrast to traditional bottom-up parsing, our scoring functions at each node do not exhibit a subtree independence property; instead, we score shapes against a set of exemplars using IDSCs, which convey global shape information over both small and large regions of the body. We also infuse the process with multiple image segmentations as a pool of shape candidates at all levels, in contrast to typical parsing which only utilizes local image features at the leaf level.

We demonstrated competitive results on the challenging task of human pose estimation, on a dataset of baseball players with substantial pose variation. To the best of our knowledge, we are the first to present both quantitative segmentation and pose estimation results on this task. Note that in general, we need not start parsing with the legs only; it would be entirely feasible to add other nodes (e.g. arms) as leaves.

Further, we use larger shapes (composed of multiple body limbs) than typical pose estimation methods. The notion of layers may also be useful in handling occlusion, as well as describing the shape relation of arms to the torso, since the arms often overlap the torso. Better grouping techniques (ones that introduce fewer hypotheses) are a good idea, since this would save substantial computation.

We also studied the introduction of more traditional features such as PB and SIFT codebook histograms and demonstrated that these features can make an important contribution. Our learning framework is well-suited to the parsing problem since unlike traditional maximum likelihood estimation, we do not explicitly require the best parse for each image. Instead, we simply require a function to provide a ground truth score for each parse hypothesis.

References

1. Belongie, S., Malik, J., Puzicha, J.: Shape matching and object recognition using shape contexts. IEEE Trans. Pattern Anal. Mach. Intell (2002)
2. Borenstein, E., Malik, J.: Shape guided object segmentation. In: CVPR (2006)
3. Cour, T., Benezit, F., Shi, J.: Spectral segmentation with multiscale graph decomposition. In: CVPR (2005)
4. Felzenszwalb, P.F., Huttenlocher, D.P.: Pictorial structures for object recognition. IJCV (2005)
5. Hua, G., Yang, M.-H., Wu, Y.: Learning to estimate human pose with data driven belief propagation. In: CVPR (2005)
6. Lee, M.W., Cohen, I.: Proposal maps driven mcmc for estimating human body pose in static images. CVPR (2004)
7. Ling, H., Jacobs, D.W.: Using the inner-distance for classification of articulated shapes. In: CVPR (2005)

8. Lowe, D.: Distinctive image features from scale-invariant keypoints. In: IJCV (2003)
9. Martin, D.R., Fowlkes, C., Malik, J.: Learning to detect natural image boundaries using local brightness, color, and texture cues. IEEE Trans. Pattern Anal. Mach. Intell (2004)
10. Mori, G.: Guiding model search using segmentation. In: ICCV (2005)
11. Mori, G., Ren, X., Efros, A.A., Malik, J.: Recovering human body configurations: combining segmentation and recognition. In: CVPR (2004)
12. Ramanan, D.: Learning to parse images of articulated bodies. In: NIPS (2007)
13. Ramanan, D., Forsyth, D.A., Zisserman, A.: Strike a pose: Tracking people by finding stylized poses. In: CVPR (2005)
14. Ren, X., Berg, A.C., Malik, J.: Recovering human body configurations using pairwise constraints between parts. In: ICCV (2005)
15. Ronfard, R., Schmid, C., Triggs, B.: Learning to parse pictures of people. In: Heyden, A., Sparr, G., Nielsen, M., Johansen, P. (eds.) ECCV 2002. LNCS, vol. 2350, Springer, Heidelberg (2002)
16. Sigal, L., Black, M.J.: Measure locally, reason globally: Occlusion-sensitive articulated pose estimation. In: CVPR (2006)
17. Zhang, J., Luo, J., Collins, R., Liu, Y.: Body localization in still images using hierarchical models and hybrid search. In: CVPR (2006)
18. Zhu, L., Yuille, A.: A hierarchical compositional system for rapid object detection. In: NIPS (2005)

Introduction to a Large-Scale General Purpose Ground Truth Database: Methodology, Annotation Tool and Benchmarks

Benjamin Yao, Xiong Yang, and Song-Chun Zhu

Lotus Hill Institute of Computer Vision and Information Sciences
EZhou City, HuBei Province, P.R. China
{st.zyyao,xyang.lhi}@gmail.com,
sczhu@stat.ucla.edu
http://www.lotushill.org

Abstract. This paper presents a large scale general purpose image database with human annotated ground truth. Firstly, an all-in-all labeling framework is proposed to group visual knowledge of three levels: scene level (global geometric description), object level (segmentation, sketch representation, hierarchical decomposition), and low-mid level (2.1D layered representation, object boundary attributes, curve completion, etc.). Much of this data has not appeared in previous databases. In addition, *And-Or Graph* is used to organize visual elements to facilitate top-down labeling. An annotation tool is developed to realize and integrate all tasks. With this tool, we've been able to create a database consisting of more than 636,748 annotated images and video frames. Lastly, the data is organized into 13 common subsets to serve as benchmarks for diverse evaluation endeavors.

Keywords: Ground truth Annotation, Image database, Benchmark, Sketch representation, Top-down/Bottom-up Labeling.

1 Introduction

The importance of having an image database containing *ground truth* annotations parsed by humans for a wide variety of images is widely recognized by the machine vision community. The goal of our project is to build up a publicly accessible annotated image database with over 1,000,000 of images and more than 200 categories of objects. Because manual annotation of millions of images is too time-consuming a task for every vision lab to do independently, we hope to compile this centralized database to serve the community's diverse training and evaluation endeavors.

The challenges are many fold, however. First, there is no standard or handy tools for general purpose annotation. We need to find answers to questions like "what to label" and "how to represent common visual knowledge", so that we can develop a suitable labeling tool. Secondly, with the potential scale of the database

A.L. Yuille et al. (Eds.): EMMCVPR 2007, LNCS 4679, pp. 169–183, 2007.
© Springer-Verlag Berlin Heidelberg 2007

in mind, we would like to devise top-down algorithms to guide and speed up annotation, yet it is a non-trivial task to organize and abstract visual knowledge from labeled images for this purpose. In addition, to make the database general enough to be used for different evaluation tasks, we need to build up benchmarks for a variety of visual patterns, thus we need to define equivalent distances over different spaces.

In this paper, we present our efforts in confronting these challenges, and show examples of data from our database. By consulting with several vision groups, we have gathered a consensus on the commonly desired information for labeling in three levels:

- *Scene Level*: Global geometry information, scene category (indoor/outdoor), events and activities;
- *Object Level*: Hierarchical decomposition, object segmentation, sketching and semantic annotation;
- *Low-middle Level*: Contours types (object boundary, surface norm change or albedo change), Amodal completions, Layered representation (2.1-D), etc.

According to the requirements listed above, we developed a novel annotation tool, which integrates several functional modules designed for specific task(s). We show that by properly combining these functions, the tool can perform customized annotation tasks blending all kinds of information. Moreover, the tool is associated with an *And-Or Graph knowledgebase*[4], which organizes and summarizes labeled visual knowledge in a universal way.

To the best of our knowledge, there has not been much previous work on building a large scale general purpose database. However, there are many special purpose databases publicly available, which provide us with some valuable insight. Here we only list those most related to our database:

- **LabelMe** database of MIT-CSAIL [Torralba et al.[11]]. This is the most similar dataset to ours. Images in this database are of natural images and cluttered scenes and contain objects under multiple views/poses. Its current limitation is that only the rough boundary of the object is annotated, as opposed to fine segmentation or hierarchical decomposition.
- **CalTech 101 and 256** [Fei-Fei, Griffinet et al.[6,7]]. These two datasets provide a great number of diverse object classes. Its limitations are that the objects are not positioned in real scenes, are centered in the image, and have a limited number of viewpoints.
- **The Berkeley Segmentation Dataset** [Martin and Fowlkes[9]]. This dataset is a pioneer effort on large scale ground truth annotation on general natural images. It makes a valuable contribution to error control and benchmark, but it is limited in regards to scale and content.
- **UA (Arizona) localized semantics dataset** [Barnard et al.[1]]. This dataset provides a good semantic annotation standard based on the data of Berkeley dataset.

2   Methodology: Representation and Organization of Labeling Information

2.1   Region Segmentation and Semantic Annotation

Segmentation is the foundation of image annotation. Common annotation task requires *object level* segmentation. For example, when we see the *water* in Figure 1, we tend to interpret it as a single *thing*, even though it is actually composed of several disconnected image areas. Therefore, in our representation framework, there are two levels of data. At the lower level, we define a *region* as an image area with closed boundary. At the higher level, we define an entity represent for object named *PO* (as short for "Physical Object"). A *PO* can represent any meaningful entity in an image, such as a scene, an object, an object part, a texture area or a text block. In Figure 1, all the regions of the road are aggregated into a single *PO* (masked with a same color). *PO* is a core element in our framework. It is composed of several lower level elements including regions and sketch graph (section 2.2). It also corresponds to a node in the parsing graph (section 2.3).

Fig. 1. This figure shows the result of object level segmentation. Sub-figure on the left is the original image. Sub-figure on the right is the segmentation mask. The *region* of an object (e.g. the road surface) may be composed of several disconnected image areas.

It is also worth mentioning that our annotation tool enables users to do "fine" segmentation and thus can output accurate object boundary, in comparison to coarse outlines provided by other datasets. This is especially important when the objects are small and in great numbers. A typical example is the annotation task of aerial images. As shown in Figure 2, the average number of *PO*s in aerial images is over one hundred. Such task can hardly be accomplished without fine segmentation.

Furthermore, to make the naming convention general and accurate, we use WordNet [10] as the reference. The following rules are adopted when a semantic

Fig. 2. This figure illustrates the segmentation and annotation results of a typical aerial image (a parking area)

Fig. 3. Sketch graph representation of object: The sketch graph can capture most essential perceptual information of an object, like the structural information of chair or the folds, sewing lines, albedo/lighting changes of cloth

name is chosen for a *PO*: (1) Words of object name correspond to their Word-Net definition. (2) The sense in WordNet (if multiple) should be mentioned as word[i], where i is the sense number in WordNet. (3) Synonym or description of same object in a different way should be given as additional entry(s) (e.g. *grass, ground*). (4) For parts of an object, add the object name as prefix (e.g. *horse:head*). (5) Descriptive words can be added to provide further information(e.g. *[white]human, [frontal view]car*).

2.2   Sketch Graph Representation

Since segmentation only provides the outer boundary of an object, we adopt a sketch graph representation like the Primal Sketch model [8] to record structural features inside the object boundary. The sketch is composed of a set of strokes (long curves) aligned through landmarks. It is not required to have a closed boundary (see the small figures on the right side of each object in Figure 3). This figure demonstrates that the sketch graphs can capture essential perceptual information of the object, such as the structural information of the chair and the folds, sewing lines, albedo/lighting changes of cloth. We believe that it is valuable to record sketch information in addition to the object boundary. Another example is showed in Figure 6.

Sketch graph representation is further augmented to include low-middle level vision elements by adding attributes to each curve (*attributed curves*) . As illustrated in Figure 4, the curves of object are classified by different colors to represent for *occlusion, surface normal change, lighting/albedo change* respectively. Besides, model/amodel completions and 2.1D layered representation are clearly defined with attributed curves (illusory contours, Figure 5).

Fig. 4. Attributed curves. "Winnie Pooh" is labeled with three types of curve attributes: *Surface*: curves generated by surface norm change; *Boundary*: curves on object boundary; *Shadow*: curves generated by shadow.

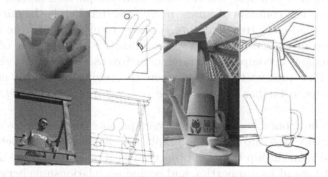

Fig. 5. Illusory contours for model/amodel completion and 2.1D layered representation. This figure shows how to label 2.1D layered representation and curve completions in a 2D image. The green line stand for the illusory(occluded) contours.

Figure 6 uses a high resolution sketch to do artistic rendering. We can see that the sketch graph image on the right welly depicts the appearance, clothing, and even expression of the little girl in the left image, with little distortion. Therefore, with such kind of *sketch*, people can easily do image rendering by filling in color patterns with all kinds of artistic styles. Moreover, the sketch graph is also a flexible and expressive tool to localize image features. As shown in Figure 11, we use the landmarks of sketch graph to represent the body pose and limb directions of human.

Fig. 6. Artistic rendering. *Sketch* captures the appearance, clothing, and expression of the child in the image with little distortion. Based on this, people can easily do image rendering with all kinds of artistic styles.

2.3   Hierarchical Decomposition and Parsing Graph

Compositionality is a common phenomenon in natural images. As shown in Figure 7(a), the image is composed of objects (car and background), while the car is composed of several parts and, if the resolution of the image were higher, we could further decompose the car body into sub-parts. Therefore it is natural to organize the contents of image in a hierarchical style. In our annotation framework, we adopt a tree structure (parsing tree) to record the hierarchical decomposition of objects (similar to the *Image Parsing* concept of Zhu et al. [12]). The image is decomposed hierarchically from scene to object then to parts and so on. This process terminates when the resolution of leaf nodes is too low to decompose further. Nodes of the parsing tree are the *PO*s mentioned in 2.1.

By adding horizontal connections between parsing tree nodes, we further augment the parsing tree into a parsing graph. The horizontal links between nodes represent the relationship between object/parts. For example, the dashed lines in Figure 7(b) stand for supporting and occluding relationships between objects in the image.

Another example of a parsing graph is shown in Figure 8. In this figure, the bike is labeled at two different scales. At the low resolution, the bike is labeled as a whole. At the high resolution, it is further decomposed into parts with details. This process represents the scaling phenomena in vision.

2.4   And-Or Graph Knowledgebase and Bottom-Up/Top-Down Labeling Procedure

To build up an annotated image database with millions of images, it is important to abstract and organize visual knowledge from labeled images. It is

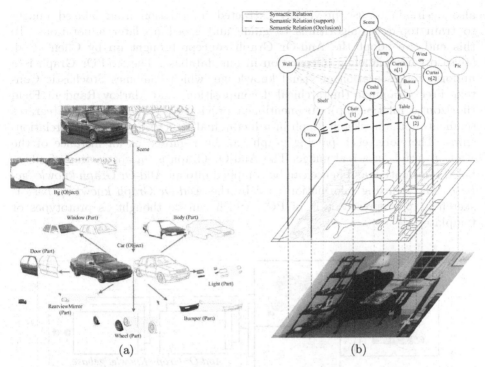

Fig. 7. (a) A parsing tree: The image is hierarchically decomposed from scene to object then to parts like a tree. The node of the parsing tree is *PO* (section 2.1), which includes both region and sketch graph. (b) A parsing graph. Solid lines represent hierarchical decomposition. Dashed lines represent spatial relationships (supporting and occluding in this figure).

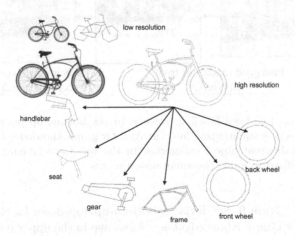

Fig. 8. Multiple scale/resolution annotation: The bike in the figure is labeled at two different scales. At low resolution, the bike is represented by a single sketch graph. At high resolution, it is further decomposed into parts.

also desirable to use previously extracted information from labeled images
to train top-down algorithms to guide and speed up later annotations. To
this end, we adopt the And-Or Graph concept brought up by Chen et al.
[4] to organize labeling information in our database. The And-Or Graph is a
uniform representation of visual knowledge, which combines Stochastic Context Free Grammar (hierarchical decomposition) and Markov Random Field
(horizontal relationship). As mentioned previously, *Parsing Graph* integrates
segmentation and sketch graph in a hierarchial structure with spatial relationships. Therefore, each parsing graph can be regarded as an instance of the
And-Or graph in a real image. The And-Or Graph is an abstraction of all labeled information and hence can be compiled into an *And-Or Graph knowledgebase*. The indexing information used by the *And-Or Graph knowledgebase* is
sketch graph representation of POs, which can be thought as prototypes or
templates.

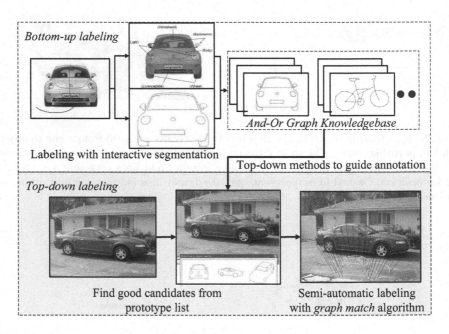

Fig. 9. Bottom-up/top-down labeling process: In the bottom-up process, labeled information of objects is summarized into the And-Or graph knowledge base and stored
as prototypes of different object categories. In the top-down labeling process, these
templates are utilized to guide the annotation process.

As shown in Figure 9, we devise a bottom-up/top-down labeling procedure
with the *And-Or Graph Knowledgebase*. As shown in the upper part of Figure 9,
when a new category or novel instance is input, the object is labeled manually
or with interactive method such as GraphCut[3]. The graph representation is
then stored into the *And-Or Graph Knowledgebase* as templates. This process

is called bottom-up labeling. When there are sufficient templates recorded to cover the inter-class variety of an object category, the templates can be utilized in a downward direction. First, good candidates are automatically selected from the template pool in the *And-Or Graph Knowledgebase*. After the best template is selected (manually or automatically), match algorithm based on sketch graph representation such as *Shape Context*[2] or *Graph Match*[1] is used to fit the template onto object. Thus, the labeling procedure is speeded up dramatically. The top-down labeling process[2] is shown in the bottom part of Figure 9. Through this bottom-up/top-down labeling procedure, visual knowledge is accumulated towards the final goal of automated image parsing and labeling.

2.5 Global 3D Geometry Information

Global 3D geometric information is very important for scene understanding. Our annotation tool provides a module to label Global coordinate frame and perspective projection parameters in 2D image. As shown in Figure 10, the pairs of yellow lines in the figure are perspective parallel lines in real world. *Vanishing points* can be computed by a group of parallel lines. The *horizon line* is easily derived by connecting two *vanishing points*. The *ground plane* can be created using two pairs of parallel lines, which is same for the vertical planes.

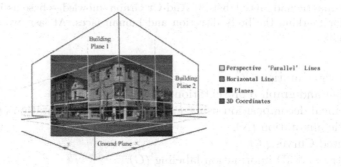

Fig. 10. This figure illustrate the labeling results of global geometry information. Yellow lines are *Perspective 'Parallel' Lines*. *Vanishing Points* can be computed by the intersection of parallel lines. Green line is *Horizon Line*. Black frame is ground plane, blue frame stands for vertical planes. Red axis is global coordinate.

3   Annotation Tool: Integrating Functional Modules

In previous section, we present the representation and organization methodology of labeling information with examples from our database. In our annotation tool, we realize the functions with seven modules:

---

[1] Discussed in a companion paper in CVPR07.

[2] The automatic algorithms exploited in top-down labeling process is detailed discussed in a another companion paper submitted to ICCV07.

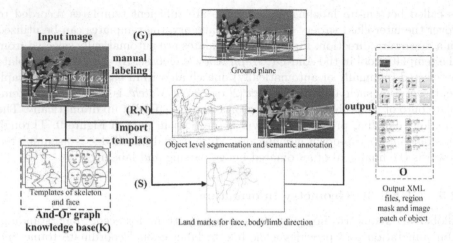

Fig. 11. Integration of functional modules. The task is to label images for sports activities. Four aspects of information are required: 1) ground plane and horizon line 2) segmentation of objects and semantic labels 3) layer labels for foreground and background 4) body direction and faces of athletes. To finish this task we integrate three modules: 1) geometry and 3D scene label module(G) for marking out ground plane on the image 2) region segmentation and annotation module(R) for labeling segmentation of objects, semantic and layer labels 3) And-Or Graph knowledge base assistant label module(S) for marking the body direction and human faces. At last, we output the labeling results.

1. Region segmentation *(R)*.
2. Sketching and graph representation *(S)*.
3. Hierarchical decomposition using *Parsing Graph* representation *(P)*.
4. Semantic annotation *(N)*.
5. Attributed Curves *(A)*.
6. Geometry and 3D information labeling *(G)*.
7. *And-Or Graph Knowledgebase (K)*.

In this section, we study the case of a practical image annotation task to illustrate how we integrate several functional modules to perform a specific annotation task. The task is to create an annotation subset for sports activities. It requires visual knowledge of following aspects:

1. Ground plane and horizon line of image
2. Segmentation of objects and semantic labels
3. Layer labels for foreground and background
4. Body direction and faces of athletes.

It is obvious that requirement (1) is a global coordinate frame and perspective projection parameters labeling issue; Requirement (2) and (3) can be reduced to region segmentation and semantic annotation. Since there are many templates of face and body skeleton in our *And-Or Graph knowledgebase*, we can fulfil requirement (4) by a top-down procedure introduced in section 2.4.

As shown in Figure 11, we integrate five modules to finish this task. First, we use the 3D Geometry module to label ground plane and horizon line on the image(G). Second, we use the segmentation and annotation to perform object level segmentation (R,N). Third, we use templates of human body skeleton and faces from And-Or Graph knowledge base to mark the body direction and faces(S,K). Integrating these five modules, we derive a customized annotation procedure.

4   Database Statistics, Subsets and Benchmarks

There are 3,927,130 *POs*, 636,748 images and video frames in our database at present, and the number is growing everyday. As illustrated in Figure 14, widespread images have been annotated.

To serve the community's dire needs for *dataset* for training and evaluation, we have organized 13 common subsets from our database to serve as benchmarks. Figure 12 illustrate the typical image collection of these subsets. Table 1 illustrates more detailed statistics of these subsets, including image number, class number, visual knowledge included (functional modules involved), etc. These subsets are:

1. *Common scene classification:* A subset for general scene classification (Row1 in Figure 12, Row1 in Table 1).

   classes: Images are categorized into 14 classes including: *bathroom, bedroom, cityview, corridor, hall, harbor, highway, kitchen, living room, office, parking, rural, seashore, street.*

   labeling information: 3D Geometric description of scene, Object-level Segmentation (objects included in a scene, such as sky, tree, pedestrians, cars), Semantic Annotations, Parsing graph is used to perform scene decomposition and record occluding relation between objects.

2. *Activity and Events classification:* A subset for activity and event classification (Row2 in Figure 12, Row2 in Table 1).

   classes: Images are categorized into 16 classes including: *dinning, harvest, lecture, meeting, shopping, badminton, bocce, croquet, high jump, hurdles, iceskate, polo, rowing, snowboarding, RockClimbing, sailing.*

   labeling information: Similar with the common scene classification subset. Since judgement of events and activities highly related with human in the scene, special annotations of human body are added (for specific, the face, body and limb directions, the case in Section 3 is an example).

3. *Aerial images:* A subset for aerial image segmentation (Row3 in Figure 12, Row3 in Table 1).

   classes: Images are grouped into 10 classes: *airport, business, harbor, industry, intersection, mountain, orchard, parking, residential, school.*

   labeling information: Segmentation of main objects in a scene, such as building roof, parking area, single car and road surface.

Bedroom   Bathroom   Office   Highway   Rural   Harbor

Meeting   Lecture   Sports   Harvest   Dinning   Shopping

School   Industry   Intersection   Business   Airport   Marina

Airplanes of multi-view   Cars of multi-view

Bicycles of multi-scale   Motorcycles of multi-scale

Backpack   Camera   Mug   Horse   Squirrel   Cat

Natural images (2.1D Sketch)

Texts in different languages   Video frames

Different age   Different pose   Different expression

Fig. 12. Exemplary images of 13 subsets for benchmarks. Subset1: scene classification (Row 1); Subset2: events and activity (Row 2); Subset3: aerial images (Row 3); Subset4-6: 20 categories of popular objects, which are in multiple views, scales and resolutions (Row 4-5); Subset7-9: Generic objects of 200 categories (Row 6); Subset10: Human faces with different age, pose, expression and etc. (Row 9); Subset11: Surveillance video frames (Row 8); Subset12: Text (Row 8); Subset13: Natural images for 2.1D layered representation.

Fig. 13. Popular object subsets: 20 categories of common objects are selected and labeled with multiple views, scales and resolution. The first and third lines of this figure are original image patches. The second and fourth lines of this figure are sketch representation of objects.

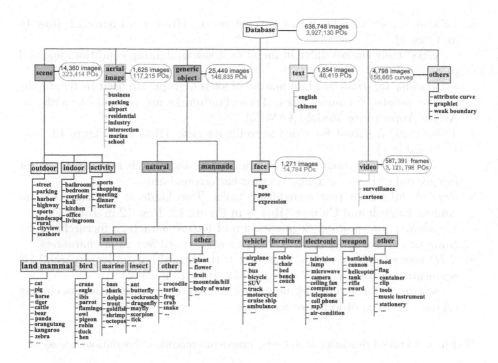

Fig. 14. This tree list is a comprehensive inventory of our dataset. From root node to leaf nodes, the entire set is decomposed into subsets and categories hierarchically. Terminal nodes without boxes are corresponding to the most detailed categories. The numbers in arc angle boxes are the statistics relatively. The *PO* in the figure means Physical Object mentioned in section 2.1.

4. *Popular objects:* Three subsets for object categorization, object bounding box localization and object outline detection. These objects are putted in multiple views and resolutions. (Row 4,5 in Figure 12, Row 4-6 in Table 1).

   classes: airplane, bicycle, bucket, chair, clock, couch, cup, frontcar, glasses, hanger, keyboard, knife, lamp, laptop, monitor, motorcycle, sidecar, table, teapot, watch.

   labeling information: Both segmentation and sketch representation of object are labeled. Objects are labeled under two or three resolutions. At the high resolution, object is decomposed hierarchically into parts and sub-parts.

   Figure 13 shows the segmentation and sketch representation of both high and low resolution of this subset.

5. *Generic object:* Three subsets for object categorization, bounding box localization and outline detection. (Row 6 in Figure 12, Row 7-9 in Table 1).

   labeling information: Similar with popular objects, except that only one resolution and single view is labeled. Because many objects are rarely seen, thus it is very hard to collect enough images for different views and resolutions.

6. *Face:* A subset for human face categorization. (Row 9 in Figure 12, Row 10 in Table 1).

   class: Four classes differ in facial expression, lighting condition, age and pose.

   labeling information: Landmarks of sketch graph are used to record the feature points of human faces. These landmarks are compatible with the Active Appearance Model (AAM)[5].

7. *Video clips:* A subset for video surveillance task. (Row 8 in Figure 12, Row 11 in Table 1).

   labeling information: Both segmentation and sketch representation for foreground objects. Segmentation for background areas.

8. *Text:* A subset for text recognition tasks. Two kinds of languages are included: English and Chinese (Row 8 in Figure 12, Row 12 in Table 1).

   labeling information: Segmentation of letter(character), hierarchical decomposition from *text block* to *lines* to *words* until *letters or characters*.

9. *2.1D layered representation:* A subset with natural images for general 2.1D segmentation tasks (Row 7 in Figure 12, Row 13 in Table 1).

   labeling information: Segmentation and sketch representation, occluding relation between objects are recorded.

Table 1. Detailed Statistics of Subsets, functional module abbreviations see sec. 3

Subsets		Class Num	Functional Modules							Image Num
			R	S	P	K	N	A	G	
Scene Classification	Common scene	14	V		V		V		V	9637
	Activity	16	V	V	V	V	V		V	4723
	Aerial	10	V		V	V	V			1625
Popular Object	Categorization	20					V			9585
	Bounding Box	20	V				V			9585
	Outline	20	V	V	V	V	V	V		9585
Generic Object	Categorization	200					V			15864
	Bounding Box	200	V				V			15864
	Outline	200	V	V		V	V	V		15864
Face		4	V	V	V	V	V			1271
Video		1		V	V		V			587391
Text		2	V	V	V	V	V			1854
2.1D Sketch		1		V					V	1446

5   Conclusions and Future Works

In this paper, we present a new large-scale general purpose ground truth image database. We bring up the representation and organization methodology of generally desired labeling information. We also demonstrate that, by properly combining the functional modules of our annotation tool, one can perform annotation tasks blending all kinds of desired information. Besides, a bottom-up/top-down labeling framework is proposed using the And-Or Graph knowledgebase to speed up labeling process. Lastly, thirteen subsets of labeled data are organized to serve as standard *Benchmarks*. Further investigations are needed on the automatic algorithms related with bottom-up/top-down labeling procedure to realize the long term goal of semi-automatic labeling and automatic labeling. Besides, to set up benchmarks for image understanding (rather than simple classification), further investigation are needed on defining equivalent distance over diverse visual spaces.

Acknowledgements. The project is supported by: National 863 project, Contact No. 2006AA01Z121; National Natural Science Foundation of China, Contact No. 60672162; National Natural Science Foundation of China, Contact No. 60673198; National 863 project, contact No. 2006AA01Z339.

References

1. Barnard, K., Fan, Q., et al.: Evaluation of localized semantics: Data, methodology, and experiments. University of Arizona, Computing Science, Technical Report,TR-05-08. (September 2005)
2. Belongie, S., Malik, J., Puzicha, J.: Shape matching and object recognition using shape contexts. IEEE Trans. Pattern Recognition and Machine Intelligence, 509–522 (April 2002)
3. Boykov, Y., Veksler, O., Zabih, R.: Fast approximate energy minimization via graph cuts. IEEE Transactions on Pattern Analysis and Machine Intelligence 11, 1222–1239 (2001)
4. Chen, H., Xu, Z.,J., Zhu, S.: Composite templates for cloth modeling and sketching. In: CVPR'2006, pp. 943–950 (2006)
5. Cootes, T.F., Taylor, C.J.: Active appearance models. In: ECCV'1998 (1998)
6. Fei-Fei, L., Fergus, R., Perona, P.: One-shot learning of object categories. IEEE Trans. Pattern Recognition and Machine Intelligence, pp. 594–611 (April 2006)
7. Griffin, G., Holub, A., Perona, P.: The caltech 256. Caltech Technical Report
8. Guo, C., Zhu, S., Wu, Y.: Primal sketch: Integrating texture and structure. Computer Vision and Image Understanding (2006)
9. Martin, D., Fowlkes, C., et al.: A database of human segmented natural images and its application to evaluating segmentation algorithms and measuring ecological statistics. In: ICCV'2001, p. 416 (2001)
10. Miller, F.C., Tengi, R., Wakefield, P., et al.: Wordnet - a lexical database for english (1990)
11. Russel, B.C., Torralba, A., Murphy, K.P.: Labelme: a database and web-based tool for image annotation, M.I.T., C.S. and A.I. Lab Techinical Report, MIT-CSAIL-TR-2005-056 (September 2005)
12. Tu, Z., Chen, X., Yuille, A.L., Zhu, S.-C.: Image parsing: Unifying segmentation, detection and recognition. Int'l. J. of Computer Vision, Marr Prize Issue (2005)

An Automatic Portrait System Based on And-Or Graph Representation

Feng Min[1,4], Jin-Li Suo[2,4], Song-Chun Zhu[3,4], and Nong Sang[1]

[1] IPRAI, Huazhong University of Science and Technology, China
[2] Graduate University of Chinese Academy of Sciences, China
[3] Departments of Statistics and Computer Science, University of California
[4] Lotus Hill Institute for Computer Vision and Information Science, China
fmin.lhi@gmail.com, jlsuo.lhi@gmail.com, sczhu@stat.ucla.edu,
nsang@hust.edu.cn

Abstract. In this paper, we present an automatic human portrait system based on the And-Or graph representation. The system can automatically generate a set of life-like portraits in different styles from a frontal face image. The system includes three subsystems, each of which models hair, face and collar respectively. The face subsystem can be further decomposed into face components: eyebrows, eyes, nose, mouth, and face contour. Each component has a number of distinct sub-templates as a leaf-node in the And-Or graph for portrait. The And-Or graph for portrait is like a "mother template" which produces a large set of valid portrait configurations, which is a "composite templates" made of a set of sub-templates. Our approach has three novel aspects:(1) we present an And-Or graph for portrait that explains the hierarchical structure and variability of portrait and apply it into practice; (2) we combine hair, face and collar into a system that solves a practical problem; (3) The system can simultaneously generate a set of impressive portraits in different styles. Experimental results demonstrate the effectiveness and life-likeness of our approach.

Keywords: And-Or Graph, Non-Photorealistic Rendering, Face.

1 Introduction

A portrait is a concise yet expressive representation of each person. A life-like portrait should not only resemble the appearance of an individual, but also capture the spirit of an individual.

It is a difficult and challenging task to automatically generate a life-like portrait from a given face image. There have been a few attempts to interactively or automatically generate a stylistic facial sketch by observing images drawn by artists. For example, a few template-based facial caricature systems were developed by Koshimizu et al [1] and Li et al[2], which simply linked face feature points using image processing methods and produced stiff sketches. A number of example-based approaches have been proposed for sketch. For instance, Librande[3] developed an example-based character drawing system.

A.L. Yuille et al. (Eds.): EMMCVPR 2007, LNCS 4679, pp. 184–197, 2007.
© Springer-Verlag Berlin Heidelberg 2007

Freeman et al [4] presented an example-based system for translating a sketch into different styles. With the development in texture synthesis[5] and face hallucination[6], Chen et al [7,8] developed an example-based face sketch generation system. They used inhomogeneous non-parametric sampling to capture the statistical likelihood between the sketch and original image, and fit a flexible template to generate the sketch. However, this method is the mapping from image to sketch, it is difficult to change the style of portrait.

Inspired by recent development in generative model[16,9], Xu et al [10] presented a high resolution grammatical model for face representation and sketching. They adopted three-layer generative model and coarse-to-fine computation to generate fine face sketches. Chen et al [11,12] presented a generative model of human hair for hair sketching and composite templates for cloth modeling and sketching. These models can generate vivid sketches of hair, face and cloth, however, no one combines these separate parts into one portrait system.

In this paper, we present an automatic portrait system based on the And-Or graph representation. We build an And-Or graph for portrait which can account for the variability of portraits by separating the structure and style of portrait. Additionally, we build a set of sketch dictionaries for portrait components in different styles. With the And-Or graph for portrait and sketch dictionaries, we can automatically generate a set of life-like portraits in different styles from a given face image as shown in Figure 10, 11.

The rest of this paper is organized as follows. We introduce the And-Or graph for portrait in Section 2. The automatic portrait system is presented in Section 3. Experimental results are shown in Section 4. In Section 5, we will discuss the limitations of our approach and the future work.

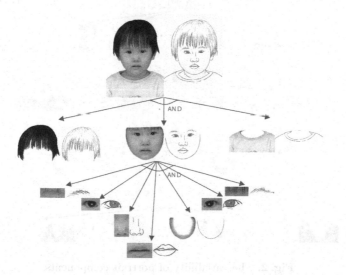

Fig. 1. Decompose portrait into components

2   The And-Or Graph for Portrait

As is shown in Figure 1, a portrait includes three parts: hair sketch, face sketch, and collar sketch. The face sketch can be further decomposed into eyebrows sketch, eyes sketch, nose sketch, and mouth sketch. All of these components form a rather fixed spatial configuration within the face contour. At the same time, variability still exists in portraits, not only globally such as different views or posture, but also locally such as open/closed mouth or different types of hair. For example, Figure 2 shows various hair, collar, eyebrows, eyes, nose, mouth, and face contour together with their corresponding sketches.

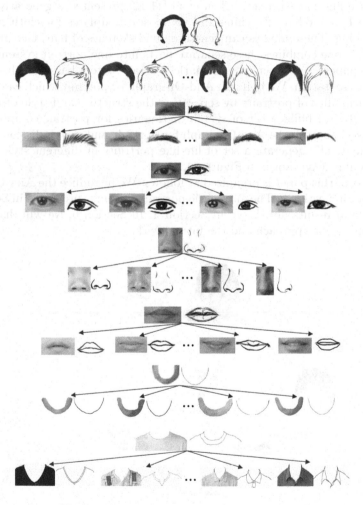

Fig. 2. The variability of portrait components

We need to take three categories of variability into account. (1) Topological configuration, such as V/T collar; (2) Geometric deformation, such as thick/thin

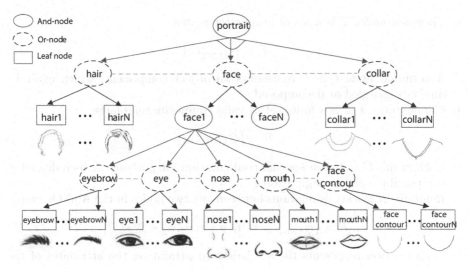

Fig. 3. The And-Or graph for portrait

eyebrows; (3) Photometric variabilities, such as light/dark hairs. In order to account for these variabilities, we propose an And-Or graph for portrait.

As is shown in Figure 3, each terminal(leaf) node represents a component or sub-template. Different sub-templates in the same category are represented by distinct subgraphs. The non-terminal nodes are either And-nodes whose children must be chosen jointly or Or-nodes of which only one child can be selected. An And-node is an instance of the semantic meaning, which expands the regularized configurations. An Or-node only has the syntax meaning, which is a switch between alternative sub-configurations. So the And-nodes, the Or-nodes, and the leaf nodes constitute an And-Or graph. The And-Or graph has horizontal dash lines to specify the spatial relations and constrains among the nodes. For example, hair is above the face, while collar is below the face. Eyebrows, eyes, nose, mouth are within the face contour. These relations and constrains help to link the components together to form a valid representation. Thus the And-Or graph is like a "mother template" which produces a set of valid portrait configurations -"composite templates" that are made of a set of sub-templates.

As a matter of fact, an And-Or Graph is a context sensitive grammar [13], which can be regarded as a 5-tuple

$$G_{and-or} = < N = U \cup V, T, \Sigma, R, A > \tag{1}$$

Each element is explained as below:

1. *Non-terminal nodes* $N = U \cup V$ includes a set of And-nodes and Or-nodes.

$$U = \{u_1, ..., u_n\}; V = \{v_1, ..., v_n\} \tag{2}$$

   An And-node $u \in U$ represents a composite template, which is composed of a set of sub-templates. An Or-node $v \in V$ is a switch pointing to a set of alternative sub-templates.

2. *Terminal nodes* T is a set of atomic templates.

$$T = \{t_1, ..., t_n\} \tag{3}$$

A terminal node $t \in T$ represents an object component, which can't be further expanded or decomposed.

3. *Configurations* Σ is a finite set of valid composite templates.

$$\Sigma = \{G_1, ..., G_n\} \tag{4}$$

Each graph $G \in \Sigma$ is a specific configuration for portrait. Σ includes all of the possible valid configurations.

4. *Relations* R is a set of relations between any two nodes in the And-Or graph.

$$R = \{r_{(n_i, n_j)} = < n_i, n_j >; n_i, n_j \in N \cup T\} \tag{5}$$

Each relation represents the statistical constraint on the attributes of the nodes.

5. *Attributes* A is a set of attributes for each node in the And-Or graph. For the terminal nodes $t_i \in T$, A is a set of photometric and geometric transforms.

$$A = \{(A^{pho}_{(t_i)}, A^{geo}_{(t_i)}); i = 1, 2, ..., n\} \tag{6}$$

3   The Automatic Portrait System

In order to automatically generate a set of life-like portraits in different styles, we need a large number of sub-templates in different styles as the terminal nodes. We asked artists to draw the portrait in different styles on top of the original image with a different layer in PhotoShop. Then we manually decompose the portrait into a set of components as is shown in Figure 1. By collecting these components, we build a large database for hair, collar, eyebrows, eyes, nose, mouth, face contour and their corresponding sketches in different styles. From this large database, we can extract different types of hair, face components, and collars to build a set of sketch dictionaries in different styles. It is also convenient to change the style of portrait by changing the sketch dictionaries.

Based on the And-Or graph for portrait, we divide the portrait system into three subsystems: hair subsystem, face subsystem and collar subsystem. The face subsystem is the key part which can be further decomposed into face components: eyebrows, eyes, nose, mouth, and face contour. We detect face rectangle using a boosted cascade of features[14]. Then we adopt a local Active Appearance Model (AAM) [15,16,17] for each face component. The hair subsystem and the collar subsystem are connected to the face subsystem. We first find the hair contour and the collar contour by the spatial relationships. Then we use shape matching to find the best matched shape with shape contexts [18]. Last we warp the best matching shape to the corresponding shape contour by the Thin Plate Spline(TPS) [19,20] model. The details of the three subsystems will be presented in the following sections.

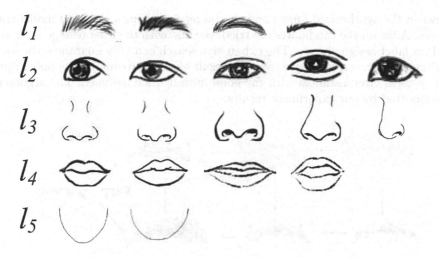

Fig. 4. The different types of facial components extracted from the database

3.1   Face Subsystem

Because of the diversity in the database, we categorize the face components into four types of eyebrows, five types of eyes, five types of nose, four types of mouth, and two types of face contour shown in Figure 4. Each type of component has its own AAM. The AAM representation includes a set of principle components for the geometric deformations and a set of principle components for the photometric variabilities after aligning the landmarks. Therefore we build a dictionary for all (4+5+5+4+2=20) components and their corresponding sketches.

$$\Delta_I^{cp} = \{I_{geo}^{cp,i}, I_{pho}^{cp,i}, i = 1, 2, ..., 20\} \tag{7}$$

$$\Delta_S^{cp} = \{S_i^{cp}, i = 1, 2, ..., 20\} \tag{8}$$

where $I_{geo}^{cp,i}$ and $I_{pho}^{cp,i}$ respectively denote the geometric and photometric models of AAM for component i. S_i^{cp} denotes the sketch of component i.

The 20 component models are learned in a supervised manner from the database. The selection of the model for each component is controlled by five switch variables $l_j \in \{1, 2, .., 20\}, j = 1, 2, ..., 5$. Because of the symmetry of the two eyes and eyebrows, there are only five variables l_1, l_2, l_3, l_4, l_5 which respectively denoted eyebrows, eyes, nose, mouth, and face contour. The inference of the face sketch can be represented as:

$$p(S^{cp}|I; \Delta_S^{cp}, \Delta_I^{cp}) = \prod_{j=1}^{5} p(l_j) \cdot \prod_{j=1}^{5} p(S_{l_j}^{cp}|I; \Delta_S^{cp}, \Delta_I^{cp}) \tag{9}$$

The inference of switch variables l_j is done through a exhaustive way. We firstly pick node j as a candidate from the AAM. Then we obtain the residue between the synthesized image and the target image until the local model converges. After all the candidates are tried, the one with the least residue is chosen and its label assigned to l_j. The exhaustive search can only guarantee the local optimum. However, we argue that the result shall approximate the global optimal in most circumstances with the good initialization assumed. The argument is supported by our experiment results.

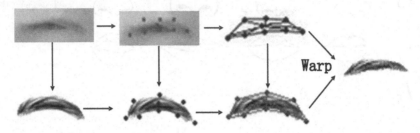

Fig. 5. Render the sketch of eyebrows through its corresponding sketch

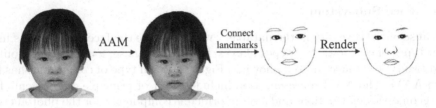

Fig. 6. The flow chart of face subsystem

Once we have determined the switch variables l_j for each component, we can render the sketch of each component through its corresponding sketch $S_{l_j}^{cp}$. Taking eyebrows as an example shown in Figure 5, we can extract the accurate shape and associated texture information using the best fitting AAM from a new eyebrow. Then we define the same triangular mesh over the landmarks and warp each triangle separately from source to destination to get the vivid sketch of eyebrow.

The flow chart of face subsystem is shown in Figure 6. Firstly, we infer switch variables l_j through a exhaustive way. Secondly, we extract the accurate shape using the AAM for component l_j and connect landmarks. Finally, we render the sketch of each component through its corresponding sketch $S_{l_j}^{cp}$.

3.2   Hair and Collar Subsystem

There are various types of collar and hair. we select some simple and typical types of collar and hair as templates from the database as is shown in Figure 7, 8. We use Δ_S^h and Δ_S^c to represent the various sketch of hair and collar respectively.

Fig. 7. The different types of collars extracted from the database

Fig. 8. The different types of hairs in two styles extracted from the database

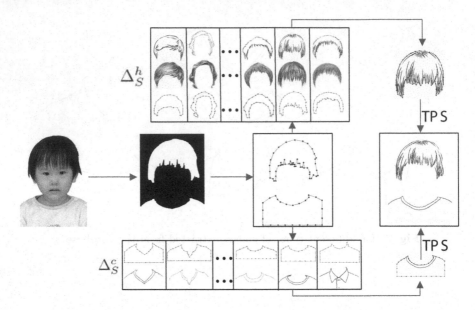

Fig. 9. The flow chart of the hair and collar subsystem

Hair and collar cannot be handled in the same way as the face due to two reasons. (1) Because they have many styles and are not structured in the same regular way as the faces, building a model is not feasible. (2) There is no clear correspondence between source shape contours and destination shape contours. Without a correspondence, the triangular warp is not valid. Therefore, we adopt the method of shape contexts[18] to solve these problems. Shape contexts can measure similarity between shapes and find the correspondence between two similar shapes. Shapes are represented by a set of points sampled from the shape contours. In fact, the shape context is a descriptor for a reference point that captures the distribution of the remaining points with respect to the reference. As the corresponding points on similar shapes will have similar shape contexts, we can find the correspondence between two similar shapes by solving an optimal assignment problem. To measure similarity between shapes, shape context distance is defined as the symmetric sum of shape context matching costs over best matching points.

In order to choose the closest template using shape contexts, we should find the shape contour of hair and collar from a given image first. We can find the overall face region by skin detection. To reduce the influence of luminance on skin color, we transform the given image from RGB space to YCbCr space and only use Cb and Cr to detect skin. Because we have found face rectangle in face subsystem, we can get the approximate mean of skin color by calculating the mean Cb and Cr in the region of face rectangle. Using the skin color, we can quickly and effectively detect the the overall face region. If the background of the given image is clear and static, we can easily segment out the background by color detection. If the background is complex, we can use graph cuts [21,22] to segment out the background. Therefore, we can obtain the region of hair and collar after we segment out the overall face region and background from the given image.

Searching the closest template is the process of shape matching. More formally, $D(p, q)$ denotes shape context distance between shapes p and q, $P = \{p_1, ..., p_n\}$ denotes a set of templates, P^* denotes the closest template for a given shape Q, thus

$$P^* = \arg \min_{p_k \in P} D(p_k, Q) \tag{10}$$

The flow chart of hair and collar subsystem is shown in Figure 9. We firstly obtain the shape contour of hair and collar by segmenting out the background and face. Then we get a set of points sampled from the shape contour of hair and collar and find the closest template by minimizing the shape context distance. Finally, we use a regularized TPS to map the closest template onto the corresponding shape contour.

3.3   Multiple Style Rendering

For each portrait component, we always have a corresponding graph representation shown in Figure 4, Figure 7 and Figure 8. We call them the sketch dictionaries

$$\Delta_S = \{\Delta_S^h, \Delta_S^{cp}, \Delta_S^c\} \tag{11}$$

The Δ_S represents a special style. The inference of portrait can be represented as:

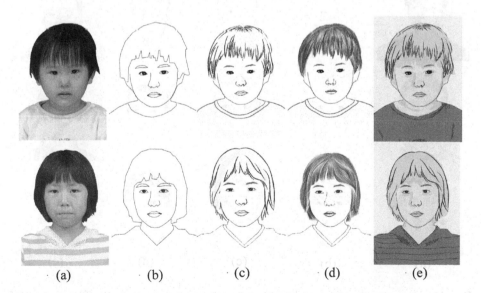

(a)   (b)   (c)   (d)   (e)

Fig. 10. Multiple style rendering. (a) Input image; (b) The portrait of primitive sketch; (c) The portrait of literary sketch; (d) The portrait of pencil sketch; (e) The colored portrait.

$$p(S|I; \Delta_S)$$
$$=p(S^h|I; \Delta_S^h) \cdot p(S^{cp}|I; \Delta_S^{cp}, \Delta_I^{cp}) \cdot p(S^c|I; \Delta_S^c)$$
$$=p(S^h|I; \Delta_S^h) \cdot \prod_{j=1}^{5} p(l_j) \cdot \prod_{j=1}^{5} p(S_{l_j}^{cp}|I; \Delta_S^{cp}, \Delta_I^{cp}) \cdot p(S^c|I; \Delta_S^c) \tag{12}$$

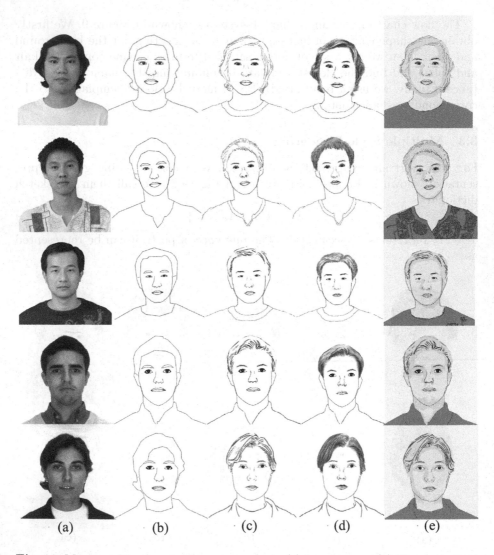

    (a)   (b)   (c)   (d)   (e)

Fig. 11. More results generated by our approach. (a) Input image; (b) The portrait of primitive sketch; (c) The portrait of literary sketch; (d) The portrait of pencil sketch; (e) The colored portrait.

It is easy to generate another style by replacing the Δ_S with another sketch dictionary $\Delta_{S'}$. Additionally, we can get a colored portrait by tinging the region of each portrait component. The colored portrait is more expressive. We can see these effect in Figure 10. With a larger sketch dictionary, we can generate more styles.

4 Experiments

To verify the framework we proposed, experiments were conducted based on 225 frontal face images chosen from different genders, ages and races. Some images are taken from the AR data set [23]. All the images are resized to the same resolutions: 500×500. We take 125 images for training and 100 images for testing. The training data satisfy following prerequisites:

1. Frontal view only(no hats and glasses).
2. Plain style, no exaggeration.
3. Each pair of image and portrait matches perfectly.
4. One image corresponds a set of portraits in different styles.

Figure 11 shows some results generated by our approach. It takes about 5 seconds on a Pentium IV 2.93 GHZ PC to generate a 500×500 portrait. We conclude that our approach is convenient to change the style of portrait and has good expansibility.

5 Discussion and Conclusions

We have presented an approach to automatically generating a set of life-like portraits in different styles. The And-Or graph for portrait is employed to account for the variability of portraits and separate the structure and style of the portraits. Our approach benefits from large sketch dictionaries in different styles. By replacing the sketch dictionaries, it is convenient to change the style of portrait. However our approach is not able to handle the old people because the wrinkles are not taken into account.

Our approach is aimed at a number of applications, such as low bit portrait communication in wireless platforms, cartoon sketch and canvas in non-photorealistic rendering, portrait editing and make-up on the Internet. In future work, we will add richer features including wrinkles, mustache, and lighting variabilities. We'd also like to extend our approach to cartoon sketch, side face and motion.

Acknowledgement

This work is done when the author is at the Lotus Hill Research Institute. The author thanks ZiJian Xu for extensive discussions. The project is supported by the National Natural Science Foundation of China under Contract 60672162.

References

1. Koshimizu, H., Tominaga, M., Fufiwara, T., Murakami, K.: On kansei facial processing for computerized facial caricatruing system picasso. In: IEEE International Conferece on Systems, Man and Cybernetics, vol. 6, pp. 294–299 (1999)
2. Li, Y., Kobatake, H.: Extraction of facial sketch based on morphological processing. In: IEEE international conference on image processing, vol. 3, pp. 316–319 (1997)
3. Librande, S.E.: Example-based character drawing, Masters thesis. MIT, Cambridge, MA (1992)
4. Freeman, W.T., Tenenbaum, J.B., Pasztor, E.: An example-based approach to style translation for line drawings, Technical Report 11, MERL Technical Report, Cambridge, MA (1999)
5. Efros, A.A., Leung, T.K.: Texture synthesis by nonparametric sampling. In: Seventh International Conference on Computer Version (1999)
6. Baker, S., Kanade, T.: Hallucinating faces, AFGR00 (2000)
7. Chen, H., Xu, Y.Q., Shum, H.Y., Zhu, S.C., Zheng, N.N.: Example-based facial sketch generation with non-parametric sampling. ICCV 2, 433–438 (2001)
8. Chen, H., Liu, Z.Q., et al.: Example-based composite sketching of human portraits, NPAR, 95–102 (2004)
9. Jones, M.J., Poggio, T.: Multi-dimensional morphable models: a framework for representing and matching object classes. IJCV 2(29), 107–131 (1998)
10. Xu, Z.J., Chen, H., Zhu, S.C.: A high resolution grammatical model for face representation and sketching. CVPR 2, 470–477 (2005)
11. Chen, H., Zhu, S.C.: A generative model of human hair for hair sketching. CVPR 2, 74–81 (2005)
12. Chen, H., Zhu, S.C.: Composite templates for cloth modeling and sketching. CVPR 1, 943–950 (2006)
13. Rekers, J., Schurr, A.: A parsing algorithm for context sensitive graph grammars, TR-95-05, Leiden Univ (1995)
14. Viola, P., Jones, M.: Rapid object detection using a boosted cascade of simple features, CVPR (2001)
15. Cootes, T.F., Taylor, C.J., Cooper, D., Graham, J.: Active shape models-their training and application. Computer Vison and Image Understanding 61(1), 38–59 (1995)
16. Cootes, T.F., Edwards, G.J., Taylor, C.J.: Active appearance models, proceedings of ECCV (1998)
17. Davies, R.H., Cootes, T.F., Twining, C., Taylor, C.J.: An Information theoretic approach to statistical shape modelling. In: Proc. British Machine Vision Conference, pp. 3–11 (2001)
18. Belongie, S., Malik, J., Puzicha, J.: Shape matching and object recognition using shape contexts. PAMI 24(4), 509–522 (2002)
19. Meinguet, J.: Multivariate interpolation at arbitrary points made simple. J. Applied Math. Physics (ZAMP) 5, 439–468 (1979)
20. Chui, H., Rangarajan, A.: A new algorithm for non-rigid point matching, CVPR (2000)

21. Boykov, Y., Veksler, O., Zabih, R.: Faster approximate energy minimization via graph cuts. PAMI 23(11), 1222–1239 (2001)
22. Boykov, Y., Kolmogorov, V.: An Experimental Comparison of Min-Cut/Max-Flow Algorithms for Energy Minimization in Computer Vision. PAMI 26(9), 1124–1137 (2004)
23. Martinez, A., Benavente, R.: The ar face database, Technical Report 24, CVC (1998)

Object Category Recognition Using Generative Template Boosting

Shaowu Peng[1,3], Liang Lin[2,3], Jake Porway[4], Nong Sang[1,3],
and Song-Chun Zhu[3,4]

[1] IPRAI, Huazhong University of Science and Technology,
Wuhan, 430074, P.R. China
{swpeng,nsang}@hust.edu.cn
[2] School of Information Science and Technology, Beijing Institute of Technology,
Beijing, 100081, P.R. China
linliang@bit.edu.cn
[3] Lotus Hill Institute for Computer Vision and Information Science,
Ezhou, 436000, P.R. China
[4] Departments of Statistics, University of California, Los Angeles
Los Angeles, California, 90095, USA
{jporway,sczhu}@stat.ucla.edu

Abstract. In this paper, we present a framework for object categorization via sketch graphs, structures that incorporate shape and structure information. In this framework, we integrate the learnable And-Or graph model, a hierarchical structure that combines the reconfigurability of a stochastic context free grammar(SCFG) with the constraints of a Markov random field(MRF), and we sample object configurations as training templates from this generative model. Based on these synthesized templates, four steps of discriminative approaches are adopted for cascaded pruning, while a template matching method is developed for top-down verification. These synthesized templates are sampled from the whole configuration space following the maximum entropy constraints. In contrast to manually choosing data, they have a great ability to represent the variability of each object category. The generalizability and flexibility of our framework is illustrated on 20 categories of sketch-based objects under different scales.

1 Introduction

In the last few years, the problem of recognizing object classes has received growing attention in both the fields of whole image classification[6,7,15,5] and object recognition[1,14,16]. The challenge in object categorization is to find class models that are invariant enough to incorporate naturally occurring intra-class variations and yet discriminative enough to distinguish between different classes.

In the vision literature, the majority of existing object representations for categorization use local image patches as basic features. These appearance based models achieve simple and rich visual/image representations and range from

A.L. Yuille et al. (Eds.): EMMCVPR 2007, LNCS 4679, pp. 198–212, 2007.
© Springer-Verlag Berlin Heidelberg 2007

global appearance models (such as PCA) in the 1990s (Nayar et al.)[19], to local representations using invariant feature points (Lowe et al.)[2], patches[12]
and fragments[18]. Recently graphical models have been introduced to account
for pictorial deformations[11] and shape variances of patches, such as the constellation model[13,12]. However, these models generally do not account for the
large structural or configurational variations exhibited by functional categories,
such as vehicles. Other class methods model objects via edge maps, detected
boundaries[21,3],or skeletons[16], which are essential for object representations,
but are not designed to deal with large variations, particularly at different scales.

Recently, one kind of hierarchical generative model using a stochastic attribute
grammar, known as the And-Or graph, was presented by Han[4] and Chen[8].
It is a model that combines the flexibility of a stochastic context free grammar
(SCFG) with the constraints of a Markov Random Field (MRF). The And-Or
graph models object categories as collections of parts constrained by pairwise
relationships. The SCFG component models the probability that hierarchies of
object parts will appear and with certain appearances, while the MRF component ensures these parts are arranged meaningfully. More recently, Zhu[17] and
Porway[9] defined a probability model on the And-Or graph that allows us to
learn its parameters in a unified way for each category.

Our categorization framework is built upon a sketch representation of objects
along with a learned And-Or graph model for each object category. We sample
object configurations from the And-Or graph offline as training templates from
which a dictionary of graphlets is learned. Then, four steps of discriminative
approaches are adopted to prune candidate categories online, and one graph
matching algorithm[10] is developed for top-down verification.

The main contribution of this paper is using sampled object templates from
the learned And-Or graph model instead of collecting training data manually,
and presenting an efficient computing framework for object categorization, integrated with discriminative approaches and top-down verification.

Fig. 1 (a) shows the And-Or graph for the bicycle category. The Or-nodes
(dashed) are "switching variables", like nodes in an SCFG, for possible choices
of the sub-configurations, and thus account for structural variance.Only one child
is assigned to each Or-node during instantiation. The And-nodes (solid) represent pictorial composition of children with certain spatial relations. The relations
include *butting, hinged, attached, contained, cocentric, colinear, parallel, radial*
and others constraints. [9] describes how these relations can be effectively pursued in a minimax entropy framework with a small number of training examples.
Assigning values to the Or-nodes and estimating the variables on the And-nodes
produces various object instances as generated templates. Some examples of the
bicycle template are shown in Fig. 1 (b), which can be used as training data in
our framework. These templates are synthesized following the maximum entropy
principle and are much more uniform than manually collected data would be.

In order to show the generalizability and power of categorization using
shape and structure information, we use hand-drawn objects, so-called "perfect
sketches", as the testing instances.

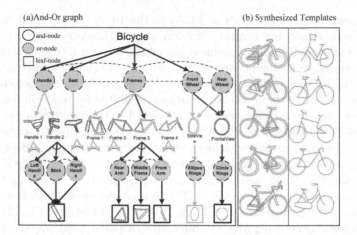

Fig. 1. A specific object category can be represented by an And-Or graph.(a) shows one example for the bicycle category. An Or-node (dashed) is a "switching variable" for possible choices of the components and only one child is assigned for each object instance. An And-node (solid) represents a composition of children with certain relations. The bold arrows form a sub-graph (also called a parsing graph) that corresponds to a specific object instance (a clock) in the category. The process of object recognition is thus equivalent to assigning values to these Or-nodes to form a "parsing graph". (b) Several synthesized bicycle instances using the And-Or graph to the left as training data.

The remainder of this paper is arranged as follows. We present the principle of synthesizing templates from the And-Or graph model in Section 2. We then follow with a description of the inference schemes, involving discriminative approaches and a top-down matching algorithm[10], along with detailed experiments with comparisons in Section 3. The paper is concluded in Section 4 with a discussion of future work.

2   Generative Template Synthesis

We begin by briefly reviewing the And-Or graph model and then describe the principle of sampling instances from the learned And-Or graph model. The detailed And-Or Graph learning work for each object category was proposed in [9].

2.1   The Learnable And-Or Graph

The And-Or graph can be formalized as the 6-tuple

$$G = < S, V_N, V_T, R, P, \Sigma >$$

where S is the root node, V_N are non-terminal nodes for objects and parts, V_T are terminal nodes for the atomic description units, R are a set of pairwise relationships, P is the probability model on the graph, and Σ is the set of all valid configurations producible from this graph.

In Fig. 1 (a), the root node S is an And node, as bicycle must be expanded into *handle, seat, frame, front wheel, rear wheel*, while the handle node is an Or node, as only one appearance of handle should exist for each instance of bicycle. Each Or Node V_i^{OR} has a distribution $p(\omega_i)$ over which of its $\omega = \{1, 2, ..., N(\omega_i)\}$ children it will be expanded into. $V_T = \{t_1, t_2, ..., t_{T_n}\}$ represents the set of terminal nodes. In our model, the terminal nodes are low-level sketch graphs for object components. They are combined to form more complex non-terminal nodes, as illustrated in Fig. 1 (a).

$R = \{r_1, r_2, ..., r_{N(R)}\}$ in the formulation represents the set of pairwise relationships defined as functions over pairs of nodes $v_i, v_j \in V_T \cup V_N$.

$$r^\alpha = \psi^\alpha(v_i, v_j)$$

each relationship is a function of each node's attributes, for example the distance between the centers of the two nodes. These relationships are defined at all levels of the tree.

$P = p(G, \Theta)$ is the probability model over the graph structure. The deriving process of $P = p(G, \Theta)$ and the detailed learning process of θ from training set was described in [9].

As the And-Or graph embeds an MRF in a SCFG, it borrows from both of their formulations. We first define the *parsing graph* as a valid traversal of an And-Or graph, and we then collect a set of parsing graphs labeled from a category of real images as training data to learn the structure and related parameters of the And-Or graph for this category. Each parsing graph will consist of the set of non-terminal nodes visited in the traversal, $V = v_1, v_2, ..., v_{N(v)} \in V_N$, a set of resulting terminal nodes $T = t_1, t_2, ..., t_{N(t)} \in V_T$, and a set of relationships observed between parts, $R \in \mathcal{R}$.

Following the SCFG[23], the structural components of the And-Or graph can be expressed as a parsing tree, and its prior model follows the product of all the switch variables ω_i at the Or nodes visited.

$$p(T) = \prod_{i \in V} p_i(\omega_i)$$

Let $p(\omega_i)$ be the probability distribution over the switch variable ω_i at node V_i^{OR} , θ_{ij} be the probability that ω_i takes value j, and n_{ij} the number of times we observe this production, we can rewrite $p(T)$ as

$$p(T) = \prod_{i \in V^{OR}} \prod_{j=1}^{N(\omega_i)} \theta_{ij}^{n_{ij}}$$

The MRF is defined as a probability on the configurations of the resulting parts of the parsing tree. It can be written in terms of the pairwise energies between parts. We can extend these energies to include constraints on the singleton nodes, e.g. singleton appearance constraints.

$$p(C) = \frac{1}{Z}\exp^{-\sum_{i \in T} \phi(t_i) - \sum_{<i,j> \in V} \psi(v_i, v_j)}$$

where $\phi(t_i)$ denotes the singleton constraint function corresponding to a singleton relationship, and $\psi(v_i, v_j)$ denotes the pairwise constraint corresponding to a pairwise relationship.

Following the deriving process of [9], we obtain the final expression of $P = p(G, \Theta)$. Suppose the number of times we observe this production is n_{ij}, R_N^1 is the number of singleton constraints, and R_N^2 is the number of pairwise constraints,

$$p(G, \Theta) = \frac{1}{Z}(\Theta)p(T)\exp\{-E(g)\}$$

$$E(g) = \log(p(T)) + \sum_{i \in T}\sum_{a=1}^{R_N^1} \alpha_i^a(\phi^a(t_i)) + \sum_{<i,j> \in V}\sum_{b=1}^{R_N^2} \beta_{ij}^b(\psi^b(v_i, v_j))$$

where $\Theta = (\theta, \alpha, \beta)$ are related parameters of the probability model and can be learned from a few parsing graphs, as proved in [9]. Intuitively, for each object category, the basic structural components (object parts or sub-parts) and corresponding relationships are essential and finite, like the basic words and grammar rules, and thus can be learned from a few typical instances, as illustrated in Fig. 2 (a),(b). Furthermore, a huge number of object configurations can be synthesized from the learned And-Or graph that are representative of the in-class variability of each object category, as illustrated in Fig. 2 (c).

2.2   Template Synthesise

To generate sample templates from the learned And-Or graph, we first sample the tree structure $p(T)$ of the And-Or graph. This is done by starting at the root of the And-Or graph and decomposing each node into its children nodes recursively. And Nodes are decomposed into all of their children, while an Or node V_i^{OR} selects one of its children to decompose into according to the $p(\omega_i)$ learned in [9]. This process is similar to generating a sentence from a SCFG[23] model. This continues until all nodes have decomposed into terminal nodes. The output from this step is a parsing graph of the hierarchy of object parts, though no spatial constraints have been imposed yet. For example, we may have selected certain templates for the wheels, frame, seat, and handle bars of a bike, but we have not yet constrained their relative appearances.

To arrange the parts spatially, we use *Gibbs* sampling to sample new part positions and appearances for each pair of parts. At each iteration, we update the value of each of the relationships that exists between each pair of nodes at the same level in the parsing graph. The parameters for each relationship are inherited from the And-Or graph. A relationship R for a given pair of nodes (v_i, v_j) is represented as a histogram in our framework [9], and we can thus calculate the change in energy that would occur if this pair were to take on every value in R. For each r that R can assume, we arrange the parts so that their response for this relationship is r. We then compute the total change in energy for the tree by observing the new values of any relationships R' that were altered between any pairs (v_x, v_y) by this change, including (v_i, v_j) and R themselves. This energy is easily calculated by plugging the affected relationship responses into our prior model $p(G)$. We record this energy and then arrange according to the next r.

For example, given a parsing graph of a car, we would need to update the relative position between the wheels and the body. This would be just one of the relationships between this pair of parts that we would update during this iteration. To do this, we set the size of the wheels to each possible value that the relative position can take on, and then compute the resulting change in energy of the parsing graph. Any relationships between the body and wheels that depend upon the relative position would be affected, as well as any relationships between other parts that depend on the position of the wheels.

Once we have a vector of energies for every possible r for a given pair P under R, we exponentiate to get a probability distribution from which we sample the next value v for this P under R. We then move on to the next pair or next relationship. This process continues for 50 iterations in our experiment, at which point we output the final parsing graph, which contains the hierarchy we selected as well as the spatial arrangements of the parts. This process is described in the algorithm below.

1. Sample the tree structure $p(T)$ to get a parsing graph.
2. Compute the energy of this tree by calculating the energy of each relationship between each pair of terminals. These begin uniformly distributed.
3. For each pair of parts P
   (a) For each relationship R between P
       i. Arrange P so that its response to R is r.
       ii. For every other pair that shares a part with P and every relationship affected by R, compute new responses.
       iii. Calculate the change in energy ΔE by plugging these new responses into our energy term.
       iv. Repeat steps i-iii for each value R can take.
       v. Normalize the ΔE's to get a probability distribution. Sample a new value y from it for this P under R.
4. Update all pairs of parts under each relationship to their new values y.

This sampling procedure thus selects the new relationship values for the pairs of parts proportional to how much they lower the energy at each iteration. As lowering the energy corresponds to increasing the probability of this tree, we are sampling configurations proportional to their probabilities from our learned model.

Once we have a parsing graph with all of the leaf nodes appropriately arranged, we must impose an ordering on the terminal nodes to create an appropriately occluded object. This is a logistic issue that is needed to transform

(a) Instances for And-Or graph learning   (b) And-Or graph for each category   (c) New Object configurations

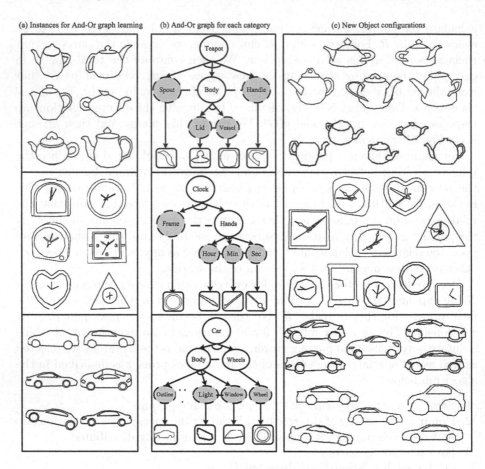

Fig. 2. Examples of And-Or graphs for three categories (in (b)), and their selected training instances (in (a)) and corresponding samples (in (c)). And The samples (in (c)) contain new object configurations, compared with the training instances (in (a)). Note that, for the sake of space, the And-Or graphs (in (b)) have been abbreviated. For example, the terminal nodes show only one of the many templates that could have been chosen by the Or Node above.

our object from overlapping layers of parts into one connected structure. In our experiment we hard-coded these layer values by part type. For example, teapot spouts are always in front of the teapot base. We are currently experimenting with learning an occlusion relationship between pairs so that we can sample this ordering in the future as well. Once the ordering is determined, intersection points between layers are determined and the individual leaf templates are flattened into one final template.

By the end of this process we can produce samples that appear similar to the training data, but vary in the arrangements and configurations of their parts. Figure 2 (c) shows examples of these samples at both high and low resolution, along with the corresponding And-Or graph for that category (Figure 2 (b)). Note that, a few new configurations are synthesized, compared with the training instances (Figure 2 (a)). For our experiments, we produced 50 samples for each category.

3   Inference and Experiments

Using the synthesized templates as training data, we illustrate our framework on classifying sketch graphs into 20 categories. In classifying a given sketch graph g, we utilize four stages of discriminative tests to prune the set of candidate categories it can belong to. These discriminative methods are useful as they are quite computationally fast. Finally, a generative top-down matching method is used for verification. All discriminative approaches work in a cascaded way, meaning each step keeps a few candidate categories and candidate templates from each candidate category for the next step. These discriminative approaches were adopted due to their usefulness in capturing object-specific features, local spatial information, and global information respectively. The verification by top-down template matching operates on this final pruned candidate set. This inference framework will be described in the next subsections.

We collect 800 images at multiple scales from 20 categories as testing images from the Lotus Hill Institute's image database[24], as well as the ground truth sketch graphs for each image. We show a few typical instances in Fig. 3 and measure our categorization results with a confusion matrix.

3.1   Discriminative Prune

We first create a dictionary of atomic graph elements, or graphlets. Graphlets are the most basic components of graphs and consist of topologies such as edges, corners, T-junctions, etc. To create a dictionary of graphlet clusters, we first collect all the sketch templates from all categories. Starting with the edge element as the most basic element, we count the frequency of all graphlets across all graphs. A TPS (Thin Plate Spline) distance is used in the counting process to determine the presence of different graphlets. To consider the detectability of each graphlet, we then use each graphlet as a weak classifier in a compositional boosting algorithm[22].

$$\triangle_{Glet} = \{(g_i, \omega_i), i = 1, ..., 13\}$$

According to their weights, we select the top 13 detectable graphlets, and their distributions in each category are plotted in Fig. 4. Other graphlets are either very rare (occurrence frequency is less than 1.0%) or uniform in every category, and thus are ignored.

To prune candidate categories for a testing instance, we integrate four steps of discriminative approaches. We use confusion matrix to estimate the true positive rate of categorization. In each step, we keep candidates and plot the confusion

Fig. 3. Selected testing images with sketch graphs from Lotus Hill Institute's image database[24]

matrix of top N candidate categories, which describes whether the true category is in the top N candidate categories. The number N of candidates we keep is empirically determined in the training stage and it helps to guarantee high true positive rate and narrow the computing range of the next step. Estimating this number is straightforward and its description is thus ignored in this paper.

Step 1 Category histogram. The histogram of graphlets for each category is learned by counting graphlet frequency over the training templates. Each testing graph can then be converted into a sparse histogram and the likelihood of its histogram against each category histogram can be calculated for classification.

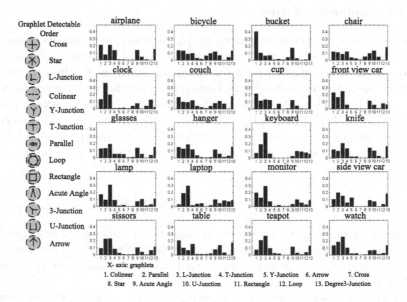

Fig. 4. The graphlets distributions in each category. The graphlets are listed in order according to detectability.

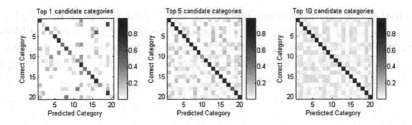

Fig. 5. Step 1 discriminative prune. Three confusion matrix show results with the top 1, top5 and top10 candidate categories. We keep the top 10 candidates for step 2.

The results of step 1 are shown in Fig. 5 and the top 10 candidates are selected for step 2.

Step 2 Nearest Neighbor. Each instance in the training data and our testing graph are converted into sparse vectors. Each vector represents the frequency of each graphlet along with its weight. Suppose we have M training templates in each category, then each vector is

$$V_j = \{\omega_i * N_i, i = 1, ..., 13\}, j = 1, ..., M$$

Where ω_i and N_i are graphlet g_i's weight and occurrence frequency in template j.

To get a distance between two sparse vector, we use a modified Hamming Distance, the L1-Norm of two vectors' subtraction. Firstly we calculate the distance between our testing instance and each training instance from the 10 candidate categories kept from step 1. Then we find the 8 shortest distances within each category and calculate the average distance between them. The candidate categories can then be ordered by the length of these average distances. We keep the 8 closest candidate categories and 10 closest candidate templates within these categories for step 3, the results of which are shown in Fig. 6.

Step 3 Nearest Neighborhood via Composite Graphlet. We next introduce spatial information into our features. To capture informative attributes of a graph, adjacent graphlets are composed into cliques of graphlets. The training and testing templates can be decomposed into these graphlet cliques and then

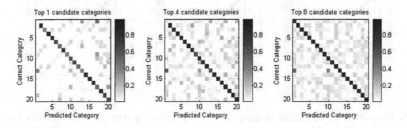

Fig. 6. Step 2 discriminative prune. Three confusion matrix show results with the top 1, top4 and top8 candidate categories. We keep the top 8 candidates for step 3.

vectorized for further pruning. The composite graphlets selection work is similar to single graph selection, and the top 20 detectable composite graphlets are shown in Fig. 7. The distance metric is computed in a similar fashion to step 2. We then keep the top 6 candidate categories and top 8 candidate templates from our remaining set, as shown in Fig. 8.

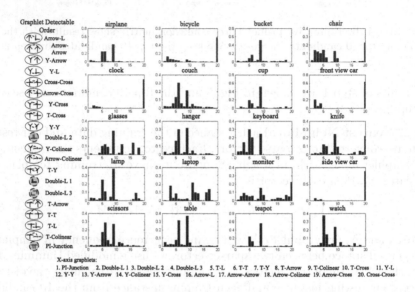

Fig. 7. Top 20 detectable composite graphlets and their distributions

Step 4 Nearest Neighborhood via Shape Context. Shape context[20] can be considered as one kind of global shape descriptor. It can be used to further prune the candidates proposed by previous steps. We keep the top 5 candidate categories and top 5 candidate templates from each category in this step, as shown in Fig. 9.

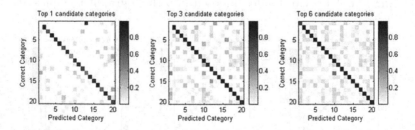

Fig. 8. Step 3 discriminative prune. Three confusion matrices show results with the top 1, 3, 6 candidate categories. We keep the top 8 candidate templates for step 4.

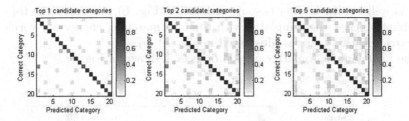

Fig. 9. Step 2 discriminative prune. Three confusion matrix show results with the top 1, top3 and top5 candidate categories. We keep the top 5 candidates for top-down verification.

3.2 Top-Down Verification

Given the results of our cascade of discriminative tests, the template matching process is then activated for top-down verification. The matching algorithm we adopted is a stochastic layered matching algorithm with topology editing that tolerates geometric deformations and some structural differences [10]. As shown in Fig. 10, the final candidate categories and templates for testing objects are listed, and top-down matching is illustrated. The input testing image and its perfect sketch graph are shown in Fig. 10 (a), in which there are two testing instances, a cup and a teapot. After the discriminative pruning, there are two candidate categories left (cup and pot) with candidate templates for the testing

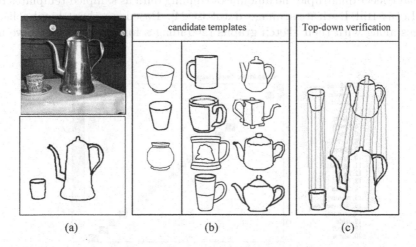

Fig. 10. The top-down verification of candidate templates. The input testing image and its perfect sketch graph are shown in (a), where there are two testing instances, a cup and a teapot. After the discriminative prune there are two candidate categories remaining (cup and teapot) with candidate templates for the testing cup, as well for the testing teapot, as listed in (b). After the top-down matching process, the templates with the best matching energy are verified, as illustrated in (c).

cup, as well for the testing teapot, as listed in Fig. 10 (b). After the top-down matching process, the templates with the best matching energy are verified for the categorization task, as illustrated in Fig. 10 (c).

The final confusion matrix on 800 objects is shown in Fig. 11.

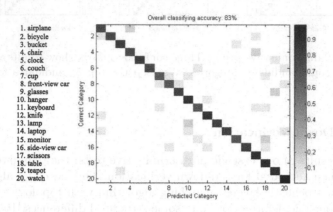

Fig. 11. The final confusion matrix on 800 objects

3.3 Comparison

For comparison purposes, we discard the generative And-Or graph model and instead collect quadruple the amount of training data as sampled templates from the Lotus Hill Institute's image database[24]. Based on these hand-collected images along with their sketch graphs (200 images for each category), we infer

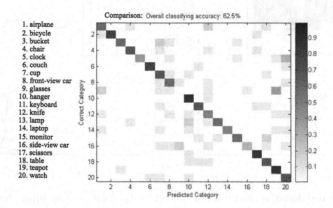

Fig. 12. Confusion matrix showing the overall classification accuracy of the test categories using training data from hand-collected images. Compared with Fig. 11, it shows numerically that the manually collected training data are of less powerful in representing varied object configurations.

the same testing instances for categorization as we described above. We just plotted the final confusion matrix for illustration, as shown in Fig. 12. With the same inference steps and testing instances, the overall classification accuracy is 21.5% lower than inference using the generative And-Or graph model. The essential reason for comparison results is that most categories include a nearly infinite number of configurations, which are impossible to cover completely by hand-collected data.

4 Summary

In this paper, a framework composed of a generative And-Or graph model, a set of discriminative tests and a top-down matching method, is proposed for a sketch-based object categorization task. Instead of collecting training data manually, we synthesize object configurations as object templates from the And-Or graph model. In the computational process, four steps of discriminative pruning are adopted to narrow down possible matches, and a template matching algorithm is used on the final candidates for verification. We show detailed experiments of classifying testing sketch graphs into 20 categories step by step. In order to show the generalizability and power of this method, we use the human-annotated sketch graphs as testing instances. With the experimental results and comparisons, the synthesized templates show their ability to represent object with large variability in appearance, and the compositional inference shows their efficiency and accuracy with the experiments results.

To get a better performance on words clustering and generative model, we currently focus on objects in form of manually drawn sketch graph. In future, further work about imperfect sketch recognition will be done on pixel images, new model of sketch extraction and object detection will be consider.

Acknowledgment

This work is done when the authors are at the Lotus Hill Research Institute. The project is supported by Hi-Tech Research and Development Program of China (National 863 Program, Grant No. 2006AA01Z339 and No. 2006AA01Z121) and National Natural Science Foundation of China (Grant No. 60673198 and No. 60672162).

References

1. Berg, A., Berg, T., Malik, J.: Shape Matching and Object Recognition using Low Distortion Correspondence, CVPR (2005)
2. Lowe, D.G.: Distinctive image features from scaleinvariant keypoints. IJCV 60(2), 91–110 (2004)
3. Estrada, F., Jepson, A.: Perceptual Grouping for Contour Extraction, ICPR (2004)
4. Han, F., Zhu, S.C.: Bottom-up/top-down image parsing by attribute graph grammar, ICCV, 2 (2005)

5. Jurie, F., Triggs, B.: Creating Efficient Codebooks for Visual Recognition, ICCV (2005)
6. Csurka, G., Dance, C., Fan, L., Willamowski, J., Bray, C.: Visual Categorization with Bags of Keypoints. In: SLCV workshop in conjunction with ECCV (2004)
7. Dorko, G., Schmid, C.: Selection of Scale-Invariant Parts for Object Class Recognition, ICCV (2003)
8. Chen, H., Xu, Z., Liu, Z., Zhu, S.C.: Composite Templates for Cloth Modeling and Sketching. CVPR 1, 943–950 (2006)
9. Porway, J., Yao, Z., Zhu, S.C.: Learning an and-or graph for modeling and recognizing object categories, submitted to CVPR 2007, NO. 1892 (2007)
10. Lin, L., Zhu, S.C., Wang, Y.: Layered Graph Match with Graph Editing, submitted to CVPR 2007, NO. 2755 (2007)
11. Fischler, M., Elschlager, R.: The representation and matching of pictorial structures. IEEE Transactions on Computers 22(1), 67–92 (1973)
12. Weber, M., Welling, M., Perona, P.: Towards automatic discovery of object categories, CVPR (2000)
13. Felzenszwalb, P., Hut tenlocher, D.: Pictorial Structures for Object Recognition. IJCV 61(1), 55–79 (2005)
14. Viola, P., Jones, M.: Rapid Object Detection using a Boosted Cascade of Simple Features, CVPR (2001)
15. Fergus, R., Perona, P., Zisserman, A.: Object class recognition by unsupervised scale- invariant learning, CVPR (2003)
16. Zhu, S.C., Yuille, A.L.: Forms: A flexible object recognition and modeling system. IJCV 20(3), 187–212 (1996)
17. Zhu, S.C., Mumford, D.: Quest for a Stochastic Grammar of Images, Foundations and Trends in Computer Graphics and Vision (to appear, 2007)
18. Ullman, S., Sali, E., Vidal-Naquet, M.: A Fragment-Based Approach to Object Representation and Classification. In: Proc. 4th Intl. Workshop on Visual Form, Capri, Italy (2001)
19. Nayar, S.K., Murase, H., Nene, S.A.: Parametric Appearance Representation. In: Nayar, S.K., Poggio, T. (eds.) Early Visual Learning (1996)
20. Belongie, S., Malik, J., Puzicha, J.: Shape matching and object recognition using shape contexts. PAMI 24(4), 509–522 (2002)
21. Ferrari, V., Tuytelaars, T., Van Gool, L.: Object Detection by Contour Segment Networks, ECCV (2006)
22. Tu, Z.W.: Probabilistic Boosting Tree: Learning Discriminative Models for Classification, Recognition, and Clustering, ICCV (2005)
23. Chi, Z., Geman, S.: Estimation of probabilistic context-free grammars, Computational Linguistics, 24(2) (1998)
24. Yao, Z., Yang, X., Zhu, S.C.: An Integrated Image Annotation Tool and Large Scale Ground Truth Database, submitted to CVPR 2007, NO. 1407 (2007)

Bayesian Inference for Layer Representation with Mixed Markov Random Field

Ru-Xin Gao[1,2], Tian-Fu Wu[2], Song-Chun Zhu[2,3], and Nong Sang[1]

[1] IPRAI, Huazhong University of Science and Technology, China
[2] Lotus Hill Research Institute, China
[3] Departments of Statistics and Computer Science, UCLA
rxgao.lhi@gmail.com, tfwu.lhi@gmail.com, sczhu@stat.ucla.edu,
nsang@hust.edu.cn

Abstract. This paper presents a Bayesian inference algorithm for image layer representation [26], 2.1D sketch [6], with mixed Markov random field. 2.1D sketch is an very important problem in low-middle level vision with a synthesis of two goals: segmentation and 2.5D sketch, in other words, it is to consider 2D segmentation by incorporating occulision/depth explicitly to get the partial order of final segmented regions and contour completion in the same layer. The inference is based on Swendsen-Wang Cut (SWC) algorithm [4] where there are two types of nodes, instead of all nodes being the same type in traditional MRF model, in the graph representation: atomic regions and their open bonds desribed by address variables. These makes the problem a mixed random field. Therefore, two kinds of energies should be simultaneously minimized by maximizing a joint posterior probability: one is for region coloring/layering, the other is for the assignments of address variables. Given an image, its primal sketch is computed firstly, then some atomic regions can be obtained by completing some sketches into a closed contour. At the same time, T-junctions are detected and broken into terminators as the open bonds of atomic regions after being assigned the ownership between them and atomic regions. With this graph representation, the presented inference algorithm is performed and satisfactory results are shown in the experiments.

Keywords: Layer Representation, 2.1D Sketch, Bayesian Inference, Contour Completion, Mixed Markov Field, Swendsen-Wang Cut, MCMC.

1 Introduction

This paper presents a Bayesian inference algorithm for image layer representation [26], that is 2.1D sketch [6], with mixed Markov random field. The general goal of 2.1D sketch is to resume the occluded structure part of object and find the occlusion relation (partial order) of all objects in a scene. 2.1D sketch is a very important issue in low-middle level vision tasks and remains a challenging problem yet in the literature. Solving 2.1D sketch is a critical step for scene understanding in both still images and video, such as foreground/background separation, 2.5D sketch, motion analysis, etc.

A.L. Yuille et al. (Eds.): EMMCVPR 2007, LNCS 4679, pp. 213–224, 2007.
© Springer-Verlag Berlin Heidelberg 2007

Layer representation is firstly presented by Wang and Adelson for motion analysis [26]. Based on this idea, the concept of 2.1D sketch was then proposed firstly by Nitzberg and Mumford [6]. In their works, T-junctions detected on region's boundaries provide the cue information of occlusion relation, and an energy function is minimized to get 2.1D sketch with experiments on some simple images.

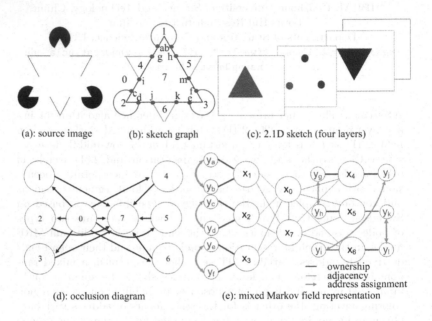

(a): source image   (b): sketch graph   (c): 2.1D sketch (four layers)

(d): occlusion diagram   (e): mixed Markov field representation

Fig. 1. Illustration of the 2.1D sketch. (a) is the original image, Kanizsa figure. (b) is the sketch graph computed by the primal sketch model after interactively labeling. there are 8 atomic regions and 12 terminators broken form T-junctions. (c) is the 2.1D sketch. There are 4 layer and the contour completion is performed using Elastica rules. (d) is the Hasse diagram for partial order relation, such as $< 7, 1 >$ means region 7 occludes region 1. (e) is the graph representation with mixed Markov random field. Each big circle denotes a vertex of atomic region, each blue bar denotes one terminator, each little circle denotes a vertex of open bound described as address variable. Each region may have two or more terminators. The blue line segments connect the two neighboring regions. The red two-way arrows connect two terminators and each terminator is assigned another terminator's address variable.

As Fig. 1 shown, it is an illustration of the 2.1D sketch for the well known Kanizsa figure. The final layer representation with contour completion is shown in Fig. 1(c). There are 8 atomic regions in total as shown in Fig. 1(b), denoted by $0, 1, ..., 7$ in Fig. 1(d) for showing the diagram of the partial order and $x_0, x_1, ..., x_7$ in Fig. 1(e) for showing the the graph representation with mixed random field. There are 12 T-junctions in total and they are break into terminators $a, b; c, d; e, f; g, h; i, j; k, m$ as shown in Fig. 1(b). These terminators are called open bonds described as address variables for corresponding atomic regions, denoted by $y_a, y_b; ...; y_k, y_m$ in Fig. 1(e). Both the atomic region vertices and the corresponding address variables vertices yield the graph representation and address variables would reconfigure the graph structure. And it is these address variables that make the problem a mixed random field:

(1) The vertices are inhomogeneous with different degrees of connections and should be inferred from the images;

(2) The neighborhood of each vertex is no longer fixed but inferred as address variables, which yields and would reconfigure the graph structure.

The compatibility between any two terminators is a function defined on different cues such as appearance, geometric properties (eg. Elastic rules), etc. The compatibility function is then used as the weight of edge in the graph representation. For the simple case shown in Fig. 1, Fig. 2 shows the inferential computing procedure using Gibbs sampler with anneal simulating starting from initial temperature $T = 20$.

In the literature, there are some related works on 2.1D sketch. Stella [8] etc proposed a model for Figure-Ground segregation based on hierarchical Markov random field. They used clique potentials in MRF to encode local logical decision rules and demonstrated a system that automatically integrating sparse local relative depth cues arisen from T-junctions over long distance into a global ordering of relative depths. Eseddoglu and Riccardo March [7] proposed to segment image with depth but without detecting junctions. They described a technique in the variational formulation of minimizing Mumford-Shah's energy function that avoided explicit detection/connection of T-junction. These works did not consider the the problem of address variable explicitly.

In contrast, this paper formulates 2.1D sketch using mixed random field, and presents a Bayesian inference algorithm based on Swendsen-Wang Cut algorithm. According to the authors' knowledge, this is the first time to solve 2.1D sketch problem inferencing based on mixed random field.

Given an image, its primal sketch is firstly computed using the primal sketch model [2], then some atomic regions can be obtained by completing some sketches into a closed contour using some interactive operations for the time being as shown in Fig. 3 and other results in experiments. From sketch to atomic region is not the main issue of this paper, so, currently we use some interactive operations, and this is also a common assumption in the literature. At the same time, T-junctions are detected using a method developed in the author's group recently [14] as shown in Fig. 3 and other results in experiments. T-junctions are then broken into terminators as the open bonds or address variables of atomic regions after being assigned the ownership between them and atomic regions. Each address variable should be assigned with another one to make contour completion to get larger segmented regions consisting of some atomic regions. Then a graph representation can be obtained and consists of two kinds of vertices: atomic regions and its corresponding address variables. So there are two kinds of energies that should be minimized at the same time: region coloring/layering and address variable assignment, by maximizing a joint posterior probability. Some results are shown in Section 2 and Section 3.

The remainder of the paper is organized as following. Section 2 formulates the problem under Bayesina framework and defines some used concepts and variables. Section 3 presents the inference algorithm based on Swendsen-Wang Cut algorithm with experiments in Section 4. Some conclusions and discussions are come at in Section 5.

2   Problem Formulation of the 2.1D Sketch

2.1   Graph Representation

Given an image defined on lattice Λ with occlusion relations among the objects in it, eg. Fig. 1 and other examples in this paper, its sketch graph is computed using the primal sketch model [1,2], followed by some manually interactive operations to get initial partition of the image domain with some atomic regions. Because the step from primal sketch to initial atomic regions are not the main issue of this paper and it is another research subject in the literature, we simplify this step by using manually interactive operations, and this is a common assumption in dealing with 2.1D sketch for the time being. These atomic regions are one kind of vertex in the graph. For each atomic region, there are some open bonds, described as address variables, which are obtained by breaking the T-junctions. Each T-junction is broken into a line segment and a terminator, and the terminator is described by address variable treated as the other kind of vertex in the graph. Address variables should be assigned to another one, inferring from the image. Some examples are shown in Fig. 3, etc. With this graph representation, 2.1D sketch can be formulated as graph partition problem. Swendsen-Wang Cut(SWC) is a recently developed algorithm to solve general graph partition problem. In the following, some definitions of the problem domain are given, bayesian formulation and the inference algorithm based on SWC are derived.

2.2   Definition of the Problem Domain

In this section, we define the elements in the 2.1D sketch representation. As stated above, it is a mixed random field with two kinds of vertices in the graph representation: a set of connected regions and a set of terminators broken from T-junctions. The regions set are defined as:

$$\Omega_x = \{R_1, R_2, \cdots, R_n\} \tag{1}$$

the regions will be divided into an unknown number of K layers.

$$\Omega_l = \{1, 2, \cdots, K\}, K \in |\Omega_x| \tag{2}$$

each region $R_i, i \in [1, n]$ has a label for its layer, $l_{R_i} \in X$ defined below. The terminators are also called open-bonds to be completed after inference, which can be defined as:

$$\Omega_a = \{a, b, \cdots, N\} \tag{3}$$

on Ω_x we define the layer information:

$$X = (x_1, x_2, \cdots, x_n), x_i \in \Omega_l, \forall i \in \Omega_x \tag{4}$$

region R_i is said to occlude R_j if $x_i < x_j$, region R_i and R_j are in the same layer and can be connected into one region if $x_i = x_j$.

On Ω_a, a set of address variables are defined for the contour completion information of open bonds:

$$Y = \{y_a, y_b, \cdots, y_N\}, y_a \in \Omega_a, \forall a \in \Omega_a \tag{5}$$

this will tell to whom the terminator a is connected. A terminator connects to itself when the T-junction is not broken. An example is shown in Fig. 1.

2.3 Bayesian Formulation

According to the above definitions, 2.1D sketch is then represented by a hidden vector variable W, which describes the world state for generating the image I.

$$W = (X, Y) \tag{6}$$

The solution vector W includes the following information in two graphical representation: (1) Partial orders represented by an Hasse diagram as shown in Fig. 1 (d), where each arrow indicates an occlusion relation. The arrow between two nodes i, j often decide the ownership of the boundary (contour) between regions R_i and R_j. For example, if region 2 occludes region 0, then region 2 owns the red boundary and consequently, it owns the two terminators c and d. Similarly, if region 4 occludes region 0, then region 4 owns the green contour. Thus region 4 own terminator g and i. This is not a trivial problem. In some rare accidental cases, two regions may co-own the contour between them; and (2) Graph representation with mixed Markov random field as shown in Fig. 1 (e).

In a Bayesian framework, we make inference about W from input image I over a solution space Ω:

$$W^* = (X, Y)^* = \text{arg max } p(X, Y|I) = \text{arg max } p(I|X, Y)p(X, Y) \tag{7}$$

or preserving multiple interpretations, each with a probability:

$$W_1, W_2, \cdots, W_k \sim p(I|X, Y)p(X, Y) \tag{8}$$

As we mentioned before, there are two kinds of energies that should be minimized simultaneously and described in the above joint probability distribution.

The prior probability $p(W)$ is a product of the following probabilities.

1. An exponential model for the number of layers $p(K) \propto \exp\{-\lambda_0 K\}$.

2. A model for the size of each "larger" region R, $A = |R|$, consisted of some atomic regions after contour completion in each layer. The prior encourage large and connected regions. We take a distribution $p(A) \propto \exp\{-\gamma A^c\}$, where $c = 0.9$ is a constant in this paper.

3. The prior for the layer of a region $p(l_R)$ is assumed to be uniform, and the prior for the address variable is also uniform.

In this paper, we adopt a mixture of Gaussians for generative region model,

$$J(x, y) = \prod_{v \in R} [\sum_{i=1}^{m} (\alpha_i G(I_v - \mu_i; \Sigma_i))]. \tag{9}$$

The image likelihood $p(I|X, Y)$ is defined as:

$$p(I|X, Y) \propto \exp\{-\sum_{i=1}^{K} \sum_{(x,y) \in D_i} |I(x, y) - J(x, y)| - \sum_{i=1}^{K} \sum_{m=1, n=1}^{N} T(m, n)\} \tag{10}$$

where K is the total number of layer, D_i is the domain of region in the layer i ,$I(x, y)$ is color value in RGB space for any $(x, y) \in \Lambda$. N is the total number of open bonds, $T(m, n)$ denote the link energy of terminator m and terminator n when the edge between the two terminators are 'on'.

3 Inference

The presented inference algorithm is based on Swendsen-Wang Cut algorithm [3,4,23] which generalized a well accepted cluster sampling algorithm for solving graph partition problem. Generally speaking, there are three steps in performing a SWC algorithm: (1) initialization of an adjacency graph by computing local discriminative probabilities for each edge based on image data and the prior; (2) graph clustering under a given partition by removing all edges between vertices of different labels, probabilistically turning on/off of each of the remaining edges according to its weight; (3) graph flipping of the label for a selected connected component with a probability driven by the posterior. Details of the SWC algorithm are referred to reference [3,4]. In the following, how to design the discriminative probabilities as weights of edges is given firstly, then we describe the inference algorithm for 2.1D sketch.

3.1   Discriminative Probabilities on Edges

After being broken from T-junctions, terminators or open bonds need to find a match by testing the compatibilities on different cues such as appearance (eg. profile matching), geometric properties (eg. smoothness), etc. These cues define the discriminative probabilities as the weights of edges in the graph. By sampling these probabilities, we can get new configurations of the graph.

Given an adjacency graph, for each open bond of each atomic region (vertex in initial graph), we extract a patch from the terminator and compute a 3D histogram with 15-bin in each dimension then normalize the histogram. For each edge $e = < v_i, v_j >$, we assign a binary random variable $\mu_e \in \{on, off\}$ and define the discriminative probability on edge as following:

$$q_e = p(\mu)e^{-E_{terminator}} \tag{11}$$

$$E_{terminator} = \alpha_0 E_{profile} + \alpha_1 E_{elastica} \qquad (12)$$

Where α_0 and α_1 are constant, they will be adjusted in experiments.

$$E_{profile} = (KL(p_i\|p_j) + KL(p_j\|p_i)/2 \qquad (13)$$

$$E_{elastica} = \int_{\Gamma} (\nu + \alpha k^2) ds \qquad (14)$$

Where ν and α are constant, Γ is the boundary. By minimizing the energy, the Elastica function considered the orientation consistency as well as the shortest pathway.

3.2   Algorithm

We summarize the presented algorithm in the following table.

Bayesian Inference for 2.1D Sketch

Input an image and set the parameters ($\alpha_0, \alpha_1, \nu, \alpha$, sweep number and temperature value for anneal simulating):
1.   Computing primal sketch for input image and interactively completing the sketches into close contour to get atomic regions
2.   T-junctions detection and obtaining the open bond for each atomic region.
3.   Initialization of adjacency graph.
4.   Computing the discriminative probabilities on edges.
5.   Select a region and select a layer(label) to it as in SWC algorithm;
6.   Sample the open bonds according to the edge probabilities in each layer;
7.   Connect terminators with curves according to open bonds' link status and generate new regions if some terminators and line segments can be combined into contours;
     Repeat 5-7 within given sweep number and at the same time reduce the temperature every several steps.

At each step, the acceptance probability is:

$$\alpha(A \to B) = \arg\min\{1, \frac{q(B \to A)}{q(A \to B)} \cdot \frac{p(B)}{p(A)}\} \qquad (15)$$

Where A and B are the two different states in node flipping process. $q(B \to A)$ is the posterior probability from state B to state A and $q(A \to B)$ is from A to B, let π denote state A or B, then the likelihood is in equation 10.

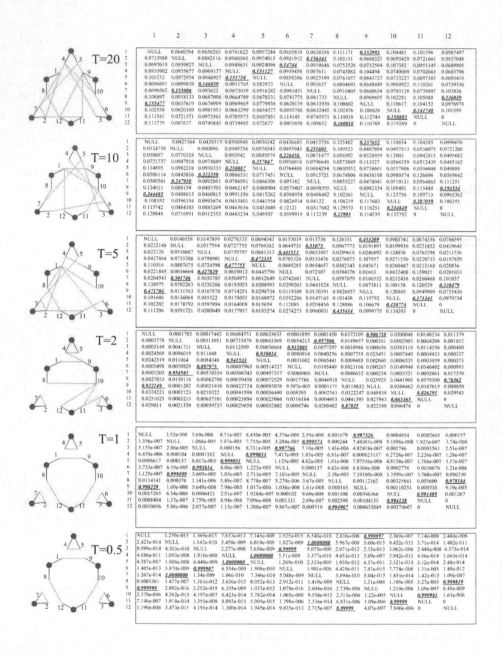

Fig. 2. Illustration of inferential Computing procedure. Each red node denotes a region, black line segment denotes terminator and the green line segment shows the inferred connection or assignment of address variable. Inference starts from a initial temperature $T = 20$, $(a) \sim (h)$ are the results in different temperatures. After $T = 1$, the result is right as in the figure.

the transition probability is:

$$q(A \to B) = q(A \to B)_{layer} * q(A \to B)_{connect} \tag{16}$$

and

$$q(A \to B)_{layer} = \frac{p(B \to A)}{p(A \to B)} = \frac{n_A + 1}{n_B + 1} \tag{17}$$

$$q(\pi) = \exp\{-\sum_{i=1}^{K} \sum_{(x,y) \in A_i} |I(x,y) - J(x,y)| - \sum_{i=1}^{K} \sum_{m=a,n=a}^{N} T(m,n)\} \tag{18}$$

where A_i is the terminator's patch domain. all pixels' profile can get from the primitive's photometric attributes. $I(x,y)$ is the profile getting from the primitive dictionary, $J(x,y)$ is the color value from a color model, here we use gaussian model. N is the total number of open bonds, $T(m,n)$ denote the link energy of terminator m and terminator n when the edge between the two terminators is 'on'.

Fig. 3 is an experiments result for the algorithm. The image is divided into 3 layers: the foreground layer, middle layer and background layer. The gaps between terminators are filled by curves which come from minimizing the Elastica function. The area are filled with a kind of color, the initial boundaries condition are from the corresponding primitive patch, the primitives are selected from a primitive dictionary.

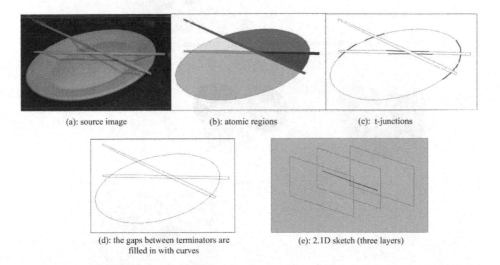

(a): source image   (b): atomic regions   (c): t-junctions

(d): the gaps between terminators are
filled in with curves

(e): 2.1D sketch (three layers)

Fig. 3. One running example. There are three layers in the original image (a). (b) is atomic regions obtained by completing some sketched into a close contour; (c) shows the T-junction detection results based on the primal sketch.(d) shows the contour completion result after inference and (e)is the final layer representation.

4   Experiments

The presented inference algorithm is tested on a variety of natural images.

As shown in Fig. 3, the boundaries of the plate in the image are occluded by a chopstick and the chopstick is still occluded by another chopstick and the plate and chopstick's regions can be cut into several atomic regions. The algorithm connects these regions in each layer with some curves cued by the terminator's boundaries' information and completed with boundaries' profile using heat diffusion. The background sketchable also are filled with some curves and the area was completed using colors.

More results are shown in Fig. 4.

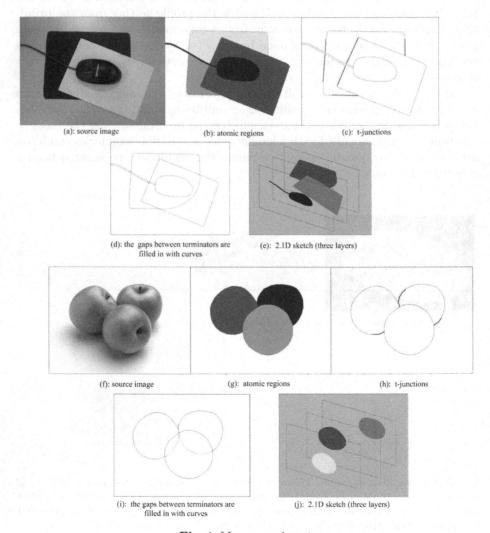

(a): source image

(b): atomic regions

(c): t-junctions

(d): the gaps between terminators are filled in with curves

(e): 2.1D sketch (three layers)

(f): source image

(g): atomic regions

(h): t-junctions

(i): the gaps between terminators are filled in with curves

(j): 2.1D sketch (three layers)

Fig. 4. More experiments

5 Discussion

This paper presents a Bayesian inference algorithm for image layer representation with mixed Markov random field. We conducted several experiments to demonstrate the algorithm by satisfactory results. The key contribution of this paper is that it is the first time to formulate the 2.1D sketch using mixed random field and present an inference algorithm to solve region coloring/layering and assignments of open bonds simultaneously.

Acknowledgements

This work is done when the authors are at the Lotus Hill Research Institute. The project is supported by : National 863 project Contact No. 2006AA01Z121, National Science Foundation China Contact No. 60672162, National Science Foundation China Contact No. 60673198, National 863 project contact No. 2006AA01Z339.

References

1. Guo, C.E., Zhu, S.C., Wu, Y.N.: Modeling visual patterns by integrating descriptive and generative models. IJCV 53(1), 5–29 (2003)
2. Guo, C.E., Zhu, S.C., Wu, Y.N.: Primal sketch: integrating texture and structure. In: Proc. Int'l. Conf. on Computer Vision (2003)
3. Barbu, A., Zhu, S.C.: Graph Partition by Swendsen-Wang Cuts. In: Proc. Int'l. Conf. on Computer Vision (2003)
4. Barbu, A., Zhu, S.C.: Generalizing Swendsen-Wang to Sampling Arbitrary Posterior Probabilities. IEEE Trans. on PAMI 27, 1239–1253 (2005)
5. Marr, D.: Vision. Freeman Publisher, San Francisco (1983)
6. Nitzberg, M., Shiota, T., Mumford, D.: Filtering, Segmentation and Depth. LNCS, vol. 662. Springer, Heidelberg (1993)
7. Eseddoglu, S.: Segment Image With Depth but Without Detecting Junction. Journal of Mathematical Imaging and Vision. 18 (2003)
8. Yu, S.X., Lee, T.S., Kanade, T.: A Hierarchical Markov Random Field Model for Figure-Ground Segregation. In: Figueiredo, M., Zerubia, J., Jain, A.K. (eds.) EMMCVPR 2001. LNCS, vol. 2134, pp. 118–133. Springer, Heidelberg (2001)
9. Chan, T., Shen, J.: Mathematical Models for Local Nontexture Inpaintings. SIAM Journal of Applied Mathematics 62, 1019–1043 (2002)
10. Bertalmio, M., Sapiro, G., Ballester, C.: Image Inpainting.Computer, Graphics, SIGGRAPH (2000)
11. Joyeux, L., Buisson, O., Besserer, B.: Detection and Removal of Line Scratches in Motion Picture Films. In: Proceedings of CVPR'99. In: IEEE Int. Conf. on Computer Vision and Pattern Recognition, FortCollins (1999)
12. Oshi, S., Srivastava, A., Mio, W.: Hierarchical Organization of Shapes for Efficient Retrieval. In: Proc. ECCV 2004, pp. 570–591 (2004)
13. Kumar, M.P., Torr, Zisserman, P.H.S.: Obj. Cut. In: Proceedings of the IEEE Conference on Computer Vision and Pattern Recognition. vol. 3 (2005)

14. Authors from the same group: Compositional boosting for computing hierarchical image structures. CVPR2007 (submitted) (2007)
15. Kimia, B.B., Frankel, I., Popescu, A.M.: Euler spiral for shape completion. International journal of computer vision 54, 159–182 (2003)
16. Mumford, D., Shah, J.: Optimal approximations of piecewise smooth functions ans associated variatioanl problems. comm. in pure and appl. Math 42, 577–685 (1989)
17. Saund, E.: Perceptual organization of occluding contours generated by opaque surfaces. CVPR 19999, 624–630 (1999)
18. Shum, H.: Prior, Context and Interactive Computer Vision. The Microsoft Research Asia Technical Report (2006)
19. Horn, B.K.P.: The curve of least energy. ACM Transactions on Mathematical Software 9, 441–460 (1983)
20. Ballester, C., Bertalmio, M., Caselles, V.: Filling-In by Joint Interpolation of Vector Fields and Gray Levels. IEEE Transactions on Image Processing 10, 1200–1211 (2001)
21. Kolmogorov, V., Zabih, R.: What Energy Functions Can Be Minimized via Graph Cuts. IEEE Transactions on Pattern Analysis and Machine Intelligence 26, 147–159 (2004)
22. Fridman, A.: Mixed Markov models. Applied mathematics. PNAS 100(14), 8092–8096 (2003)
23. Gilks, W.R., Richardson, S., Spiegelhalter.: Markov Chain Monte Carlo In practive. Chapman and Hall, Sydney (1996)
24. Saund, E.: Perceptual organization of occluding contours generated by opaque surfaces. In: Proceedings of the 1999 Conference on Computer Vision and Pattern Recognition, pp. 624–630 (1999)
25. Geiger, D., Kumaran, K., Parida, L.: Visual organization for figure/ground separation. In: Proc. IEEE Conf. on Computer Vision and Pattern Recognition, pp. 155–160 (1996)
26. Adelson, E.A., Wang, J.Y.A.: Representing Moving Images with Layers. IEEE Trans. on Image Processing 3, 625–638 (1994)
27. Wang, J., Gu, E., Betke, M.: MosaicShape: Stochastic Region Grouping with Shape Prior. Computer Vision and Pattern Recognition 1, 902–908 (2005)
28. Efros, A.A., Freeman, W.T.: Image Quilting for Texture Synthesis and Transfer. In: Proceedings of SIGGRAPH '01, Los Angeles, California, (August 2001)

Dichromatic Reflection Separation from a Single Image

Yun-Chung Chung[1], Shyang-Lih Chang[2], Shen Cherng[3], and Sei-Wang Chen[1]

[1] Department of Computer Science and Information Engineering
National Taiwan Normal University, Taiwan
[2] Department of Electronic Engineering, St. John's University, Taiwan
[3] Department of Electrical Engineering, Cheng Shiu University, Taiwan
schen@csie.ntnu.edu.tw

Abstract. A feature-based technique for separating specular and diffuse components of a single image is presented. In the proposed approach, Shafer's dichromatic reflection model is utilized, which assumed a light reflected from a surface point is linearly composed of diffuse and specular reflections. The major idea behind the proposed method is to classify the boundary pixels of the input image to be specular-related or diffuse-related. A fuzzy integral process is proposed to classify boundary pixels based on their local evidences, including specular and diffuse estimation information. Based on the classification result of boundary pixels, an integration method is evoked to reconstruct the specular and diffuse components of the input image, respectively. Unlike previous researches, the proposed method has no color segmentation and iterative operations. The experimental results have demonstrated that the proposed method can perform dichromatic reflectance separation effectively with small misadjustments and rapid convergence.

Keywords: dichromatic reflection separation, specular and diffuse components, fuzzy integral classification.

1 Introduction

When a light hits a surface of opaque objects, three situations, absorbing, scattering, and reflecting may occur as shown in Fig. 1. The reflected light from the surface is called the reflecting light. Part of light may pierce through the surface. Some of the piercing light is absorbed, and the other returns back to the air after a number of back and forth bouncing between molecules. The latter light is called scattering light. The scattering light together with reflecting light forms the reflection light in this study.

The reflecting of the incoming light is influenced by albedo of the surface, and we denote albedo as $\rho(\theta_i, \theta_o)$, where θ_i is the angle between incoming light and surface normal, and θ_o is the angle between reflecting light and surface normal. The reflecting light, E_R, can be modeled as $E_R = \rho(\theta_i, \theta_o) E_I$, where E_I is the incoming light. For rough surfaces, the reflecting may uniformly scatter at all possible angle θ_o. But for smooth surfaces, a major reflection direction often appears at a particular angle θ_o that depends on both the angle of the incoming irradiance θ_i and the surface normal, which

A.L. Yuille et al. (Eds.): EMMCVPR 2007, LNCS 4679, pp. 225–241, 2007.
© Springer-Verlag Berlin Heidelberg 2007

can be often observed as highlight reflection. For imaging, we call the strong reflection to be the specular reflection, and all the remaining reflection to be the diffuse reflection. Shafer [12] proposed a dichromatic reflection model to describe the combination of the two reflections, which assumes the reflection be linearly composed of the specular and diffuse components.

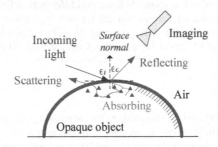

Fig. 1. The interaction of a light hits a surface of opaque objects

The specular reflection is often an obstacle to computer vision applications. Specular reflection on a smooth surface generates undesired highlight area and also changes color properties which annoy object recognition processes. Most of computer vision applications assumed perfect diffuse surface and ignored the specular reflection effect. However, specular reflection is inevitable in natural scenes, and the method to separate specular and diffuse reflections is desired. In addition, the separated specular and diffuse reflection images can provide extra information about the types of surface materials and smoothness condition of the observed objects.

Many techniques have been developed to separate specular and diffuse reflections from images. They can be broadly categorized into three classes: multiple images, segmented color surfaces in a single image, and single image using pixel-based analysis. First, in early researches, multiple images were used. Wolff and Boult [4] proposed a polarization-based method to separate reflection components. A set of polarization filters with different polarization angles are utilized to generate several gray images with different polarizing effects. According to the difference among multiple images, highlights are located. Later, Nayar et al. [13] extended their work by integrating color information with polarizing filters. Criminisi et al. [1] uses a collection of images taken under different locations of light sources to create a spatial-temporal EPI plane. EPI-strips are extracted and each individual EPI-strip is analyzed and decomposed it into its specular and diffuse components. Lin and Shum [10] proposed a neutral interface reflection model for separating the diffuse and specular reflection components in color images. From two photometric images, the RGB intensities of the two reflection components are computed for each pixel using a linear model of surface reflectance. Many other researches using multiple images to separate specular and diffuse components are in [11][15][17]. All of them have successful separation results on multiple images; however, using several input images is not applicable under many circumstances.

Later, researchers tried to solve the separation problem using segmented color surfaces in a single image, and analyzed one segmented color surface at a time. To analyze colors, the dichromatic reflection model is proposed by Shafer [12] and used by Klinker et al. [2] and Bajscy et al. [7] to develop specular and diffuse separation methods. They analyzed the color distribution on the single color object surface, and transformed to their specified color space to separate specular and diffuse pixels based on the color information. One common limitation of these methods is they all required color segmentation beforehand to mark a single color area before performing color analysis. Color segmentation techniques have been applied to help categorizing color regions; however, texture surfaces and complex scenes are very common in real world scenes, and the applications of their methods are quite limited.

Recently, researchers tried to use pixel-based approach to avoid the above limitation of single color surfaces. Tan and et al. [6] successfully use image inpainting technique to remove highlight based on local information of pixels. Unlike traditional inpainting, occluded image regions are directly filled by neighboring pixels. Highlight color analysis and uniform illumination color are utilized to estimate the underlying diffuse color. The experimental results demonstrate their method works well. However, it requires manual interactions to mark highlight regions a priori, and image inpainting methods are iterative in nature. Mallick et al. [14] transformed image pixels to a SUV space where specular pixels have particular grouping areas. They proposed a partial differential equation (PDE) to describe the property, and the PDE iteratively erodes the specular component at each pixel to remove the specularity. For different image types, i.e., textureless and highly textured scenes, a family of PDEs with multi-scale morphological templates are designed. Tan and Ikeuchi [8] proposed a specular-free image in which diffuse colors are not kept to estimate diffuse surface. The specular-free image is utilized to perform iterative operations (include specularity reduction and diffuse verification) to decrease the specularity of pixels to get diffuse values. Their method produced satisfied separation results. However, pixel-based approaches require iterative operations, which consume a lot of time to compute the pixel-wise local information to separate specular and diffuse components.

In this paper, we propose a feature-based approach to separate specular and diffuse components from a single image. Our method utilizes specular and diffuse estimation, and a fuzzy integral classification algorithm to classify the edge pixels. The proposed method does not require a priori color segmentation, and iterative operations of pixel-based approach are not needed, either. The key idea of the proposed method is to classify edges of the input image to be specular-related or diffuse-related, and to use integration method to reconstruct specular and diffuse images. The proposed method performs satisfied separation results in a short time. In addition, we have some comparisons with the recent Tan and Ikeuch's work [8].

The rest of this paper is organized as follows: we address the separation method in Sec. 2, and the information extraction methods are given in Sec. 3. Next, Sec. 4 is devoted to the boundary classification method using fuzzy integral technique. The performance of the proposed method is demonstrated in Sec. 5. Concluding remarks and future works are finally given in Sec. 6.

2 Reflection Image Decomposition

Figure 2 shows a flowchart for the proposed approach to extracting dichromatic reflection images from a single color image. The approach consists of four major steps referred to as: boundary generation, information extraction, boundary classification, and reflection image composition, respectively. The major idea behind the proposed method is to classify the boundary pixels of the input image to be specular-related or diffuse-related edge pixels. Based on the classification results, we reconstruct the specular and diffuse components of the input image.

Let $I = (I^r, I^g, I^b)$ denote the input color image, where I^r, I^g and I^b are the red, green and blue components of the input image. From Shafer's dichromatic reflection model [12], each color component I^i ($i = r, g, b$) is modeled as $I^i = S^i + D^i$, where S^i and D^i are the specular and diffuse components of I^i, respectively. We call them the dichromatic reflection components of I^i. The first three steps of the process of Fig. 2 are applied to each of the color components. In the last step, the final results are composed of the chromatic reflectance components of the three color component, R, G, and B.

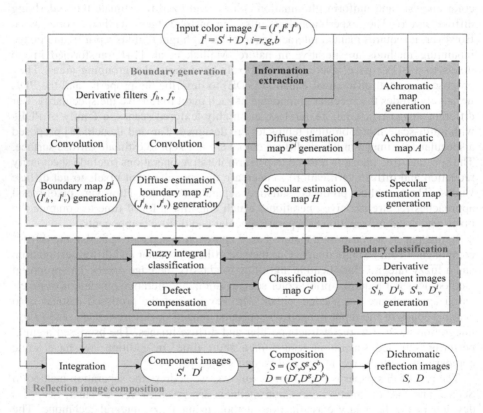

Fig. 2. Flowchart for extracting dichromatic reflection images from a single image

The decomposition process begins with the boundary generation stage, and the image I^i is convolved with a horizontal derivative filter f_h and a vertical derivative filter f_v, resulting in two derivative component images I^i_h and I^i_v. In this study, the Prewitt derivative filters are utilized. The boundary map B^i contains horizontal derivative values I^i_h and vertical derivative values I^i_v, i.e., $B^i = (I^i_h, I^i_v)$.

Next, highlight characteristics that will provide the criteria for the later classification process are extracted in the information extraction stage. Highlight characteristics are calculated as two maps, the specular estimation map, H, and the diffuse estimation map, P^i. The specular estimation map reveals the degrees of pixels belonging to highlight areas by calculating the pixel's intensity and saturation information, and the diffuse estimation map is defined as removing specular effects from the input image by properly shifting each pixel's intensity and maximum chromaticity nonlinearly. The detail processes of this step are addressed in the Sec. 3. These two maps, H and P^i, are used in the next stage for classifying the pixels of derivative components into specular-related or diffuse-related edge pixels. After the diffuse estimation map P^i is obtained, the diffuse estimation boundary map F^i is generated by convolving with the derivative filters, i.e, $F^i = (J^i_h, J^i_v) = (f_h * P^i, f_v * P^i)$.

In the boundary classification step, the diffuse estimation boundary map F^i and the specular estimation map H calculated in the previous step are utilized to classify the pixels of boundary map B^i and the result is the boundary classification map G^i. The fuzzy integral classification method is employed to classify the pixels of derivative components into specular-related or diffuse-related edge pixels. An additional defect compensation process is utilized to compensate for the classification results on the highlight corona area which may often fail in the classification process due to low edge magnitudes. The detail processes of this step are addressed in the Sec. 4.

Finally, in the reflectance image composition step, an integration [18] process is applied to the derivative components, from which the specular S^i and diffuse D^i component images are computed. From each convolution equation j, $f_j * S^i = S^i_j$, we design a reverse filter f^r_j and convolve with each equation as $f^r_j * f_j * S^i = f^r_j * S^i_j$. Summing all the equations over j, we have

$$\sum_j f^r_j * f_j * S^i = \sum_j f^r_j * S^i_j. \tag{1}$$

Next, let g be the normalization function, where $g * (\sum_j f^r_j * f_j) = \delta$, and δ is the Kronecker delta function.

The specular and diffuse components can be calculated as

$$S^i = g * (\sum_{j=h,v} f^r_j * S^i_j), \quad D^i = g * (\sum_{j=h,v} f^r_j * D^i_j). \tag{2}$$

Note that the f^r_j is a user-defined reversed function of f_j, and the g function can be obtained after f^r_j is defined. Here, the reversed function is defined as $f^r_j(p) = f_j(-p)$.

After obtaining color components, S^i and D^i, the dichromatic reflection images, S and D, of the input image I are directly composed by S^i and D^i as $S = (S^r, S^g, S^b)$ and $D = (D^r, D^g, D^b)$.

3   Information Extraction

In this section, we address how to calculate the specular estimation map, H, and the diffuse estimation map, P^i, for each image pixel. To this end, we temporarily treat pixels as edge pixels when calculating their characteristic values.

3.1 Specular Estimation Map

The specular estimation map, H, reveals the degree of a pixel belonging to the highlight area based on the pixel's intensity and saturation information. First, consider a pixel $p(x, y) \in$ image I, define an achromatic map as $A(x, y) = true$, while the color $r \approx g \approx b$. Otherwise, $A(x, y) = false$. From the definition, $A = true$ represents the R, G, B values of pixel p are very close, which means it is an achromatic pixel; otherwise it is a chromatic pixel. Next, consider only chromatic pixels based on the work of Lehmann and Palm [16], we design the specular estimation map H while $A(x, y) = false$, let $H(x, y) = (e_I(x, y), e_S(x, y))$, and

$$e_I(x, y) = \frac{1}{I_{max}} \sum_{i=r,g,b} I^i(x, y), \text{ and} \tag{3}$$

$$e_S(x, y) = 1 - \frac{1}{S_{max}} + \frac{3 \min_{i=r,g,b}(I^i(x, y))}{S_{max} \sum_{i=r,g,b} I^i(x, y)}, \tag{4}$$

where the maximum intensity and saturation in the whole image are normalization terms and denoted as

$$I_{max} = \max_{x,y} \sum_{i=r,g,b} I^i(x, y) \quad, \quad \text{and} \quad S_{max} = 1 - 3 \max_{x,y} \left\{ \min_{i=r,g,b}(I^i(x, y)) / \sum_{i=r,g,b} I^i(x, y) \right\},$$

respectively.

(a)   (b)   (c)

Fig. 3. Specular estimation map, (a) a sample image, (b) e_I map, (c) e_S map

In Eq. (3), $e_I(x, y)$ is designed to show the ratio of the intensity of pixel $p(x, y)$ to the maximum intensity in the whole image. The larger value e_I the pixels have, the more they belong to specular component in terms of intensity; i.e., more bright the pixel is, the degree of it belonging to specular is higher. An example of e_I map (normalized) of input image Fig. 3(a) is shown in Fig. 3(b).

In Eq. (4), $e_S(x, y)$ is designed to show the degree of the saturation of pixel $p(x, y)$ to the maximum saturation in the whole image. From the definition of saturation, let

$$S(x, y) = 1 - \min_{i=r,g,b}(I^i(x, y)) / [\sum_{i=r,g,b} I^i(x, y)/3].$$

Since specular areas are tend to be unsaturated, i.e. the smaller S, we define

$$S_{max} - S(x, y) = S_{max} - 1 + \frac{\min_{i=r,g,b}(I^i(x,y))}{\sum_{i=r,g,b} I^i(x,y)/3} > 0, \tag{5}$$

and the larger value the pixels have, the more they belong to specular component in terms of saturation. Normalize Eq. (5) with S_{max}, we can obtain Eq. (4). An example of e_S map of input image Fig. 3(a) is shown in Fig. 3(c).

With e_I and e_S, the specular estimation map H describes the degree of the pixels which belong to diffuse or specular components. For achromatic pixels, i.e., $\forall p(x, y)$, $A(x, y) = true$, we just let $H(x, y) = $ (NULL, NULL).

3.2 Diffuse Estimation Map

The diffuse estimation map removes specular effects from the input image by properly shifting each pixel's intensity and maximum chromaticity nonlinearly. Namely, the diffuse estimation image is the image with specular components removed from the input image, and keeps the diffuse components only. However, the surface color is not kept in the diffuse estimation image since the color of the original diffuse components is unknown and assigned randomly, and the color is changed nonlinearly. This technique is modified from [8], where it was named as specular-free image.

Refer to Fig. 4, the diffuse estimation map generation process requires the achromatic map, A, and specular estimation map, H, calculated from the previous

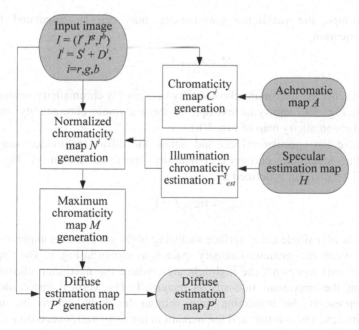

Fig. 4. Flowchart for diffuse estimation map, P^i, generation

section. The diffuse estimation map generation process will be addressed following the flowchart shown in Fig. 4. First, we calculate the chromaticity map, C^i, which considers only chromatic pixels ($A = false$,) and it is defined as

$$C^i = I^i / \sum_{i=r,g,b} I^i . \tag{6}$$

Note that the chromaticity map, C^i, is also known as the normalized RGB model, and Fig. 5(a) shows an example of the chromaticity map of the input image of Fig. 3(a). The chromaticity map reveals the degree of chromaticity of each pixel.

Next, we calculate the environmental illumination chromaticity estimation, Γ^i_{est}. We will use it later to remove the environmental illumination from the image chromaticity. To calculate Γ^i_{est}, first, the pixels of high magnitudes of the specular estimation map H are transformed to the inverse-intensity chromaticity space [9]. In this space, the correlation between image chromaticity and illumination chromaticity becomes linear. Base on the linear correlation relationship, Hough transform and histogram analysis are utilized to estimate Γ^i_{est}.

The normalized chromaticity map, N^i, is then defined as

$$N^i = C^i / \Gamma^i_{est}, \tag{7}$$

which removes environmental illumination chromaticity effect. Fig. 5(b) shows an example of the normalized chromaticity map of the chromaticity map of Fig. 5(a).

Subsequently, the maximum chromaticity map can be calculated from the following equation

$$M = \max_{i=r,g,b} N^i, \tag{8}$$

which represents the maximum chromaticity of a pixel's chromaticity among its r, g, b channels. Fig. 5(c) shows an example of the maximum chromaticity map of the normalized chromaticity map of Fig. 5(b).

To remove the specular effects and obtain the diffuse estimation map, P^i, we consider the maximum chromaticity-intensity space as shown in Fig. 6. The maximum intensity, I_M, is defined as

$$I_M = \max_{i=r,g,b} I^i / \Gamma^i_{est}. \tag{9}$$

For pixels of a single color surface including highlight area, a sample distribution on the maximum chromaticity-intensity space is as shown in Fig. 6. The upper curve (denoted as red) represents the highlight area, where the maximum chromaticity is reduced but the maximum intensity is increased. The vertical curve (denoted as orange) represents the remaining area without highlight, where the maximum chromaticity remains constant and the maximum intensity varies according to lighting intensity.

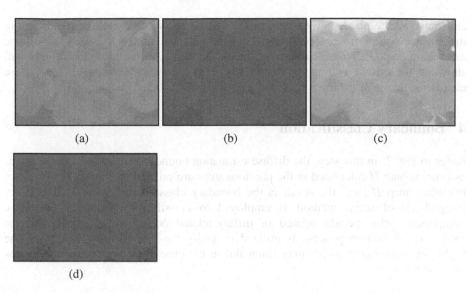

(a)   (b)   (c)

(d)

Fig. 5. The processes to calculate the diffuse estimation image of Fig. 3(a), (a) chromaticity map, C^i, (b) normalized chromaticity map, N^i, (c) maximum chromaticity map, M, and (d) diffuse estimation map, P^i with given $D_a = 0.5$

See Fig. 6, to remove the specular effect means to correct the highlight curve to be vertical non-highlight value, in this example, we can simply move the highlight pixels horizontally to the constant maximum chromaticity to eliminate the highlight effect.

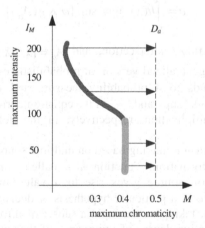

Fig. 6. A sample distribution of pixels of a single color surface (including highlight area) on the maximum chromaticity-intensity space

However, in real world scenes, multiple-color objects do not have a simple pattern like the example, i.e., there is no constant maximum chromaticity for all colors. Here, we can assign a pseudo constant diffuse maximum chromaticity, D_a ($1/3 \leq D_a \leq 1$), and move all the pixels horizontally to D_a. The highlight effect will be removed, but the surface colors are not preserved, and the adjusted image is called the diffuse estimation map, P^i. Note that given different D_a values, different surface colors will show on the diffuse estimation image. Fig. 5(d) shows an example of the diffuse estimation image of the input image of Fig. 3(a) as $D_a = 0.5$.

4 Boundary Classification

Refer to Fig. 2, in this step, the diffuse estimation boundary map F^i and the specular estimation map H calculated in the previous steps are utilized to classify the pixels of boundary map B^i, and the result is the boundary classification map G^i. The fuzzy integral classification method is employed to classify the pixels of derivative components into specular-related or diffuse-related edge pixels. In addition, the Defect compensation process is utilized to verify the classification results on the highlight corona area which may often fail in the classification process due to low edge magnitudes.

4.1 Fuzzy Integral

Fuzzy integral [3] is a nonlinear numeric approach for integrating multiple sources of uncertain information or evidence to attain a value, which expresses the degree of confidence in a particular hypothesis or decision. Let $h: X \rightarrow [0, 1]$ be a function defined on a set X and $g: P(X) \rightarrow [0, 1]$ be a set function defined over the power set of X. The fuzzy integral [3] of function $h(\cdot)$ with respect to function $g(\cdot)$ is defined as

$$e = \int_X h(x) \cdot g = \sup_{\alpha \in [0,1]} \{\alpha \wedge g(A_\alpha)\}, \tag{10}$$

where $\wedge$ denotes the fuzzy intersection, and $A_\alpha = \{x \in X \mid h(x) \geq \alpha\}$. The fuzzy integral is in a sense a generalized version of probabilistic expectation. Accordingly, function $g(\cdot)$ corresponds to a probability measure and α to its probability of occurrence. The operators "sup" and "$\wedge$" in the equation correspond to the operations of integration and multiplication, respectively, in the definition of probabilistic expectation.

In applications of decision making based on multiple sources of information, set X collects sources of information. Function $h(\cdot)$, called confidence function, after receiving an information source gives rise to a value indicating the degree of confidence of the source in a particular hypothesis or decision. Function $g(\cdot)$, called fuzzy measure function, taking as the input a subset of information sources gives a value indicating the relative degree of importance of the set of sources to the other sources. If $g(\cdot)$ has a singlet $\{x\}$ (i.e., a set contains one and only one element) as its input, $g(\{x\})$ is called the fuzzy density of element x. Let $s(\cdot)$ denote the fuzzy density function defined as $s(x) = g(\{x\})$.

In practice, fuzzy densities of information sources are readily evaluated. Fuzzy measures of subsets of information sources are derived from the fuzzy densities. Sugeno [5] provided a computational framework for calculating fuzzy measures from fuzzy densities. Let A and B be two disjoint subsets of X. The Sugeno measure function, denoted by $g_\lambda(\cdot)$, is given by

$$g_\lambda(A \cup B) = g_\lambda(A) + g_\lambda(B) + \lambda g_\lambda(A) g_\lambda(B), \tag{11}$$

with boundary conditions $g_\lambda(\phi) = 0$ and $g_\lambda(X) = 1$. For any subset S of X, the fuzzy measure of S can be computed by recursively invoking Eq. (11),

$$g_\lambda(S) = (\prod_{x_i \in S} (1 + \lambda s(x_i))) / \lambda \cdot \tag{12}$$

This equation together with the boundary conditions constrain λ to satisfy the relationship function as

$$\lambda + 1 = \prod_{x_i \in X} (1 + \lambda s(x_i)). \tag{13}$$

Referring to Eq. (10), there will be $2^{|X|}$ subsets of X needed to perform the fuzzy integral. Eq. (10) can be simplified by presorting the information sources in X. Let $X' = \{x'_1, x'_2, \cdots, x'_{|X|}\}$ be the sorted version of X, in which $h(x'_1) \geq h(x'_2) \geq \cdots \geq h(x'_{|X|})$. Eq. (10) can then be rewritten as

$$e = \int_{X'} h(x) \cdot g = \sup_{\alpha \in [0,1]} \{\alpha \wedge g(A_\alpha)\} = \bigvee_{1 \leq i \leq |X|} [h(x'_i) \wedge g(X'_i)], \tag{14}$$

where $X'_i = \{x'_1, x'_2, \cdots, x'_i\}$. The above equation greatly reduces the number of subsets required to perform the fuzzy integral from $2^{|X|}$ (by Eq. (10)) to only $|X|$.

4.2 Fuzzy Decision

We now address how to integrate available information to arrive at a concordant decision using fuzzy integral technique. Since the goal of the fuzzy integral classification algorithm is to categorize an edge pixel as being specular-related or reflection-related, the decision hypotheses are assigned as two classes as $C = \{c_1, c_2\}$. Let $c_1 = 1$ and $c_2 = 0$ represent diffuse and specular edge classes, respectively.

In addition, for each decision hypothesis c_j, we have the decision value set $V_j = \{v_1, v_2, v_3\}$. The decision value v_1 is the ratio between the original boundary map B^i and the diffuse estimation boundary map F^i, which is defined as

$$v_1(x, y) = \frac{1}{3} \sum_{i=R,G,B} \left(\sum_{j=h,v} J^i_j(x, y) / \sum_{j=h,v} I^i_j(x, y) \right), \tag{15}$$

where $J^i_j \in F^i$, and $I^i_j \in B^i$. When the considered boundary pixel has lower diffuse estimation to original boundary value, the chance of the boundary pixel belonging to specular-related category should be lower, and the v_1 value will be also lower.

The decision values v_2, v_3 are designed to describe the neighborhood support from intensity and saturation information, respectively, which are given by

$$v_2(x, y) = 1 - \frac{1}{n_I} \sum_{p \in n(x,y)} e_I(p), \text{ and} \tag{16}$$

$$v_3(x, y) = 1 - \frac{1}{n_S} \sum_{p \in n(x,y)} e_S(p), \tag{17}$$

where $n(x, y)$ is the non-boundary neighbor pixels of (x, y), and $\{e_I, e_S\} \in H$. For non-boundary neighbor pixel $p \in n(x, y)$, the edge magnitude of p satisfies $\sum_{i=R,G,B} I_h^i(p) < \xi_h$ and $\sum_{i=R,G,B} I_v^i(p) < \xi_v$, where ξ_h and ξ_v are small constants, and $I_j^i \in B^i$. Only nearby pixels within a pre-defined distance are considered, and n_I and n_S are the normalization terms which denote the numbers of non-boundary neighbor pixels of v_2, v_3, respectively.

The degree of confidence of source x_i on hypothesis c_j is defined based on the distance between the decision values and the hypothesis value as

$$h_i^j(v_i^j, c_j) = e^{-\beta|v_i^j - c_j|}, \tag{18}$$

where β is a constant and $0 \le \beta \le 1$.

Sugeno's method is employed to obtain the fuzzy integral results. Sugeno's measure function $g(.)$ in Eq. (11) can be solve by Sugeno boundary conditions constrain λ in Eq. (13), which can be solved by

$$\lambda + 1 = (1 + \lambda s_1)(1 + \lambda s_2)(1 + \lambda s_3). \tag{19}$$

Simplify Eq. (19), we obtain a quadratic equation in λ as

$$s_1 s_2 s_3 \lambda^2 + (s_1 s_2 + s_1 s_3 + s_2 s_3)\lambda + (s_1 + s_2 + s_3 - 1) = 0. \tag{20}$$

With calculated $g(.)$ and $h(.)$, the fuzzy integral results can be obtained by Eq. (14). Let G^i be the classification map, which contains the fuzzy integral classification results of all pixels. For a pixel $p(x, y)$, the value $G^i(x, y) = true$ represents p is classified as a specular-related pixel, $G^i(x, y) = false$ represents p is classified as a diffuse-related pixel, and $G^i(x, y) = NULL$ means p is not yet classified which will be determined in the next step.

4.3 Defect Compensation

In the previous step, some edge pixels may not be classified due to low magnitude of edge values, especially the corona area around specular highlights. For these pixels, we introduce the defect compensation process which incrementally determines their edge types through progressive propagations of local evidence with predefined morphological operators.

Consider the specular-related pixels of the classification map G^i obtained in the previous step, denote as G_S^i. First, we apply the binary morphological dilation operation on the specular-related pixels map G_S^i with a pre-defined small operator.

This operation extends the areas of specular-related pixels to their surrounding areas, which are the potential low magnitude corona areas, and we call the extended areas as the corona area candidate map, X^i.

Next, we examine each pixel belongs to the corona area candidate map but not contains in the specular-related pixels map, i.e. $\forall p \in X^i$ and $p \notin G^i_S$. If $B^i(p) < \xi_B$, we add p into G^i_S, (i.e., let $p \in G^i_S$), where ξ_B is a small constant. On the other hand, if $B^i(p) > \xi_B$, which means p is not possible a corona pixel, and it will be skipped. After checking all the candidate pixels in one iteration, we will go back to repeat the operations for certain times within a small exploring radius to make sure all the corona pixels are included in G^i_S.

Based on the compensated resultant classification map G^i, we generate specular-related $S^i_d = (S^i_h, S^i_v)$ and diffuse-related $D^i_d = (D^i_h, D^i_v)$ derivative component images from derivative component images I^i_h and I^i_v as

$$S^i_d = (S^i_h, S^i_v) = \{B^i(x, y) \mid G^i(x, y) = true\}, \text{ and}$$

$$D^i_d = (D^i_h, D^i_v) = \{B^i(x, y) \mid G^i(x, y) = false\}, \tag{21}$$

and we call S^i_h, S^i_v, D^i_h, and D^i_v the dichromatic reflection derivative component images.

5 Experimental Results

Both synthetic and real images are tested in our experiments. Each input image is an RGB color image of size 320×240 pixels. The output results include a specular image and a diffuse image extracted from the input image. The source code was written in C++ run on a 2.4 GHz Pentium4 based PC with 512MB RAM. The program takes about two seconds to decompose an input image into its specular and diffuse images.

Figure 7 shows a synthetic teapot image with highlight reflection on it; Fig. 7(a) is the input image, and Fig. 7(b) illustrates the classification result of specular-related pixels. We present the specular-related pixels using their original pixel values in Fig. 7(a), and for diffuse-related pixels, we denote them as black pixels. Note that there are some black pixels in the middle center of the specular highlight reflection area, which are fully saturated pixels (the intensity of red, green and blue are all at maximum value, i.e. 255), and they will be treated as achromatic pixels in A. For these pixels, we recover them as specular-related pixels in the defect compensation process, and the result is shown in Fig. 7(c). Finally, the dichromatic reflection separation results are shown in Fig. 7(d) as the specular image, and Fig. 7(e) as the diffuse image. Note that on the diffuse image Fig. 7(e), we observe discontinuous intensities on the teapot body, and it is because that the removal of highlight area in Fig. 7(d) on the teapot body makes the surface reflection information is also removed, and compared with the specular reflection, the diffuse reflection can be neglected. In addition, for better visually effect, we can propagate the surrounding pixel values to specular area to compensate the discontinuing.

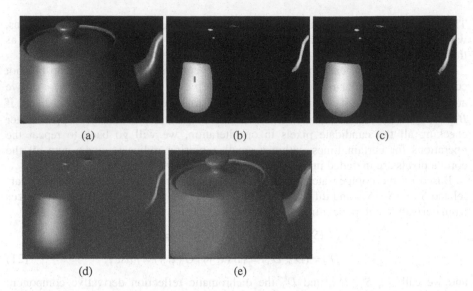

(a)   (b)   (c)

(d)   (e)

Fig. 7. Dichromatic reflection separation result (synthetic image), (a) input image, (b) fuzzy integral result, (c) defect compensation result, (d) specular image, and (e) diffuse image

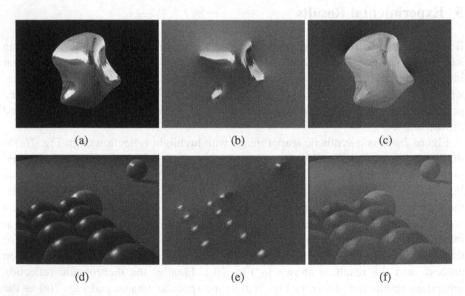

(a)   (b)   (c)

(d)   (e)   (f)

Fig. 8. Dichromatic reflection separation results (synthetic images), (a, d) input images, (b, e) specular images, and (c, f) diffuse images

Next, we test the reflection separation algorithm on two more different types of material objects. Fig. 8(a, d) are the input images, where Fig. 8(a) is mental surface, and Fig. 8(d) is plastic billiards. The separation results are shown in Fig. 8(b, e) and

Fig. 8(c, f), which are the specular images and the diffuse images, respectively. In addition, in Fig. 8, some real image examples are illustrated. Fig. 9(a, d) are the input images, and the separation results are shown in Fig. 9(b, e) and Fig. 9(c, f), which are the specular images and the diffuse images, respectively. The separation results demonstrate that our algorithm can effectively separate the input images into specular and diffuse images.

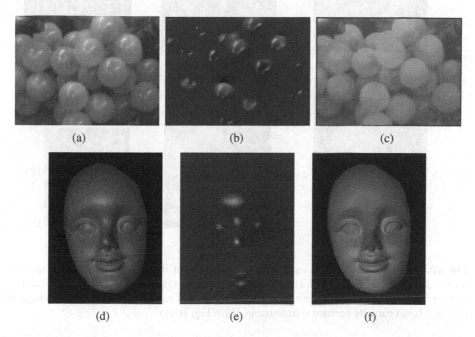

(a)   (b)   (c)

(d)   (e)   (f)

Fig. 9. Dichromatic reflection separation results (real images), (a, d) input images, (b, e) specular images, and (c, f) diffuse images

To compare our method with Tan and Ikeuchi's [8], see a plastic doll example image in Fig. 10. Fig. 10 (a) is the input image, Fig. 10 (b) is our specular image, and Fig. 10 (c) is the specular image obtained by Tan and Ikeuchi [8]. In addition, Fig. 10 (d) is our diffuse image, and Fig. 10 (e) is the diffuse image obtained by Tan and Ikeuchi [8]. To illustrate the comparison, we zoom in some part of the images. Fig. 10 (f) is a partially zoomed input image, Fig. 10 (g) is partially zoomed on our diffuse image, and Fig. 10 (h) is partially zoomed diffuse image of [8]. From the zoomed images, compare the original image Fig. 10 (f) and the result of our methods in Fig. 10 (g), Fig. 10 (g) preserves most details of the input image, while Fig. 10 (h) from [8] smoothes the image by iteratively pixel modification operations. In addition, our method has visually the same good separation results but a much faster speed since the computation time of pixel-base approach reported in [8] is several minutes, while we only require about two seconds.

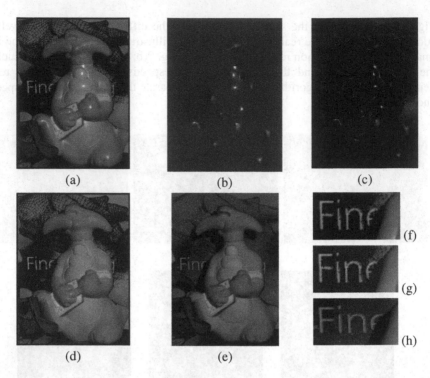

Fig. 10. Dichromatic reflection separation results, (a) input image, (b) the proposed specular image, and (c) specular image of Tan and Ikeuchi's [8], (d) the proposed diffuse image, and (e) diffuse image of [8], (f) partially zoomed input image, (g) partially zoomed diffuse image of Fig. 10 (d), (h) partially zoomed diffuse image [8] of Fig. 10 (e)

6 Concluding Remarks and Future Works

In this paper, we present a novel and fast feature-based approach to separate the specular and diffuse images from a single image. The major idea of the proposed method is to classify the boundary edge of the input image to be specular-related or diffuse-related components, and then using the integration method to decompose the specular and diffuse images. Shafer's dichromatic reflection model is utilized and local information (specular estimation and diffuse estimation maps) are integrated with a fuzzy integral classification algorithm. Compared with previous researches, the proposed method does not require color segmentation and also avoid iterative operations of pixel-based approaches. The experimental results demonstrate that the proposed method performs satisfied separation results in a short time.

The separation results are based on the classification results of the boundary pixels, and currently specular estimation and diffuse estimation information are employed to be the cues to classify the boundary pixels. Additional information can be developed to robust the classification process, such as shape information. We will continue to study related information to improve the results.

Acknowledgments. This work is supported in part by the National Science Council, Taiwan, Republic of China under contract NSC-94-2213-E-003-008.

References

[1] Criminisi, A., Kang, S.B., Swaminathan, R., Szeliski, S., Anandan, P.: Extracting Layers and Analyzing Their Specular Properties Using Epipolar Plane Image Analysis. Technical Report MSR-TR-2002-19, Microsoft Research (2002)

[2] Klinker, G.J., Shafer, S.A., Kanade, T.: The Measurement of Highlights in Color Images. Int'l. J. Computer Vision 2, 7–32 (1990)

[3] Keller, J.M., Krishnapuram, R.: Fuzzy decision models in computer vision. In: Yager, R.R., Zadeh, L.A. (eds.) Fuzzy Sets, Neural Networks, and Soft Computing. Van Nostrand Reinhold, New York, pp. 213–232 (1994)

[4] Wolff, L.B., Boult, T.: Constraining Object Features Using Polarization Reflectance Model. IEEE Trans. Pattern Analysis and Machine Intelligence 13(7), 635–657 (1991)

[5] Sugeno, M.: Fuzzy measure and fuzzy integrals: a survey. Fuzzy Automatic and Decision Processes, North Holland, Amsterdam, pp. 89–102 (1977)

[6] Tan, P., Lin, S., Quan, L., Shum, H.-Y.: Highlight removal by illumination-constrained inpainting. In: IEEE Int'l. Conf. on Computer Vision, Nice, France, pp. 164–169 (2003)

[7] Bajscy, R., Lee, S.W., Leonardis, A.: Detection of Diffuse and Specular Interface Reflections by Color Image Segmentation. Int'l J. Computer Vision 17(3), 249–272 (1996)

[8] Tan, R.T., Ikeuchi, K.: Separating reflection components of textured surfaces using a single image. IEEE Trans. on Pattern Analysis and Machine Intelligence 27(2), 178–193 (2005)

[9] Tan, R.T., Nishino, K., Ikeuchi, K.: Color Constancy through Inverse-Intensity Chromaticity Space. J Opt Soc Am A Opt Image Sci Vis 21(3), 321–334 (2004)

[10] Lin, S., Shum, H.Y.: Separation of Diffuse and Specular Reflection in Color Images. In: Proc. IEEE Conf. Computer Vision and Pattern Recognition (CVPR '01) (2001)

[11] Lin, S., Li, Y., Kang, S.B., Tong, X., Shum, H.Y.: Diffuse-Specular Separation and Depth Recovery from Image Sequences. In: Proc. European Conf. Computer Vision (ECCV '02), pp. 210–224 (2002)

[12] Shafer, S.: Using Color to Separate Reflection Components. Color Research and Applications 10, 210–218 (1985)

[13] Nayar, S.K., Fang, X.S., Boult, T.: Separation of Reflection Components Using Color and Polarization. Int. J. of Computer Vision 21(3), 163–186 (1997)

[14] Mallick, S.P., Zickler, T.E., Belhumeur, P.N., Kriegman, D.J.: Specularity Removal in Images and Videos: A PDE approach. In: European Conference on Computer Vision, Graz, Austria, vol. I, pp. 550–563 (May 2006)

[15] Lee, S.W., Bajcsy, R.: Detection of Specularity Using Color and Multiple Views. Image and Vision Computing 10, 643–653 (1992)

[16] Lehmann, T.M., Palm, C.: Color line search for illuminant estimation in real-world scene. Journal of the Optical Society of America A 18(11), 2679–2691 (2001)

[17] Sato, Y., Ikeuchi, K.: Temporal-Color Space Analysis of Reflection. J. Optics Soc. Am. A, 11 (1994)

[18] Weiss, Y.: Deriving Intrinsic Images from Image Sequences. In: IEEE Int'l. Conf. on Computer Vision, vol. 2, pp. 68–75 (2001)

Noise Removal and Restoration Using Voting-Based Analysis and Image Segmentation Based on Statistical Models

Jonghyun Park, Nguyen Trung Kien, and Gueesang Lee

Dept. of Computer Science, Chonnam National University, 300 Yongbong-dong, Buk-gu, Gwangju, Korea, Tel.: +82-62-530-0147, Fax.: +82-62-530-1917
jhpark@chonnam.ac.kr, trung_kien_kg@yahoo.com, gslee@chonnam.ac.kr

Abstract. Restoration and segmentation in corrupted text images are very important processing steps in digital image processing and several different methods were proposed in the open literature. In this paper, the restoration and segmentation problem in corrupted color text images are addressed by tensor voting and statistical method. In the proposed approach, we assume to have corruptions in text images. Our approach consists of two steps. The first one uses the tensor voting algorithm. It encodes every data point as a particle which sends out a vector field. This can be used to decompose the pointness, edgeness and surfaceness of the data points. And then noises in a corrupted region are removed and restored by generalized adaptive vector sigma filters iteratively. In the second step, density mode detection and segmentation using statistical method based on Gaussian mixture model are performed in values according to hue and intensity components in the image. The experimental results show that proposed approach is efficient and robust in terms of restoration and segmentation corrupted text images.

Keywords: corrupted image restoration, segmentation, tensor field, voting.

1 Introduction

Due to fast growing of the mobile systems, there is an increased demand for high performance and robust image processing in natural scene images. More and more mobile imaging systems, such as camera phone, PDA and digital camera, incorporate several image processing algorithms. In general, the images obtained by a digital camera are used for various purposes such as sending multimedia message, storage, printing, vision system for recognition and authentication, etc. And the images are including various signs. Especially, text information in natural scene images is useful it can convey important information even though it is simple. In this paper, we so consider the text images with corruption regions by noises such as graffito, streaks cracks, etc. Therefore, noise suppression or noise removal is an important task in previous processing for pattern recognition or computer vision from corrupted natural scene images. In general, the results of the noise removal have a strong influence on the quality of the following

A.L. Yuille et al. (Eds.): EMMCVPR 2007, LNCS 4679, pp. 242–252, 2007.
© Springer-Verlag Berlin Heidelberg 2007

image processing techniques. Several techniques for noise removal are well established in image processing. However, it is not easy to adapt these techniques to damaged images from noises such as streaks, graffito, and surface corrosion.

In this paper, we show the analysis method of features like surface elements from a corrupted image. We propose a restoration method that uses the tensor voting as the first step to remove and restore the corrupted regions in an input image. In the second step, we perform a segmentation based on statistical method with mode values detected by mean shift theory to evaluate the result of restored images. The tensor voting can not only be used for handling 2D or 3D data but also to process motion fields or stereo data. In most cases the input data is of small scale in contrast to corrupted text image data and the output is only used for visualization in pixel or voxel representation. We have a look on the tensor voting in Section III. And then we show how the output of the tensor voting can analyze input data in 3D. Fig. 1 below shows the overall framework of our proposed method for text segmentation in corrupted text image from natural scene images.

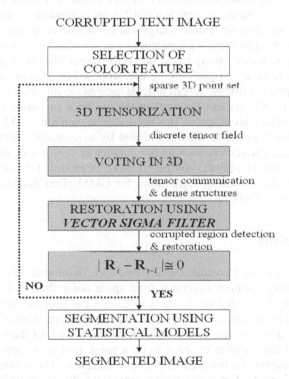

Fig. 1. The overall framework of proposed method to restore and segment in corrupted text images

2 Related Work

In general, segmentation refers to the low-level tasks of partitioning a corrupted image into disjointed and homogeneous regions which should be meaningful for a given application; this operation is usually preliminary to higher-level tasks such as object recognition, classification, and semantic interpretation.

Diverse approaches for common image segmentation have been investigated for a long time. Some segmentation algorithms only deal with gray scale images or RGB color space [1][2]. In general, the segmentation is sensitive to illumination, so results are somewhat poor. Image segmentation in the HSI color space, as proposed by C. Zhang and P. Wang, produces better results [3]. HSI space is therefore preferred in natural scenes and easily influenced by illuminations [4][5]. In general, natural scene images have diverse objects and, among them, texts are important objects since they convey important meanings for image understanding. The fact has inspired many efforts on text recognition in static images, as well as video sequences [6][7][8]. In [9], Qixiang Ye *et al.* use Gaussian mixture models (GMMs) in HSI color space with spatial connectivity information to segment characters from a complex background. They do not explicitly take into account the fact that characters in natural scenes can be severely corrupted by noises. In such cases, characters may not be segmented as separated objects due to the corruption of strokes which may cause errors when used as input in optical character recognition (OCR), as mentioned in the future work in [10]. Therefore, we propose feature analysis of chromatic or achromatic components based on tensor voting and text segmentation using separated clustering algorithm.

In this paper, the tensor voting based on vector sigma filter is used to detect or remove noises automatically. Tensor voting was proposed by Medioni *et al.* in [11][12], and has been applied to diverse research fields such as the inference of object boundaries: as a consequence, its use can explain the presence of noises based on surface saliency in the image feature space. This noise can be then removed by a vector sigma filter. The improved image is then segmented by a separated clustering algorithm. Clustering requires parameters such as the number or centroid of modes, which are generally not known as a priori. We use adaptive mean shift for automatic mode detection and use these modes as seed values for GMM. Text images are finally segmented as uniform objects.

3   Tensor Voting

Tensor voting is a unified computational framework developed by Guy and Medioni for inferring multiple salient structures from spare noisy data in 2-D or 3-D space [11][12]. The goal of the tensor voting is to extract the structure inherently given in the input data. The results of the tensor voting process are three continuous vector fields, represented by discrete grid points. The scalar a parts of these fields represent the likelihood of the location in space to be a point, part of a curve, a surface. The vector part represents the orientation of the occurrence. These three fields can be searched through to find maxima which represents the most likely location of a wanted feature.

3.1   Tensor Field in Physical Analogy

A tensor field is a very general concept of variable geometric quantity. It is used in differential geometry and the theory of manifolds, in algebraic geometry, in general relativity, in the analysis of stress and strain in materials, and in numerous applications in the physical sciences and engineering. It is a generalization of the idea of vector field, which can be considered as a vector that varies from point to point.

3.2 Tensor Encoding

The above mentioned field is a tensor field that means each point in space has an associated tensor. In our case this tensor is a second order symmetrical tensor in 3D. If we can formulate the tensor field like in (1), and it is encoded as a rotated 3D ellipsoid. We can describe the geometrical meaning of (1). The normalized vectors $(\mathbf{e}_1, \mathbf{e}_2, \mathbf{e}_3)$ are the main axes of the ellipsoid. They build a local right-handed coordinate system.

$$\mathbf{T} = \begin{bmatrix} \mathbf{e}_1 & \mathbf{e}_2 & \mathbf{e}_3 \end{bmatrix} \begin{bmatrix} \lambda_1 & 0 & 0 \\ 0 & \lambda_2 & 0 \\ 0 & 0 & \lambda_3 \end{bmatrix} \begin{bmatrix} \mathbf{e}_1^T \\ \mathbf{e}_2^T \\ \mathbf{e}_3^T \end{bmatrix} \tag{1}$$

$$\mathbf{T} = \lambda_1 \mathbf{e}_1 \mathbf{e}_1^T + \lambda_2 \mathbf{e}_2 \mathbf{e}_2^T + \lambda_3 \mathbf{e}_3 \mathbf{e}_3^T \tag{2}$$

Rearranging the eigensystem, the ellipsoid is given by

$$\mathbf{T} = (\lambda_1 - \lambda_2)\mathbf{S} + (\lambda_2 - \lambda_3)\mathbf{P} + \lambda_3 \mathbf{B} \tag{3}$$

where S defines a stick tensor, P defines a plate tensor and B defines a ball tensor:

$$\mathbf{S} = \mathbf{e}_1 \mathbf{e}_1^T, \; \mathbf{P} = \mathbf{e}_1 \mathbf{e}_1^T + \mathbf{e}_2 \mathbf{e}_2^T \text{ and } \mathbf{B} = \mathbf{e}_1 \mathbf{e}_1^T + \mathbf{e}_2 \mathbf{e}_2^T + \mathbf{e}_3 \mathbf{e}_3^T$$

This ellipsoid has the dimensions λ_1, λ_2 and λ_3 in the main axis directions. We define that $\lambda_1 > \lambda_2 > \lambda_3$. For surface inferences, surface saliency is then given by $\lambda_1 - \lambda_2$, with normal direction estimated as $\mathbf{e}_1$. Curve saliency is given by $\lambda_2 - \lambda_3$, with tangent direction estimated as $\mathbf{e}_3$, and junction saliency of curves estimated by the eigenvalue λ_3, which is encoded by the ball tensor.

3.3 Voting Communication

Every initial location sends out a tensor field and propagates it into the space in a certain neighborhood. Every other location in this neighborhood is then influenced by this field. To calculate the total influence on a certain location we simply have to summarize the tensor fields of all neighbors in a given radius. Therefore we have to look how the tensor field propagates in tensor voting space. As we have seen above, the tensor field represents three vector fields with different meanings. Thus these vector fields behave differently while propagating, we have to handle each part by itself and assemble them again afterwards (4). The reason why we formulate the three fields in a single tensor is the communication between these fields.

The voting kernel defines the most likely normal by selecting the most likely continuation curve between two points O and P in Fig. 2. The length of the normal vector at P, representing the strength of the vote, is defined by the following equation in spherical coordinates:

$$DF(s,k,\sigma) = \exp\left(\frac{s^2 + ck^2}{\sigma^2}\right) \qquad (4)$$

where is $s = (l\theta)/\sin(\theta)$ and $k = 2\sin(\theta)/l$.

The parameter s is the arc length OP, k is the curvature, c is a constant, and σ is the scale of voting field controlling the size of the voting neighborhood and the strength of votes in Fig. 2.

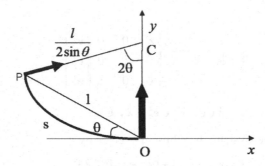

Fig. 2. Second order votes cast by a stick voting field

We have obtained normal directions at each input site, we can infer a smooth structure that connects the tokens with high feature saliencies in 3D (surface saliency is represented by $\lambda_2 - \lambda_3$). Feature extraction for analyzing input data can be achieved by casting votes to all discrete sites by the same voting fields and voting algorithm. Given this dense information, in 3D we analyze true surface points and connect them. We detail feature analysis based on tensor voting for corrupted text images in next section.

4   Noise Removal and Restoration Using Generalized Adaptive Vector Sigma Filters in Tensor Voting

4.1   Image Analysis in Tensor Voting Space

In order to analyze feature vectors in corrupted text images, we used the method proposed in [13]. This method can split color components in RGB space that is independent of illumination. Instead of thresholding on saturation, we derive a chromaticity measure based on the sum of differences of R, G and B components at each pixel (x,y). Therefore, Hue and intensity values are defined in different ranges, and the selected feature vectors are used for analyzing a corrupted image in tensor voting.

A *surface saliency map* defines surface saliency at every pixel after tensor voting as described in Fig. 3. The map is able to indicate the presence of noise on text as Fig. 3(c) by black regions. The black region can be replaced by applying a generalized adaptive vector sigma filter to the low saliency values.

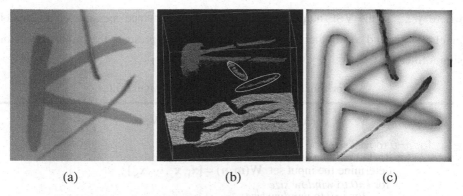

(a)   (b)   (c)

Fig. 3. Saliency image in corrupted text image: (a) corrupted image, (b) Yellow circles indicate corruption regions in tensor voting, (c) Saliency map image with surface saliency

4.2 Region Restoration Using Vector Sigma Filter

In a surface saliency map for an image restoration by tensor voting, low saliency values are primarily considered noises and should be replaced with values of the high surface saliency neighbors by a vector sigma filter. We briefly review a smoothing method for noise removal and restoration using vector sigma filter from corrupted regions of an image. In [14], R. Lukac *et al.* introduced the vector sigma filter. Let ψ_γ be the approximation of the multivariate variance of the vectors contained in a supporting window **W** with sufficiently large window size **N** that is the number of pixels in the filter window **W**, given by

$$\psi_\gamma = \frac{\mathbf{L}_{(1)}}{\mathbf{N}-1} \tag{5}$$

where $\mathbf{L}_{(1)} = \sum_{j=1}^{N} \| \mathbf{x}_{(1)} - \mathbf{x}_j \|_\gamma$ is the aggregated distance calculated by (5) and associated with the vector median $\mathbf{x}_{(1)}$. This approximation represents the mean distance between the vector median and all other saliency values contained in **W**. The division of the smallest aggregated distance $\mathbf{L}_{(1)}$ by N-1 (number of distances from $\mathbf{x}_{(1)}$ to all samples from **W**) ensures that the dispersion measure is independent of the filtering window size. Following the switching concept in [15], the Vector Sigma Filter (VSF) uses the threshold **TH** defined as follows:

$$\mathbf{TH} = \mathbf{L}_{(1)} + \lambda \psi_\gamma = \frac{\mathbf{N}-1+\lambda}{\mathbf{N}-1} \mathbf{L}_{(1)} \tag{6}$$

where $\mathbf{L}_{(1)}$ is the smallest aggregated Minkowski metric, ψ_γ is the approximated variance, and λ is the tuning parameter used to adjust the smoothing properties of the VSF. If the threshold **TH** is compared with the distance measure $\mathbf{L}_{(N+1)/2}$, it is possible to derive a simple switching rule for the replacement of noisy pixels [15]:

$$\mathbf{y}_{\text{SVMF}} = \begin{cases} \mathbf{x}_{(1)} & \text{for } \mathbf{L}_{(N+1)/2} \\ \mathbf{x}_{(N+1)/2} & \text{otherwise} \end{cases} \tag{7}$$

where $\mathbf{y}_{SVMF}$ is the VSF output, $\mathbf{L}_{(N+1)/2}$ denotes the distance measure of the center pixel $\mathbf{x}_{(N+1)/2}$, and $\mathbf{x}_{(1)}$ is the VMF output. Algorithm I presents the proposed vector sigma filter.

Algorithm 1. Algorithm of the vector sigma filter

Input data: $\mathbf{M} \times \mathbf{N}$ corrupted image, window size W;

for $a=0$ to *row size*;
    for $b=0$ to *column size*;
        determine the input set $\mathbf{W(m,n)} = \{\mathbf{x}_1, \mathbf{x}_2, \cdots, \mathbf{x}_N\}$;
        for $i=0$ to *window size*
            for $j=0$ to *window size*
                compute the distance $\mathbf{f}(\mathbf{x}_i, \mathbf{x}_j)$;
            end
            compute the sum $\zeta_i = \mathbf{f}(\mathbf{x}_i, \mathbf{x}_1) + \mathbf{f}(\mathbf{x}_i, \mathbf{x}_2) + \cdots + \mathbf{f}(\mathbf{x}_i, \mathbf{x}_N)$;
        end
            sort $\zeta_1, \zeta_2, \cdots, \zeta_N$ to the ordered set $\zeta_{(1)} \leq \zeta_{(2)} \leq \cdots \leq \zeta_{(N)}$;
            compute **TH** by (6)
        IF $\zeta_{(N+1)/2} \geq \mathbf{TH}$
            apply the same ordering to $\mathbf{x}_1(\zeta_1), \mathbf{x}_1(\zeta_1), \cdots, \mathbf{x}_N(\zeta_N)$
            store ordering sequence as $\mathbf{x}_{(1)} \leq \mathbf{x}_{(2)} \leq \cdots \leq \mathbf{x}_{(N)}$
            let the filter output $\mathbf{y(a,b)} = \mathbf{x}_{(1)}$
        ELSE
            let the filter output $\mathbf{y(a,b)} = \mathbf{x}_{(N+1)/2}$
        end
    end

Algorithm 2. Noise removal and restoration

STEP1: Compute the surface saliency in the $(w+s) \times (w+v)$ window surrounding the noise pixel : $S(x, y)$, $1 \leq x \leq w+s, 1 \leq y \leq w+v$
    (As an initial value, $W=3$ and $s=v=0$.)
STEP2: The number of high surface saliency pixel in the window:
    IF $(S(x, y) \geq TH_2)$ Count = Count+1; //
STEP3: The size change of windows for *vector sigma filter* in a corrupted region
    IF (Count<TH_3) $s=s+2$, $v=v+2$ and **GOTO STEP 1** // (here, $TH_3 = 8$.)
    ELSE **GOTO STEP 4**
STEP4: Enumerate values of pixels corresponding to high surface saliency
    in the increased window: $H(x, y)$, $1 \leq x \leq w+s, 1 \leq y \leq w+v$
STEP5: Find a value among the enumerated values and fill a value
    out of the noise pixels.

If a pixel is judged as noise, values surrounding the pixel in a **W** subwindow are initially examined to find a high saliency value. If the noise region is broad, the **W** subwindow may be insufficient to find high surface saliency defined by TH_2. The size of the subwindow $((W+s)\times(W+s))$ is therefore increased until the proper number of high surface saliency pixels is detected. The steps below represent the process of this proposed method. Then, one value among the high surface saliency pixels within the final window is selected. However, some noises may remain after a single pass of the tensor voting and vector sigma filter. We therefore repeatedly apply the filter to remove the remaining noises.

4.3 Segmentation Using Statistical Method

The number and centroid of modes selected using mean-shift algorithm proposed in [16] are used as seed values in GMM (Gaussian mixture models) based on deterministic annealing EM [17]. GMM method is then applied to values in the improved image to segment the text regions. We performed color image segmentation using deterministic annealing EM proposed in our previous work [17].

5 Experimental Results

We have tested our approach on natural scene images, which are corrupted by noise. The images are with the sizes of 256×256. After text region detection to define a region of interest, our approach is performed. In our experiment, text regions are manually detected and the selected regions are segmented using our method.

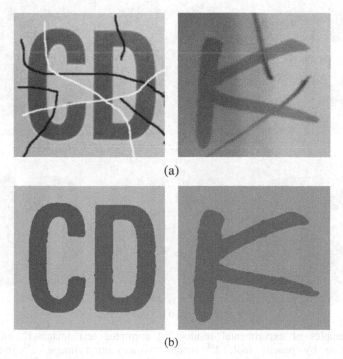

(a)

(b)

Fig. 4. Experimental results: (a) corrupted text image in natural scene images, (b) segmented image by the proposed method

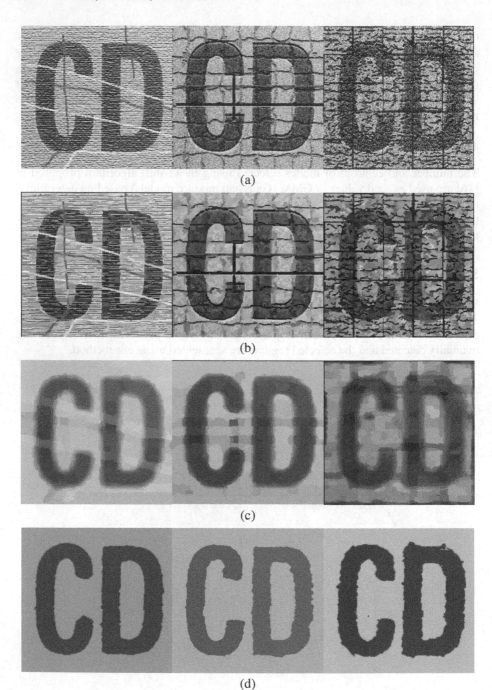

Fig. 5. Examples of experimental results: (a) corrupted text images(1st image: texturized+corruption by pencile tool, 2nd image: mosaic+cutted image, 3rd image: texturized+cutting+added noises), (b) the results by EDISON, (c) the results segmented by EDISON after median filter with optimal subwindow, (d) segmented result by the proposed method

Fig. 4 shows our experimental results. Used images contain some corruptions and confusions on text regions which can typically cause problems when use the text images for image segmentation. The results show that the proposed method considers the corrupted part in complex regions as well as removes noises.

The segmentation results for text images corrupted by Photoshop tools are shown in Fig. 5. To verify the region examination of the texts, we make in Fig 5 several artificial corruptions on the original images. The proposed method is compared with EDISON [16] and median+EDISON. In the experimental results, the background and foreground are regarded as the same region in the segmented result of the proposed method, while the segmentation of the EDISON contains many small regions.

Table 1. Performance comparison of three approaches with processing time (sec.)

	Our approach (num. of iterations)	EDISON	Median+ EDISON
1st image in Fig. 5(a) (256x256)	8.5 (2 times)	3.0	5.0
2nd image in Fig. 5(a) (256x256)	12.6 (4 times)	2.5	4.3
3rd image in Fig. 5(a) (256x256)	14.3 (4 times)	3.4	5.4

6 Conclusions

In this paper, a new approach for a restoration and segmentation has been proposed. The approach takes into account tensor voting based on vector sigma filter and statistical model using GMM based on deterministic annealing EM algorithm. The achieved results show good restoration and segmentation on preservation capabilities in corrupted regions. Especially, the vector sigma filter selects proper values to replace the noise regions, which is presented on corrupted regions. The improved image by tensor voting is segmented by statistical model based on Gaussian mixture models.

In our experiments, the proposed method can remove different kinds of noises well and segment one text as one object. The result can then contribute to improving text recognition rate as well as reducing the complexity of final step like OCR for text recognition.

References

1. Pal, N.R., Pal, S.K.: A review on image segmentation techniques. Pattern Recognition 26(9), 1277–1294 (1993)
2. Moghaddamzadeh, A., Bourbakis, N.: A fuzzy region growing approach for segmentation of color images. Pattern Recognition 30(6), 867–881 (1997)
3. Zhang, C., Wang, P.: A new method of color image segmentation based on intensity and hue clustering. In: IEEE Int. Conference on Pattern Recognition, vol.3, pp. 3617–3621 (2000)
4. Jie, X., Peng-fei, S.: Natural color image segmentation. In: IEEE Int. conference on Image Processing, vol. 1, pp. 14–17 (2003)

5. Lucchese, L., Mitra, S.K.: Color image segmentation: a state-of-the-art survey. Pro. of the Indian National Science Academy 67, 207–221 (2001)
6. Zhong, Y., Karu, K., Jain, A.K.: Locating text in complex color images. Pattern Recognition 28, 1523–1536 (1995)
7. Jain, K., Yu, B.: Automatic Text location in images and video frames. Pattern Recognition 31, 2055–2076 (1998)
8. Haritaoglu, I.: Scene text extraction and translation for handheld devices. In: IEEE Conference on Computer Vision and Pattern Recognition, pp. 408–413 (2001)
9. Ye, Q., Gao, W., Huang, Q.: Automatic text segmentation from complex background. In: IEEE Int. Conference on Image Processing, vol. 5, pp. 2905–2908 (2004)
10. Lucas, S.M., Panaretos, A., Sosa, L., Tang, A., Wong, S., Young, R.: ICDAR: 2003 robust reading competitions. In: IEEE Int. Conference on Document Analysis and Recognition, pp. 682–687 (2003)
11. Medioni, G., Lee, M.S., Tang, C.K.: A Computational Framework for Segmentation and Grouping. Elsevier, Amsterdam (2000)
12. Lee, M.-S., Medioni, G.: Grouping.,-,→, into regions, curves, and junctions. Computer Vision and Image Understanding 76(1), 54–69 (1999)
13. Lim, J.G., Park, J.H., Medioni, G.G.: Text segmentation in color images using tensor voting. Image and Vision Computing 25, 671–685 (2007)
14. Lukac, R., Smolka, B., Plataniotis, K.N., Venetsanopoulos, A.N.: Vector sigma filters for noise detection and removal in color images. Journal of Visual Communication and Image Representation 17, 1–26 (2006)
15. Lukac, R., Plataniotis, K.N., Venetsanopoulos, A.N., Smolka, B.: A statistically-switched adaptive vector median filter. J. Intell. Robotic System 42, 361–391 (2005)
16. Comaniciu, D., Meer, P.: Mean shift: a robust approach towards feature space analysis. IEEE Tran. on Pattern Analysis and Machine Intelligence 24(5), 1–18 (2001)
17. Park, J.H., Cho, W.H., Park, S.Y.: Deterministic annealing EM and its application in natural image segmentation. In: Zhang, J., He, J.-H., Fu, Y. (eds.) CIS 2004. LNCS, vol. 3314, pp. 639–644. Springer, Heidelberg (2004)

A Boosting Discriminative Model for Moving Cast Shadow Detection

Yufei Zha[1], Ying Chu[2], and Duyan Bi[1]

[1] Signal and Information Processing Lab, Engineering College of Air Force
Engineering University, xi'an, China
[2] Institute for Pattern Recognition and Artificial Intelligence, Huazhong University of
Science and Technology, China
yfzha.lhi@gmial.com, yingchu.lhi@gmail.com, biduyan@126.com

Abstract. Moving cast shadow causes serious problem while segmenting and extracting foreground from image sequences, due to the misclassification of moving shadow as foreground. This paper proposes a boosting discriminative model for moving cast shadow detection. Firstly, color invariance subspace and texture invariance subspace are obtained by the color and texture difference between current image and background image; then, boosting is selected based on theses subspaces to discriminate cast shadow from moving objects; finally, temporal and spatial coherence of shadow and foreground is employed on Discriminative Random Fields for accurate image segmentation through graph cut. Results show that the proposed method excels classical method both in indoor and outdoor scene.

Keywords: Shadow detection, boosting, Discriminative Random Fields, graph cut.

1 Introduction

Detection and tracking of moving objects is an important topic in dealing with image sequences. In real world, cast shadow causes serious problem while segmenting and extracting foreground, due to the misclassification of shadow as foreground. The difficulties associated with cast shadow detection arise since shadow and foreground share some same motional features. The performance of effective foreground detection would be seriously degraded due to this problem. A reliable and efficient shadow detection algorithm is required.

1.1 Related Work

Many methods have been proposed for detecting moving cast shadow. Horprasert *et al.*[1] exploits the Lambertian hypothesis that considering color as a product of irradiance and reflectance and classifies a pixel into four categories according to distortion of the brightness and the chrominance. Mikic *et al.*[2]uses a linear transformation to measure color change under changing illumination and each

A.L. Yuille et al. (Eds.): EMMCVPR 2007, LNCS 4679, pp. 253–266, 2007.
© Springer-Verlag Berlin Heidelberg 2007

color channel is approximately multiplied by a single overall multiplicative factor. Cucchiara *et al.*[3] eliminates shadow in HSV color space because casting shadow does not change significantly its hue and saturation information. Salvador *et al.*[4] exploits invariant color features in a scene for shadow detection. Photometric color invariants are demonstrated to be invariant to a change in the imaging conditions, such as viewing direction, objects surface orientation and illumination conditions. Elgammal *et al.*[5]detects shadow through comparing the appearance value of a point in the current image with the corresponding points in background image in normalized rgb color space. Those methods are based on the assumption that shadow does not change the color information of the background covered. However, not all the shadow points are justified that assumption. In addition, some foreground points may have the same color with shadow when comparing with background.

Texture information is also employed for shadow detection in some literatures. Stauder *et al.*[6] develops an edge detector. Shadow is distinguished from the background discontinuities through heuristic rules. Gradient features are exploited in dynamic conditional random field in Wang *et al.*[7]. These methods can not work well in some case. For instance, foreground and shadow have the some texture for uniform regions both in moving pixels and background image.

Otherwise,Nadimi *et al.*[8]uses dichromatic reflection model and accounts for both the sun and the sky illuminations, however some assumption is right only in outdoor. Porikli *et al.*[9] models shadow using multivariate Gaussians in RGB color space, and it needs training for a long time. Zhao *et al.*[10] proposes a geometrical approach for shadow detection. Any foreground pixel which lies in the shadow ellipse and whose intensity are lower than that of the corresponding pixel in the background image by a threshold is classified as a shadow pixel. That needs prior knowledge.

1.2   Proposed Methods

Above methods have only exploit some cues, for example, the foreground points which darker than background image but have the some color with the background wills misclassification as shadow. If the foreground region has uniform texture information with the background texture information, error wills appearance.

In orderto addressing the problem and achieve robust performance, all these cues need to be exploited simultaneously in a uniform framework for detecting moving cast shadow. This paper proposes a boosting discriminative model to eliminate cast shadow on Discriminative Random Fields (DRFs). The method combines different features for boosting to discriminate cast shadow from moving objects, then temporal and spatial coherence of shadow and foreground are incorporated on Discriminative Random Fields (DRFs); finally, the problem can be solved by graph cut[11][12][13]. The flowchart is shown in Fig. 1. Firstly, moving objects are obtained by background subtraction; secondly, shadow candidates can be derived through pre-processing moving objects, in terms of the shadow physical property; thirdly, color information and texture information is derived by comparing shadow and foreground points in current image with corresponding points in background image, which are selected as features for boosting; finally, temporal and spatial coherence of shadow and foreground is employed on Discriminative Random Fields and discriminate shadow and foreground by graph cut accurately.

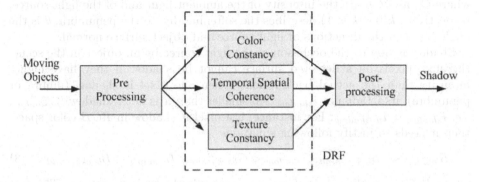

Fig. 1. Flowchart of proposed shadow detection. Firstly, moving objects are obtained by background subtraction; secondly, shadow candidates can be derived through pre-processing moving objects, in terms of the shadow physical property; thirdly, color information and texture information are derived by comparing shadow and foreground points in current image with corresponding points in background image, which are selected as features for boosting; finally, temporal and spatial coherence of shadow and foreground is employed on discriminative random fields for accurate image segmentation and discriminate shadow and foreground by graph cut accurately.

2   Shadow Model

We aim to obtain accurate foreground by detecting cast shadow in moving objects which is derived through background subtraction [5]. The shadow candidates from moving objects can be obtained through analyzing the physical properties of shadow in this section.

Shadow is generated due to a relative absence of light. Therefore, their brightness and color will change with the changes in the cast surface. The spectral characteristics of shadow depend on the characteristics of the light that illuminates the shadow compared to the light that would illuminate the same area if there is no obstruction.

According to[14], the appearance of a pixel I_r can be modeled as:

$$I_r = i_r \rho_r \tag{1}$$

where I_r is the image luminance of pixel at the position r. ρ_r is the reflectance of the object surface and i_r is the irradiance. Assumes that light source is far from the object, the distance between light source and surface is constant, the light source emits parallel light rays and the observation point is fixed. In this case, the irradiance i_r can be approximated as[15]:

$$i_t^r = \begin{cases} C_A + C_p \cos\theta & \text{no shadow} \\ C_A + kC_p \cos\theta & \text{penumbra} \\ C_A & \text{umbra} \end{cases} \qquad (2)$$

where C_A and C_P are the intensity of the ambient light and of the light source, respectively. $k(0 < k < 1)$ describes the softening due to the penumbra, θ is the angle between the directions of light source and object surface normal.

Shadow is due to the occlusion of the light source by an object in the scene, therefore pixels on a detected surface cannot be shadow if they have higher intensity than the actual background. When one pixel is shadow (umbra or penumbra), its irradiance i_{shadow} will smaller than it is not a shadow $i_{noshadow}$, i.e. $i_{shadow} < i_{noshadow}$. For instance, if a pixel is shadow in RGB color space, then it needs to justify following equations:

$$R_{shadow} < R_{background}, G_{shadow} < G_{background}, B_{shadow} < B_{background} \qquad (3)$$

where R, G, B are the values of each component in the RGB color space. This test does not reduce the shadow areas, but can successfully reduce foreground areas, thus reducing the binary mask and computational load. Otherwise, pixels nearing the black vertex of the RGB space may generate unstable invariant features values in presence of noise. As mentioned in[4], those pixels are extracted whose RGB values are below the value of 30 (on a range of 256 levels). It is therefore possible to discard these pixels from shadow analysis. The works of [3] arrive at the same conclusions.

3   Discriminative Random Fields for Cast Shadow Detection

We use the Discriminative Random Fields (DRFs) [16] to describe the features and coherence in the shadow and foreground, respectively. Discriminative Random Fields is a discriminative framework for the classification of image regions by incorporating neighborhood interactions in the labels as well as the observed data.

Given an input sequence of images $I_{1:t}$ represent observed values of images at time $1, 2, \cdots, t$ and the corresponding background values are $B_{1:t}$ which is obtained by background subtraction. Let $L = \{shadow = -1, foreground = 1\}$ be the labels of detected moving objects points. We define the conditional probability of the class label L given image sequence $I_{1:t}$ and background sequence $B_{1:t}$:

$$p(L_i|I_{1:t}, B_{1:t}) = \frac{1}{Z} \exp(\sum_{r \in V} A_r(l_r, I_{1:t}, B_{1:t}) +$$

$$\sum_{r \in V} \sum_{(r,s) \in N} SM_{rs}(l_r, l_s, I_{1:t}, B_{1:t})) \qquad (4)$$

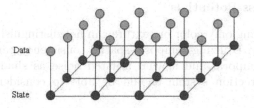

Fig. 2. Discriminative Random Fields for cast shadow detection. The upper level is observed data, which includes image I and background B, in different feature subspace; the lower level is hidden states, *i.e.* the label of each pixel. Vertical lines represent the data potentials and the smoothness potentials are represented by the horizontal plane lines.

Where Z is a normalizing constant known as the partition function, V is the sets of shadow candidates obtained from section 2, N is the sets of all adjacent pixel pairs, $-A_r(l_r, I_{1:t}, B_{1:t})$ and $-SM_{rs}(l_r, l_s, I_{1:t}, B_{1:t})$ are the data and smoothness potentials, respectively, and $l_r = \{-1, 1\}$. $(1 : t$ will be omitted for simple writing in following sections). It is shown in figure 2, the data potentials is determined by the observed data in different feature subspace in each pixel and the smoothness potentials is determined by the neighborhoods.

3.1 Data Potentials

In DRFs frameworks, $A_r(l_r, I_{1:t}, B_{1:t})$ is modeled using a local discriminative model that outputs the association of the site r with class l_r. In this research, the data potentials $A_r(l_r, I_{1:t}, B_{1:t})$ is obtained by boosting to represent the color and texture constancy of shadow. These features and the boosting procedure used to discriminate cast shadow from moving objects are described in section 4 in detail. As mentioned in [16], the local class posterior can be modeled using logical function:

$$p(l_r|I, B) = \frac{1}{1 + \exp(1 + exp(-\alpha l_r \sum_{m=1}^{M} \lambda_m h_m(I, B))} \tag{5}$$

where   is the constant which determines the smoothness of probability distribution, h_m is the weak classifier and λ_m is the corresponding weight which is determined by boosting in the color and texture feature subspace, respectively.

The data potentials can be written as:

$$A_r(l_r, I, B) = \log(p(l_r|I, B))$$

$$= -\log(1 + exp(-\alpha l_r \sum_{m=1}^{M} \lambda_m h_m(I, B))) \tag{6}$$

3.2 Smoothness Potentials

Methods considering only color or texture can not distinguish shadow and foreground accurately. Utilizing temporal-spatial consistency can prevent wrongly
classifying the temporal and spatial isolated noise as shadow regions. A reliable shadow detection system should be able to consider all these factors
simultaneously.

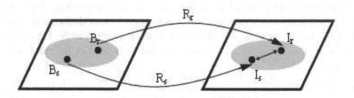

Fig. 3. Temporal-Spatial coherence. Neighborhoods not only have similar value, but
also have similar change with background pixels.

As noticed in [8], temporal-spatial ratio is used to help detect shadow and
foreground. As show in figure 3, neighborhoods not only have similar value, but
also have similar change with background. Define the temporal ratio as:

$$R_r = \frac{I_r - B_r}{I_r - B_r}, R_s = \frac{I_s - B_s}{I_s - B_s}, (r,s) \in N \qquad (7)$$

In which I is the current image at time t, B is the background image at time
$t-1$. If two neighboring pixels belong to the same surface, their temporal ratios
will be close to each other.
   Define the temporal-spatial ratio as:

$$R_{tem\_spa} = \frac{1}{C} exp(\frac{R_r - R_s}{R_r + R_s}), (r,s) \in N \qquad (8)$$

where C is normalized const.
   The temporal-spatial relation of two neighborhood pixels will be close to zero,
if they belong to the same region, *i.e.* they have the same label. The smoothness
potentials can be denoted as follows:

$$SM_{rs}(l_r, l_s, I, B) = R_{tem\_spa}$$
$$= \frac{1}{C} exp(\frac{R_r - R_s}{R_r + R_s}) \delta(l_r, l_s) \qquad (9)$$

where,

$$\delta(l_r, l_s) = \begin{cases} 1, & \text{if } l_r = l_s \\ 0, & \text{if } l_r \neq l_s \end{cases} \quad (r,s) \in N$$

4   Boosting Learning for Data Potentials

The observation is that shadow does not change seriously the color and texture of background covered, which are combined for shadow detection. The most important part of DRFs is the data potentials, which is obtained by boosting based on the color and texture feature subspace.

4.1   Color Invariance Subspace

Shadow has the same color with background covered. The spectral property of shadow can be derived in the hypothesis by considering illumination invariants color space. Color invariants subspace is functions which describe the color configuration of each image point discounting shading, shadow, and highlights. These functions are demonstrated to be invariant to the change in the image conditions, such as viewing direction, object surface orientation and illumination conditions. Examples of photometric color invariants are normalized rgb, $hue(H)$, $saturation(S)$, $c_1c_2c_3$ and $l_1l_2l_3$ [4]. Normalized rgb color space is adopted in this study for its simple computation.

Comparing pixel-wise color information between current image and background image can help to detect shadow. Define F as one of the above mentioned photometric color invariants features. F_{shadow} is assumed the value of a point in shadow, and $F_{background}$ is the value of the same point in background image. Ideally, the differences of them is zero, $i.e.$ $|F_{shadow} - F_{background}| = 0$.

In the normalized rgb color space, define the differences as:

$$D_r^{color} = \frac{1}{\alpha}|I_r^{rgb} - B_r^{rgb}|^2, \alpha = \sum_{r \in V}|I_r^{rgb} - B_r^{rgb}|^2 \tag{10}$$

where V is the set of shadow candidates.

Using following equation to compute the probability of the differences:

$$P_{color}(I_r^{rgb}, I_r^{rgb}|l_r) \propto \exp(-D_r^{color}) \tag{11}$$

Through formulation (11), illumination invariance color probability map of shadow is established, as shown in figure 4(c). The brighter points of image correspond to the smaller probability.

4.2   Texture Invariance Subspace

Shadow neither changes the color of the background covered, nor changes the texture of the background covered. The ratio between current pixel and its neighborhoods is employed to describe texture of the pixel. This also can help to detect the shadow.

Define the ratio of current pixel and its neighborhoods as:

$$R_{r,s} = \frac{I_r}{I_s}, r \in V, (r, s) \in N \tag{12}$$

where V is the set of shadow candidates obtained from section 2, and N is the sets of all adjacent pixel pairs.

Assuming the neighboring pixels receive the same irradiance, *i.e.* $i_r = i_s, (r, s) \in N$. Considering the formulation (1), it can write as follows:

$$R_{(r,s)} = \frac{I_r}{I_s} = \frac{i_r \rho_r}{i_s \rho_s} = \frac{\rho_r}{\rho_s}, r \in V, (r, s) \in N \tag{13}$$

Similarly, the ratio of the neighboring pixels on the background image B can be calculated as follows:

$$R'_{(r,s)} = \frac{B'_r}{B'_s} = \frac{i'_r \rho'_r}{i'_s \rho'_s} = \frac{\rho'_r}{\rho'_s}, r \in V, (r, s) \in N \tag{14}$$

They have the same reflection, *i.e.* $\rho = \rho'$, due to the pixel in both current image and background image are projected by the same point. Therefore, the difference between current image pixels and background pixels is zero when it is shadow in ideal case, *i.e.* $R_{(r,s)} - R'_{(r,s)} = 0$.

Define the differences as:

$$D_r^{texture} = (-\beta \sum_{(r,s) \in N} |R_{(r,s)} - R'_{(r,s)}|^2) \tag{15}$$

where, $r \in V, \beta = \frac{1}{\sum_{r \in V} \sum_{(r,s) \in N} |R_{(r,s)} - R'_{(r,s)}|^2}$, V is the set of shadow candidates obtained from section 2. Like color constancy, the probability can be obtained as following:

$$P_{texture}(I_r^{rgb}, I_r^{rgb} | l_r) \propto exp(-D_r^{texture}), r \in V \tag{16}$$

At the same time the corresponding texture difference map can established as color probability map which is shown in figure 4(d)

(a)   (b)   (c)   (d)

Fig. 4. Color and texture probability map of differences. (a) is original image; (b) is the ground truth, blue is foreground and red is shadow, and (c) and (d) are the illumination invariance color probability map and texture probability map, respectively. The brighter points of image correspond to the smaller probability.

4.3   Boosting for Data Potentials

The AdaBoost algorithm is introduced by Freund and Schapire[17] and it reweighs the training samples instead of re-sampling them. The algorithm takes as input a train set $\{(x_i, l_i)|i = 1, 2, \cdots, K\}$, where each x_i belongs to some domain or instance space X, and each label set Y. The cast shadow removal is a binary classification. We assume we have M classifiers, and then each pixel label can be obtained through AdaBoost as:

$$l_i = sign(\sum_{m=1}^{M} \lambda_m h_m(x_i)) \tag{17}$$

Freund and Schapire [18] have proved strong bounds on the training and generalization error of AdaBoost. For the case of binary classification the training error drops exponentially fast with respect to the number of boosting rounds M (*i.e.* number of weak classifiers).

The strong classifier can be obtained based on ground truth through discrete AdaBoost. It is combined by many weaker classifiers with different weights, which is shown in figure 5. Then the data potentials can be derived trough formulation (5) and (6). Finally, we can classify the foreground and shadow from moving objects.

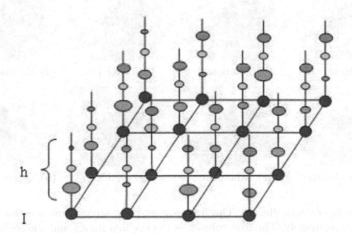

Fig. 5. Boosting for data potentials. The lower level is the image pixels and upper levels is the weaker classifiers, the size of points represent the corresponding weights. The strong classifier if composed of these weaker classifiers.

5   Experiments Result

Typical sequences are used in experiments, in which 'HighwayI' sequence are outdoor scenes, 'Intelligent room' and 'Laboratory' are indoor scenes. The ground truth of each video is manual labeled. Blue and red colors correspond to the true foreground and true shadow pixels, respectively. The frame is 320×240 video on a 2.4GHz desktop PC.

5.1   Performance Evaluate

In order to evaluate the proposed method quantitatively, we use the foreground detection rate ζ and shadow detection rate η defined in [15]:

$$\zeta = \frac{TP_F}{TP_F + FN_F} \tag{18}$$

$$\eta = \frac{TP_S}{TP_S + FN_S} \tag{19}$$

TP_F: the foreground pixels correctly detected;
FN_F: the foreground pixels detected as shadow.
TP_S: the shadow pixels correctly detected;
FN_S: the shadow pixels detected as foreground;

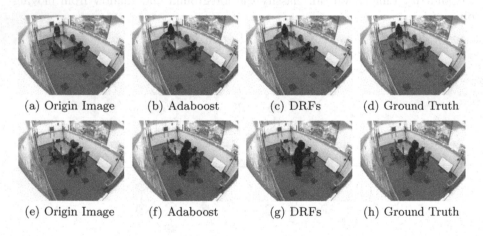

| (a) Origin Image | (b) Adaboost | (c) DRFs | (d) Ground Truth |

| (e) Origin Image | (f) Adaboost | (g) DRFs | (h) Ground Truth |

Fig. 6. Intelligentroom Result. The first row and the second row are the 198th and 278th frame separately. The first column is the original image and the second column is detected result by Adaboost, the third column is detected result in DRFs and the last is the ground truth.

5.2   Experiments

We test the proposed method on these videos and compare with the previous methods. All results are shown in figure 6 to figure 8. Figure 9 is comparing with original image, ground truth detected result. Intelligent room result is shown in

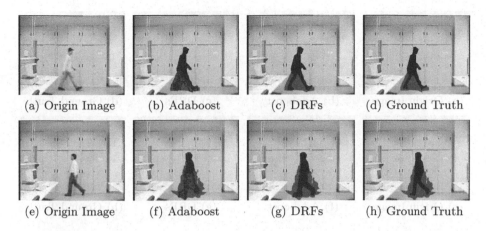

(a) Origin Image   (b) Adaboost   (c) DRFs   (d) Ground Truth

(e) Origin Image   (f) Adaboost   (g) DRFs   (h) Ground Truth

Fig. 7. Laboratory Result. The first row and the second row are the 108th and 155th frame separately. The first column is the original image and the second column is detected result by Adaboost, the third column is detected result in DRFs and the last is the ground truth.

Table 1. shadow detection rate ζ and shadow discrimination rate η

	HighwayI		laboratory		Intelligentroom	
	η	ζ	η	ζ	η	ζ
SNP	0.8159	0.6376	0.8403	0.9235	0.7282	0.8890
SP	0.5959	0.8470	0.6485	0.9539	0.7627	0.9074
DNM1	0.6972	0.7693	0.7626	0.8987	0.7861	0.9029
DNM2	0.7549	0.6238	0.6034	0.8157	0.6200	0.9398
Proposed Method	0.6586	0.8236	0.7886	0.9588	0.8552	0.9210

figure 6 and the first row and the second row are the 198th and 278th frame, respectively. Figure 7 is the laboratory result with the 108th and 155th frame. HighwayI result is shown in figure 8 with 119th and 328th frame. The first column is the original image and the second column is detected result by Adaboost, the third column is detected result in DRFs and the last is the ground truth. Blue color is the foreground and the red represent shadow. Figure 9 is the shadow accurate detection (SR) and foreground accurate detection (FR) of intelligent room, laboratory and HighwayI, respectively. Horizontal axis is the frame and vertical axis is the detection probability. The triangle is the shadow accurate detection (SR) and the square is foreground accurate detection (FR). Quantitative comparison with the methods referred in the survey and the results are shown in table 1. The results show that the proposed method excels the classical method in both indoor and outdoor scene.

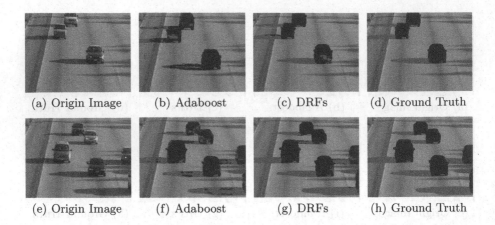

(a) Origin Image   (b) Adaboost   (c) DRFs   (d) Ground Truth

(e) Origin Image   (f) Adaboost   (g) DRFs   (h) Ground Truth

Fig. 8. HighwayI Result. The first row and the second row are the 119*th* and 328*th* frame separately. The first column is the original image and the second column is detected result by Adaboost, the third column is detected result in DRFs and the last is the ground truth.

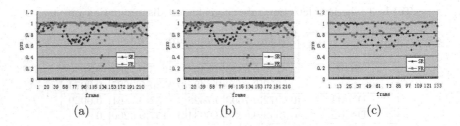

(a)   (b)   (c)

Fig. 9. Detection Rate of Result. Horizontal axis is the frame and vertical axis is the detection probability. The triangle is the shadow accurate detection (SR) and the square is foreground accurate detection (FR).

6   Conlusions

This paper proposes a boosting discriminative model for moving cast shadow detection. The color and texture invariance subspace are selected as features for boosting, then, temporal and spatial coherence of foreground and shadow are incorporated on Discriminative Random Fields for accurate image segmentation.

The results show that the proposed method can work well in both indoor and outdoor scene. Comparing with the classical methods, our method has good performance both in shadow accurate detection and foreground accurate detection. However, high level knowledge of shadow is needed for more accurate detection.

Acknowledgements

This work is done when the authors are at the Lotus Hill Research Institute. The authors would like to thank Xiong Yang, Xun Wang and Zheng Li for their invaluable help.

References

1. Horprasert, T., Harwood, D., Davis, L.: A statistical approach for real-time robust background subtraction and shadow detection. In: presented at Proc. of IEEE ICCV Frame-rate Workshop (1999)
2. Mikic, I., Cosman, P., Kogut, G., Trivedi, M.: Moving shadow and object detection in traffic scenes. In: Proceedings of the 15th International conference on Pattern Recognition, Barcelona, vol. 1, pp. 321–324 (2000)
3. Cucchiara, R., Grana, C., Piccardi, M., Prati, A.: Detecting moving objects, ghosts and shadow in video streams. IEEE Transactions on Pattern Analysis and Machine Intelligence 25, 1337–1342 (2003)
4. Salvador, E., Cavallaro, A., Ebrahimi, T.: Cast shadow segmentation using invariant color features. Computer Vision and Image Understanding 95, 238–259 (2004)
5. Elgammal, A., Harwood, D., Davis, L.S.: Elgammal a.background and foreground modeling using nonparametric kernel density estimation for visual surveillance. Proceedings of IEEE 90, 1151–1163 (2002)
6. Stauder, J., Mech, R., Ostermann, J.: Detection of moving cast shadows for object segmentation. IEEE Transactions on Multimedia 1, 65–76 (1999)
7. Wang, Y., Loe, K.F., Wu, J.K.: A dynamic conditional random field model for foreground and shadow segmentation. IEEE Transactions on Pattern Analysis and Machine Intelligence 28, 279 289 (2006)
8. Nadimi, S., Bhanu, B.: Physical models for moving shadow and object detection in video. IEEE Trans on Pattern Analysis and Machine Intelligence 26, 1079–1087 (2004)
9. Porikli, F., Thornton, J.: Shadow flow: A recursive method to learn moving cast shadow. In: IEEE International Conference on Computer Vision (ICCV), Beijing (2005)
10. Zhao, T., Nevatia, R.: Tracking multiple humans in complex situations. IEEE Transactions on Pattern Analysis and Machine Intelligence 26, 1208–1221 (2004)
11. Boykov, Y., Kolmogorov, V.: An experimental comparison of min- cut/max-flow algorithms for energy minimization in vision. IEEE Transactions on Pattern Analysis and Machine Intelligence 26, 1124–1137 (2004)
12. Boykov, Y., Veksler, O., Zabih, R.: Fast approximate energy minimization via graph cuts. IEEE Transactions on Pattern Analysis and Machine Intelligence 23, 1222–1239 (2001)
13. Kolmogorov, V., Zabih, R.: What energy functions can be minimized via graph cuts? IEEE Transactions on Pattern Analysis and Machine Intelligence 26, 147–159 (2004)
14. Klinker, G., Shafer, A., Kanada, T.: A physical approach to color image understanding. International Journal of Computer Vision 4, 7–38 (1990)
15. Prati, A., Mikic, I., Trivedi, M., Cucchiara, R.: Detecting moving shadow: Algorithms and evaluation. IEEE Transactions on Pattern Analysis and Machine Intelligence 25, 918–923 (2003)

16. Kumar, S., Hebert, M.: Discriminative random fields: A discriminative framework for contextual interaction in classification. In: IEEE International Conference on Computer Vision (ICCV) (2003)
17. Schapire, R.: The boosting approach to machine learning: An overview. In. In: Proc. MSRI workshop on Nonlinear Estimation and Classification (2001)
18. Freund, Y., Schapire, R.: A decision-theoretic generalization of on-line learning and an application to boosting. Journal of Computer and System Sciences 55(1), 119–139 (1997)

An *a Contrario* Approach for Parameters Estimation of a Motion-Blurred Image

Feng Xue[1,2], Quansheng Liu[1,3], and Jacques Froment[1]

[1] LMAM, Université de Bretagne Sud, Campus de Tohannic - Y. Coppens, BP573,
56017 Vannes, France
[2] Sense Technology Ltd, 3F, Tower D, Tian Ji Plaza Tian An Cyber Industrial Park
518040 Shenzhen, P.R. China
[3] School of Mathematics and Computing Science, Changsha University of Science
and Technology, Changsha, Hunan, 410076, P.R. China

Abstract. The recovery of a motion-blurred image is an important ill-posed inverse problem. But this subject has not recently received lot of attention. We propose a probabilistic method for the estimation of motion parameters based on the geometrical characteristic of the Fourier spectrum. Indeed, the Fourier spectrum of the blurred image is made by the product of the original Fourier spectrum with an oriented cardinal sine function. The estimation of the parameters reduces to the detection of the direction and of the gap between oscillations of the Fourier spectrum. Using the Helmholtz principle, the maximum meaningful parallel alignments are detected in the frequency domain, and then the direction and the extent of the blur are identified by an adapted K-means cluster algorithm. Simulation results show that the approach is very promising.

1 Introduction

There are plenty of situations where an image may be degraded by a blurred motion. The problem is common when the camera is in a moving vehicle or, in low shutter-speed situation, when the camera is held by human hands. However, the literature rarely addresses the problem of a specific motion blur while, by assuming this kind of degradation, one gets enough information from the degraded image to recover a far better image than the one reconstructed by a general deblurring method.

For the case of a uniform linear camera motion, Cannon [1] used the power cepstrum to identify the blur parameters. Yitzhaky [2] utilized the fact that image characteristics along the blur direction are different from those along other directions, and emphasized the Point Spread Function (PSF) characteristics by filtering the blurred image. Zhang and Wen [3] employed blur invariant moments as normalization constraints to normalize the image to a canonical form, and obtained the motion parameters from the normalizing transformation matrices. None of these approaches explicitly uses the visual geometric information given by the oriented cardinal sine function, due to the difficulty of automatically extracting this information from the Fourier spectrum.

A.L. Yuille et al. (Eds.): EMMCVPR 2007, LNCS 4679, pp. 267–279, 2007.
© Springer-Verlag Berlin Heidelberg 2007

There are also many other methods which can deblur the image and estimate the PSF simultaneously in the restoration process ([4], [5], [6]). These methods are more complicated, require more computations and cannot give very precise results. They should be dedicated to cases where the type of the blur is unknown.

In this paper, we develop a probabilistic method to estimate the motion parameters from a blurred image by analyzing the geometric characteristics of the Fourier Spectrum.

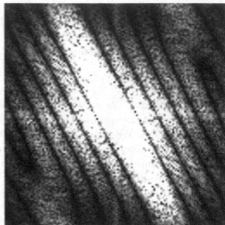

Fig. 1. Simulated motion blurred image and its Fourier spectrum, with blur length 12 and blur direction 28 degree

We assume that the motion blur degradation is uniform linear and space invariant. From the Fourier Spectrum (Fig.1), we see that the blur direction is perpendicular to the most perceptual parallel lines in the spectrum. The reason is that the motion blur effectively performs a low-pass filter on the image along the blur direction by multiplying frequency components with a cardinal sine function. An intuitive way to estimate the PSF parameters is to analyze the characteristics of parallel lines generated by the zero-crossings of the sine function.

According to the Gestalt theory and the Helmholtz principle ([7],[8]), an observed geometric structure is perceptually meaningful if its number of occurrences would be very small in a random situation: observed objects are "meaningful events" which should not happen by accident in an image according to probabilistic estimates. In this paper, we propose to use the Helmholtz principle to detect meaningful parallel segments *a contrario* to a null hypothesis. Then, we utilize an adapted K-means cluster algorithm to estimate the PSF parameters from the detected parallel segments.

As observed by Cannon [1], the technique of searching for zero's of the Fourier transform usually fails for the blur identification, although this technique seems to be the most natural one. The point is that the location of zero-crossings is often very noisy and does not constitute true straight line, however, with our eyes we can observe some parallel lines in the Fourier spectrum. Although these parallel lines are very obscured, we are able to find them with very good precision using the Helmholtz principe, at least on simulated motion-blurred images. The blur parameters can then be determined from the detected parallel lines.

Compared with blind deconvolution methods, our approach for the identification of blur parameters have some advantages. Sometimes we are more interested in the parameters than in the original image. For example, this could enable us to find the exact physical phenomenon responsible for the observed image. Also, once the exact nature of the blur has been determined, there exist more efficient methods to perform the deconvolution.

The paper is organized as follows. The motion blur degradation problem is described in Section 2. The Helmholtz principle is introduced in section 3. In Section 4, we explain how this principle is used to detect "meaningful parallel segments". In Section 5, an adapted K-mean cluster algorithm is presented, and is applied to determine the blur length and the blur direction. In Section 6 we give some experimental results showing that our approach is efficient on simulated motion-blurred images. Section 7 concludes the paper.

2   The Model

An image degraded by a uniform motion blur can be modeled by

$$g(x,y) = \frac{1}{T} \int_0^T f(x - x(t), y - y(t))dt + n(x,y), \tag{1}$$

where $x(t) = (a\cos\theta)t/T$, $y(t) = (a\sin\theta)t/T$; T, θ and a represent respectively the duration of the exposure, the motion blur direction and blur length along θ; $n(x,y)$ is a Gaussian noise with mean 0 and variance σ^2. We need to search for the original image f from the observed image g.

The Fourier transform of (1) is given by

$$G(u,v) = H(u,v)F(u,v) + N(u,v), \tag{2}$$

where F is the Fourier transform of f, N is that of n, and

$$H(u,v) = \int_0^T \exp[-j2\pi(ux(t) + vy(t))]dt \tag{3}$$

is the transfer function of the system.

If we consider the uniform motion blur only in the x direction, i.e. $y(t) = 0$, and if there in no noise, then (3) yields

$$H(u,v) = \int_0^T \exp[-j2\pi u\frac{at}{T}]\,dt$$
$$= \frac{T}{\pi ua}\sin(\pi ua)e^{-j\pi ua} \tag{4}$$

It is evident that $H(u,v)$ vanishes at $u = m/a$, where m is an integer. Because of this property, we can observe parallel lines along the blur direction with equal distance $\Delta u = 1/a$ in the spectrum of the degraded image. Hence, the problem is to identify the parallel lines which give values of the blur direction and the blur length: the blur direction is $\theta = \alpha - \pi/2$, where $\alpha \in (0, \pi)$ is the direction of the parallel lines, and the blur length is $a = 1/\Delta u$.

The general case with an arbitrary blur direction θ can be reduced to the preceding case where $\theta = 0$. When there is a noise, the geometric properties of the Fourier spectrum still hold.

3   The Helmholtz Principle

The Helmholtz principle is a general perception law. It states that an event is significant if its number of occurrences in a random situation is very small. Therefore significant events represent large deviations from randomness.

According to Gestalt theory, "grouping" is the general law of visual perception. Its main idea is that whenever objects have a common characteristic, they get grouped and form a new larger visual object, called a "Gestalt". The Helmholtz principle gives computational grouping thresholds associated with gestalt quality. It can be stated in the following generic way [9].

Assume that objects $O_1, O_2, ..., O_n$ are present in an image, and that k of them, say $O_1, ..., O_k$, have common property, such as: same color, same orientation, position etc... We are then facing the dilemma: is this common feature happening by chance, or is it significant and enough to group $O_1, ..., O_k$ to form a new, large visual object (a "Gestalt")? Desolneux, Moisan and Morel [9] introduced the notion of "ϵ-meaningful event" to answer this question. We assume *a priori* that the considered quality has been randomly and uniformly distributed on all objects $O_1, O_2, ..., O_n$. Then we (mentally) assume that the observed position of objects in the image is a random realization of this uniform process. We say that an event of type "such configuration of geometric objects has such property" is ϵ-meaningful if the expectation of number of occurrences of this event is less then ϵ [9]. The smaller ϵ, the more meaningful the event.

The Helmholtz principle has been applied recently to many image feature detection problems, see for example Desolneux et al. ([9], [7],[8]), and Cao [10].

4 ϵ- Meaningful Parallel Segments

In order to solve our problem, the Helmholtz principle will be applied twice: we first use it to obtain maximal meaningful segments in the spectrum of the degraded image, and we then use it to detect maximal meaningful parallel segments with which we will identify the blur parameters.

We consider the spectrum of the degraded image as a gray-level image of size $N \times N$ (that is a regular grid of N^2 pixels). At each point we compute a direction, which is the direction of the level line passing by the point calculated on a 2×2 pixels neighborhood. The direction at (i, j) is computed by rotating by $\frac{\pi}{2}$ the direction of the gradient, calculated by the order 2 interpolation at the center of the 2×2 window made of pixels $(i, j), (i + 1, j), (i, j + 1), (i + 1, j + 1)$[7]:

$$dir(i, j) = (d_1, d_2)/\sqrt{(d_1^2 + d_2^2)},$$

where $d_1 = -\frac{\partial u}{\partial y}(i,j) = -\frac{1}{2}[u(i,j) + u(i,j+1)] + \frac{1}{2}[u(i,j) + u(i+1,j)]$, $d_2 = \frac{\partial u}{\partial x}(i,j) = \frac{1}{2}[u(i+1,j) + u(i+1,j+1)] - \frac{1}{2}[u(i,j) + u(i,j+1)]$. We say that two points X and Y have the same direction with precision $p = 1/n$ if

$$Angle\ (dir(X), dir(Y)) \leq 2\pi/n,$$

where we take $n > 2$. According to the Helmholtz principle, we assume that the direction at each point is a random variable uniformly distributed on $[0, 2\pi)$, so that p is the probability for a given point to have a given direction with the given accuracy p. The directions of any two pixels with a distance ≥ 2 are independent random variables.

Let A be a (straight) segment in the image made of l independent points $x_1, ..., x_l$. For convenience we shall say that the segment A has length l, although its real length is $2l$ or $2l - 1$. Let θ_0 be a given direction. Let X_i be the random variable defined by $X_i = 1$ if the point x_i has the given direction θ_0 (the direction of the segment), and $X_i = 0$ otherwise, and set

$$S_l = X_1 + ... + X_l.$$

Then S_l is the number of points among $x_1, ..., x_l$ having the same direction θ_0, and has the binomial distribution with parameters l and p:

$$P(S_l = k) = C_l^k p^k (1 - p)^{l-k}, \ 0 \leq k \leq l.$$

Definition 1 (ϵ-meaningful segment [7]). *A segment of length l is called ϵ-meaningful in a $N \times N$ image if it contains at least $k(l)$ points having their direction aligned with that of the segment, where $k(l)$ is given by*

$$k(l) = \min\{k \in \mathbf{N}, P[S_l \geq k] \leq \frac{\epsilon}{N^4}\}. \tag{5}$$

Notice that N^4 is approximately the total number of segments (N^2 choices for the first point of a segment and N^2 for the last one). By the definition, it can be easily checked (see [9]) that the expectation of the number of ϵ-meaningful segments is less than ϵ, so that the definition is justified.

The notion of ϵ-meaningful is closely related to the classical "α-significance" in statistics, where α is simply ϵ/N^4. The main advantage to use the expectation instead of probability is that the linearity of the expectation does not need the independence of related variables.

If A is a segment of length l with k points having the (given) direction of A, we define

$$NFA(A) = N^4 P(S_l \geq k)$$

as the *number of false alarms* of A. Therefore A is ϵ- meaningful if and only if $NFA(A) \leq \epsilon$.

An ϵ-meaningful segment may contain sub-segment which is also ϵ-meaningful. We therefore need the notion of *maximal ϵ-meaningful segment*. A segment A is called maximal if it does not contain a strictly more meaningful segment (i.e. $\forall B \subset A, NFA(B) \geq NFA(A)$), and if it is not contained in a more meaningful segment (i.e. $\forall B \supset A, NFA(B) > NFA(A)$). A segment is said to be maximal ϵ-meaningful if it is maximal and ϵ-meaningful.

Fig.2 shows the maximal meaningful segments found in the spectrum of a simulated motion-blurred image, with blur length 10 and horizontal blur direction.

Fig. 2. Simulated motion-blurred image with blur length 10 and blur direction 0, and its Fourier Spectrum with detected maximal meaningful segments

To find parallel segments from detected maximal meaningful segments, we proceed in a similar way. Let $L_1, ..., L_l$ be a group of segments. This time we define $X_i = 1$ if the direction of i-th segment L_i coincides with that of the group (a given direction θ_0), and $X_i = 0$ otherwise. We can still suppose that X_i are independent random variables with the same law $P(X_i = 1) = p$ and $P(X_i = 0) = 1 - p$, because the direction of a segment is the common direction of all of its points. For example, in the case where L_1 contains a point which is of distance ≥ 2 from a point of L_2, then the directions of L_1 and L_2 are independent since so are the directions of the two mentioned points. The contrary case occurs hardly and can be neglected. Therefore we can still suppose that $S_l = X_1 + ... + X_l$ has the binomial law with parameters l and p.

Definition 2 (ϵ-meaningful parallel segments). *A group of l segments $L = (L_1, ..., L_l)$ is called ϵ-meaningful in a $N \times N$ image if it contains at least $k(l)$ segments having their direction aligned with that of the group (a given direction), where $k(l)$ is given by (5).*

Just as in the case of meaningful segments, we define in a similar way the number of false alarms of a group L of segments and the notion of maximal ϵ-meaningful parallel segments. For example, if $L = (L_1, ..., L_l)$ is a group of l- segments with k segments having the (given) direction of L, we define

$$NFA(L) = N^4 P(S_l \geq k)$$

as the *number of false alarms* of L.

With the previous definitions, we can now describe our algorithm for the detection of ϵ-meaningful parallel segments as follows:

1. Detect the maximal ϵ- meaningful segments $L = (L_1, ... L_l)$ with the algorithm in [7];
2. Calculate the tail probability of the binomial distribution :

$$P[S_l \geq k] = \sum_{i=k}^{l} \binom{l}{i} p^i (1-p)^{l-i} \tag{6}$$

where $0 \leq k \leq l$;

3. For $1 \leq i \leq j \leq l$, calculate $k_{i,j}$, the number of segments in the group $L_{i,j} = (L_i, ..., L_j)$ having the same direction as the group $L_{i,j}$;
4. Calculate $P = P[S_{j-i+1} \geq k_{i,j}]$;
5. if $P \leq \frac{\epsilon}{N^4}$, the group $L_{i,j}$ of segments is ϵ-meaningful parallel.

To demonstrate our algorithm, we performed it with Fig.2. The detected parallel meaningful segments are shown in Fig.3.

We now need to identify the most possible value for the equidistance of the detected parallel segments. To this end we design a K-means cluster algorithm.

5   Adapted K-Means Cluster Algorithm

Due to the very obscured characteristic of the Fourier spectrum, we should be very careful to identify the real distance between segments. If all were perfect, the consecutive segments would have equal distance $\Delta = \Delta u$ but for the two segments around the origin whose distance would be 2Δ. However, as segments in the Fourier transform are hardly visible, it is very possible that some of detected segments are not the real ones. For example we could detect two segments which might represent the same segment in the Fourier spectrum. Due to the characteristic of the Fourier transform, it is reasonable to classify the distances of consecutive segments into three clusters, say C_1, C_2 and C_3, where C_1 consists of distances near 0, C_2 of distances near the desired value Δ, and C_3 of distances near 2Δ. We need to identify these three clusters. To this end we use the K-means cluster algorithm.

This algorithm aims to partition a set of n inputs into k clusters in a reasonable way. Mathematically the problem can be stated as follows. For a set $C = (x_1, ..., x_n)$ of points of the real line $\mathbb{R}$ (or of the Euclidean space $\mathbb{R}^d$ in the general case), we search for the partition of C into k-clusters $C_1, ..., C_k$ such that

$$\sum_{i=1}^{k} \sigma^2(C_i)$$

is minimal among all possible partitions of C, where $\sigma^2(C_i) = \sum_{x \in C_i} |x - m_i|^2$, with $m_i = \frac{1}{|C_i|} \sum_{x \in C_i} x$, $|C_i|$ being the cardinality of C_i.

Fig. 3. Detected ϵ-meaningful parallel segments of Fig.2

The K-means algorithm has been widely applied to image analysis, such as pattern recognition [11]. Here we can adapt it thanks to some *a priori* information of cluster inputs. With this at hand, we can get rid of classical drawbacks of the K-means algorithm [12].

Let $L = (L_1, ..., L_{n+1})$ be the group of detected ϵ-meaningful parallel segments, arranged from left to right. Let $C = \{d_1, ..., d_n\}$ be the perpendicular distance of adjacent parallel segments.

For $1 \leq i \leq n$, let $r(d_i)$ be the number of $j = 1, ..., n$ such that $\mid d_j - d_i \mid \leq \varepsilon$. Then $r(d_i)$ is the number of repetition in C of the value d_i with precision ε. As explained in the beginning of this section, the value of the equal distance is approximately the most occurred value in C:

$$\Delta \approx d_{i_0}, \text{where } i_0 \text{ is such that } r(d_{i_0}) = \max_{1 \leq j \leq n} r(d_j). \tag{7}$$

We shall give a more precise value for Δ. This algorithm consists of two steps:

Step I: initialisation. The first step is a pre-processing procedure. It aims to assign the cluster number, the *initial seed point* of each cluster, and the initial partition. As was explained in the beginning of this section, the cluster number is assigned to be 3: we search for a partition of C to three clusters C_1, C_2, C_3. Let m_i be the *initial seed point* of each cluster C_i, where:

$$m_1 = \varepsilon,$$
$$m_2 = \max\{d_j, 1 \leq j \leq n\}/2, \qquad (8)$$
$$m_3 = \max\{d_j, 1 \leq j \leq n\} - \varepsilon,$$

where $1 \leq \varepsilon \leq 3$ is the precision. We usually take $\varepsilon = 2$.

As shown in Fig.3, m_1 is approximately the distances of segments which are very close each other; m_3 is about the distance of the two segments around the origin (see also Fig.4), and m_2 the distance of usual adjacent segments.

We then define C_i as the set of all $d \in C$ which are nearest from m_i compared with $m_j (j \neq i)$. That is, for $i = 1, 2, 3$,

$$C_i = \{d \in C : |d - m_i| = \min_{1 \leq j \leq 3} |d - m_j|\}.$$

Step II: iteration. The second step is the iteration process. We now replace the values of m_i by the mean of C_i:

$$m_i = \frac{1}{|C_i|} \sum_{x \in C_i} x, \quad 1 \leq i \leq 3,$$

where $|C_i|$ is the cardinality of C_i. With these new values of m_i, we define in the same way as in Step I the new clusters C_i, and so on. The process will be stopped

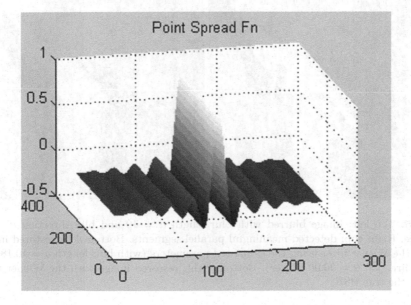

Fig. 4. Transfer function H in the frequency domain

if $|m_i^{(n+1)} - m_i^{(n)}| < \delta$ for each $i = 1, 2, 3$, where δ is the desired precision, and $m_i^{(n)}$ denotes the seed point of C_i at the n-th iteration.

The final value of m_2 can be considered as the desired value of Δ. But in practice, to reduce errors, we calculate the value of Δ as the following mean value: $\Delta = (x_{n+1} - x_1)/(|C_2| + 2)$, if $(x_i, 0)$ is the intersection point of the i-th segment L_i with the x-axis. (We have added the number 2 because the middle interval with length 2Δ is not considered in C_2.)

Fig. 5. Top left: image blurred with blur length $a = 10$ and blur direction $\theta = 90$ degree. Right top: detected meaningful parallel segments. Bottom left: restored image with the Wiener filter using the estimated parameters (with blur length $a = 10.18$ and blur direction $\theta = 90.00$ degree). Bottom right: restored image with the Wiener filter using the true PSF.

6   Experimental Results

In this section, numerical experiments are presented to test the proposed algorithm to identify PSF parameters of motion-blurred image.

To check the validity of the method, the experiments are performed on a simulated blurred image. The Lena image is blurred with different blur lengths and directions, and Gaussian noise is add to the blurred image. The procedure of our experiments is as follows:

1. blur the image with given length and direction;
2. detect the maximal ϵ-meaningful segments in the spectrum of blurred image;

Fig. 6. Top left: image blurred with blur length $a = 7$ and blur direction $\theta = 28$ degree. Right top: detected meaningful parallel segments. Bottom left: restored image with the Wiener filter using the estimated parameters (with blur length $a = 7.23$ and blur direction $\theta = 28.11$ degree). Bottom right: restored image using the true PSF.

3. detect the ϵ-meaningful parallel segments;
4. assign the *initial seed points*;
5. cluster the ϵ-meaningful parallel segments;
6. calculate the blur length and blur direction from the clustering result.

Two examples are shown in Fig.5 and Fig.6. The length and direction are always very accurately estimated.

7   Conclusion

We have presented a new method for motion-blur parameters detection based on the most perceptual parallel lines in the blurred image spectrum.

By our method, the blur direction and length can be estimated with good precision, via the Fourier spectrum, using the characteristic of PSF. The Helmholtz principle is applied to get meaningful parallel segments; an adapted K-means algorithm is used to cluster the detected segments. The blur length and blur direction are obtained from the clustering result.

At that time, experiments have been performed using simulated motion-blurred images only. The difficulty with natural blurred images is to define a protocol that ensures images to satisfy the uniform linear and space invariant motion blur assumption. We expect to present such experiments on natural images by the time of the conference. The next step will be to generalize the approach so that more general motion blurs would be included, resulting in an effective new algorithm to restore motion-blurred photographs.

References

1. Cannon, M.: Blind deconvolution of spatially invariant image blurs with phase. IEEE Trans. Acoust. Speech Signal Process 24 (1976)
2. Yitzhaky, Y., Kopeika, N.: Identification of blur parameters from motion blurred images. Graphical Models and Image Processing 59(5), 310–320 (1997)
3. Zhang, Y., Wen, C., Zhang, Y.: Estimation of motion parameters from blurred images. Pattern Recogniton Letters 21, 425–433 (2000)
4. Schafer, R.W., Mersereau, R.M., Richards, M.A.: Constrained iterative restoration algorithm. Proc. IEEE 69, 432–450 (1981)
5. Lagendijk, R., Biemond, J., Mastin, G.: Systematic approach to two-dimentional blind deconvolution by zero-sheet separation. IEEE Trans. Acoust, Speech, Signal Processing 38(7) (1990)
6. Kundur, D.: Blind deconvolution of still images using recursive inverse filtering. Proceedings of the IASTED International Conference on Signal and Image Processing 59(5), 310–320 (1997)
7. Desolneux, A., Moisan, L., Morel, J.: Edge detection by helmholtz principle. Mathematical Imaging and Vision 14(3), 271–284 (2001)
8. Desolneux, A., Moisan, L., Morel, J.: A grouping principle and four applications. IEEE Trans. on Pattern analysis and Machine Intelligence 25(4), 508–513 (2003)
9. Desolneux, A., Moisan, L., Morel, J.: Meaningful alignements. International Journal of Computer Vision 40(1), 7–23 (2000)

10. Cao, F.: Application of the gestalt principles to the detection of good continuations and corners in image level lines. Computing and Visualisation in Science 7, 3–13 (2004)
11. Chen, C., Luo, J., Parker, K.: Image segmentation via adaptive k-mean clustering and knowledge-based morphological operations with biomedical applications. IEEE Trans. Image Processing 7(12), 1673–1683 (1998)
12. Cheung, Y.: k^*-means: a new generalized k-means clustering algorithm. Pattern Recognition Letters 24, 2883–2893 (2003)

Improved Object Tracking Using an Adaptive Colour Model

Zezhi Chen[1] and Andrew M. Wallace[2]

[1] School of Mathematics and Computer Sciences
[2] School of Engineering and Physical Sciences
Joint Research Institute in Signal and Image Processing
Heriot-Watt University, Edinburgh, UK, EH14 4AS
{Zc19,A.M.Wallace}@hw.ac.uk

Abstract. We present the results of a study to exploit a multiple colour space model (*CSM*) and variable kernels for object tracking in video sequences. The basis of our work is the mean shift algorithm; for a moving target, we develop a procedure to adaptively change the *CSM* throughout a video sequence. The optional *CSM* components are ranked using a similarity distance within an inner (representing the object) and outer (representing the surrounding region) rectangle. Rather than use the standard, Epanechnikov kernel, we have also used a kernel weighted by the normalized Chamfer distance transform to improve the accuracy of target representation and localization, minimising the distance between the two distributions of foreground and background using the Bhattacharya coefficient. To define the target shape in the rectangular window, either regional segmentation or background-difference imaging, dependent on the nature of the video sequence, has been used. Experimental results show the improved tracking capability and versatility of the algorithm in comparison with results using fixed colour models and standard kernels.

Keywords: Object tracking, video sequences, adaptive colour space models.

1 Introduction

As defined by Comaniciu et al. [1], tracking algorithms have two essential components

- Target representation and localization, primarily a bottom-up process, which has to adapt to changes in the appearance of the target.
- Filtering and data association, primarily a top-down process, incorporating the dynamics of the tracked object, learning of scene priors, and evaluation of different hypotheses.

In real time tracking of objects through video sequences, combined analysis of the shape, intensity or colour distribution, and motion of the object against an assumed background can be used in principle to differentiate (segment) and track

A.L. Yuille et al. (Eds.): EMMCVPR 2007, LNCS 4679, pp. 280–294, 2007.
© Springer-Verlag Berlin Heidelberg 2007

the object in a dynamic scene, one in which the camera may be fixed or static and the object is assumed to be in motion. Tracking has been achieved by simple Gaussian or adaptive Gaussian mixture models [2], using Kalman-filters [3] or dynamic programming [4] assuming either fixed and deformable contours. Exemplars include face tracking in a crowded scene [5] [6] [7], aerial video surveillance [8], and the detection and tracking of multiple people and vehicles in cluttered scenes using multiple cameras [9]. In many real time applications only a small percentage of the system resources can be allocated for tracking, the rest being required for the pre-processing stages or for high-level tasks such as recognition, trajectory interpretation, and reasoning. Therefore, it is desirable to keep the computational complexity of a tracker as low as possible.

The mean shift algorithm is a nonparametric statistical method to find the nearest mode of a point sample distribution. This algorithm has been adapted as an efficient technique for image segmentation [10] [11] and object tracking [1] [12]. The justification for its use as a density estimation-based non-parametric clustering method is that the feature space can be regarded as the empirical probability density function (*pdf*) of the represented parameter. Dense regions in the feature space correspond to local maxima of the *pdf*, i.e. the modes of the unknown density. Once the location of a mode is determined, the associated cluster can be delineated based on the local structure of the feature space. As usually applied, the mean shift tracking algorithm [1] is a kernel based object tracking method that uses a symmetric Epanechnikov kernel (E-kernel) within a pre-defined elliptical or rectangular window. A candidate object is defined in a first image, interactively or by automatic segmentation. The 2D region that encloses this object is modelled by a probability density function that describes the first order statistics of that region. If the image sequence is in colour then the usual choice is to employ a colour histogram. Then, the task is to find a *candidate* position for the target in subsequent images by finding the minimum distance between the model and candidate histograms using an iterative procedure.

Tracking success or failure depends primarily on how distinguishable an object is from its surroundings. The features that best distinguish between foreground and background are the best features for tracking. However, in common instances when the camera viewpoint changes so that the background and illumination parameters change, it is sensible to continuously re-evaluate these features, rather than rely on a fixed set of features that are determined a priori. For example, Collins et al. [13] presented an online feature selection mechanism for evaluating multiple features while tracking and adjusting the set of features used to improve tracking performance. They computed log likelihood ratios of class conditional sample densities from object and background are used to form a new set of candidate features tailored to the local object/background discrimination task.

In this paper, we propose to improve object tracking by using an adaptive colour model. In order to select the best colour space model (*CSM*), we sort the several candidate *CSM* components using the Bhattacharyya coefficient [14], which is an approximate measurement of the amount of overlap between the two distributions of foreground and background, instead of using a variance ratio

measure of the distribution of likelihood values. For a moving target, it is usual
to define a rectangular target window in an initial frame, and then process the
data within that window to separate the tracked object from the background by
a segmentation algorithm (e.g. mean shift segmentation). However, rather than
use the standard, Epanechnikov kernel, we have also used a kernel weighted by
the Normalized Chamfer Distance Transform (NCDT) to improve the accuracy
of target representation and localization, introduced in [15]. Experimental results
demonstrate the improved capability of the new algorithm for object tracking in
colour sequences.

2   Defining Similarity by Histograms

In tracking an object through a colour image sequence, we represent it by a discrete distribution of samples from a region in colour space, localised by a kernel
whose centre defines the current position. Hence, we want to find the maximum
in the distribution of a function, ρ, that measures the similarity between the
weighted colour distributions as a function of position (shift) in the *candidate*
image with respect to a previous *model* image. If we have two sets of parameters
for the respective densities $p_i(x), i = 1, 2$, the Bhattacharyya coefficient is an
approximate measurement of the amount of overlap, defined by

$$\rho = \int \sqrt{p_1(x)p_2(x)}dx \tag{1}$$

Referring to Bhattacharyya [14], and as described by Kailath [16], the set,
$\{\sqrt{p_i(x)}, i = 1, 2\}$, can be regarded as the direction-cosines of two vectors in the
space of $\mathbf{x}$. Alternatively, they may be thought of defining two points on the unit
sphere (since $\int p_i(x)dx = 1, i = 1, 2$) of $\mathbf{x}$. The Bhattacharyya coefficient ρ can
be regarded as the cosine of the angle between these two vectors, i.e., $\rho = cos\theta$,
where θ is the angle between the lines with direction cosines $\{\sqrt{p_i(x)}\}$. The
angle θ must clearly be between 0 and $\pi/2$ and therefore $0 \leq \rho \leq 1$. Hence, it
is natural to also consider the distance between the points $\{\sqrt{p_i(x)}, i = 1, 2\}$ on
the unit sphere

$$d = \int \left[\sqrt{p_1(x)} - \sqrt{p_2(x)} \right]^2 dx \tag{2}$$

Hence,

$$d = 4sin^2\frac{\theta}{2} = 2(1 - cos\theta) = 2(1 - \rho) \tag{3}$$

Since we are dealing with discretely sampled data from colour images, we use
discrete densities stored as m-bin histograms in both the *model* and *candidate*
image. The discrete density of the model is defined as

$$q = \{q_u\}, u = 1, 2, \cdots, m \qquad \sum_{u=1}^{m} q_u = 1 \tag{4}$$

Similarly, the estimated histogram of a candidate at a given location y in a subsequent frame is

$$p(y) = \{p_u(y)\}, u = 1, 2, \cdots, m \qquad \sum_{u=1}^{m} p_u = 1 \qquad (5)$$

According to the definition of Equations (1) and (3), the sample estimate of the Bhattacharyya coefficient is given by

$$\rho(y) = \rho[p(y), q] = \sum_{u=1}^{m} \sqrt{p_u(y)q_u} \qquad (6)$$

The distance between two distributions can be defined as

$$d(y) = \sqrt{1 - \rho(y)} \qquad (7)$$

The distance $d(y)$ lies between zero and unity, obeying the triangle inequality. In a discrete space, $x_i, i = 1, 2, \cdots, n$ are the pixel locations of the model, centred at a spatial location $\mathbf{0}$, which is defined as the position of the window that we want to track in a preceding frame. A function b: $\mathbf{R}^2 \longrightarrow \{1, 2, \cdots, m\}$ associates to the pixel at location x_i the index $b(x_i)$ of the histogram bin corresponding to the value of that pixel. Hence a normalized histogram of the region of interest can be formed (using q as an example)

$$q(u) = \frac{1}{n} \sum_{i=1}^{n} \delta[b(x_i) - u], \qquad u = 1, 2, \cdots, m \qquad (8)$$

where δ is the Kronecker delta function.

3   Selecting the Best Colour Space Model

Numerous *CSM*s can used be for segmentation and tracking of objects in a scene [7] [13] [17] [18] [19]. Most represent colours in a three coordinate colour space, and conversion formulas are used to transform from one space to another. Our main objective is to achieve the best discrimination of an object of a certain colour distribution from the background. In our context, we need to measure the similarity between the two histograms of the internal (target) and external (background) regions. We use the distance criterion Eq.(7) to measure this dissimilarity. As tracking proceeds, the best colour space is selected by finding the *CSM* with maximum distance value between the foreground and background. This distance measure based on the Bhattacharyya coefficient is simple and effective, but we have considered alternatives. For example, the augmented variance ratio (AVR) is the ratio of the between class variance of the feature to the within class variance of the feature [20]. The earth mover's distance (EMD) is based on the minimal cost that must be paid to transform one distribution into the other, in a precise sense [21]. In the color index method [22], the index contains the complete color distributions of the image in the form of cumulative color histograms.

We use a "center-surround" approach to sample pixels from object and background. A rectangular set of pixels covering the object is chosen to represent the object pixels, while a larger surrounding ring of pixels is chosen to represent the background. For the internal rectangle of size $h \times w$ pixels, the outer margin of width $(\sqrt{2} - 1)hw/2$ pixels forms the background sample. The foreground and background have the same number of pixels if $h = w$. Each potential feature set can have many tunable parameters and therefore the full number of potential features that could be used for tracking is enormous. We construct 16 single frame *CSM* components from 5 different colour spaces ($R.G.B$, $L.a.b$, $H.S.V$, $Y.I.Q/Y.Cb.Cr$, $C.M.Y.K$). All the values of pixels are normalized to 0 to 255, yielding feature histograms with 256 bins.

Fig.1(a) shows a sample object. The distance measure between the foreground and background of each CSM components are shown in Fig.1(b) and the set of all 16 candidate images after rank-ordering the feature according to the criterion formula (7) are shown in Fig.1(c). The image with the most discriminative feature (best for tracking) is at the upper left. The image with the least discriminative feature (worst for tracking) is at the lower right.

4   Normalized Chamfer Distance Transform

The mean shift procedure is equivalent to gradient ascent on the similarity function for kernels that are radially symmetric, non-negative, non-increasing and piecewise continuous over the profile [23]. The most popular choice for a kernel, K, is the optimal E-kernel that has a uniform derivative of $G = 1$. However, when tracking an object through a video sequence, there is no reason to suppose that the target has radial symmetry, and even if an elliptical kernel is used, i.e. there is a variable bandwidth, the background area that is being sampled for the colour distribution will change. If the background is uniform this will not affect the colour *pdf*, and hence the gradient ascent will be exact. However, if it is not uniform, but varies markedly and in a worst case has similar properties to the target, as we shall see in the next section, then multiple modes will be formed in the *pdf* and the mean shift is no longer exact.

Therefore, we use a *distance transform* (DT), matched to and changing with the shape of the tracked object, as a kernel function. For the DT each foreground pixel is given a value that is a measure of the distance to the nearest edge pixel. The edge and background pixels are set to zero. We use the normalised Chamfer distance transform (NCDT) rather than the true Euclidean distance, as it is an efficient approximation [24]. A 3×3 pixel neighborhood is used. The two local distances in 3×3 neighborhood are the distance between horizontal/vertical neighbors and between diagonal neighbors. The NCDT kernel is a better representation of the colour distribution of the tracked target because it takes account of the shape of the enclosing contour, but it retains the more reliable centre weighting of the radially symmetric kernels. We aim to show

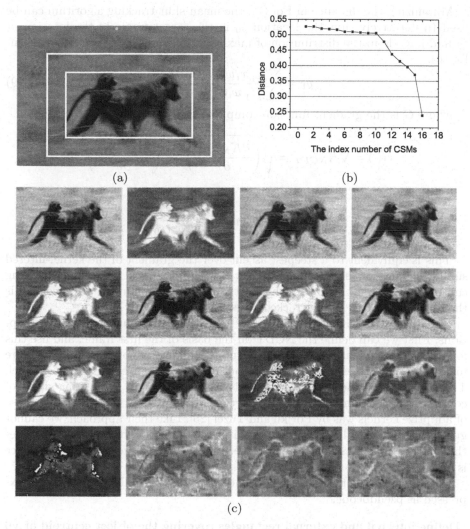

Fig. 1. (a) A sample image with concentric boxes delineating the object and background. (b). The similarity distance of each *CSM* component. (c) Rank-ordered 16 images.

that this weighting increases the accuracy and robustness of representation of the *pdf*'s as the target moves. The target area is separated from the background by a mean shift segmentation [10]. Therefore, we are investigating whether the anticipated gain in excluding background pixels from the density estimates and weighting more substantially those more reliable pixels towards the centre of the tracked object, will outweigh the possibility of forming false modes because of the shape of the NCDT, bearing in mind that the radially symmetric kernels may produce false modes due to badly defined densities.

Minimizing the distance in Eq. (7), the mean shift tracking algorithm can be used to iteratively shift the location y_0 in the target frame to the location y_1 to find a mode in the distribution of maximization the Bhattacharya coefficient Eq.(6).

$$y_1 = \frac{\sum_{i=1}^{n} x_i w_i G(y_0 - x_i)}{\sum_{i=1}^{n} w_i G(y_0 - x_i)} \tag{9}$$

where G is the gradient function computed on NCDT-kernel.

$$G(\cdot) = \nabla K_{NCDT} = \sqrt{\left(\frac{\partial K_{NCDT}}{\partial x}\right)^2 + \left(\frac{\partial K_{NCDT}}{\partial y}\right)^2}$$

$$w_i = \sum_{u=1}^{m} \delta\left[b(x_i) - u\right] \sqrt{\frac{q_u}{p_u(y_0)}}$$

This is equivalent to a steepest ascent over the gradient of the kernel-filtered similarity function based on the colour histograms. Local mean shift tracking can be used with each CSM, such that the y with the minimum distance $d(y)$ is the optimal location of the tracked object.

In applying the NCDT kernel to the mean shift procedure, we have two options. First we can define the NCDT on the basis of the first frame, and use this for the whole sequence. Second we can update the kernel on each frame, before mean shifting in the subsequent frame but retaining the previous frame's kernel. The advantage of the latter technique is that it adapts to a changing shape of contour (e.g. as a human is tracked) as well as to a changing CSM. A previous use of adaptive kernels was by Porikli and Tuzel [25]. Like their approach, we cannot guarantee theoretical convergence conditions [10], although the NCDT satisfies these in part with its decreasing profile. Practical tests show that convergence is achieved.

To summarise our discussion, the adaptive tracking algorithm can be expressed as pseudocode,

Define internal and external rectangles covering the object centroid at y0 in the first image.
Sort CSMs by similarity distance criterion Eq.(7).
Choose preferred CSMs.
Apply the mean shift segmentation algorithm to separate foreground (target) and background in the external rectangle.
Compute NCDT-kernel using Chamfer distance to form model histogram, q, in selected CSMs.
Repeat
    Input the next image.
    Repeat
        Compute candidate histogram p(y0) in preferred CSM using NCDT-kernel.
        Find next location of candidate using Eq.(9)
        Compute error, $e = \|y_1 - y_0\|$

Set $y_0 = y_1$ **(y_0 is the new location)**
Until $e \leq \epsilon$**, error threshold or maximum iteration reached.**
Choose preferred CSM**s.**
Compute NCDT-kernel and the histogram, q, in selected CSM**s.**
Until end of input sequence

5 Experimental Evaluation

In this section, we present an evaluation of adaptive mean shift tracking using the NCDT-kernel and an adaptive colour model. All the tests were carried out on a Pentium 4 CPU 3.40 GHz with 1GB RAM. The code was implemented in Matlab, so that it would be reasonable to assume a considerable increase in processing speed if re-implemented in another language. Even so, real-time operation is possible.

In the first experiment, we compare the tracking of a moving monkey in a video sequence that includes 401 frames of 324 × 480 pixels. The target location was initialized by a rectangular region of size 140 × 290 pixels. Fig.2(a) shows one of the CSM components of the first frame and the tracked object. In this case a simple regional homogeneity criterion has been applied as the target had relatively uniform intensity. Fig.2(b) shows the comparison of the 3D-NCDT kernel with the E-kernel in Fig.2(c). Fig.3 shows some sample images of the tracked monkey, frames 11, 142, 286 and 375, from the whole sequence.

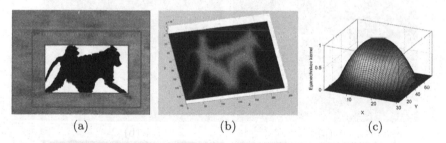

(a)   (b)   (c)

Fig. 2. (a) The tracked object in the best CSM component of the first frame. (b) The 3-D NCDT kernel. (c) The E-kernel.

Fig.4 shows another sample sequence (187 frames). These images have much lower contrast with a car travelling through patches of forest and grass. Fig.4(a) shows the first frame with a labelled object (red box) and background (white box). Fig.4(b), (c) and (d) show the CSM components with highest, median and lowest similarity distance between the foreground and background. Fig.5 shows some tracking results as the car moves through the whole sequence. These are frames 14, 36, 126 and 171. The car changes noticeably in size and projected shape, as well as colour distribution as the sequence progresses. Using the same sequence, we illustrate in Fig.6 what happens if we use the standard E-kernel mean shift tracking algorithm, without adapting the CSMs. As can be seen,

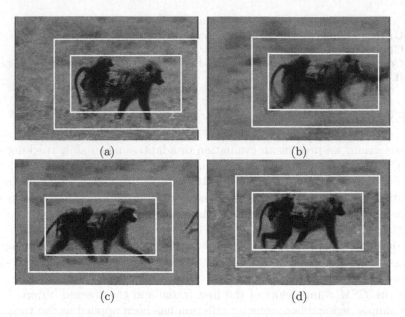

(a)   (b)

(c)   (d)

Fig. 3. Tracking the moving monkey (frames 11, 142, 286 and 375)

(a)   (b)

(c)   (d)

Fig. 4. Some samples of ranked *CSM* components. (a) The first frame with labeled object (red box) and background (white box). (b)-(d) Ranked *CSM*s.

(a)   (b)

(c)   (d)

Fig. 5. Tracking results of frames 14, 36, 126 and 171 using the adaptive tracking algorithm

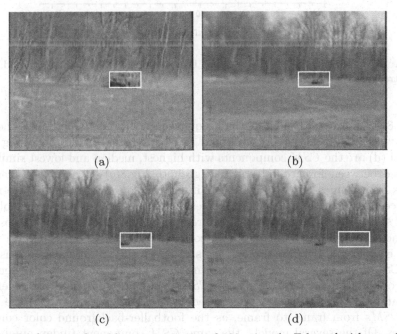

(a)   (b)

(c)   (d)

Fig. 6. Tracking results of frames 1, 80, 98 and 108 using the E-kernel without adaptive *CSM*

after the 98th frame, the target is lost. Frames 1, 80, 98 and 108 are illustrated. Fig.7 shows the variation of the distance between the real centre and the estimated centre of the tracked car, using manual positioning of the centroid as ground truth. The results show that the adaptive algorithm is more robust in this instance, and the corresponding sharp rise in error as the car is lost when using the E-kernel.

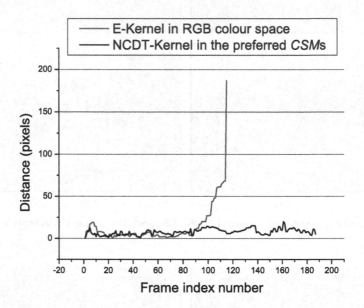

Fig. 7. A comparison results between the E-kernel and NCDT kernel with adaptive *CSM* tracking

Fig.8 shows frames from a football sequence (129 frames). Fig.8(a) is the first frame with the labeled object (red box) and background (white box). Fig.8(b), (c) and (d) are the *CSM* components with highest, median and lowest similarity distance between foreground and background. Fig.8(e)-(f) show the tracking results from frames 16, 36, 69 and 127 in the whole sequence. There are many instances in which the footballer changes shape and crosses with other players of the same or the other side. The source images are low contrast videos, but the tracker is still able to maintain its correct position.

For each frame, the best three of the sixteen possible components are used to form CSM. Fig.9 presents a trace showing that, during the football sequence of Fig.8, only five of the possible sixteen *CSM* components are used. However, the adaptation of the algorithm is illustrated by the changing ranking of the three best *CSM*s from frame to frame, as the footballer-background color contrast changes. Of the several models, the same *CSM* component (index number 12 (Q component)) is selected as either first, second or third throughout the video sequence.

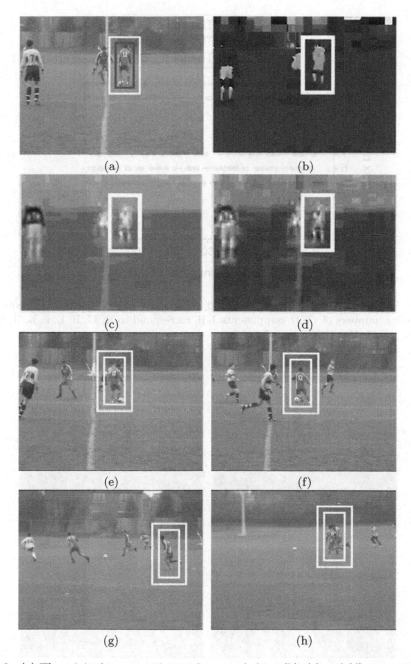

Fig. 8. (a) The original image with initial rectangle box. (b), (c) and (d) some samples of ranked *CSM* components. (e)-(h) tracking results of frames 16, 36, 69 and 127.

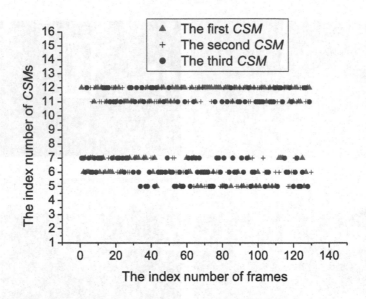

Fig. 9. Trace of *CSM* components selected to track the object in a football sequence. The index numbers of *CSM* components 1-16 correspond to R, G, B, L, a, b, H, S, V, Y, I, Q, C, M, Y and K in our definition.

Fig. 10. Tracking a walking man with added Gaussian noises of mean zero with variance 0.03

Finally, to further test the robustness of the new algorithm, as requested by a referee, we added Gaussian white noise of mean zero with variance 0.03 (intensities ranging from 0 to 1) to a video of a moving male pedestrian sequence. Some samples (frames 4, 34, 52 and 64) of tracking results are shown in Fig.10. Again, the approach is successful.

6   Conclusions

We have presented two modifications to the mean-shift tracking approach, first to continuously evaluate a multiple colour space model while tracking, and second to use a normalised Chamfer distance kernel as a weighting and constraining function. The algorithm maximises the similarity between model and candidate distributions in each of the selected CSMs. The results presented on a number of video sequences show that the new algorithm performs well in maintaining a stable and robust estimate of the target position, when compared with the non-adaptive methods.

Acknowledgements

The work reported in this paper was funded by the Systems Engineering for Autonomous Systems (SEAS) Defence Technology Centre. The authors would like to acknowledge the provision of data sequences from the Caviar project at the University of Edinburgh.

References

1. Comaniciu, D., Ramesh, V., Meer, P.: Kernel-based object tracking. IEEE Transactions on Pattern Analysis and Machine Intelligence 25(5), 564–575 (2003)
2. McKenna, S.J., Raja, Y., Gong, S.: Tracking colour objects using adaptive mixture models. Image and Vision Computing 17, 225–231 (1999)
3. Chen, Y., Rui, Y., Huang, T.: JPDAF-based HMM for real-time contour tracking. In: Proc. IEEE Conf. on Computer Vision and Pattern Recognition, vol. I, pp. 543–550. IEEE Computer Society Press, Los Alamitos (2001)
4. Geiger, D., Gupta, A., Costa, L.A., Vlontzos, J.: Dynamic programming for detecting, tracking, and matching deformable contours. IEEE Transactions on Pattern Analysis and Machine Intelligence 17(3), 294–302 (1995)
5. Hsu, R.L., Abdel-Mottaleb, M., Jain, A.K.: Face Detection in Color Images. IEEE Transactions on Pattern Analysis and Machine Intelligence 24(5), 696–706 (2002)
6. Zarit, B.D., Super, B.J., Quek, F.K.H.: Comparison of five color models in skin pixel classification. In: Proc. on International Workshop on Recognition, Analysis, and Tracking of Faces and Gestures in Real-Time Systems, pp. 58–63 (1999)
7. Sobottka, K., Pitas, I.: Segmentation and tracking of faces in color images. In: Proceedings of the Second International Conference on Automatic Face and Gesture Recognition, October 1996, pp. 236–241 (1996)
8. Kumar, R., Sawhney, H., et al.: Aerial video surveillance and exploitation. Proceedings of the IEEE 89(10), 1518–1539 (2001)

9. Cohen, I., Medioni, G.: Detecting and tracking moving objects in video surveillance. In: Proceedings of the IEEE Conference on Computer Vision and Pattern Recognition, vol. 2, pp. 319–325 (1999)

10. Comaniciu, D., Meer, P.: Mean shift: A robust approach toward feature space analysis. IEEE Transactions on Pattern Analysis and Machine Intelligence 24(5), 603–619 (2002)

11. Singh, M., Ahuja, N.: Regression based bandwidth selection for segmentation using Parzen windows. In: ICCV. Proceedings of International Conference of Computer Vision, vol. 1, pp. 2–9 (2003)

12. Chen, H., Meer, P.: Robust computer vision through kernel density estimation. In: Heyden, A., Sparr, G., Nielsen, M., Johansen, P. (eds.) ECCV 2002. LNCS, vol. 2350, pp. 236–250. Springer, Heidelberg (2002)

13. Collins, R.T., Liu, Y., Leordeanu, M.: Online selection of discriminative tracking features. IEEE Transactions on Pattern Analysis and Machine Intelligence 27(10), 1631–1643 (2005)

14. Bhattacharyya, A.: On a measure of divergence between two statistical populations defined by their probability distributions. Bulletin of the Calcutta Mathematics Society 35, 99–110 (1943)

15. Chen, Z., Husz, Z.L., Wallace, I., Wallace, A.M.: Video object tracking based on a Chamfer distance transform. In: Proceedings of IEEE International Conference on Image Processing, San Antonio, Texas, September 2007, IEEE Computer Society Press, Los Alamitos (2007)

16. Kailath, T.: The divergence and Bhattacharyya distance measures in signal selection. IEEE Transactions on Communication Technology 15(1), 52–60 (1967)

17. Stern, H., Efros, B.: Adaptive color space swiching for face tracking in multi-coloredlighting environments. In: Fifth IEEE International Conference on Automatic Face and Gesture Recognition, pp. 236–241. IEEE Computer Society Press, Los Alamitos (2002)

18. Quek, F., Mysliwiec, T.A., Zhao, M.: FingerMouse: A Freehand Computer Pointing Interface. In: Proc. of Int'l Conf. on Automatic Face and Gesture Recognition, Zurich, Switzerland, pp. 372–377 (1995)

19. Yang, J., Lu, W., Waibel, A.: Skin-color modeling and adaptation. In: Proceedings of the 3rd Asian Conference on Computer Vision, vol. 2, pp. 687–694 (1998)

20. Liu, Y., Schmidt, K., Cohn, J., Mitra, S.: Facial asymmetry quantification for expression invariant human identification. Computer Vision and Image Understanding 91(1/2), 138–159 (2003)

21. Rubner, Y., Tomasi, C., Guibas, L.J.: The earth mover's distance as a metric for image retrieval. International Journal of Computer Vision 40(2), 99–121 (2000)

22. Stricker, M., Orengo, M.: Similarity of color images. In: SPIE Conference on Storage and Retrieval for Image and Video Databases III, vol. 2420, pp. 381–392 (1995)

23. Cheng, Y.Z.: Mean shift, mode seeking, and clustering. IEEE Transactions on Pattern Analysis and Machine Intelligence 17(8), 790–799 (1995)

24. Borgefors, G.: Hierarchical chamfer matching: A parametric edge matching algorithm. IEEE Transaction on Pattern Analysis and Machine Intelligence 10(6), 849–865 (1988)

25. Porikli, F.M., Tuzel, O.: Human body tracking by adaptive background models and mean-shift analysis. In: IEEE International Workshop on Performance Evaluation of Tracking and Surveillance (2003)

Vehicle Tracking Based on Image Alignment in Aerial Videos*

Hong Zhang** and Fei Yuan

Image Processing Center, BeiHang University, Beijing 100083,
China
dmrzhang@buaa.edu.cn, flyuan@sa.buaa.edu.cn

Abstract. Ground vehicle tracking is an important component of Aerial Video Surveillance System (AVS). We address the problem of real-time and precise vehicle tracking from a single moving airborne camera which faces the challenges of congestion, occlusion and so on. We track a set of point features of the selected vehicle by the technique of image alignment. An edge feature-based outlier rejection criterion is proposed to eliminate the outlier caused by congestion and occlusion. Large motion and total occlusion is handled by a Kalman filter. Furthermore, a reappearance verification program is used to ensure the tracker gets back the right object. Experimental results on real aerial videos show the algorithm is reliable and robust.

Keywords: Vehicle tracking, image alignment, outlier rejection, Kalman filter, reappearance verification.

1 Introduction

Aerial Video Surveillance System has important military and civilian uses [1]. Especially, video-based tracking across an airborne platform has extensive uses in monitoring enemy vehicles and criminal-driving vehicles. For example, we have to track and recognize military vehicles or runaway criminal vehicles until they are destroyed or cornered. So a robust, efficient, and reliable tracking algorithm is desired to maintain precise and durative tracking of the selected vehicle. However, it is not an easy thing. The main difficulties lie in the following:

1. The background is complex and unpredictable. The object may appear in any situation.
2. The image resolution is low. In order to get higher resolution, it is generally necessary to use a telephoto lens, with a narrow field of view [1].
3. The hostile vehicle may be interspersed with civilian traffic, and encounter the problem of congestion and occlusion.
4. The tracked vehicle has maneuverability, which may cause big motion between frames.

---

* Supported by national key laboratory (51476010105HK0101).
** Corresponding author.

A.L. Yuille et al. (Eds.): EMMCVPR 2007, LNCS 4679, pp. 295–302, 2007.
© Springer-Verlag Berlin Heidelberg 2007

5.   The tracked vehicle will leave the field of view because of inaccuracies in platform pointing directions and long-time occlusion. The tracker needs to get back and verify the former tracked object.

We have to overcome all these difficulties to develop a reasonable good system. In our research, we track a vehicle from arbitrary viewpoint, not limiting to vertical viewpoint. The size of vehicles varies from 5 x 10 to 40 x 100.

The tracking algorithm presented in this paper is based on the alignment between the selected vehicle and input images. The Inverse Compositional Algorithm [2] is used and extended. First, we propose an edge feature-based outlier rejection criterion, which can eliminate the outliers caused by congestion and occlusion. Second, a Kalman filter estimator is introduced, which predicts the future position and velocity of the tracked vehicle. In virtue of the Kalman filter, the estimated initial value is located near to the real value of object state, thereby decreasing the iteration times and making the algorithm robust to large motion. Third, when the object leaves the field of view because of inaccuracies in platform pointing directions and long-time occlusion and reappears, we get back it by correlation matching and verify it by combining feature-based quasi-rigid alignment with flexible local template matching.

The rest of this paper is organized as follows. Section 2 outlines the tracking algorithm based on image alignment. Section 3 describes edge feature-based outlier rejection criterion. Section 4 covers how the Kalman filter is combined with the tracking algorithm. We describe how the occlusion is handled in Section 5, and finally reach the experiment and conclusion in Section 6.

2   Tracking Algorithm Based on Image Alignment

Image alignment has successful application in tracking [3] ~ [6]. The basic idea of alignment-based tracking algorithm is: Establishing a general parametric model for the set of allowable image motions and deformations of the target region, then considering the tracking problem as finding the best set of parameter values of the parametric model. Traditional alignment-based tracking algorithms, such as Lucas-Kanade algorithm [3], need to calculate the Jacobian and the Hessian matrix in every iteration, thereby having low efficiency. The Inverse Compositional Algorithm [2] is an improved edition of Lucas-Kanade algorithm, which switches the role of the image and the template, thereby allows pre-computing the time-consuming Hessian matrix, and has good efficiency.

The goal of traditional Lucas-Kanade algorithm [3] is to minimize the sum of squared error between the template T and the image I. That is to minimize:

$$\sum_x [I(W(x; p)) - T(x)]^2. \tag{1}$$

Where, $x = (x, y)^T$ is a column vector containing the pixel coordinates, $W(x; p)$ denotes the parameterized set of allowed warps, e.g. considering affine warps with 6 parameters:

$$W(x\ ;\ p)=\begin{pmatrix}(1+p_1)*x & +p_3*y & +p_5 \\ p_2*x & +(1+p_4)*y & +p_6\end{pmatrix}=\begin{pmatrix}(1+p_1) & p_3 & p_5 \\ p_2 & (1+p_4) & p_6\end{pmatrix}\begin{pmatrix}x \\ y \\ 1\end{pmatrix}\quad\text{and}$$

$$p=(p_1,p_2,p_3,p_4,p_5,p_6)^T\ .$$

The Inverse Compositional Algorithm [2] inverses the role of the template T and the image I. That is,

$$\sum_x [T(W(x;\Delta p))-I(W(x;p))]^2\ . \tag{2}$$

Δp can be estimated by Gauss-Newton algorithm:

$$\Delta p=H^{-1}\sum_x [\nabla T\frac{\partial w}{\partial p}]^T[I(W(x\ ;p))-T(x)]\ . \tag{3}$$

Where H is Hessian matrix of the template T : $H=\sum_x [\nabla T\frac{\partial w}{\partial p}]^T[\nabla T\frac{\partial w}{\partial p}]$.

Then updates the warp:

$$W(x;p)\leftarrow W(x;p)\circ W(x;\Delta p)^{-1}\ . \tag{4}$$

(2) and (4) are iterated until the estimates of the parameters p converge.

3 Outlier Rejection

Image alignment is actually a form of template matching. When the object is occluded, or the brightness constancy assumption is invalid, it needs a robust and efficient algorithm to eliminate the outliers. Normalized cross-correlation [6] and normalized error value of the image and the template [7] is used to determine the outliers. However, when the gray-value of the discrepancy between the image and the template is small, the above algorithms fall easily. Considering the low resolution of aerial images, and edges being the most dominant features for many man-made objects such as vehicles and buildings, we develop an edge feature-based outlier rejection criterion. First, by the characteristic of spatial coherence of outliers, the template is subdivided into a set of rectangular blocks ($B_1,B_2,...,B_k$, k is the number of the blocks). Then, we deal the template T and the image I with Canny operator [11], and score up the number of pixels whose values are 1, noting $T_1,T_2,...,T_k$ and $I_1,I_2,...,I_k$. Finally, we compute the ratio of T_i and I_i , noting E_i , $E_i=\dfrac{T_i}{I_i}$, $i=1,2,...,k$. When E_i is above a threshold τ_1 or under a threshold τ_2 , block B_i is classified as containing outlier pixels.

The block size may be 5x5 or 10x10 pixels, which performs finely in the experiments with the block size varies from 1 x 1 pixels to 20 x 20 pixels. This is a result of the slight smoothing in the computation of the Hessian. The threshold τ_1 and τ_2 is related with the edge feature of the template T. When the template T has abundant edge feature, τ_1 and τ_2 can be chosen near to 1, and vice versa.

4 Kalman Filter

As Takahiro Ishikawa pointed out, large motion can easily result in the Inverse Compositional Algorithm falling into a local minimal [7]. Pyramid technique is often used to alleviate this problem [8]. In [8], large motion is resolved by building pyramids from the support vectors and using a coarse-to-fine approach in the classification. This method is complex and can't completely eliminate the problem. Furthermore, the appropriate initial value is important. When the estimated initial value is close to the real value of object state, the iteration times is small, and the precision of optimization is high, and vice versa.

To account for these two problems, we introduce a Kalman filter estimator. By establishing precise motion model of the object, we predict the future position and velocity of the object, which makes the initial value locate near to the real value of object state.

Suppose that the motion is constrained on a two-dimension plane. The state vector is $X = [x, y, \dot{x}, \dot{y}, \ddot{x}, \ddot{y}]$, and the observation vector is $Z = [x, y]$. Where, $(x, y), (\dot{x}, \dot{y}), (\ddot{x}, \ddot{y})$ denotes the displacement, velocity and acceleration of the object. The state model [10] is:

$$\begin{cases} X(k+1) = \Phi(k+1,k)X(k) + W(k) \\ Z(k+1) = H(k+1)X(k+1) + V(k+1) \end{cases}. \tag{5}$$

Where, $\Phi(k+1,k) = \begin{bmatrix} 1 & 0 & 1 & 0 & 0.5 & 0 \\ 0 & 1 & 0 & 1 & 0 & 0.5 \\ 0 & 0 & 1 & 0 & 1 & 0 \\ 0 & 0 & 0 & 1 & 0 & 1 \\ 0 & 0 & 0 & 0 & 1 & 0 \\ 0 & 0 & 0 & 0 & 0 & 1 \end{bmatrix}$,

$H(k+1) = H = \begin{bmatrix} 1 & 0 & 0 & 0 & 0 & 0 \\ 0 & 1 & 0 & 0 & 0 & 0 \end{bmatrix}$, $W(k)$ is a zero-mean, white random noise, and $V(k+1)$ is a zero-mean, white observation noise.

Since the main motion of the object is the translation portion, we predict it by the velocity estimator. That is,

$$\begin{cases} p_5 \leftarrow p_5 + \dot{x} \\ p_6 \leftarrow p_6 + \dot{y} \end{cases} \tag{6}$$

Table 1. The comparison of the average iteration times before prediction and after prediction

Test	Average iteration times before prediction	Average iteration times after prediction
1	13.64	5.85
2	12.4	4.9
3	15.6	8.36

5 Occlusion Treating

5.1 Occlusion Detection

Occlusion detection is done by robust outlier rejection criterion proposed in Section 3. After performing outlier identification, we calculate the ratio of the occluded blocks. When the ratio of the occluded blocks is under a threshold T_1, we consider that the object is in partial occlusion, and let these blocks be away from calculation. When the ratio of the occluded blocks is above a threshold T_2, we consider that the object is in total occlusion. Total occlusion is resolved by updating the position of the object using the velocity estimation before total occlusion with the help of the Kalman filter proposed in Section 4.

5.2 Reappearance Detection

When the object is in total occlusion, we still estimate the number of blocks containing the outliers between the forecasting region and the template using robust outlier rejection criterion proposed in Section 3. When the ratio of the occluded blocks is under the threshold T_2, we consider the object appear again, and go to the step of reappearance verification. When the object is lost because of inaccuracies in platform pointing directions and long-time occlusion, we get back it by Kalman filter-based correlation matching. The details are as follows.

1. Searching region: Take the predicting value of the object as the center of searching region. The length and width of searching region is $K\rho_x$ and $K\rho_y$, respectively. Where, ρ_x and ρ_y is the predicting error in X and Y direction, K is searching coefficient, choosing $K = 2$ at first.

2.   Perform correlation matching in searching region. If a potential object is hunt, go to step 3. Or else, enlarge the searching region by updating $K = K + 1$ and go on searching until a potential object is hunt.

3.   When a potential object is hunt, go to the step of reappearance verification. If the potential object is the previously tracked object, stop. Or else, update $K = K + 1$ and go back to step 2.

5.3 Reappearance Verification

When the object reappears or the tracker gets back a potential object, we have to decide if the potential object is the previously tracked object. The difficulties are that the reappeared object may have change in pose, aspect and appearances. To handle this problem, a feature-based quasi-rigid alignment and flexible local template matching algorithm [9] is adopted. In feature-based quasi-rigid alignment step, first, detect lines by performing canny edge detection, and classify them into two groups(X-edges and Y-edges). Then, establish the line correspondences and their end point correspondences, and define one image flow field for each line. Finally, use the interpolated flow field to align the potential object and the template. After alignment, calculate the matching score between the potential object and the template by flexible local template matching. If the score is above a threshold, we assume that the potential object is the previously tracked object.

6   Experiment Results and Conclusion

Experiments on real images were conducted to investigate the performance of our tracking system. Fig. 1 shows results of the proposed outlier rejection criterion. In Fig. 1 the car being tracked is occluded by a pillar whose gray-value is very close to the tracked car. It can be seen that our criterion plus normalized error value criterion is more robust than normalized error value criterion. It is able to exclude more than ninety percent of outlier pixels and continues to track the car, while normalized error value criterion fails to exclude most of the outlier pixels, causing unsuccessful tracking.

We tested our algorithm in the presence of total occlusion, as is in Fig. 2. The tracked vehicle is occluded by trees for several seconds. When it reappears, the tracker gets back it successfully.

We extend the Inverse Compositional Algorithm, and successfully apply it in video-based vehicle tracking across airborne platform. First, we propose an edge feature-based outlier rejection criterion, which can eliminate the outliers caused by congestion and occlusion. Then, a Kalman filter estimator is introduced to decrease the iteration times and make the algorithm robust to large motion. Finally, a reappearance verification program is used to ensure the tracker get back the right object. Experimental results on real aerial videos shows the algorithm behaves well. Future work including vehicle detecting and multi-vehicle tracking will be done.

The proposed criterion plus normalized error value criterion

Normalized error value criterion

Reference Frame

Frame A

Frame B

Fig. 1. Two example frames from the sequence. The car being tracked is occluded by a pillar whose gray-value is very close to the tracked car. The yellow rectangle shows the location of the tracked car. In the top left of reference frame we show the template, and in the top left of frame A and frame B we show the tracked car region with the excluded blocks highlighted. In the middle left of each frame in the left column we show the template and the tracked car region after performing canny operator. Our criterion plus normalized error value criterion is able to exclude more than ninety percent of outlier pixels and so track the car successfully. The normalized error value criterion fails to exclude most of the outlier pixels and looses the track.

a. Before total occlusion   b. After total occlusion

Fig. 2. The results of our algorithm in total occlusion

References

1. Kumar, R., Sawhney, H., Samarasekera, S., Hsu, S., Tao, H. (eds.): Aerial Video Surveillance and Exploitation. In: Proceedings of the IEEE Special Issue on Third Generation Surveillance Systems, pp. 1518–1539 (2001)
2. Baker, S., Matthews, I.: Equivalence and efficiency of image alignment algorithms. In: Proceedings of the IEEE Conference on Computer Vision and Pattern Recognition, pp. 1090–1097 (2001)
3. Lucas, B., Kanade, T.: An iterative image registration technique with an application to stereo vision. In: Proceedings of the International Joint Conference on Artificial Intelligence, pp. 674–679 (1981)
4. Chen, H., Varshney, K., Slamani, M.-A.: On registration of regions of interest (ROI) in video sequences. In: Proceedings of the IEEE Conference on Advanced Video and Signal Based Surveillance (AVSS'03), pp. 313–318 (2003)
5. Singh, M., Arora, H., Ahuja, N.: Robust Registration and Tracking Using Kernel Density Correlation. In: Proceedings of the IEEE Conference on Computer Vision and Pattern Recognition Workshops (CVPRW'04), pp. 174–182 (2004)
6. Jin, H., Favaro, P., Soatto, S.: Real-time feature tracking and outlier rejection with changes in illumination. In: Proceedings of the IEEE Conference on Computer Vision, pp. 684–689 (2001)
7. Ishikawa, T., Matthews, I., Baker, S.: Efficient Image Alignment with Outlier Rejection. Technical Report CMU-RI-TR-02-27,CUM Robotics Institute (2002)
8. Avidan, S.: Support Vector Tracking. IEEE Transactions on Pattern Analysis and Machine Intelligence, 1064–1072 (2004)
9. Guo, Y., Hsu, S., Shan, Y.H. (eds.): Vehicle Fingerprinting for Reacquisition & Tracking in Videos. In: International Conference on Computer Vision & Pattern Recognition (CVPR05), pp. 761–768 (2005)
10. Tekalp, A.M.: Digital Video Processing. Prentice Hall, Englewood Cliffs (1995)
11. Canny, J.: A computational approach to edge detection. IEEE Transactions on Pattern Analysis and Machine Intelligence, 679–698 (1986)

Probabilistic Fiber Tracking Using Particle Filtering and Von Mises-Fisher Sampling

Fan Zhang[1], Casey Goodlett[2], Edwin Hancock[1], and Guido Gerig[2]

[1] Dept. of Computer Science, University of York, York, YO10 5DD, UK
{zfan,erh}@cs.york.ac.uk
[2] Dept. of Computer Science, University of North Carolina at Chapel Hill,
NC 27599-3175, USA
{gcasey,gerig}@cs.unc.edu

Abstract. This paper presents a novel and fast probabilistic method for white matter fiber tracking from diffusion weighted magnetic resonance imaging (DWI). We formulate fiber tracking on a nonlinear state space model which is able to capture both smoothness regularity of fibers and uncertainties of the local fiber orientations due to noise and partial volume effects. The global tracking model is implemented using particle filtering. This sequential Monte Carlo technique allows us to recursively compute the posterior distribution of the potential fibers, while there is no limitation on the forms of the prior and observed information. Fast and efficient sampling is realised using the von Mises-Fisher distribution on unit spheres. The fiber orientation distribution is theoretically formulated by combining the axially symmetric tensor model and the formal noise model for DWI. Given a seed point, the method is able to rapidly locate the global optimal fiber and also provide a connectivity map. The proposed method is demonstrated both on synthetic and real-world brain MRI dataset.

1 Introduction

Diffusion tensor MRI (DTI) has become a popular tool for non-invasive exploration of the anatomical structure of the white matter in vivo. It endows each voxel with a 3×3 symmetric positive-definite matrix, which characterises the local water diffusion process. It is based on a Gaussianity assumption concerning water molecule motion. White matter fiber tracking or "tractography" estimates the fiber paths by tracing the local fiber orientations, which are corresponding to the principal eigenvectors of diffusion tensors. However, the fiber orientations measured by DTI are not completely reliable due to both noise and ambiguity at voxels where multiple fibers cross or branch. In the latter case, the diffusion process within the voxels is no longer Guassian, and the diffusion tensor is an incomplete model of the diffusion signal. Thus, traditional streamline tracking approaches [1] are generally susceptible to errors in orientation measurement.

To deal with the uncertainty in fiber orientation measurements, probabilistic fiber tracking methods have received considerable interest recently [2,3,4,5]. Instead of reconstructing the fiber pathways, they aim to measure the probability of connectivity between brain regions. These methods can be described in terms of two steps. Firstly, they model the uncertainty in DTI measurements at each voxel using a probability density function (PDF) of the fiber orientations [3,5]. Behrens et al. [3] was the first to

A.L. Yuille et al. (Eds.): EMMCVPR 2007, LNCS 4679, pp. 303–317, 2007.
© Springer-Verlag Berlin Heidelberg 2007

formalise the PDF of local fiber orientations using a Bayesian model. Friman et al. [5] proposed an alternative Bayesian method based on a simplified tensor model that is more tractable. However, their PDF does not consider the uncertainty of partial volume effects. On the other hand, this PDF can also be determined from other sophisticated diffusion imaging modalities which can represent multi-fiber orientations at each voxel, such as high angular resolution imaging [6] and q-ball imaging. However, these new imaging methods require much more data acquisition time than that of DTI. Here, we focus on estimating the PDF of fiber orientations from DTI. Secondly, the probabilistic tracking algorithms simply repeat a streamline propagation process (typically $1000 \sim 10000$ times) with propagation directions randomly sampled from the PDF of fiber orientations. The fraction of the streamlines that pass through a voxel provides an index of the strength of connectivity between that voxel and the starting point. Most previous methods estimate the connectivity map by sampling directly from the PDF for fiber orientations. The sampling process is difficult, thus it is necessary to resort to Markov Chain Monte Carlo (MCMC) methods [3] or to evaluate the PDF discretely with low angular resolution. The intrinsic drawback of the previous methods is their computational complexity (often more than several hours on a high-end PC [3,5]), and this is unacceptable in practice.

In this paper, we propose a new probabilistic method for white matter fiber tracking. Our contributions are threefold: First, we formulate fiber tracking using a nonlinear state space model and recursively compute the posterior distribution using particle filtering [7,8,9]. This is a technique which has successfully been used in computer vision for visual tracking [10], contour extraction [11], etc. This sequential Monte Carlo technique provides a sound statistical framework for propagating a sample-based approximation of the posterior distributions. There is almost no restriction on the type of model that can be used. The proposed model can capture both smoothness regularity of the fibers and the uncertainties of the local fiber orientations. Since samples from the posterior path distribution are maintained at each propagation step, different decision criteria can be used to identify the true fiber. This procedure is similar to the active testing and tree pruning method for maximum a posteriori (MAP) road tracking developed by Geman et al. [12]. Given a seed point, our method is able to rapidly locate the global optimal fiber path and also provide the connectivity map. Our second contribution concerns the PDF modeling of local fiber orientations from DTI. To do so, we group diffusion tensors in white matter into two classes, namely prolate tensors and oblate tensors, according to the shape metrics in [13]. We then build PDF for each class separately. For prolate tensors, we characterise the uncertainty of fiber orientation based on the axially symmetric model of diffusion tensors [6] and a formal noise model for diffusion weighted MRI (DWI) [14]. For oblate tenors, since the fiber orientations are not collinear with the principal eigenvectors of diffusion tensors due to partial volume effects, we model the PDF using the normal distribution of the angle with the smallest eigenvectors of diffusion tensors. Finally, we use the von Mises-Fisher distribution to model prior and the importance density for sampling particles. This spherical distribution provides a natural way to model noisy directional data, and it can be efficiently sampled from using the Wood's simulation algorithm [15].

Section 2 of this paper formally develops the global fiber tracking model based on particle filtering. Section 3 describes the various ingredients of the tracking model. We have tested the algorithm both on synthetic data and a real diffusion MRI dataset, and the experimental results are presented in Section 4. Section 5 concludes the paper.

2   Tracking Algorithm

The problem of fiber tracking from a 3D diffusion MRI volume is to extract the fiber pathway from a predefined seed point. Contrary to the standard tracking problem where the data arrive one bit after another as time progresses, we receive the whole set of data before tracking begins. Thus, at each step of propagation, we set the observation set as the data visible only from the current position. In this sense, the fiber tracking problem is similar to the contour extraction [11] and road tracking [12] in computer vision. However, it is a more challenging problem because there are numerous fiber paths in white matter that often crosses each other.

We formulate the fiber tracking problem using a state space model. Given the prior probability densities that characterise the properties of the expected fiber paths and the observation densities that characterise the uncertainty of local fiber orientations, a posterior distribution of the target fiber can be estimated. Because of the complex geometry of the fiber paths and the various uncertainties of the orientation measurements, both the prior density and the observation density in our case are non-Gaussian. Thus, standard linear state space techniques such as the Kalman filter are inappropriate here, and a nonlinear filter is necessary. In contrast to the work of Gossl et al. [16] which used the Kalman filter to locate the optimal path with regard to smoothness constraint for the fibers, our method deals with both smoothness regularity and uncertainties of fiber orientations induced by noise and partial volume effects.

2.1   Global Tracking Model

A white matter fiber path P can be modeled as a sequence of points in the image space $\Omega \subset R^3$, i.e. $P_{n+1} = (x_0, x_1, ..., x_{n+1})$, as shown in Fig. 1. Thus, starting from a seed point x_0, a progressive growing of a path in discrete time can be described as

$$x_{i+1} = x_i + \rho_i \hat{v}_i, \tag{1}$$

where ρ_i and $\hat{v}_i$ are respectively the step size and the direction of propagation (unit vector) at step i. As most previous methods, we set the step size as a constant, i.e. $\rho_i = \rho, i = 0, ..., n$. Thus, the dynamics of a path growing is only determined by the propagation directions $\hat{v}_i$. In the following, we denote a path as a sequence of unit vectors $P_{n+1} = \hat{v}_{0:n} = \{\hat{v}_0, ..., \hat{v}_n\}$. Let $\mathcal{Y}$ be the set of observation or image data of a 3D DWI volume. The image data observed at $\hat{v}_i$ is $y_i = \mathcal{Y}(\hat{v}_i) = \mathcal{Y}(x_i)$. Our goal is to propagate a sequence of unit vectors that best estimates the true fiber path based on prior density $p(\hat{v}_{i+1}|\hat{v}_{0:i})$ and the observation model $p(\mathcal{Y}|\hat{v}_{0:i})$.

We assume that the tracking dynamics forms a Markov chain, so that

$$p(\hat{v}_{i+1}|\hat{v}_{0:i}) = p(\hat{v}_{i+1}|\hat{v}_i). \tag{2}$$

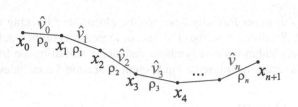

Fig. 1. Path representation

This means the new state is conditioned directly only on the immediately preceding state, independent of the past. Thus, the prior of the fiber path is

$$p(\hat{v}_{0:n}) = p(\hat{v}_0) \prod_{i=1}^{n} p(\hat{v}_i|\hat{v}_{i-1}). \qquad (3)$$

Another assumption is that the observations or diffusion measurements are conditionally independent given $\hat{v}_{0:n}$, i.e $p(\mathcal{Y}|\hat{v}_{0:n}) = \prod_{r\in\Omega} p(\mathcal{Y}(r)|\hat{v}_{0:n})$. We also assume that the diffusion measurement at a point does not depend on any points in the history of the path, i.e $p(y_i|\hat{v}_{0:i}) = p(y_i|\hat{v}_i)$. Using Equation (3), the posterior distribution $p(\hat{v}_{0:n}|\mathcal{Y})$ can be expanded to

$$p(\hat{v}_{0:n}|\mathcal{Y}) = p(\hat{v}_0|\mathcal{Y}) \prod_{i=1}^{n} p(\hat{v}_i|\hat{v}_{i-1}, \mathcal{Y}). \qquad (4)$$

Applying Bayes theorem, we have

$$p(\hat{v}_i|\hat{v}_{i-1}, \mathcal{Y}) = \frac{p(y_i|\hat{v}_i)p(\hat{v}_i|\hat{v}_{i-1})}{p(y_i)}. \qquad (5)$$

In our tracking model we do not assume any prior information about the diffusion measurements, thus we can simply consider $p(y_i)$ as a constant regularity factor, so that

$$p(y_i) = \int_{\hat{v}_i} p(y_i|\hat{v}_i)p(\hat{v}_i|\hat{v}_{i-1}). \qquad (6)$$

Most previous methods [2,4,3] for probabilistic fiber tracking estimate the posterior $p(\hat{v}_{0:n}|\mathcal{Y})$ by sampling $1000 \sim 10000$ streamline paths from $p(y_i|\hat{v}_i)$. The sampling is difficult and time consuming. Moreover, these methods do not take into account the smoothness constraint for fibers. On the other hand, Friman et al. [5] estimate the posterior by sampling from $p(\hat{v}_i|\hat{v}_{i-1}, \mathcal{Y})$. This sampling is again difficult and requires the computation of the integral in Equation (6) over the new state. To avoid these difficulties, they discretise the problem using a finite set of several thousand directions for propagating paths from $\hat{v}_{i-1}$ to $\hat{v}_i$. Thus, sampling the discretised version of $p(\hat{v}_i|\hat{v}_{i-1}, \mathcal{Y})$ becomes possible, and the integral becomes a sum. In addition to introducing errors, this discretised sampling is still very time consuming, since all discretised directions must be evaluated at all locations in the volume. Moreover, previous simple sequential sampling methods may degenerate as the number of propagation steps becomes large [9].

2.2 Recursive Posterior Using Particle Filtering

We wish to estimate iteratively in time the posterior distribution. By inserting Equation (5) into Equation (4), we have

$$p(\hat{v}_{0:n}|\mathcal{Y}) = p(\hat{v}_0|\mathcal{Y}) \prod_{i=1}^{n} p(\hat{v}_i|\hat{v}_{i-1}) \prod_{i=1}^{n} \frac{p(y_i|\hat{v}_i)}{p(y_i)}, \tag{7}$$

where $p(\hat{v}_0|\mathcal{Y})$ is predefined. The modeling of the transition probability $p(\hat{v}_i|\hat{v}_{i-1})$ and the distribution $p(y_i|\hat{v}_i)$ will be detailed in the next section. We recast the problem of tracking the expected fiber path as that of approximating the MAP path from the posterior distribution.

It is straightforward to obtain the following recursive formula for the posterior from Equation (7)

$$p(\hat{v}_{0:i+1}|\mathcal{Y}) = p(\hat{v}_{0:i}|\mathcal{Y}) \frac{p(\hat{v}_{i+1}|\hat{v}_i)p(y_{i+1}|\hat{v}_{i+1})}{p(y_{i+1})}. \tag{8}$$

Since the denominator of this expression requires the evaluation of a complex high-dimensional integral, it is infeasible to locate the maximum likelihood path analytically. Like above methods, we evaluate the posterior using a large number of samples which efficiently characterise the required posterior. Thus, the statistical quantities, such as the mean, variance and maximum likelihood, can be approximated based on the sample set. Since it is seldom possible to obtain samples from the posterior directly, here we use the particle filtering (sequential Monte Carlo technique) to recursively compute a finite set of sample paths from the posterior based on the Equation (8).

To sample a set of K paths, we set K particles at the starting location and allow them to propagate as time progresses. Given the states of the set of particles $\{\hat{v}_{0:i}^{(k)}, k = 1, ..., K\}$ at time i, the process of sequentially propagating the particles to the next time step $i+1$ can be described in three stages. These are referred to as *prediction, weighting* and *selection*. Let $\pi(\hat{v}_{0:i}|\mathcal{Y})$ be a so-called importance function which has a support including that of the posterior $p(\hat{v}_{0:i}|\mathcal{Y})$. For our sequential importance sampling, suppose that we choose an importance function of the form [8]

$$\pi(\hat{v}_{0:n}|\mathcal{Y}) = \pi(\hat{v}_0|\mathcal{Y}) \prod_{i=1}^{n} \pi(\hat{v}_i|\hat{v}_{i-1}, \mathcal{Y}). \tag{9}$$

In the first *prediction* stage, each simulated path $\hat{v}_{0:i}^{(k)}$ with index k is grown by one step to be $\hat{v}_{0:i+1}^{(k)}$ through sampling from the importance function $\pi(\hat{v}_{i+1}^{(k)}|\hat{v}_i^{(k)}, \mathcal{Y})$. The new set of paths generally is not an efficient approximation of the posterior distribution at time $i + 1$. Thus, in the second *weighting* stage, we measure the reliability of the approximation using a ratio, referred to as the importance weight, between the truth and the approximation

$$w_{i+1}^{(k)} = \frac{p(\hat{v}_{0:i+1}^{(k)}|\mathcal{Y})}{\pi(\hat{v}_{0:i}^{(k)}|\mathcal{Y})\pi(\hat{v}_{i+1}^{(k)}|\hat{v}_i^{(k)}, \mathcal{Y})} \tag{10}$$

We are more interested in the normalised importance weights, given by

$$\tilde{w}_{i+1}^{(k)} = \frac{w_{i+1}^{(k)}}{\sum_{l=1}^{K} w_{i+1}^{(l)}}. \tag{11}$$

Inserting Equation (8) and Equation (10) into the above expression, it goes as

$$\tilde{w}_{i+1}^{(k)} \propto \tilde{w}_i^{(k)} \frac{p(\hat{v}_{i+1}^{(k)}|\hat{v}_i^{(k)})p(y_{i+1}|\hat{v}_{i+1}^{(k)})}{\pi(\hat{v}_{i+1}^{(k)}|\hat{v}_i^{(k)}, \mathcal{Y})}. \tag{12}$$

The choice of importance function plays an important role for the performance of particle filtering. This will be detailed in the next section. At this point the resulting weighted set of paths provides an approximation of the target posterior. However, the distribution of the weights $\tilde{w}_{i+1}^{(k)}$ may becomes more and more skewed as time increases. The purpose of the last *selection* stage is to avoid this degeneracy. We measure the degeneracy of the algorithm using the effective sample size N_{eff} introduced in Liu [17],

$$N_{eff} = \frac{1}{\sum_{k=1}^{K}(\tilde{w}_{i+1}^{(k)})^2}. \tag{13}$$

When N_{eff} is below a fixed threshold N_s, then resampling procedure is used [7,9]. The key idea here is to eliminate the paths or particles with low weights $\tilde{w}_{i+1}^{(k)}$ and to multiply offspring particles with high weights. We obtain the surviving particles by resampling K times from the discrete approximating distribution according to the importance weight set $\{\tilde{w}_{i+1}^{(k)}, k = 1, .., K\}$.

Both fiber reconstruction and connectivity map can be easily solved based on the discrete distribution of the posterior conveyed by the importance weight set. The MAP estimate of the true fiber path from starting point x_0 is the path with the maximal importance weight. In order to set up the connectivity map, the algorithm records the full tracking history of all the particles at each time step. The probability of connectivity between x_0 and points in the history of the paths is determined by the importance weight at that point before the resampling step.

3   Algorithm Ingredients

In this section, we give the details of the local ingredients of the global tracking model.

3.1   Observation Density

We commenced by showing how to set up $p(y_i|\hat{v}_i)$, which encodes the uncertainty in local fiber orientation due to both noise and partial volume effects. Our observation density function is constructed using single diffusion tensor model. Despite its weakness in capturing complex fiber architecture, DTI is still the most widely used diffusion MRI modality. Formally, the diffusion-weighted intensity s_j is related to the diffusion tensor D by the Stejskal-Tanner equation [18]

$$s_j = s_0 e^{-b_j \hat{g}_j^T D \hat{g}_j}, \tag{14}$$

where gradient direction the $\hat{g}_j$ and the b-value b_j are the scanner parameters for data acquisition, and, s_0 is the intensity with no diffusion gradients applied.

Let $\lambda_1 \geq \lambda_2 \geq \lambda_3 \geq 0$ be the decreasing eigenvalues of D and $\hat{e}_1, \hat{e}_2, \hat{e}_3$ be the corresponding normalized eigenvectors. The degree of anisotropy of water diffusion at a voxel can be characterised using the fractional anisotropy (FA) [19] of its diffusion tensor. We can classify the diffusion tensors in white matter into two groups: prolate tensors $(\lambda_1 > (\lambda_2 \approx \lambda_3))$ and oblate tensors $((\lambda_1 \approx \lambda_2) < \lambda_3)$. This can be done using the prolate shape metric proposed by Westin et al. [13], i.e.

$$c_l = \frac{\lambda_1 - \lambda_2}{\sqrt{\lambda_1^2 + \lambda_2^2 + \lambda_3^2}}. \tag{15}$$

In this work, we distinguish prolate tensors and oblate tensors by using a threshold $\tau = 0.25$.

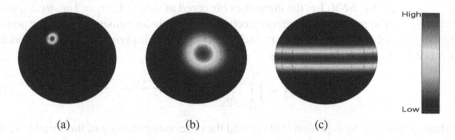

(a)   (b)   (c)

Fig. 2. Example of the observation density of three tensors from the brain white matter. (a) a prolate tensor with $FA = 0.9299, c_l = 0.9193$. (b) a prolate tensor with $FA = 0.3737, c_l = 0.3297$. (c) an oblate tensor $FA = 0.7115, c_l = 0.2157$. For voxels with prolate tensors, the larger the value of FA and c_l, the more focused the fiber orientation distribution. For voxels with oblate tensors, fiber orientations are focused on the plane spanned by the two leading eigenvector.

In the case of prolate tensors $(c_l > \tau)$, it can be assumed that a single dominant diffusion direction, $\hat{e}_1$, is collinear with underlying fiber orientation. Borrowing ideas from Anderson [6], we suppose the prolate diffusion tensor within a voxel is a single axially-symmetric tensor. Let us set up a local coordinate system for the diffusion tensor D so that it can be written as a diagonal matrix $diag(\lambda_\perp, \lambda_\perp, \lambda_\parallel)$ where the axis corresponding to $\lambda_\parallel$ is associated with the fiber orientation. In spherical coordinates, let θ be the polar angle from the $\lambda_\parallel$-axis and ψ be the azimuth angle from one of $\lambda_\perp$-axis. Then, a gradient direction $\hat{g}_j$ with (θ, ψ) has local coordinates $g(\theta, \psi) = [\sin\theta\cos\psi, \sin\theta\sin\psi, \cos\theta]$. Thus, diffusion along $\hat{g}_j$ is

$$\begin{aligned} \hat{g}_j^T D \hat{g}_j &= g(\theta, \psi) \cdot diag(\lambda_\perp, \lambda_\perp, \lambda_\parallel) \cdot g(\theta, \psi)^T \\ &= \lambda_\perp + \cos^2\theta \cdot (\lambda_\parallel - \lambda_\perp). \end{aligned} \tag{16}$$

Suppose the trace of the diffusion tensor $tr(D)$ is known and varies unsignificantly over the white matter. The mean of the three main diffusivities is $\bar{\lambda} = tr(D)/3$. It also follows that $\lambda_\parallel = 3\bar{\lambda} - 2\lambda_\perp$. Let $\hat{v}$ be the true fiber orientation, then $\cos\theta = \hat{v} \cdot \hat{g}_j$. Therefore, Equation (16) can be written as

$$\hat{g}_j^T D\hat{g}_j = \lambda_\perp + 3(\hat{v} \cdot \hat{g}_j)^2(\bar{\lambda} - \lambda_\perp). \tag{17}$$

By inserting Equation (17) into Equation (14), we have

$$s_j = s_0 e^{-b_j(\lambda_\perp + 3(\hat{v} \cdot \hat{g}_j)^2(\bar{\lambda} - \lambda_\perp))}. \tag{18}$$

The intensity measured along any gradient direction is subject to two unknown parameters $\hat{v}$ and $\lambda_\perp$.

Due to noise, the intensity u_j in the DWI measured by the scanner is a noisy observation of the true intensity s_j. It is well known that the noise in MRI can be described accurately by a Rician distribution. Salvador et al. [14] formally investigated the errors in the logarithm of the intensity. They showed that the error distribution conforms closely to a normal distribution with a zero-mean and standard deviation equal to the inverse of the signal-to-noise ratio (SNR), i.e. $\epsilon_j = \log(u_j) - \log(s_j) \sim N(0, \varrho_j^{-1})$, where $\varrho_j = s_j/\sigma_j$ is the SNR. Let the intensities observed at voxel i be $y_i = \{u_0, u_1, ..., u_M\}$ where M is the number of gradient directions used in data acquisition. Then for prolate tensors, our observation density at voxel i is found by multiplying the error distribution for DWIs of all gradient directions

$$p(y_i|\hat{v}_i) = \prod_{j=1}^{M} \frac{\varrho_j}{\sqrt{2\pi}} e^{-\frac{\varrho_j^2(\log u_j - \log s_j)^2}{2}}, \tag{19}$$

where s_j is given by Equation (18). To find the observation density of the variable $\hat{v}_i$ for fiber orientations, we need to solve the three unknown parameters, i.e. $\bar{\lambda} = tr(D)/3$, $\lambda_\perp$ and ϱ_j, in Equation (19). The value of $tr(D)$ and $\lambda_\perp$ can be estimated using the method in [6]. Here, we simply set $tr(D)$ as the trace of the diffusion tensor D estimated using the linear least squares estimation [18], and set $\lambda_\perp = (\lambda_2 + \lambda_3)/2$. The SNR is estimated using the weighted least squares method in [14]. Panels (a) and (c) of Fig. 2 show two examples of the fiber orientation distribution calculated using Equation (19). The figure tells that the orientation distribution of a prolate tensor is very concentrated when its FA and c_l is relatively large.

In the case of oblate tensors ($c_l \leq \tau$), the dominant direction of diffusion is ambiguous and Equation (19) is inappropriate. It is possible that diffusion in the plane defined by $\hat{e}_1$ and $\hat{e}_2$ contains two or more significant non-collinear diffusion directions, each corresponding to a separate fiber tract. This situation may be indicative of fiber crossing and branching. This is a weakness of DTI. In this case, we set up a local coordinate system with $\hat{e}_3$ as its Z axis and represent the fiber orientation $\hat{v}$ in spherical coordinates. Let θ' be the polar angle from the $\hat{e}_3$-axis, i.e. $\theta' = \arccos(\hat{v} \cdot \hat{e}_3)$, and ψ' be the azimuth angle (relative to an arbitrary reference direction in the plane spanned by $\hat{e}_1$ and $\hat{e}_2$). The vector $\hat{v}$ is mainly distributed on the plane spanned by $\hat{e}_1$ and $\hat{e}_2$. Hence, we choose the distribution of the polar angle θ' to be normal with mean $\pi/2$ and standard deviation σ. The azimuth ψ' is assumed to have a uniform distribution over the interval $[0, 2\pi]$. Thus, our fiber orientation distribution for oblate tensors is given by

$$p(y_i|\hat{v}) = \frac{1}{\sigma\sqrt{2\pi}} \exp(-\frac{(\arccos(\hat{v} \cdot \hat{e}_3) - \pi/2)^2}{2\sigma^2}) \cdot \frac{1}{2\pi}. \tag{20}$$

Here, $\hat{e}_3$ is the eigenvector of the diffusion tensor D estimated using linear least squares. Panel (c) of Fig. 2 shows an example of the observation density of the fiber orientation of an oblate tensor in white matter.

3.2 Prior Density

The transition density $p(\hat{v}_{i+1}|\hat{v}_i)$ specifies a prior distribution for the change in fiber direction between two successive steps. Here, we adopt a model of the prior density based on the von Mises-Fisher (vMF) distribution over a unit sphere. This is one of the simplest parametric distribution for directional data.

For a d-dimensional unit random vector $\mathbf{x}$, the probability density function for the vMF distribution is given by

$$f_d(\mathbf{x}; \mu, \kappa) = \frac{\kappa^{d/2-1}}{(2\pi)^{d/2} I_{d/2-1}(\kappa)} \exp(\kappa \mu^T \mathbf{x}), \tag{21}$$

where $\kappa \geq 0$, $\|\mu\| = 1$, and $I_{d/2-1}(\cdot)$ denotes the modified Bessel function of the first kind and order $d/2 - 1$. The density $f_d(\mathbf{x}; \mu, \kappa)$ is parameterised by the mean direction vector μ and the concentration parameter κ. The greater the value of κ the higher the concentration of the distribution around the mean direction μ. In particular, when $\kappa = 0$, the distribution is uniform over the sphere, and as $\kappa \to \infty$, the distribution tends to a point density. The distribution is rotationally symmetric around the mean μ, and is unimodal for $\kappa > 0$.

In our case, the directions are defined on a two directional unit sphere in R^3, i.e. $d = 3$. Thus, we choose our prior density as the vMF distribution with mean $\hat{v}_i$ and concentration parameter κ, i.e.

$$p(\hat{v}_{i+1}|\hat{v}_i) = f_3(\hat{v}_{i+1}; \hat{v}_i, \kappa). \tag{22}$$

The value of the concentration parameter κ here controls the smoothness regularity of the tracked paths. The value of κ is set manually to optimally balance the prior constraints on smoothness against the evidence of v_{i+1} observed from the image data.

3.3 Importance Density Function

As discussed in Doucet et al. [8], the optimal importance density, which minimises the variance of the importance weight $\tilde{w}_{i+1}$ conditional upon $\hat{v}_i$ and $\mathcal{Y}$, is $p(\hat{v}_{i+1}|\hat{v}_i, \mathcal{Y})$. However, as discussed above, the optimal density suffers from two major drawbacks. In our case it is both difficult to sample from $p(\hat{v}_{i+1}|\hat{v}_i, \mathcal{Y})$ and to evaluate the integral analytically over the new state. Thus, our aim is to devise an suboptimal importance function that best represents $p(y_{i+1}|\hat{v}_{i+1})p(\hat{v}_{i+1}|\hat{v}_i)$ subject to the constraint that it can be sampled from.

A most popular choice is to use the prior distribution as the importance function, i.e

$$\pi(\hat{v}_{i+1}|\hat{v}_i, \mathcal{Y}) = f_3(\hat{v}_{i+1}; \hat{v}_i, \kappa). \tag{23}$$

The von Mises-Fisher distribution in Equation (22) can be efficiently sampled from using the simulation algorithm developed by Wood [15]. In this case, $\hat{v}_{i+1}$ is predicted from $\hat{v}_i$ and the importance weight is updated using $\tilde{w}_{t+1} = \tilde{w}_t p(y_{i+1}|\hat{v}_{i+1})$. However, the prior importance function is not very efficient. Since no observation information is used, the generated particles are often outliers of the posterior distributions. As a result, the weights may exhibit large variations and the estimation results may be poor. Indeed, if the diffusion tensor at $\hat{v}_i$ is prolate, then the movement to the state v_{i+1} is mainly attributable to the fiber orientation distribution. Thus, the posterior distribution is more strongly influenced by the observation density. For prolate tensors, we believe that the observation density in Equation (19) is a good choice as the importance function. However, it is difficult to sample from. To overcome this problem, we model the observation density in Equation (19) using the von Mises-Fisher distribution. Since we use an axially symmetric tensor model, the distribution of fiber orientations in Equation (19) is also rotationally symmetric around the direction of largest probability, as shown in Fig. 2. We use the leading eigenvector, $\hat{e}_1^i$, of the diffusion tensor D_i at $\hat{v}_i$ estimated using the linear least squares as the mean direction of the fiber orientation distribution. We have found experimentally that the leading eigenvector $\hat{e}_1^i$ of D_i is almost identical to the direction of maximum probability for the distribution in Equation (19). This is based on a test of 1000 prolate tensors from the brain MRI dataset in experimental section. The average difference between the two directions is less than $2°$. The concentration parameter ν_i at each state $\hat{v}_i$ is set to $\nu_i = 90 \times c_l(D_i)$. This choice is based on empirical trial and error. A more theoretically justifiable choice can be deduced from the shape of the ellipsoid of the tensor D_i. Also, we can estimate ν_i by fitting the von Mises-Fisher distribution to Equation (19) using the algorithm in [20]. However, this will increase the computational complexity of the algorithm. Moreover, it is not necessary for particle filtering that the importance density is exactly the same as the observation density. Therefore, for prolate tensors we set the importance density as

$$\pi(\hat{v}_{i+1}|\hat{v}_i, \mathcal{Y}) = f_3(\hat{v}_{i+1}; \hat{e}_1^i, \nu_i). \tag{24}$$

For oblate tensors, since the observation density in Equation (20) is wide, in this case we still use the prior as the importance density given in Equation (23).

3.4 Algorithm Outline

To summarise, the iteration steps of the algorithm are as follows:

- given K particles at step i: $\hat{v}_{0:i}^{(k)}, k = 1, ..., K$
- compute diffusion tensor $D_i^{(k)}$ for each particle k using linear least square fitting
- **Prediction**: for $k = 1, ..., K$
  - if $D_i^{(k)}$ is a prolate tensor, sample $\hat{v}*_{i+1}^{(k)}$ using Equation (24)
  - if $D_i^{(k)}$ is a oblate tensor, sample $\hat{v}*_{i+1}^{(k)}$ using Equation (23)
- **Weighting**: for $k = 1, ..., K$
  - if prolate tensor, compute $\tilde{w}_{i+1}^{(k)}$ from Equation (12) using Equation (19), (22) and (24)

- if oblate tensor, compute $\tilde{w}_{i+1}^{(k)}$ from Equation (12) using Equation (20), (22) and (23)
- normalise all these weights
- **Selection**: Evaluate N_{eff} using Equation (13).
  - If $N_{eff} \geq N_s$, then for $k = 1, ..., K$, $\hat{v}_{i+1}^{(k)} = \hat{v^*}_{i+1}^{(k)}$
  - If $N_{eff} < N_s$, then for $k = 1, ..., K$, sample an index $z(k)$ from discrete distribution $\{\tilde{w}_{i+1}^{(k)}\}_{k=1,...,K}$, and set $\hat{v}_{i+1}^{(k)} = \hat{v^*}_{i+1}^{(z(k))}, \tilde{w}_{i+1}^{(k)} = \frac{1}{N}$

4 Experimental Results

We have tested the algorithm both on synthetic tensor fields and a real-world brain MRI dataset. Since our particles propagate in a continuous domain, an interpolation issue arises for the diffusion data that is acquired only on a discrete grid. Here, we choose the trilinear interpolation method introduced in [21]. It is computationally very cheap and can preserve the positive-definite constraint of tensors.

4.1 Synthetic Dataset

We have evaluated the performance of the algorithm on two artificial data. Our first example demonstrates the robustness of the algorithm under noise. To do this, we first generate a synthetic tensor field with $128 \times 128 \times 40$ voxels, in-plane resolution 2×2 milimeter (mm) and slice thickness $2mm$. The data contains a single cylinder, and the principal diffusion directions of the voxels within the cylinder form a concentric vector field, as shown in panel (a) Fig. 3. Each voxel is visualised by a ellipsoid whose axes are the three orthogonal eigenvectors of the tensor, and the lengths of the axes represent the eigenvalues. Then, we add different levels of i.i.d noise to each component of the tensor field. The proposed particle filter algorithm is then used to track the global optimal fiber (MAP path) from a seed point using 1000 particles for 650 propagation steps with step size $1mm$. Our result is compared with that of the standard local streamline method (FACT) [22] and the Bayesian method of Friman et al. [5]. For Friman's method, we sample 1000 paths from the seed point using their discrete sampling technique with 2562 predefined directions on the unit sphere, and choose the path with maximal probability as the optimal fiber. The results of all three methods and the ground truth are plotted in panel (b) and (c) of Fig. 3. The figure shows that under mild noise (10% noise) both our method and the Friman's method well reconstruct the true fiber, however, our method runs significantly faster than the Friman method. For instance, the MATLAB implementation of our method takes less than 100 seconds for 1000 samples to propagate 100 steps on a PC with P4 CPU. Our MATLAB implementation of the Friman's method requires more than 2 hours to sample 1000 paths with the same length. Additionally, the MCMC method of Behrens et al. runs much more slowly according to their report [3]. On the other hand, when the noise is large (25% noise), our method performs better than the Friman's method as shown in (c) of Fig. 3. This demonstrates that our algorithm samples paths more effectively due to the continuous

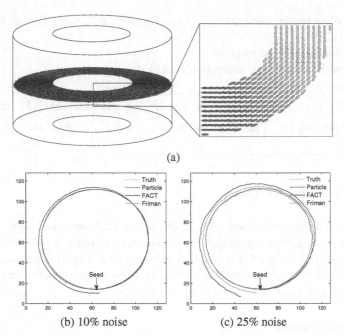

(a)

(b) 10% noise   (c) 25% noise

Fig. 3. (a) Synthetic data consisting of a cylinder and a sample slice with a zoomed view. (b) Results of our method, streamline method (FACT) [22] and the Friman method [5] under 10% noise. (c) Results under 25% noise.

simulation of Von Mises-Fisher distribution and the resampling technique of particle filtering. The results also reveal that the streamline method FACT is sensitive to noise, and that it performs more poorly than our method and Friman's method under both mild and heavy noise.

The second example is to show the behavior of the algorithm under both noise and partial volume effects. We again generate a synthetic tensor field with $128 \times 128 \times 40$ voxels and the same voxel size as the first example. In this simulation, right angle crossing of horizontal and vertical fibers are imitated (as shown in the top row of Fig. 4). Additionally, we also add 10% noise to the data set. In the crossing area, the tensor's first and second eigenvalues are assumed to be equal. As a consequence, the diffusion ellipsoids are oblate tensors in this area, in contrast to the prolate tensors in regions without fiber crossing. We then apply our method to track the fiber from a seed by propagating 1000 particles for 170 steps. The global optimal MAP path of the particle trajectories are computed and plotted in the bottom panel of Fig. 4. We also show the mean path of all the final particle trajectories. Although the principal eigenvector of the oblate tensors are not collinear with the fiber orientations, the result shows that our method still work fairly well under fiber crossing. The algorithm can interpolate over gaps in the transition evidence, and allows the prior density predominantly controls the propagation of particles at crossing areas. Since the aforementioned methods haven't specifically considered the modeling of oblate tensors, in this case we don't compare our result with others.

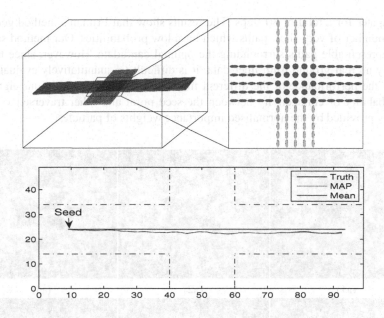

Fig. 4. Top: synthetic data with fiber crossing. Bottom: tracking result of our method.

4.2 Brain Diffusion MRI

Real diffusion MRI data was acquired from a healthy adult volunteer using a Siemens Allegra 3T head only scanner. A $128 \times 128 \times 58$ volume image was acquired with $2 \times 2 \times 2mm$ resolution. A six-direction gradient scheme was used with 10 repetitions per-image, $b = 1000s/mm^2$ for the gradient directions, and $b = 0s/mm^2$ for the baseline image. Repetitions were aligned via rigid registration of the baseline images.

We first set a seed point in the Corpus Callosum, and then applied our algorithm to track fibers from both sides of the principal direction of the seed point. A step length of 1mm and 2000 particles were used. The trajectories of all particles are fully recorded. A particle is eliminated and resampled when it enters into gray matter. A threshold of 0.2 of the FA is used to distinguish voxels in white matter and gray matter. We set a maximum length of 60 steps in each side of the seed to terminate the tracking process. Panel (a1) of Fig. 5 shows the global optimal MAP path from the seed point at the final step. Panel (b1) shows 300 sample paths generated by the particles. The result is compared with that of the Friman's method [5] with 300 sampled paths. In our method, particles with low probability of existence are eliminated during resampling stage. Thus, the sampled paths are most concentrated around the final optimal fiber. In contrast, the sampled paths of the Friman's method are more dispersed. On the other hand, the sampled paths in (a2) reveals that our probabilistic algorithm is able to handle splitting fiber bundles and ambiguous neighborhoods. Another tracking example is shown in the bottom row of Fig. 5. The algorithm is seeded at a point in the Cingulum bundles, and it propagates for a total of 80 steps. The results show that Friman's method generates a large number of very short paths which have low probabilities. Our method samples more representable paths surrounding the optimal candidate. However, since there is currently no real ground truth fiber data, it is difficult to quantitatively evaluate or to validate the performance of the different fiber tracking algorithms. In our algorithm, the probability of connectivity between the seed point and other traversed voxels is naturally provided by the normalised importance weights of particles.

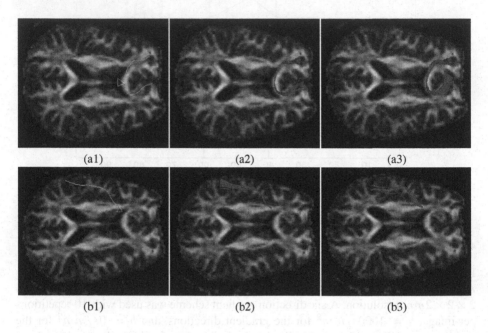

(a1)	(a2)	(a3)
(b1)	(b2)	(b3)

Fig. 5. Column 1: Global optimal fibers of our method. Column 2: 300 path samples of our method. Column 3: 300 path samples of Friman's method. [5]

5   Conclusion

We have presented a new method for probabilistic white matter fiber tracking. The global tracking model is formulated using a state space framework, which is implemented by applying particle filtering to recursively estimate the posterior distribution of fibers and to locate the global optimal fiber path. Each ingredient of the tracking algorithm is detailed. Fiber orientation distribution is formulated in a theoretical way by combining the axially symmetric tensor model and a noise model for DWI. Fast and efficient sampling is realised using the von Mises-Fisher distribution. As a consequence, there is no need to apply MCMC sampling [3] or to discretise the state space [5] to sample paths from the fiber orientation distribution. Unlike previous methods [3,5] which are computationally expensive, given a seed point our method is able to rapidly locate the global optimal fiber and compute the connectivity map for the seed point.

References

1. Mori, S., van Zijl, P.: Fiber tracking: principles and strategies - a technical review. NMR in Biomed. 15, 468–480 (2002)
2. Parker, G., Haroon, H., Wheeler-Kingshott, C.: A Framework for a Streamline-Based Probabilistic Index of Connectivity (PICo) Using a Structural Interpretation of MRI Diffusion Measurements. J. Magn. Reson. Imag. 18, 242–254 (2003)
3. Behrens, T., Woolrich, M., Jenkinson, M., Johansen-Berg, H., Nunes, R., Clare, S., Matthews, P., Brady, J., Smith, S.: Characterization and Propagation of Uncertainty in Diffusion-Weighted MR Imaging. Magn. Reson. Med. 50, 1077–1088 (2003)
4. Bjornemo, M., Brun, A., Kikinis, R., Westin, C.: Regularized Stochastic White Matter Tractography Using Diffusion Tensor MRI. In: Dohi, T., Kikinis, R. (eds.) MICCAI 2002. LNCS, vol. 2489, pp. 435–442. Springer, Heidelberg (2002)
5. Friman, O., Farneback, G., Westin, C.: A Bayesian Approach for Stochastic White Matter Tractography. IEEE Trans. on Med. Imag. 25, 965–978 (2006)
6. Anderson, A.: Measurement of Fiber Orientation Distributions Using High Angular Resolution Diffusion Imaging. Magn. Reson. Med. 54, 1194–1206 (2005)
7. Gordon, N., Salmond, D., Smith, A.: Novel approach to nonlinear non-Gaussian Bayesian state estimation. IEE Proceedings F 140, 107–113 (1993)
8. Doucet, A., Godsill, S., Andrieu, C.: On sequential Monte Carlo sampling methods for Bayesian filtering. Statistics and Computing 10, 197–208 (2000)
9. Doucet, A., de Freitas, N., Gordon, N. (eds.): Sequential Monte Carlo Methods in Practice. Springer, Heidelberg (2001)
10. Isard, M., Blake, A.: CONDENSATION-Conditional Density Propagation for Visual Tracking. International Journal of Computer Vision 29, 5–28 (1998)
11. Perez, P., Blake, A., Gangnet, M.: JetStream: Probabilistic Contour Extraction with Particles. In: Proc. IEEE ICCV, pp. 524–531 (2001)
12. Geman, D., Jedynak, B.: An Active Testing Model for Tracking Roads in Satellite Images. IEEE Trans. Pattern Anal. Mach. Intell. 18, 1–14 (1996)
13. Westin, C., Maier, S., Mamata, H., Nabavi, A., Jolesz, F., Kikinis, R.: Processing and visualization for diffusion tensor MRI. Medical Image Analysis 6, 93–108 (2002)
14. Salvador, R., Pena, A., Menon, D., Carpenter, T., Pickard, J., Bullmore1, E.: Formal Characterization and Extension of the Linearized Diffusion Tensor Model. Human Brain Mapping 24, 144–155 (2005)
15. Wood, A.T.A.: Simulation of the von Mises-Fisher distribution. Communications in Statistics. Simulation and Computation 23, 157–164 (1994)
16. Gossl, C., Fahrmeir, L., Putz, B., Auer, L., Auer, D.: Fiber Tracking from DTI Using Linear State Space Models: Detectability of the Pyramidal Tract. NeuroImage 16, 378–388 (2002)
17. Liu, J.: Metropolized independent sampling with comparison to rejection sampling and importance sampling. Statistics and Computing 6, 113–119 (1996)
18. Basser, P., Mattiello, J., LeBihan, D.: MR diffusion tensor spectroscopy and imaging. Biophysical Journal 66, 259–267 (1994)
19. Basser, P., Pierpaoli, C.: Microstructural and physiological features of tissues elucidated by quantitative-diffusion-tensor MRI. J. Magn. Reson, Series B 111, 209–219 (1996)
20. Hill, G.: Algorithm 571: Statistics for von Mises' and Fisher's Distributions of Directions. ACM Trans. on Math. Software 7, 233–238 (1981)
21. Zhukov, L., Barr, A.: Oriented Tensor Reconstruction: Tracing Neural Pathways from Diffusion Tensor MRI. In: Proc. IEEE Visualization, pp. 387–394 (2002)
22. Mori, S., Crain, B., Chacko, V., van Zijl, P.: Three-Dimensional Tracking of Axonal Projections in the Brain by Magnetic Resonance Imaging. Annals of Neurology 5, 265–269 (1999)

Compositional Object Recognition, Segmentation, and Tracking in Video

Björn Ommer and Joachim M. Buhmann*

Institute of Computational Science, ETH Zurich
8092 Zurich, Switzerland
{bjoern.ommer,jbuhmann}@inf.ethz.ch

Abstract. The complexity of visual representations is substantially limited by the compositional nature of our visual world which, therefore, renders learning structured object models feasible. During recognition, such structured models might however be disadvantageous, especially under the high computational demands of video. This contribution presents a compositional approach to video analysis that demonstrates the value of compositionality for both, learning of structured object models and recognition in near real-time. We unite category-level, multi-class object recognition, segmentation, and tracking in the same probabilistic graphical model. A model selection strategy is pursued to facilitate recognition and tracking of multiple objects that appear simultaneously in a video. Object models are learned from videos with heavy clutter and camera motion where only an overall category label for a training video is provided, but no hand-segmentation or localization of objects is required. For evaluation purposes a video categorization database is assembled and experiments convincingly demonstrate the suitability of the approach.

1 The Rational for Compositionality

Combined tracking, segmentation, and recognition of objects in videos is one of the long standing challenges of computer vision. When approaching real world scenarios with large intra-category variations, with weak supervision during training, and with real-time constraints during prediction, this problem becomes particularly difficult. By establishing a compositional representation, the complexity of object models can be reduced significantly and learning such models from limited training data becomes feasible. However, a structured representation might entail disadvantages during recognition, especially given the high computational demands of video. We present a compositional approach to video analysis that performs near real-time and demonstrates how the key concept of compositionality can actually be exploited for both, rendering learning tractable and making recognition computationally feasible.

Our compositional video analysis system unites category-level, multi-class object recognition, segmentation, and tracking in the same probabilistic graphical

---

* This work was supported in part by the Swiss national fund under contract no. 200021-107636.

A.L. Yuille et al. (Eds.): EMMCVPR 2007, LNCS 4679, pp. 318–333, 2007.
© Springer-Verlag Berlin Heidelberg 2007

model. Moreover, this Bayesian network combines compositions together with object shape. Learning object models requires only a category label for the most prominent object in a complete video sequence, thereby even tolerating distracting clutter and other objects in the background. Category specific compositions of local features are automatically learned so that irrelevant image regions can be identified and discarded without supervision. As a result tedious hand-segmentations, object localizations, or initializations of a tracker become superfluous. Since there has been only very little work on category-level segmentation and recognition in video we have started assembling a video categorization database for evaluation purposes that consists of four object categories (bicycle, car, pedestrian, and streetcar). Videos have been recorded in their natural outdoor environment and show significant scale variation, large intra-category variability, camera panning, and background clutter.

Compositionality (e.g. [11]), which serves as a foundation for this contribution, is a general principle in cognition and can be especially observed in human vision [3]. Perception exhibits a strong tendency to represent complex entities by means of comparably few, simple, and widely usable parts together with relations between them. Rather than modeling an object directly based on a constellation of its parts (e.g. [9]), the compositional approach learns intermediate groupings of parts. As a consequence, compositions bridge the semantic gap between low level features and high level object recognition by modeling category-distinctive subregions of an object, which show small intra-category variations compared to the whole object. The robustness of compositions to image changes can be exploited for tracking and grouping them over consecutive video frames. This temporal grouping of compositions improves the compositional image representation and enhances object segmentation and recognition. To be able to simultaneously recognize multiple objects in a video, we have incorporated a model selection strategy that automatically estimates the correct model complexity based on a stability analysis.

2   Related Work

Category-level recognition, segmentation, and tracking of objects in videos is related to a number of subtasks. First, motion information can be exploited by selecting relevant features for tracking (e.g. [22]) and establishing correspondences between frames, e.g. using the method of Lucas and Kanade [15]. Second, most methods for recognition describe objects based on local descriptors such as *SIFT* features [14] or flow histograms [7], and template-based *appearance patches* (e.g. [1,13]). Combining local features in an object model can then proceed along several lines. A simple approach is to compute descriptors on a regular grid and concatenate all cells to obtain a joint model [7]. More complex representations of the spatial object structure are *constellation models* [9], hough voting strategies [13], many-to-many feature correspondences [8], image parsing graphs [24], and compositional models [18]. Viola and Jones have proposed a real-time recognition system for faces that is based on a cascade of classifiers [25]. Another

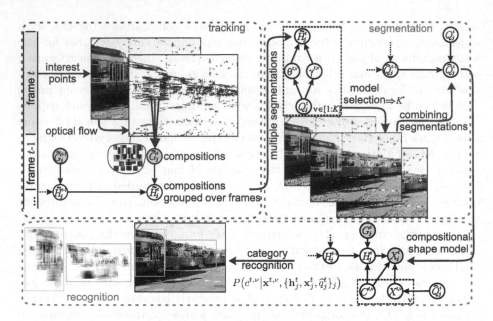

Fig. 1. Sketch of the processing pipeline for tracking, segmentation, and category-level object recognition in video

object class that many vision systems have been specifically developed for are pedestrians, e.g. [26,10]. Tracking algorithms have been studied for instance in [4] as well as [23], where the latter also presents a query-by-example approach to recognition that searches for regions which are similar to a user selected one. In contrast to tracking of a user specified region [5,2], Goldberger and Greenspann [12] propose a method for using segmentations of previous video frames to obtain a segmentation for the next.

3   Compositional Approach to Video Analysis

The following gives an overview of our compositional approach to video analysis (illustrated in Figure 1) before presenting the details in later sections. A novel video is analyzed sequentially in a frame-by-frame manner, while the underlying statistical model is propagating information over consecutive frames. Once a new frame is available, optical flow is estimated at interest points. These points and their motion pattern constitute the atomic parts of the composition system. Since interest points and their optical flow cannot be computed reliably, tracking individual points through a whole image sequence becomes error-prone. Therefore, we establish compositions of parts which are represented by probability distributions over their constituent parts. As a result, compositions are invariant with respect to individual missing parts and can be tracked reliably through a video by considering the optical flow distribution of all their constituents. The correspondence of compositions in consecutive frames is used to

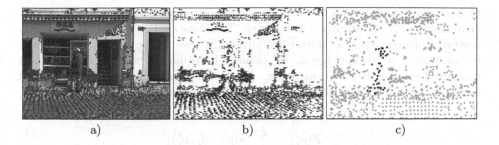

a)   b)   c)

Fig. 2. a) Detected interest points. b) Estimated optical flow at interest points. c) Locations $\mathbf{x}_j^t$ of established compositions $\tilde{\mathbf{h}}_j^t$. Brightness encodes the index of the closest codebook vector. See text for details.

group a composition over time. Subsequently, multiple segmentations are established. Therefore, compositions are clustered into different numbers of segments. To find an appropriate segmentation we incorporate a model selection strategy that analyzes the stability of the proposed segmentations over the preceding frames. The model with highest stability is then selected and combined with models from previous video frames to segment the current frame. Recognition of objects in the individual segments is then based on an extension of the *compositional shape model* from [18] which couples all compositions belonging to the same segment in a Bayesian network. The object category label is then obtained using probabilistic inference based on this model. In conclusion, tracking object constituents, segmenting objects from another, and recognizing the object category are all captured by the same statistical model, namely the graphical model illustrated in Figure 4. In this model, object representations of consecutive frames are linked together by a *Markov backbone* that connects segmentations of subsequent frames. Learning the underlying structured object model for a category proceeds in an unsupervised manner without requiring hand-segmentations or localization of objects in training videos.

3.1   Atomic Compositional Constituents

Based on the method of Shi and Tomasi [22] interest points are detected in every video frame, see Figure 2 a). Interest points from a preceding frame are then tracked into the next one using the Lucas-Kanade tracking algorithm [15]. This registration of points in consecutive frames yields an estimate of the optical flow $\mathbf{d}_i^t$ at interest point i in frame t, i.e. the displacement vector, see Figure 2 b). The interest points constitute the atomic parts of the composition system.

Codebook-Based Representation of Atomic Parts: Compositions can have different numbers of constituents and are, therefore, represented by a distribution over a codebook of atomic parts. Let $\mathbf{e}_i^t$ denote a feature vector that represents an atomic part i in frame t. The codebook that is used for representing compositions is then obtained by performing k-means clustering on all feature vectors

$\mathbf{e}_i^t$ from the training data. This vector quantization yields a common codebook of atomic parts for all object categories. To robustify the representation, each feature is described by a Gibbs distribution [27] over the codebook rather than by its nearest prototype: Let $d_\nu(\mathbf{e}_i^t)$ denote the squared euclidean distance of a measured feature $\mathbf{e}_i^t$ to a centroid $\mathbf{a}_\nu$. The local descriptor is then represented by the following distribution of its cluster assignment random variable F_i,

$$P(F_i = \nu | \mathbf{e}_i^t) := Z(\mathbf{e}_i^t)^{-1} \exp\left(-d_\nu(\mathbf{e}_i^t)\right),$$
$$Z(\mathbf{e}_i^t) := \sum_\nu \exp\left(-d_\nu(\mathbf{e}_i^t)\right). \tag{1}$$

Local Descriptors: We use two different kinds of local features to represent local parts. The first type simply represents the optical flow at an interest point, whereas the second is based on *localized feature histograms* [17] of a small surrounding region. As optical flow has to be estimated for tracking in any case, this representation has the advantage that no extra feature detector needs to be computed at each interest point and, therefore, saves computation time. For each interest point i in frame t we use its optical flow $\mathbf{d}_i^t$, giving a 2-dimensional feature vector $\mathbf{e}_i^t = \mathbf{d}_i^t$.

The second local descriptor is formed by extracting quadratic image patches with a side length of 20 pixels at interest points. Each patch is divided up into four equally sized subpatches with locations fixed relative to the patch center. In each of these subwindows marginal histograms over edge orientation and edge strength are computed (allocating four bins to each of them). Furthermore, an eight bin color histogram over all subpatches is extracted. All these histograms are then combined in a common feature vector $\mathbf{e}_i^t$.

A separate codebook is established for both types of features (optical flow features are quantized with a 10 dimensional codebook, the localized feature histograms are represented by a 60 dimensional codebook). Tracking of compositions and object segmentation is then based on the optical flow alone. Only the final inference of the object category based on the compositions in a foreground segment uses the complex, second descriptor type.

3.2 Compositions of Parts

In the initial frame of a video ($t = 0$), a random subset of all detected interest points is selected. Each of these points is then grouped with the atomic parts in its local neighborhood (radius of 25 pixel) yielding compositions of atomic parts. A composition in frame t is then represented by a mixture distribution (with uniform mixture weights) over the densities (1) of its constituent parts (cf. [18]). Let $\Gamma_j^t = \{\mathbf{e}_1^t, \ldots, \mathbf{e}_m^t\}$ denote the grouping of parts represented by features $\mathbf{e}_1^t, \ldots, \mathbf{e}_m^t$. The multivariate random variable G_j^t does then represent the composition consisting of these parts. A realization $\mathbf{g}_j^t \in [0,1]^k$ of this random variable is a multivariate distribution over the k-dimensional codebook of atomic parts

$$\mathbf{g}_j^t \propto \sum_{i=1}^m \left(P(F_i = 1 | \mathbf{e}_i^t), \ldots, P(F_i = k | \mathbf{e}_i^t)\right)^T. \tag{2}$$

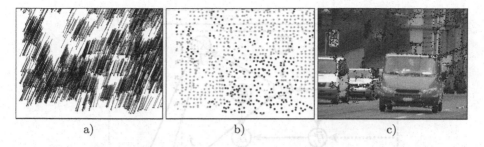

a)   b)   c)

Fig. 3. Using compositions to obtain reliable segmentations despite strong camera panning. a) Estimated optical flow at interest points. b) Compositions $\widetilde{\mathbf{h}}_j^t$. c) Segmentation.

Finally, each of the k dimensions is independently standardized to zero mean and unit variance across the whole training set, giving *z-scores*. This mixture model has the favorable property of robustness with respect to variations in the individual parts. As we are having two types of features $\mathbf{e}_i^t$ we obtain two representations of a composition j: $\mathbf{g}_j^t$ is the representation based on localized feature histograms, whereas $\widetilde{\mathbf{g}}_j^t$ builds on optical flow.

Compositions are tracked throughout a video based on the average flow estimated at their constituent parts. Given the position $\mathbf{x}_j^t$ of a composition in frame t and the optical flow vectors of its parts $\mathbf{d}_i^t$, its predicted position in the next frame is

$$\mathbf{x}_j^{t+1} = \mathbf{x}_j^t + \frac{1}{m} \sum_{i=1}^m \mathbf{d}_i^t. \tag{3}$$

In the next frame $t + 1$ the assignment of parts to compositions is updated since new interest points are computed. Therefore, all parts $\mathbf{e}_i^{t+1}$ in the local neighborhood of a composition $\mathbf{g}_j^{t+1}$ are assigned to this composition.

3.3 Temporal Grouping of Compositions

Whereas the preceding grouping was a spatial one (based on proximity) the following will present a grouping of compositions over time. By grouping compositions over consecutive frames, compositions of compositions can be formed that are more robust with respect to measurement errors in individual frames such as incorrect flow estimates. A temporal grouping of the j-th composition over consecutive frames yields the higher-order composition $\mathbf{h}_j^t$ which is represented by the distribution

$$\mathbf{h}_j^t \propto \begin{cases} \eta \mathbf{g}_j^t + (1 - \eta)\mathbf{h}_j^{t-1}, & \text{if } t > 1, \\ \mathbf{g}_j^t, & \text{else}. \end{cases} \tag{4}$$

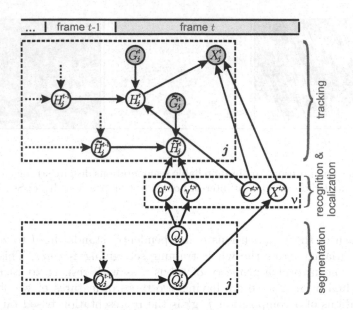

Fig. 4. Graphical model that unites category-level object recognition, segmentation, and tracking in the same statistical framework. Shaded notes denote evidence. The graph shows the dependencies between the three processes for frame t as well as the connection with the preceding frame. The involved random variables are the following: compositions represented with localized feature histograms, G_j^t, and with optical flow, $\widetilde{G}_j^t$. Temporal groupings of compositions: H_j^t and $\widetilde{H}_j^t$. Location of j-th composition: X_j^t. Assignment of compositions to segments: Q_j^t. Combining multiple segmentations over consecutive frames: $\widehat{Q}_j^t$. Segment priors: $\gamma^{t,\nu}$. Segment prototypes: $\theta^{t,\nu}$. Classification of object in segment ν: $C^{t,\nu}$. Localization of the object: $X^{t,\nu}$.

The flow representation of compositions is computed according to the same recursion formula, i.e. $\widetilde{\mathbf{h}}_j^t \propto \eta \widetilde{\mathbf{g}}_j^t + (1 - \eta)\widetilde{\mathbf{h}}_j^{t-1}$, and the mixture weight is chosen to be $\eta = 1/2$. The corresponding transition probability of the graphical model in Figure 4 is defined as

$$p(\mathbf{h}_j^t | \mathbf{g}_j^t, \mathbf{h}_j^{t-1}) := \mathbf{1}_{\{\mathbf{h}_j^t \propto \text{Eq.}(4)\}} \in \{0, 1\}. \tag{5}$$

In Figure 2 c) the centers $\mathbf{x}_j^t$ of compositions $\widetilde{\mathbf{h}}_j^t$ are displayed. As described in Section 3.2, each composition is represented by a probability distribution over a codebook of atomic parts. The brightness of the circle at $\mathbf{x}_j^t$ encodes the index of the codebook vector that has received most probability mass. In Figure 3, strong camera panning results in unreliable optical flow estimates at interest points. Compositions, however, can compensate for this difficulty and establish the foundation for an accurate segmentation. In conclusion, the visualizations show that compositions are actually valuable for a subsequent segmentation of objects.

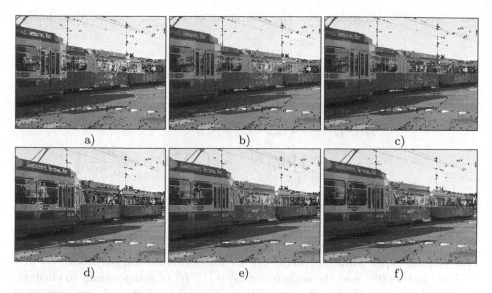

Fig. 5. Multiple segmentation hypotheses established for two frames. a) and d) show a 2 cluster solution. b) and e) 3 clusters. c) and f) 4 clusters. The segmentation with 3 clusters features the highest stability over the two frames and is, therefore, chosen by model selection.

3.4 Obtaining Multiple Segmentation Hypotheses

Subsequently, several initial hypotheses for the locations and shapes of objects that are present in a video frame are to be derived from the compositions. Since there is no prior information regarding the number of objects that are present in a scene, we have to address a difficult model selection problem. Therefore, several segmentations with varying numbers of segments are established. Model selection is then performed to retrieve the most reliable segmentation. Each segmentation partitions compositions in the optical flow feature space, $\widetilde{\mathbf{h}}_j^t$, into K segments using histogram clustering (e.g. see [19]): Compositions defined by (4) are represented as multivariate distributions over the k-dimensional part codebook, $\widetilde{\mathbf{h}}_j^t = (\widetilde{\mathbf{h}}_{j,1}^t, \ldots, \widetilde{\mathbf{h}}_{j,k}^t) \in [0,1]^k$ with $\sum_{l=1}^k \widetilde{\mathbf{h}}_{j,l}^t = 1$. The aim of clustering is then to represent $\widetilde{\mathbf{h}}_j^t$ by a mixture of K clusters $\boldsymbol{\theta}^{t,1}, \ldots, \boldsymbol{\theta}^{t,K} \in [0,1]^k$ with $\sum_{l=1}^k \theta_l^{t,\nu} = 1$ and mixture weights or class priors $\gamma^{t,1}, \ldots, \gamma^{t,K} \in [0,1]$,

$$p(\widetilde{\mathbf{h}}_j^t | \boldsymbol{\theta}^{t,1}, \ldots, \boldsymbol{\theta}^{t,K}, \gamma^{t,1}, \ldots, \gamma^{t,K}) = \sum_{\nu=1}^K \gamma^{t,\nu} \, p(\widetilde{\mathbf{h}}_j^t | \boldsymbol{\theta}^{t,\nu}). \tag{6}$$

The individual mixture components are approximated by multinomial distributions, i.e. for large $N \in \mathbb{N}$ the distribution of $\widetilde{\mathbf{h}}_{j,l}^t \cdot N$ is multinomial with parameter $\theta^{t,\nu}$. Transforming the definition of the multinomial distribution yields

$$p\left(\widetilde{\mathbf{h}}_j^t\middle|\theta^{t,\nu}\right) = \frac{N!}{\prod_l \lfloor \widetilde{\mathbf{h}}_{j,l}^t N \rfloor!} \prod_l (\theta_l^{t,\nu})^{\widetilde{\mathbf{h}}_{j,l}^t N} \tag{7}$$

$$= \frac{N!}{\prod_l \lfloor \widetilde{\mathbf{h}}_{j,l}^t N \rfloor!} \frac{\exp(\sum_l \widetilde{\mathbf{h}}_{j,l}^t \log \widetilde{\mathbf{h}}_{j,l}^t)}{\exp(\sum_l \widetilde{\mathbf{h}}_{j,l}^t \log \widetilde{\mathbf{h}}_{j,l}^t)} \exp\left\{\sum_l \widetilde{\mathbf{h}}_{j,l}^t \log(\theta_l^{t,\nu})^N\right\} \tag{8}$$

$$= \frac{N!}{\prod_l \lfloor \widetilde{\mathbf{h}}_{j,l}^t N \rfloor!} \prod_l (\widetilde{\mathbf{h}}_{j,l}^t)^{\widetilde{\mathbf{h}}_{j,l}^t} \cdot \exp\left\{-\sum_l \widetilde{\mathbf{h}}_{j,l}^t \log \frac{\widetilde{\mathbf{h}}_{j,l}^t}{(\theta_l^{t,\nu})^N}\right\} \tag{9}$$

$$= \frac{N!}{\prod_l \lfloor \widetilde{\mathbf{h}}_{j,l}^t N \rfloor!} \prod_l (\widetilde{\mathbf{h}}_{j,l}^t)^{\widetilde{\mathbf{h}}_{j,l}^t} \cdot \exp\left\{-D_{KL}\left(\widetilde{\mathbf{h}}_j^t\|(\theta^{t,\nu})^N\right)\right\}. \tag{10}$$

Here $D_{KL}(\cdot\|\cdot)$ denotes the Kullback Leibler distance between compositions and cluster prototypes while the prefactors are for normalization purposes.

The clusters $\theta^{t,\nu}$ and the assignment $Q_j^t \in [1:K]$ of compositions to clusters, i.e. $P(Q_j^t = \nu) := Prob\{j$-th composition assigned to cluster $\nu\}$, are computed by iterating an *expectation-maximization algorithm* [16]. In the *expectation-step*, assignment probabilities of compositions to segments are computed conditioned on the current estimate of clusters,

$$P\left(Q_j^t = \nu\right) := \frac{\gamma^{t,\nu} p\left(\widetilde{\mathbf{h}}_j^t\middle|\theta^{t,\nu}\right)}{\sum_\nu \gamma^{t,\nu} p\left(\widetilde{\mathbf{h}}_j^t\middle|\theta^{t,\nu}\right)}. \tag{11}$$

In the *maximization-step*, class priors $\gamma^{t,\nu}$ and cluster prototypes $\theta^{t,\nu}$ are updated conditioned on the assignment probabilities

$$\gamma^{t,\nu} := \frac{\sum_j P(Q_j^t = \nu)}{\sum_{j,\nu'} P(Q_j^t = \nu')}, \quad \theta_l^{t,\nu} := \frac{\sum_j P(Q_j^t = \nu)\widetilde{\mathbf{h}}_{j,l}^t}{\sum_j P(Q_j^t = \nu)}. \tag{12}$$

After convergence of the EM-algorithm, the cluster assignment probabilities $P(Q_j^t = \nu)$ of compositions represent the segmentation of a video frame into K segments. Since background is surrounding objects, the segment that covers most of the frame border is labeled as background, $\nu = BG$. Figure 5 shows segmentations with 2, 3, and 4 segments for two video frames. Interest points in the different segments are displayed in distinct color (black is used for the background segment).

3.5   Model Selection to Identify Reliable Segmentation Hypotheses

As there is no prior information regarding the number of objects that are present in a scene we pursue a model selection strategy to estimate the number of object segments. Therefore, segmentations $Q_j^t(K)$ for different numbers K of segments are established in each frame (currently we use $K = 2, \ldots, 5$). Bipartite matching

[6] is performed to make the current segmentation comparable with the one of the previous frame, i.e. labels are permuted so that they fit best to the preceding segment labeling. We then combine multiple segmentations of consecutive video frames into a single, more robust one $\widehat{Q}_j^t(K)$, with

$$P\big(\widehat{Q}_j^t(K) = \nu \big| Q_j^t(K), \widehat{Q}_j^{t-1}(K)\big) \tag{13}$$
$$\propto \begin{cases} \eta P(Q_j^t(K) = \nu) + (1 - \eta) P\big(\widehat{Q}_j^{t-1}(K) = \nu \big| Q_j^{t-1}(K), \widehat{Q}_j^{t-2}(K)\big), & \text{if } t > 1, \\ P(Q_j^t(K) = \nu), & \text{else}. \end{cases}$$

This dependency between segmentations of consecutive frames constitutes the Markov backbone that is represented at the bottom of the graphical model in Figure 4. It propagates segmentation hypotheses from previous frames into the current one.

An inappropriate model complexity is likely to yield unstable segmentations that change even when the input data varies only slightly. By observing the fluctuations of segmentations over multiple frames we can estimate their stability (cf. [20]) and select the most appropriate model complexity (see Figure 5 for an illustration). The stability $\zeta^t(K)$ of a K cluster segmentation is measured by the entropies $\mathcal{H}$ of the segment assignments

$$\zeta^t(K) := \sum_j \mathcal{H}\big(\widehat{Q}_j^t(K)\big) = -\sum_j \sum_{\nu=1}^K P\big(\widehat{Q}_j^t(K) = \nu\big) \log P\big(\widehat{Q}_j^t(K) = \nu\big). \tag{14}$$

The optimal number of segments is determined by selecting the K^* that minimizes this stability measure and we use the abbreviation $P(\widehat{q}_j^t) := P\big(\widehat{Q}_j^t(K^*) = \widehat{q}_j^t\big)$.

The location of the ν-th segment center $\mathbf{x}^t(\nu)$ is estimated as the center of mass of all compositions assigned to this segment ($j \in \mathcal{A}_\nu^t$)

$$\mathbf{x}^{t,\nu} := \frac{1}{|\mathcal{A}_\nu^t|} \sum_{j \in \mathcal{A}_\nu^t} \mathbf{x}_j^t, \quad \mathcal{A}_\nu^t := \{j : \nu = \operatorname*{argmax}_{\nu'} P\big(\widehat{Q}_j^t(K) = \nu'\big)\}. \tag{15}$$

3.6 Compositional Shape Model for Object Recognition

In every frame of a novel test video, the objects that are present in the individual segments have to be recognized. Therefore, all compositions $\mathbf{h}_j^t$, $j \in \mathcal{A}_\nu^t$ that are assigned to a segment ν are coupled in the graphical model shown in Figure 4. This statistical model is founded on the *compositional shape model* from [18]. The category $c^{t,\nu} \in \mathcal{L}$ of the object in segment ν can then be inferred from its posterior distribution

$$P\big(c^{t,\nu} \big| \mathbf{x}^{t,\nu}, \{\mathbf{h}_j^t, \mathbf{x}_j^t, \widehat{q}_j^t\}_j\big)$$
$$= \frac{p\big(\{\mathbf{h}_j^t, \mathbf{x}_j^t, \widehat{q}_j^t\}_{j \in \mathcal{A}_\nu^t}, \{\mathbf{h}_j^t, \mathbf{x}_j^t, \widehat{q}_j^t\}_{j \notin \mathcal{A}_\nu^t} \big| c^{t,\nu}, \mathbf{x}^{t,\nu}\big) P(c^{t,\nu} | \mathbf{x}^{t,\nu})}{p\big(\{\mathbf{h}_j^t, \mathbf{x}_j^t, \widehat{q}_j^t\}_j | \mathbf{x}^{t,\nu}\big)} \tag{16}$$

by applying Bayes' formula. Now the denominator can be skipped because it is independent of $c^{t,\nu}$. Furthermore, the category of an object should be independent of its absolute position in a frame and there should be no bias on any category, i.e. all classes are a priori equally likely. Therefore, $P(c^{t,\nu}|\mathbf{x}^{t,\nu})$ can be discarded as well.

$$\ldots \propto p\left(\{\mathbf{h}_j^t, \mathbf{x}_j^t, \widehat{q}_j^t\}_{j\in\mathcal{A}_\nu^t}, \{\mathbf{h}_j^t, \mathbf{x}_j^t, \widehat{q}_j^t\}_{j\notin\mathcal{A}_\nu^t} \big| c^{t,\nu}, \mathbf{x}^{t,\nu}\right). \tag{17}$$

Since the category of segment ν determines only compositions that have been assigned to this segment (i.e. $j \in \mathcal{A}_\nu^t$), all other compositions are independent of $c^{t,\nu}$ and can be skipped. Moreover, an assignment to segment ν implies $\widehat{q}_j^t = \nu$. Therefore $\widehat{q}_j^t$ can be dropped as well for $j \in \mathcal{A}_\nu^t$ and we obtain

$$\ldots \propto p\left(\{\mathbf{h}_j^t, \mathbf{x}_j^t\}_{j\in\mathcal{A}_\nu^t} \big| c^{t,\nu}, \mathbf{x}^{t,\nu}\right). \tag{18}$$

Compositions are conditionally independent, conditioned on the object model parameters $c^{t,\nu}$ and $\mathbf{x}^{t,\nu}$. Therefore, the likelihood factorizes and we can apply Bayes' formula again to obtain

$$\ldots \propto \prod_{j\in\mathcal{A}_\nu^t} \frac{P\left(c^{t,\nu}|\mathbf{x}^{t,\nu}, \mathbf{h}_j^t, \mathbf{x}_j^t\right) \cdot p\left(\mathbf{h}_j^t, \mathbf{x}_j^t|\mathbf{x}^{t,\nu}\right)}{P(c^{t,\nu}|\mathbf{x}^{t,\nu})}. \tag{19}$$

The factor $p\left(\mathbf{h}_j^t, \mathbf{x}_j^t|\mathbf{x}^{t,\nu}\right)$ does not depend on the object category and can be omitted. Moreover, the category of an object should be independent of its absolute position in a frame and there should be no bias on any category. Therefore, $P(c^{t,\nu}|\mathbf{x}^{t,\nu})$ can again be left out and we obtain

$$\ldots \propto \prod_{j\in\mathcal{A}_\nu^t} P\left(c^{t,\nu}|\mathbf{h}_j^t, S_j^{t,\nu} = \mathbf{x}^{t,\nu} - \mathbf{x}_j^t\right) \tag{20}$$

$$= \exp\left[\sum_{j\in\mathcal{A}_\nu^t} \ln P\left(c^{t,\nu}|\mathbf{h}_j^t, S_j^{t,\nu} = \mathbf{x}^{t,\nu} - \mathbf{x}_j^t\right)\right]. \tag{21}$$

Here the relative position of a composition with respect to the object center is represented by the shift $\mathbf{s}_j^{t,\nu} = \mathbf{x}^{t,\nu} - \mathbf{x}_j^t$. *Nonlinear kernel discriminant analysis* (NKDA) [21]) is used to estimate the distribution in (21). Therefore, probabilistic two-class kernel classifiers are trained on compositions extracted form the training data. These classifiers are coupled in a pairwise manner to solve the multi-class problem (see [21]). During recognition, an object can be recognized *efficiently* by applying the classifier to all compositions $\mathbf{h}_j^t$ and computing (21).

4   Evaluation

For still-image categorization large benchmark databases are available and the compositional approach has been shown (see [18]) to yield competitive performance compared to state-of-the-art methods in this setting. However, for the weakly supervised video analysis task that is pursued in this contribution there are, to the best of our knowledge, no comparable benchmark datasets available. Therefore, we have assembled a database for category-level object recognition in video consisting of 24 videos per category (categories car, bicycle, pedestrian, and streetcar). As can be seen from the examples in the figures, videos feature large intra-category variation (cf. Figure 6a and 7e), significant scale and viewpoint variation (e.g. Figure 7a, g), camera panning (cf. Figure 3), and background clutter. In the following, experiments are performed using 10-fold cross-validation. For each cross-validation step a random sample of 16 videos per category is drawn for training keeping the remainder for testing. Learning proceeds then on a randomly selected subset of 15 frames per video, while testing is performed on each frame. To avoid a bias towards categories with more test frames we average the retrieval rates for each category separately before averaging these scores over all frames. This evaluation approach has become the standard evaluation procedure in image categorization (e.g. see [18]).

4.1 Evaluating the Building Blocks of the Composition System

The following experiments evaluate the gain of the individual building blocks of the presented composition system for video analysis. In this first series of experiments only the most prominent object in a frame is to be detected. All key components are discarded in a first experiment before adding individual components in later experiments. The comparison of retrieval rates underlines the importance of each individual part of the compositional approach.

Baseline Performance of a Bag of Parts Approach: The compositional approach establishes an intermediate representation that is based on compositions of parts and the spatial structure of objects. In a first experiment this hidden representation layer is neglected to evaluate the gain of compositionality. A frame is then represented based on all detected localized histogram features $\mathbf{e}_i^t$ by a bag of parts $\mathbf{b}^t$ (cf. Section 3.1),

$$\mathbf{b}^t \propto \sum_i \Big(P(F_i = 1|\mathbf{e}_i^t), \ldots, P(F_i = k|\mathbf{e}_i^t) \Big)^T. \tag{22}$$

To categorize a frame, the category with highest posterior $P(c^t|\mathbf{b}^t)$ is selected. The posterior is again learned from the training data using NKDA. This approach yields a retrieval rate of $\mathbf{53.1 \pm 5.5\%}$.

Compositional Segmentation and Recognition w/o Shape Model: This experiment shows the benefit of combining segmentation with recognition in a compositional model. Therefore, compositions are established as described in Section 3.2 and Section 3.3. The prominent object in a video frame is then segmented from background clutter by establishing a 2-class segmentation as described in Section 3.4. Since only a single segmentation hypothesis is established no model selection is required. All compositions that are not assigned to the background, $\nu \neq$ BG, are then taken into account to recognize the most prominent object. Therefore, these compositions are combined using a bag of compositions descriptor $\tilde{b}^t \propto \sum_{j \in \mathcal{A}^t_{\nu \neq BG}} h^t_j$. Frames are then categorized without the compositional shape model by simply selecting the category with highest posterior $P(c^t|\tilde{b}^t)$. The combination of foreground segmentation with recognition based on compositions improves the retrieval rate to **64.5 ± 5.5%**.

Segmentation, Recognition, and Compositional Shape Model: In contrast to the previous experiment we now use the full compositional shape model of Section 3.6 to recognize the foreground object. As a result, the retrieval rate is further increased to **74.3 ± 4.3%**. The category confusion table is presented in Table 1. Another setting is to categorize video sequences as a whole and not individual frames. For this task the category hypothesis that is most consistent with all frames of a video is chosen. Here the compositional model achieves a retrieval rate of **87.4 ± 5.8%**. Obviously, this is an easier setting since information from an ensemble of frames can be used simultaneously.

By agglomerating atomic parts of limited reliability in compositions that can be tracked reliably, information has been condensed and the robustness of object representations has been improved. The underlying statistical inference problem can then be solved efficiently. As a result the compositional model segments, tracks, and recognizes objects in videos of full PAL resolution (768× 576 pixel) at the order of 1 fps on an ordinary desktop computer.

Table 1. Category confusion table (percentages) for the complete composition system

True classes →	bicycle	car	pedestrian	streetcar
bicycle	**70.1**	5.5	15.1	4.0
car	10.0	**87.6**	16.1	12.4
pedestrian	15.9	2.5	**61.4**	5.5
streetcar	4.0	4.4	7.4	**78.2**

4.2   Multi-object Recognition

In the following experiment multiple objects that appear simultaneously in a video are to be recognized. Therefore, the model selection strategy of Section 3.5 is applied to find the correct number of objects. A frame is then correctly classified if all the present objects have been found. Missing an object or detecting an object that is not present counts as an error. Given this significantly harder task our full compositional model classifies 68.1 ± 4.9% of all frames correctly.

4.3   Analyzing the Relevance of Compositions

To analyze what individual compositions contribute to object recognition the category posterior

$$P\big(c^{t,\nu}\big|\mathbf{x}^{t,\nu}, \mathbf{h}^t_j, \mathbf{x}^t_j, \widehat{q}^t_j\big)\big|_{c^{t,\nu} \,=\, \text{True Category}} \tag{23}$$

Fig. 6. Visualizing the contributions of compositions $\mathbf{h}_j^t$ to object recognition. Dark circles correspond to compositions with high posterior from (23). The gap between a) and c) and between e) and g) is both times 60 frames. i) and k) have a gap of 10 frames. Class labels are placed at the location of the segment center $\mathbf{x}^{t,\nu}$ (c:car, p:pedestrian).

is evaluated for each composition. In Figure 6 and 7 the category posterior is then encoded i) in the darkness of a circle around the composition center and ii) in the opaqueness of the underlying image region, i.e. alpha blending is used for visualization. Moreover, Figure 6 shows the propagation of an object segmentation over several frames.

5   Discussion

In this contribution we have presented a compositional approach that combines category-level object recognition, segmentation, and tracking in the same graphical model without user supervision. The compositional representation of objects

a)   b)   c)   d)

e)   f)   g)   h)

Fig. 7. Visualization of compositions. See Figure 6 for details.

is automatically learned during training. A model selection strategy has been pursued to handle multiple, simultaneously appearing objects. By agglomerating ensembles of low-level features with limited reliability, compositions with increased robustness have been established. As a result an intermediate object representation has been formed that condenses object information in informative compositions. Recognition has been formulated as an efficiently solvable, statistical inference problem in the underlying Bayesian network. Therefore, compositionality not only improves the learning of object models but also enhances recognition performance so that near real-time video analysis becomes feasible.

References

1. Agarwal, S., Awan, A., Roth, D.: Learning to detect objects in images via a sparse, part-based representation. PAMI, 26(11) (2004)
2. Avidan, S.: Ensemble tracking. In: CVPR (2005)
3. Biederman, I.: Recognition-by-components: A theory of human image understanding. Psychological Review, 94(2) (1987)
4. Brostow, G.J., Cipolla, R.: Unsupervised bayesian detection of independent motion in crowds. In: CVPR (2006)
5. Comaniciu, D., Ramesh, V., Meer, P.: Kernel-based object tracking. PAMI, 25(5) (2003)
6. Cormen, T.H., Leiserson, C.E., Rivest, R.L., Stein, C.: Introduction to Algorithms, 2nd edn. MIT Press, Cambridge (2001)

7. Dalal, N., Triggs, B., Schmid, C.: Human detection using oriented histograms of flow and appearance. In: ECCV (2006)
8. Demirci, F., Shokoufandeh, A., Keselman, Y., Bretzner, L., Dickinson, S.: Object recognition as many-to-many feature matching. IJCV, 69(2) (2006)
9. Fergus, R., Perona, P., Zisserman, A.: Object class recognition by unsupervised scale-invariant learning. In: CVPR (2003)
10. Gavrila, D.M., Giebel, J., Munder, S.: Vision-based pedestrian detection: The protector+ system. In: IEEE Intelligent Vehicles Symposium (2004)
11. Geman, S., Potter, D.F., Chi, Z.: Composition Systems. Quarterly of Applied Mathematics, 60 (2002)
12. Goldberger, J., Greenspann, H.: Context-based segmentation of image sequences. PAMI, 28(3) (2006)
13. Leibe, B., Schiele, B.: Scale-invariant object categorization using a scale-adaptive mean-shift search. In: Rasmussen, C.E., Bülthoff, H.H., Schölkopf, B., Giese, M.A. (eds.) Pattern Recognition. LNCS, vol. 3175, Springer, Heidelberg (2004)
14. Lowe, D.G.: Distinctive image features from scale-invariant keypoints. IJCV, 60(2) (2004)
15. Lucas, B.D., Kanade, T.: An iterative image registration technique with an application to stereo vision. In: IJCAI (1981)
16. McLachlan, G.J., Krishnan, T.: The EM Algorithm and Extensions. John Wiley & Sons, Chichester (1997)
17. Ommer, B., Buhmann, J.M.: Object categorization by compositional graphical models. In: Rangarajan, A., Vemuri, B., Yuille, A.L. (eds.) EMMCVPR 2005. LNCS, vol. 3757, Springer, Heidelberg (2005)
18. Ommer, B., Buhmann, J.M.: Learning compositional categorization models. In: Leonardis, A., Bischof, H., Pinz, A. (eds.) ECCV 2006. LNCS, vol. 3951, Springer, Heidelberg (2006)
19. Puzicha, J., Hofmann, T., Buhmann, J.M.: Histogram clustering for unsupervised segmentation and image retrieval. Pattern Recognition Letters, 20 (1999)
20. Roth, V., Lange, T.: Adaptive feature selection in image segmentation. In: Rasmussen, C.E., Bülthoff, H.H., Schölkopf, B., Giese, M.A. (eds.) Pattern Recognition. LNCS, vol. 3175, Springer, Heidelberg (2004)
21. Roth, V., Tsuda, K.: Pairwise coupling for machine recognition of hand-printed japanese characters. In: CVPR (2001)
22. Shi, J., Tomasi, C.: Good features to track. In: CVPR (1994)
23. Sivic, J., Schaffalitzky, F., Zisserman, A.: Object level grouping for video shots. IJCV, 67(2) (2006)
24. Tu, Z.W., Chen, X.R., Yuille, A.L., Zhu, S.C.: Image parsing: Unifying segmentation, detection and recognition. IJCV, 63(2) (2005)
25. Viola, P., Jones, M.: Rapid object detection using a boosted cascade of simple features. In: CVPR (2001)
26. Viola, P., Jones, M.J., Snow, D.: Detecting pedestrians using patterns of motion and appearance. In: ICCV (2003)
27. Winkler, G.: Image Analysis, Random Fields and Markov Chain Monte Carlo Methods—A Mathematical Introduction, 2nd edn. Springer, Heidelberg (2003)

Bayesian Order-Adaptive Clustering for Video Segmentation

Peter Orbanz, Samuel Braendle, and Joachim M. Buhmann

Institute of Computational Science, ETH Zurich
{porbanz,jbuhmann}@inf.ethz.ch

Abstract. Video segmentation requires the partitioning of a series of images into groups that are both spatially coherent and smooth along the time axis. We formulate segmentation as a Bayesian clustering problem. Context information is propagated over time by a conjugate structure. The level of segment resolution is controlled by a Dirichlet process prior. Our contributions include a conjugate nonparametric Bayesian model for clustering in multivariate time series, a MCMC inference algorithm, and a multiscale sampling approach for Dirichlet process mixture models. The multiscale algorithm is applicable to data with a spatial structure. The method is tested on synthetic data and on videos from the MPEG4 benchmark set.

1 Introduction

Clustering algorithms usually operate on a fixed set of data. When clustering is applied to perform segmentation, the input data might e.g. be a digital image (group the image into spatially coherent segments) or a time series (decompose the series into coherent segments along the time axis, such as speaker clustering). In this article, we consider a different problem arising from the formalization of video segmentation as a clustering problem: Given is a time series of fixed-size data frames, each of which has a spatial structure, i.e. the 2D structure of the frame image. The series is to be decomposed into a sequence of spatially coherent segmentations of the frames, which should reflect the temporal smoothness of the sequence.

Clustering problems of this type have been actively studied in video segmentation [1]. For example, [2] proposes a parametric mixture model for optical flow features with neighborhood constraints. The number of clusters is selected by a likelihood heuristic. Temporal context is modeled implicitly by using differential motion features. Explicit context models include designs based on HMMs [3] or frame-to-frame model adaptation [4]. A method which approaches the problem's time series structure in a manner similar to Bayesian forecasting has recently been suggested in [5]. The authors propose a Gaussian mixture model to represent image rather than motion features. Temporal context is incorporated by using the estimate obtained on a given frame in the sequence as prior information for the following frame.

A.L. Yuille et al. (Eds.): EMMCVPR 2007, LNCS 4679, pp. 334–349, 2007.
© Springer-Verlag Berlin Heidelberg 2007

We propose a Bayesian method capable of addressing both temporal context and estimation of the number of clusters by a single model. The distribution of each cluster in feature space is described by an exponential family model, which is estimated under its respective conjugate prior [6]. The components are combined in a Bayesian mixture model to represent a segmentation. For each component, the prior is defined by the component's posterior estimated during the previous time step. Due to the "chaining" properties of conjugate pairs, this results in a closed model formulation for the entire time series. The mixture proportions and number of components are controlled by a Dirichlet process (DP) prior [7]. As we will argue, the conjugate nature of the DP leads to a chaining property analogous to the exponential family case. This property is used to propagate cluster structure along the time series in a similar manner as the conjugate component distributions propagate parameter information. Inference of the model is conducted by an adaptation of the Gibbs sampler for DP mixture models [8] to the time series model. To facilitate application of our model to the large amounts of data arising in video segmentation, we (i) show how the efficiency of the Gibbs sampler can be substantially increased by exploiting temporal smoothness and (ii) introduce a multiscale sampling method to speed up processing of individual frames. Just as the model, the multiscale algorithm is based on the properties of exponential family distributions.

2   Background

Method overview. Our approach to video segmentation is based on local features extracted from each image frame (where the "image" may be the original frame image, one of its color space dimensions, or a filter response image). A local window is positioned around each point of an equidistant grid within the image, and the pixel values within the window are extracted as feature values. The data vectors described in the following may, for example, be histograms of the pixels within local windows. They will generally be denoted as $\mathbf{x}_i^t$, where t is a time index (frame index) and i indexes a window position within the frame image. Image segments are modeled as clusters. Each cluster k is modeled by a distribution $F(\mathbf{x}_i^t | \theta_k^t)$. That is, the parameter vectors θ_k^t describe the segments, and constitute the target variables of the estimation problem. Priors on the parameters will generally be denoted G. For a given time t, the individual cluster models are combined into an overall segmentation solution by joining them in a Bayesian mixture model. A DP is used adapt the model order (i.e. the number of clusters). DPs define distributions on the mixture weights of a mixture model such that the total number of clusters is a random variable. In the video or time series context, they are suitable for the definition of clustering models that allow the number of clusters to change between time steps, as segments appear in or disappear from the scene. This section will review the basic ingredients of the model: Mixture models, conjugate prior distributions and Dirichlet processes (including MCMC inference). These concepts will be used in Sec. 3 to define a model for clustering in time series. Estimation algorithms for the model are developed in Sec. 4.

2.1   Clustering with Mixture Models

A *finite mixture model* is a probability density representable as a convex combination

$$p(\mathbf{x}|\Theta) = \sum_{k=1}^{N_C} c_k F(\mathbf{x}|\theta_k) \tag{1}$$

of component densities $F(.\,|\theta_k)$. In the clustering context, each component density represents the distribution of a single cluster. For a given data set $\mathbf{x}_1, \ldots, \mathbf{x}_n$, a *clustering solution* is an assignment of each observation $\mathbf{x}_i$ to one cluster $k \in \{1, \ldots, N_C\}$. Assigning an observation expresses the assumption that it was generated by the respective density $F(.\,|\theta_k)$. We encode such a solution by a set of *assignment variables* $S_1, \ldots, S_n$, one for each observation, where $S_i = k$ if $\mathbf{x}_i$ is assigned to cluster k. Clustering solutions are obtained either by expectation-maximization (EM) algorithms [9], or by Markov chain Monte Carlo sampling in a Bayesian regime. Both approaches rely on a latent variable structure, by treating the assignment variables S_i as random quantities that are estimated from the data along with the parameters.

2.2   Conjugate Models

Several aspects of this work build on exponential family models and the concept of conjugacy [6]. We provide a brief summary of these models and those properties of importance for our approach. A distribution of a random variable $\mathbf{X}$ with domain Ω_x is called an *exponential family model* if its density can be written as

$$F(\mathbf{x}|\theta) := h(\mathbf{x}) \exp(\langle s(\mathbf{x})|\theta\rangle - \phi(\theta)) . \tag{2}$$

Here, $\theta \in \Omega_\theta$ is a parameter vector, $\langle .\,|\,.\rangle$ denotes the scalar product on the parameter space Ω_θ, and the function ϕ is the normalization term $\phi(\theta) := \log \int h(\mathbf{x}) \exp(\langle s(\mathbf{x})|\theta\rangle)d\mathbf{x}$. Of particular interest is the *sufficient statistic* function $s : \Omega_x \to \Omega_\theta$, which effectively defines all properties of the model relevant for parameter estimation. If a density F and a prior G are used in a Bayesian estimation framework, they are called *conjugate* if the resulting posterior is a distribution of the same type as G, i.e. prior and posterior differ only in their parameters. The concept of conjugacy is inherently tied to exponential families: It can be shown that, on the one hand, any exponential family model has a conjugate model (which is also an exponential family). On the other hand, only exponential family models have non-trivial conjugate models [6]. If F is a model of the form (2), then $G(\theta|\lambda, y) := \frac{1}{K(\lambda, y)} \exp(\langle \theta|y\rangle - \lambda\phi(\theta))$ is a conjugate prior, and the posterior under n independent observations $\mathbf{x}_i$ is

$$G(\theta|\lambda + n, \mathbf{y} + \sum_{i=1}^{n} s(\mathbf{x}_i)) . \tag{3}$$

The average $\sum s(x_i)$ carries all information the observation sample contains about the model parameter, i.e. it is sufficient. Conjugacy is a key property for a Bayesian time series model that uses the posterior of a previous step as prior for the present one: If a non-conjugate pair is used, the derived prior will change from time step to time step.

2.3 Dirichlet Process Mixture Models

Model order selection. For mixture model estimation, the overall number N_C of clusters is an input parameter. Techniques for estimation of this quantity from data are referred to as *model order selection*. Non-Bayesian solutions are usually based on regularity assumptions: The model's capability to explain the given data, measured by the likelihood, is traded off against a measure of model complexity to avoid overfitting. Model complexity is typically measured as a function of the model and the sample size (information criteria approaches such as AIC, BIC and MDL; cf [10] for an overview). The *stability* method [11] measures model complexity as stability of the solution under random perturbations of the data. All these methods proceed in an exhaustive search manner, i.e. scores for a range of models are computed and the best model is chosen. Bayesian DP-based methods [8] represent the number of clusters as a random variable. An estimate is obtained by sampling the model posterior under the observed data. Whereas scoring methods aim at identifying the "true" number of clusters under a chosen set of assumptions, DP methods provide a user parameter that controls the qualitative behavior of the random variable N_C. As will be discussed below, this makes them a natural choice for problems in which the number of model clusters has to change adaptively over a range of instances.

The Dirichlet process approach. In our clustering model, individual cluster components will be represented by parametric exponential family models, which are combined in a mixture model. To control the structure of the mixture, we use a Dirichlet process mixture (DPM) model [8,12]. For the sake of brevity, we will forego the model's derivation as a stochastic process and only describe its practical properties. Loosely speaking, a DPM is a mixture model with a "complexity control" term. The extra term governs the creation of new components. In the finite parametric mixture (1), different components of the mixture are defined by different values of the parameter vector θ_k. If the mixture is estimated from data, the mixture weights are chosen in proportion to the size of classes, i.e. $c_k = \frac{n_k}{n}$ if n_k out of n total data points are assigned to component k. The model (1) can be regarded as the result of integrating the distribution function F against

$$G_{\mathrm{MM}}(\theta) := \sum_{k=1}^{N_C} c_k \delta_{\theta_k}(\theta), \tag{4}$$

where δ_{θ_k} denotes the Dirac function on Ω_θ centered at θ_k. DPM models augment the expression (4) by an extra term,

$$G_{\mathrm{DP}}(\theta) := \sum_{k=1}^{N_C} c_k \delta_{\theta_k}(\theta) + \alpha G_0(\theta) . \tag{5}$$

The function G_0 is a probability distribution on the domain Ω_θ of mixture parameters, and $\alpha \in \mathbb{R}_+$ a positive scalar parameter. The dynamics of this model can be illustrated by considering a data generation process: A new datum $\mathbf{x}$ is generated by drawing from a standard mixture or DPM, which we assume to have been estimated from data $\mathbf{x}_1, ..., \mathbf{x}_n$. When drawing from a mixture model (corresponding to (4)), $\mathbf{x}$ will be drawn from one of the N_C parametric mixture components in a two-stage manner by sampling $\theta \sim G_{\mathrm{MM}}$, then $\mathbf{x} \sim F(\,.\,|\theta)$. In the DPM case (5), $\mathbf{x}$ is drawn either according to one of the N_C parametric mixture components, or (with a probability proportional to α) from a new mixture component $F(\,.\,|\theta^*_{N_C+1})$, where the new parameter value $\theta^*_{N_C+1}$ is sampled from G_0. The DPM model constitutes a Bayesian approach to model order selection in mixture models, due to its ability to generate as many clusters as required to explain the observed data.

The standard Gibbs sampling algorithm for DPM models [13] encodes assignments of data points to mixture components by means of latent variables, in a manner similar to the EM algorithm. For each $\mathbf{x}_i$, the discrete index variable S_i specifies the index of the mixture component to which $\mathbf{x}_i$ is assigned. Within the algorithm, a value of $S_i = 0$ indicates the generation of a new mixture component from G_0 as a model for $\mathbf{x}_i$. The S_i are determined by computing mixture proportions $\tilde{q}_{ik}$ as

$$\tilde{q}_{i0} := \int_{\Omega_\theta} F(\mathbf{x}_i|\theta)G_0(\theta)d\theta \tag{6}$$

$$\tilde{q}_{ik} := n_k^{-i}F(\mathbf{x}_i|\theta_k) \qquad \text{for } k = 1, ..., N_C,$$

where n_k^{-i} is the number of data points assigned to component k with $\mathbf{x}_i$ removed from the data set. The proportions are normalized to obtain mixture probabilities $q_{ik} := \frac{\tilde{q}_{ik}}{\sum_{l=0}^{N_C} \tilde{q}_{il}}$, for $k = 0, \ldots, N_C$. The sampling algorithm repeats the following steps:

Assignment step: For all $i = 1, ..., n$,

1. Compute $q_{i0}, ..., q_{iN_C}$.
2. Sample $S_i \sim (q_{i0}, ..., q_{iN_C})$.
3. If $S_i = 0$, create a new component: Sample $\theta_{N_C+1} \sim F(\mathbf{x}_i|\theta_{N_C+1})G_0(\theta_{N_C+1})$.

Estimation step: For each $k = 1, ..., N_C$, sample

$$\theta_k \sim G_0(\theta_k) \prod_{i|S_i=k} F(\mathbf{x}_i|\theta_k) \, . \tag{7}$$

In the conjugate case, the integral in (6) has an analytic solution and the component posteriors in (7) are distributions of the same type as G_0. Thus, if G_0 can be sampled, the component posterior can be sampled as well.

3   Order-Adaptive Clustering in Time Series

The present article considers the problem of obtaining a clustering solution for each time step in a multivariate time series. The clustering solutions are required to exhibit a temporal coherence. In a video sequence, each time step corresponds to a single frame image. The overall clustering solution then consists of a segmentation for each frame. If the number of clusters can change between frames, a suitable clustering method must be *order-adaptive*, i.e. capable of adjusting the model order over time. Order-adaptive methods require (i) automatic model order selection and (ii) a meaningful way to match clusters in different frames. If clustering solutions are obtained independently on each frame, the latter must be addressed by matching heuristics. Any principled approach requires the use of context information, i.e. the clustering solution for a given frame has to be obtained in a manner conditional on the solutions for the previous frame. In this section, we discuss how cluster structure can be propagated along a time series if the clustering solutions on individual frames are controlled by a DP prior.

An Order-Adaptive Clustering Model. We consider a multivariate time series $\mathbf{x}^t := (\mathbf{x}_1^t, \ldots, \mathbf{x}_n^t)$ that, for each time step $t = 1, 2, \ldots$, generates a set of n outputs $\mathbf{x}_i^t$. For each t, a single clustering solution $\mathbf{S}^t := (S_1^t, \ldots, S_n^t)$ is obtained. For temporal coherence, we require that, if $S_i^t = k$, then also $S_i^{t+1} = k$ with high probability, unless the corresponding observations $\mathbf{x}_i^t$ and $\mathbf{x}_i^{t+1}$ differ significantly. For the video segmentation problem, this reflects the assumption that size and location of segments change slowly on the time scale of frame renewal. The standard Bayesian approach to address temporal coherence requirements in time series models is to encode context with priors. The posterior distribution of the model parameter vector θ^t at a given time is used as prior distribution θ^{t+1}. This requires a conjugate model, i.e. a model for which prior and posterior belong to the same family of distributions (otherwise, the type of the prior distribution would change between time steps). Though we are ultimately interested in DP priors, let us first exemplify the approach for a parametric model. Let $F(\mathbf{x}|\theta)$ be a density modeling observations at a single time step (that is, a likelihood). We use the *ab initio* form $G(. |\lambda, \mathbf{y})$ of the prior, with G conjugate to F. The prior for the parameter vector θ^{t+1} is the corresponding posterior under the previous observation $\mathbf{x}^t$,

$$G(\theta^t | \lambda + 1, \mathbf{y} + s(\mathbf{x}^t)) \propto F(\mathbf{x}^t | \theta^t) G(\theta^t | \lambda, \mathbf{y}) . \tag{8}$$

One may variate upon this strategy to accumulate observations over time,

$$G\left(\theta^{t+1} \Big| \lambda + \tau, \mathbf{y} + \sum_{t-\tau < r \le t} s(\mathbf{x}_r)\right) \propto \prod_{t-\tau < r \le t} F(\mathbf{x}^r | \theta^r) G(\theta^{t-\tau+1} | \lambda, \mathbf{y}) \tag{9}$$

resulting in a process with a τ-step memory. A similar mechanism is applicable to DP models. The DP has a natural conjugate property, implicit in Ferguson's [7] characterization of the DP posterior: If $\theta_1, \ldots, \theta_n \sim \mathrm{DP}(\alpha G_0)$, then $\theta_{n+1} \sim \mathrm{DP}\left(\alpha G_0 + \sum_{i=1}^n \delta_{\theta_i}\right)$. For convenience, we adopt the following symbolic notation: Observe that the Dirac sum $\hat{G}_n := \sum_{i+1}^n \delta_{\theta_i}$ of n draws from G can be regarded as a finite draw from an infinite categorial distribution (i.e. as a partial, finite observation from an "infinite histogram"). Just as a finite histogram can be generated by a multinomial distribution parameterized by a draw from a

Dirichlet (its conjugate prior), the measure $\hat{G}_n$ can be explained as a draw from a multinomial process (MP) parameterized by a draw from a DP. The DP prior and posterior can then be linked explicitly in a Bayesian formula

$$\mathrm{DP}\left(\alpha G_0 + \hat{G}_n\right) \propto \mathrm{MP}\left(\hat{G}_n | G\right) \mathrm{DP}\left(\alpha G_0\right).$$

Although our use of this notation is purely symbolic, it can be derived in a mathematically precise manner using Kolmogorov's extension theorem for stochastic processes. In perfect analogy to the conjugate parametric case, we can construct a DP prior for an estimation problem in a time series at time $(t + 1)$ as the DP posterior at time t:

$$G^{t+1} \sim \mathrm{DP}\left(\alpha G_0 + \hat{G}_n^t\right) \propto \mathrm{MP}\left(\hat{G}_n^t | G^t\right) \mathrm{DP}\left(\alpha G_0\right).$$

The formulation immediately extends to a τ-step memory, by means of

$$\mathrm{DP}\left(G | \alpha G_0 + \sum_{t-\tau<r\leq t} \hat{G}_n^r\right) \propto \prod_{t-\tau<r\leq t} \mathrm{MP}\left(\hat{G}_n^r | G\right) \mathrm{DP}\left(G | \alpha G_0\right). \tag{10}$$

Note that, for the cluster propagation problem, the draws from the DP are parameters θ_i rather than observations. The observed data $\mathbf{x}^{t+1}$ at time $(t + 1)$ is then generated as

$$\mathbf{x}_i^{t+1} \sim F(. | \theta_i^{t+1}) \tag{11}$$
$$\theta_i^{t+1} \sim G^{t+1}$$
$$G^{t+1} \sim \mathrm{DP}\left(\alpha G_0 + G^t\right) .$$

Observations from multiple channels. In image and video processing applications, the input data usually consists of multiple channels. For standard color videos, three channels correspond to the three color space dimensions. Additional channels may include other features in the form of transformed data or filter responses. For multiple data channels indexed $c = 1, \ldots, C$, multiple observations $(\mathbf{x}_{i,c}^t)_{c=1,\ldots,C}$ are obtained for each frame t and location i. These are represented in the model as a product of likelihoods. That is, the generative model is obtained by substituting suitable product distributions $\prod_{c=1}^{C} F(\mathbf{x}_{i,c}^{t+1} | \theta_{i,c}^{t+1})$ and $\prod_{c=1}^{C} G_c^{t+1}$ for F and G_0 in (11). This does not increase the complexity of the clustering problem. Cluster assignments are still encoded by one variable S_i^t per location. The product model defines C parallel estimation problems, which are coupled by the assignment variables. Since the latter are the actual target variables of the clustering problem, the approach effectively increases the amount of observational data per estimated solution variable. If the features are chosen well, this can significantly increase the quality of estimates. Choosing a large number of uninformative channels may have a converse effect: The assignment probabilities q_{ik}^t computed by the inference algorithm are derived from the product model as averages over channels. Multiple uninformative features may thus obfuscate information provided by informative ones.

A note on exchangeability. A well-known fact in statistics is that data generated from a DP is *exchangeable*, i.e. its probability is invariant under permutations of the order [6]. This raises some questions about the applicability of such models to time series, where the order of observations is crucial. The model described above derives a prior for step $(t+1)$ based on two inputs, the initial parameters $(\lambda, \mathbf{y})$, which apply uniformly in all steps, and the previous observation $\mathbf{x}^t$. The implicit assumption is that pairs of adjacent steps are exchangeable. This local exchangeability is, in turn, equivalent to assuming processing of the time series to be invariant under inversion of the time axis. In other words, inference results should not depend on whether the time series is processed in increasing or decreasing index order. Such an assumption is justifiable for some time series, and in particular for the video segmentation problem. Segmentation as a mid-level vision problem is oblivious to semantic and high-level content (such as whether a falling object moves downwards or upwards). However, it requires the neighborhood relations of frames to be preserved. Hence, we may invert the order, but not shuffle the frames. The argument does not apply to a process with a multi-step memory as in (10), which would require exchangeability of more than two frames. Such a model should therefore be regarded as an approximation. Note that many models assume independence of data points, which is a much stronger assumption than exchangeability. Such models often perform well even in applications where the generative process of the data will clearly result in dependent observations. We therefore believe that, in many cases, approximation by multi-step exchangeability may be beneficial, and the ability of such a model to draw on multiple observations will outweigh the loss of descriptive precision incurred by the violation of the model assumption.

4   MCMC Inference for High-Throughput Problems

In this section, we discuss inference techniques for the model described in Sec. 3. Available methods for DPM inference are extended to address two issues: Time series structure and efficiency. Existing hidden-variable methods can be modified for time series inference by initializing inference for a given time step by the model state estimated for the previous step. To increase the sampler efficiency for individual time steps, we propose a multi-scale sampling scheme exploiting the spatial structure of frame sequence data. Both approaches can be combined to obtain a efficient sampling algorithm.

4.1   Sampling in Time Series

Parameter inference for the time series clustering problem estimates the cluster parameters θ_k^t and the states of the assignment variables S_i^t, for each $t = 1, \ldots$. Estimates are obtained by sampling the relevant posteriors with a Gibbs sampler. To derive a suitable algorithm, we note that, for a given time index t, recovering the states S_i^t and parameters θ_k^t given the current observations $\mathbf{x}_i^t$ is a DPM mixture inference problem with prior $\mathrm{DP}(\alpha G_0 + \hat{G}_n^t)$. The history of the process enters via the prior parameter, i.e. the measure $(\alpha G_0 + \hat{G}_n^t)$. Full conditionals for the Gibbs sampler are immediately obtained from the standard sampling algorithm, by substituting $(\alpha G_0 + \hat{G}_n^t)$ for αG_0. (A sampler for a series with a τ-step memory can be obtained by substituting the corresponding posterior parameter in (10)). Estimates for the whole time series can be computed by running the Gibbs sampler for the appropriate posterior at each time step. The parameter estimates (summarized by $\hat{G}_n^t$) are then substituted into the DP prior of the subsequent step. The algorithm is an online method, as it only performs a single pass over the time series. The cluster correspondence problem is solved implicitly, by propagating information from one time step to the next through the DP base measure. Initially, the same clusters as in the previous step are available for assignment, and their indices are preserved. Classes may be newly generated by drawing from the continuous component G_0 of the DPM, or deleted if no longer supported by the data. Gibbs sampling is potentially time-consuming, and performing a full run of a DPM Gibbs sampler for each time step in the series is computationally prohibitive. A substantial speed-up is achieved by exploiting temporal smoothness. If changes in the data occur slowly w. r. t. to the time scale (frame rate) of the time series, the model state estimated at time t provides an almost-perfect initialization for sampling at time $(t+1)$. The algorithm therefore obtains an initial estimate at time $t = 1$ by performing a full run of the Gibbs sampler. For $t \geq 2$, the Gibbs sampler is initialized by the previous model state, and then run only for a few steps, to allow the model to adapt to changes in the data.

4.2 Multiscale Sampling

For data exhibiting a spatial neighborhood structure, DPM inference algorithms that are more efficient then the standard Gibbs sampler can be derived using a multiscale approach. Multiscale methods attempt to increase the efficiency of iterative algorithms by replacing the original input problem with a compressed replacement problem (*coarsening*). This reduced-size problem is solved, and the solution transformed into a solution of the larger input problem (*refinement*). The compression operation exploits neighborhood structures in the data (such as spatial or sequential neighborhoods). In images, adjacent pixels are grouped into blocks, and each block B is compressed by computing a summary variable $\mathbf{x}_B$. The coarsened problem is given by the set of summary variables for all blocks. The coarsening operation therefore has to be designed to limit the loss of relevant information under compression, and to result in a coarse-scale problem to which the processing algorithm in question is applicable.

Coarsening. Our aim is the design of MCMC sampling algorithms. The information to be preserved under coarsening is therefore the information relevant to statistical parameter estimation. A simple averaging approach is not suitable in general, as it will only preserve moment information of first order. For the models considered in our work, a suitable coarsening approach can be derived from the properties of sufficient statistics. For an exponential family density as in (2), all information relevant to parameter estimation is contained in the sufficient statistic $s(\mathbf{x})$. Furthermore, for multiple observations $\mathbf{x}_1, \ldots, \mathbf{x}_n$, the sum $\hat{s}_n := \sum_{i=1}^{n} s(\mathbf{x}_i)$ is sufficient. Given a data block B, consisting of the observations $\{\mathbf{x}_{b_1}, \ldots, \mathbf{x}_{b_N}\}$, the summary variable $\mathbf{s}_B$ is computed as

$$\mathbf{s}_B := \sum_{i=1}^{N} s(\mathbf{x}_{b_i}) \, . \tag{12}$$

If $\mathbb{R}^d$ data, for example, is modeled by a Gaussian distribution, the summary variable will be the pair $\mathbf{s}_B = \left(\sum_{i=1}^{N} \mathbf{x}_{b_i}, \sum_{i=1}^{N} \mathbf{x}_{b_i} \mathbf{x}_{b_i}^{T} \right)$. Coarsening is therefore performed by averaging in parameter space, in contrast to the standard multiscale schemes used by many computer vision algorithms, which average in the data domain. A spatial partition of the input data into blocks $B_1, \ldots, B_m$ will result in a set of summary variables $\mathbf{x}_{B_1}, \ldots, \mathbf{x}_{B_m}$. The DPM sampling algorithm described in Sec. 2.3 is directly applicable to this replacement data, by substituting summary variables $\mathbf{x}_B$ for sufficient statistic values $s(\mathbf{x}_i)$.

The coarsening operation is *perfect* in the sense that it does, by the properties of sufficient statistics, preserve all information relevant for estimation purposes. More precisely, assume that $\mathbf{x}_{b_1}, \ldots, \mathbf{x}_{b_N} \sim F(\,.\,|\theta)$, with a conjugate prior $G(\theta|\lambda, \mathbf{y})$ on the parameter. Then the posterior Π satisfies the invariance

$$\Pi(\theta|\mathbf{s}_B; \lambda, \mathbf{y}) = \Pi(\theta|\mathbf{x}_{b_1}, \ldots, \mathbf{x}_{b_N}; \lambda, \mathbf{y}) \, , \tag{13}$$

since

$$\Pi(\theta|\mathbf{x}_{b_1}, \ldots, \mathbf{x}_{b_N}; \lambda, \mathbf{y}) = G\left(\theta \Big| \lambda \mathbf{y} + \sum_{i=1}^{N} s(\mathbf{x}_{b_i}) \right) = G(\theta|\lambda \mathbf{y} + \mathbf{s}_B) = \Pi(\theta|\mathbf{s}_B; \lambda, \mathbf{y}) \, .$$

Intuitively, a parameter estimated from data is a valid description of the data on the fine scale or *any* coarsened scale.

Refinement. The coarsening strategy described above and subsequent sampling on the coarse scale will result in a DPM clustering solution defined by cluster parameters $\theta_1^*, \ldots, \theta_k^*$. Because these parameters also define an admissible clustering solution on the fine (original) scale of the problem, it is not necessary to explicitly propagate coarse-scale assignments to the fine scale. Instead, a solution of the fine-scale problem is obtained by substituting the estimates θ_k^* into the fine-scale model and performing a single assignment step. We note that, as another consequence of the parametric description applying simultaneously over different coarsening scales, the method is capable of incorporating locally adaptive coarsening/refinement strategies.

Coarse-scale artifacts. When applied to clustering, the multiscale sampler may erroneously produce additional classes at a coarse scale, if a block summarized by a single variable during coarsening overlaps the boundary between two segments. The resulting mixed distribution may distinctively differ from the average distribution of both segments, and thus produce an additional cluster. Such

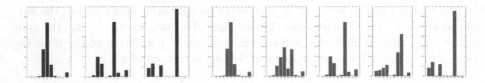

Fig. 1. Erroneous creation of classes during coarse-scale sampling: Average histograms of the three input classes (blue/left) and parameter vectors of the five classes estimated by coarse-scale sampling (red/right)

errors can be corrected by the fine-scale assignment step. To illustrate the behavior of the sampler, Fig. 1 shows estimation results obtained on a simple artificial image consisting of three block segments arranged in sequence (i.e. there are two boundaries between adjacent segments). Local windowed grayscale histograms are extracted as features. As clustering model, we apply a DPM model, with a multinomial likelihood F to account for the histograms. The cluster parameters θ_k^* can be interpreted as average histograms of the respective cluster. These average histograms are plotted for the true (generative) model on the left in Fig. 1. The multiscale algorithm is run on the data with a coarsening coefficient of 2, and the coarse-scale solution is compared to the artificial ground-truth. When sampling on a coarse scale, the algorithm models five classes, for which the class parameter vectors θ_k^* are plotted on the right. Clusters 2 and 4 are due to mixing of histograms at the segment boundaries. When assignments are performed for the fine-scale histograms to the coarse-scale class parameters, all histograms are correctly assigned to their original three classes, and clusters 2 and 4 remain empty. That is, coarse-scale artifacts vanish during the fine-scale assignment. The algorithm benefits from the ability of the DP to create new clusters for the boundary points, without distorting the remaining cluster structure.

Multiscale approaches to Markov Chain sampling have been considered e.g. in [14]. These methods are based on the idea that suitable coarse-scale formulations of a Markov chain may mix faster than the original chain, and use a coupled formulation integrating both chains. The aim is to reduce the number of iterations required for the algorithm to converge, while retaining accuracy. In contrast, our approach aims at reducing the execution time of individual iterations. Keeping in mind the large amounts of data arising in visual processing and video, we trade in accuracy and statistical guarantees for speed. Though the coarsening operation is perfect for individual distributions, it will lose in accuracy when the coarsening blocks overlap segment boundaries, as shown in the example. In practice, we should not expect the fine-scale assignment step to correct all errors. The rationale for risking a loss of accuracy is that, for vision applications, we put more emphasis on speed and plausible results than on statistical guarantees.

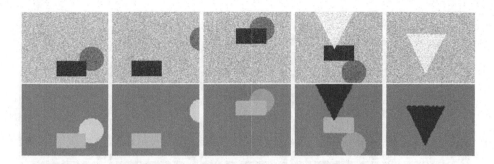

Fig. 2. Synthetic noisy data (top) and segmentation results (bottom). Clusters are correctly created or deleted as objects enter or leave the scene.

5   Experiments

This section provides experimental results for the application of our model to video segmentation. Experiments were performed on both synthetic data and real-world data (sequences from the MPEG4 benchmark set).

Processing pipeline. Features are extracted from each frame image by placing an equidistant grid within the image. A local window is placed around each grid node i, pixel values are extracted from within the window, and collected in a histogram (denoted $\mathbf{x}_i^t$ in the previous sections). For color images, the method is applied individually to each color channel. The resulting set of features for each frame is a list of multiple histograms, indexed by their position i within the image. On this data, the inference algorithm described in Sec. 4 is applied. Inference is conducted single-pass, and is hence capable of online processing. For each time step t, the assignment variables S_i^t describe the estimated segmentation (i.e. S_i^t is interpreted as the segment index of site i). In the examples shown in Figs. 2-5, segment assignments have been color-coded.

Results. Synthetic data experiments were conducted to verify the method's capability to adjust the number of clusters. The artificial data consists of simple geometric objects with additive Gaussian noise moving at random within a scene. Objects may newly appear or disappear, but only by entering or leaving the scene from the border (i.e. temporal changes are smooth). A sample experiment is shown in Fig. 2. Features used are local gray-scale histograms. In all experiments conducted, the algorithm consistently assigns each object to the same cluster over the whole running time of the sequence. The cluster number only changes if an object vanishes temporarily, either by occlusion or because it leaves the scene and reappears (as is the case for the disc in Fig. 2). As a mid-level vision algorithm, the method cannot (and should not) distinguish between initial appearance and reappearance of an object. Results on real video data were obtained on sequences from the MPEG4 benchmark set. Five feature channels where used: Four are histograms, representing the three RGB color

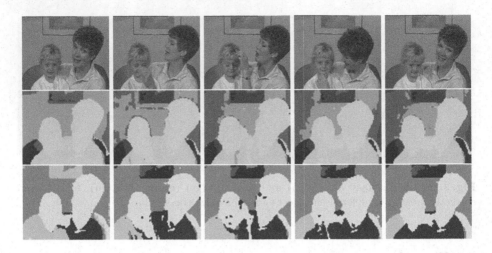

Fig. 3. "Mother and child" test sequence (top). Results are shown for color and saturation histogram features (middle), and with additional location features modeled by Gaussian distributions (bottom).

Fig. 4. "Table tennis" test sequence, with histogram and location features

Fig. 5. More difficult data: "Coastguard" test sequence (top) and segmentation results (bottom), with histogram and location features

channels plus saturation, and described by a multinomial likelihood F. In addition, we used a location feature, i.e. the center position of the local window in the image, represented by a two-dimensional Gaussian likelihood. The overall model likelihood is a product likelihood as described in Sec. 3. The scatter parameters of the parametric priors and the scatter parameter α of the DP, which control the level of cluster resolution, can be adjusted on the first few frames of the sequence. The key parameter of the feature extraction is the size of the local windows, which has to trade off sample size against precision: Large windows, to their advantage, contain many pixel examples, which results in stable histogram estimates and reduces scatter in feature space. Their drawback is a lack of precision: Large windows overlapping a cluster boundary generate histograms that represent a mixture of the two cluster distributions. Such mixtures tend to differ significantly from the individual distributions of the clusters, and hence cause additional clusters to appear at the segment boundaries. All results shown here were obtained using window sizes of 5×5 or 9×9 pixels. Sample results are shown in Figs. 3-5. In Fig. 3, results shown in the middle row where obtained using only the four histogram features (color and saturation). The background is split up into incoherent segments. Results are improved by the additional use of location features. Modeling these with a Gaussian in the spatial domain favors spatially coherent solutions, improving the segmentation of the background (bottom row). Likewise, all five features where used in the computation of results shown in Fig. 4. A more difficult sequence is shown in Fig. 5. In this case, local segmentation features provide poor information. Color differences within some segments (e.g. the large boat) are more significant than those between segments. With the size of the windows chosen sufficiently large during feature extraction to obtain stable input histograms, a boundary cluster effect is observable (note the two boats being split into an internal and a boundary segment). Results may possibly be improved by including additional motion features (such as histograms of frame differences). In general, the choice of features proves crucial for the performance of the segmentation algorithm. The parametric components of the clustering model (e.g. Gaussian and multinomial) are location-scatter type models, which represent "clouds" in their respective feature spaces. Like most mixture models, the method relies on the feature extraction step to map the segments to groups in feature space that are sufficiently well-separated to be resolved by the probabilistic model.

Average running times for our experiments on different video test sequences (300 frames at resolution 144×176 each) were: ~ 190 seconds at full resolution, ~ 110 seconds using a multi-scale sampler with coarsening coefficient 2, and ~ 35 seconds with a coarsening coefficient of 4. This does not include the feature extraction, i.e. the extraction of the fine-scale input data from the image sequence. The running time of the algorithm scales, in addition to the obvious dependence on the amount of input data, with the number of segments. The averages above were measured for relatively small numbers of classes ($N_C \leq 10$). If a large number of clusters is required (i.e. an over-segmentation), longer running times will have to be expected.

6 Conclusions

We have presented a clustering model for multivariate time series that represents temporally coherent sequences of clustering solutions. At each time index, the cluster structure is represented by a Bayesian mixture model, with a DP prior controlling the number of clusters. A conjugate model structure is used to propagate information along the time axis in a Bayesian framework. Two such structures are present in the model, one between parametric components (carrying parameter information of the mixture components), and one between the nonparametric DP components (carrying cluster structure information). To estimate the model from data, we have introduced an efficient Gibbs sampling approach which draws on both temporal context between frames and a multiscale approach for individual frames to increase efficiency. Applied to video segmentation, the model dynamically adjusts the number of clusters (segments) when new objects appear in or vanish from the scene. The achieved processing times of the sampler are just one order of magnitude below real time for videos in half-PAL format.

We have not provided convergence results for the sampling algorithm, since our analysis of the coarsening operation implies that, for individual component distributions, results for standard Gibbs samplers carry over to the multiscale approach. Any inaccuracies are due to coarsening blocks overlapping segment boundaries, a problem hard to capture by mathematical analysis. Furthermore, our estimation results are necessarily approximate, since the sampler is only run for a few steps on each frame image. Such an algorithm, for which the observed data changes (smoothly) in regular intervals raises some interesting questions. On the one hand, frame changes may pose a problem, if the model has not yet been sufficiently adapted to the current data. On the other hand, small data perturbations may help to avoid local minima. In-depth analysis of the algorithm is beyond the scope of this article.

The results presented here were computed on low-level features, such as color and saturation histograms, which necessarily results in limited precision. Since the model applies to both Gaussian and multinomial feature distributions, it is directly applicable to a wide range of features. Tracking applications, for example, require robustness but no coherent partition of the image. Hence, interest point features could be extracted on each frame and grouped with our model using a Gaussian likelihood supported on the frame image. Since DP models can be constrained by Markov random fields [15], the model itself can be extended by spatial smoothing, as advocated for video segmentation in [2]. Such an extension probably comes at the price of an increase in computation time, as more iterations per frame would be required for the smoothing to take effect.

The DP approach models the number of clusters as a random variable. The model order as an input parameter is replaced by a control parameter that allows the user to adjust the approximate level of cluster resolution. For fixed, static data sets, the approach may not constitute a practical advantage, as it arguable replaces one input parameter by another. We face a different situation for dynamic data, such as videos, where the number of clusters may change over time and the model has to adapt. Adaptation *requires* either a random description of the model order, or a transition heuristic (such as BIC scoring, or reversible jump in a Bayesian framework). Bayesian methods in general, and DP approaches in particular, are often regarded as inapplicable to data-intensive problems due to their computational costs. In our view, the reported results convincingly demonstrate that algorithmic efficiency need not pose an obstacle to the practical application of DP models, if temporal and spatial structure in the data can be exploited.

References

1. Tekalp, A.M.: Video segmentation. In: Bovik, A.C. (ed.) Handbook of Image and Video Processing, pp. 383–399 (2000)
2. Weiss, Y., Adelson, E.H.: A unified mixture framework for motion segmentation: Incorporating spatial coherence and estimating the number of models. In: Proc. of IEEE CVPR 1996 (1996)
3. Bregler, C.: Learning and recognizing human dynamics in video sequences. In: Proc. of IEEE CVPR 1997 (1997)
4. Khan, S., Shah, M.: Object based segmentation of video, using color, motion and spatial information. In: Proc. of IEEE CVPR 2001 (2001)
5. Goldberger, J., Greenspan, H.: Context-based segmentation of image sequences. IEEE Trans. on Pattern Analysis and Machine Intelligence 28(3), 463–468 (2006)
6. Bernardo, J.M., Smith, A.F.M.: Bayesian Theory. Wiley, Chichester (1994)
7. Ferguson, T.S.: A bayesian analysis of some nonparametric problems. Annals of Statistics 1(2) (1973)
8. Neal, R.M.: Markov chain sampling methods for Dirichlet process mixture models. Journal of Computational and Graphical Statistics 9, 249–265 (2000)
9. McLachlan, G.J., Krishnan, T.: The EM Algorithm and Extensions. Wiley, Chichester (1997)
10. Stoica, P., Selen, Y.: Model order selection: A review of information criterion rules. IEEE Signal Processing, 36–47 (July 2004)
11. Lange, T., Roth, V., Braun, M., Buhmann, J.M.: Stability-based validation of clustering solutions. Neural Computation 16(6), 1299–1323 (2004)
12. Walker, S.G., Damien, P., Laud, P.W., Smith, A.F.M.: Bayesian nonparametric inference for random distributions and related functions. Journal of the Royal Statistical Society B 61(3), 485–527 (1999)
13. MacEachern, S.N.: Estimating normal means with a conjugate style Dirichlet process prior. Communications in Statistics: Simulation and Computation 23, 727–741 (1994)
14. Higdon, D., Lee, H., Holloman, C.: Markov chain Monte Carlo-based approaches for inference in computationally intensive inverse problems. In: Bernardo, J.M., et al. (eds.) Bayesian Statistics vol. 7, pp. 181–198 (2003)
15. Orbanz, P., Buhmann, J.M.: Nonparametric Bayesian image segmentation. Int. J. of Computer Vision (to appear)

Dynamic Feature Cascade for Multiple Object Tracking with Trackability Analysis

Zheng Li[1,2], Haifeng Gong[2,3], Song-Chun Zhu[2,4], and Nong Sang[1]

[1] Institute for Pattern Recognition and Artificial Intelligence, Huazhong University of Science and Technology, China
[2] Lotus Hill Institute for Computer Vision and Information Science, China
[3] National Laboratory of Pattern Recognition, Institute of Automation, Chinese Academy of Sciences, China
[4] Departments of Statistics and Computer Science, University of California, Los Angeles
lizheng.lhi@gmail.com, hfgong.lhi@gmail.com, sczhu@stat.ucla.edu,
nsang@hust.edu.cn

Abstract. In multiple object tracking, the confusion caused by occlusion and similar appearances is an important issue to be solved. In this paper, trackability is proposed to measure how well given features can be used to find the correspondence of any given object in videos with multiple objects. Based on the analysis of trackability and computational complexity of the features under various occlusion conditions, a dynamic feature method cascade is presented to match the objects in consecutive frames. The cascade is composed of three tracking features: appearance, velocity and position. These features are enabled or disabled online to reduce computational complexity while obtaining similar trackability.

Experiments are conducted on 27062 frame occlusion objects, in the cases of good trackability, our experiments can obtain high succussful tracking rate with low computation burden, and in the cases of poor trackability, our estimation of trackability and confusion matrix can explain why they can not be tracked well.

Keywords: Trackability, Multiple Object Tracking, Feature Cascade, Confusion Matrix.

1 Introduction

Multiple object tracking is a challenging task in many computer vision applications such as smart surveillance. During occlusion, the correspondences of objects in consecutive frames may become uncertain. Fig. 1 shows an examples of occlusion objects tracking. How to measure the confusion among the multiple objects in tracking is an open problem now. In this study, object trackability is proposed to represent how well the object can be tracked using given features. Trackability is defined as the entropy of probability of the object in the current frame corresponding to each object in the next frame. For example, there are n objects in frame $t-1$ and m objects in frame t. One object in frame $t-1$

A.L. Yuille et al. (Eds.): EMMCVPR 2007, LNCS 4679, pp. 350–361, 2007.
© Springer-Verlag Berlin Heidelberg 2007

has correspondence probability to every object in frame t. If the distribution of the correspondence probability is dispersing, the entropy of the correspondence probability will be large, and the confusion degree is high. On the contrary, if the probability is converged, the entropy will be small, and one correspondence is dominant, the confusion is reduced and the trackability increases. So, the entropy is a good measure of the trackability.

The trackability is closely related to occlusion and features. Generally speaking, more features are introduced, and the objects are more distinguishable and more trackable, but more features need more computing. On the other aspect, as the degree of occlusion increases, the trackability decreases. But there are also interaction between features and degree of occlusion, because many feature such as appearance may be no longer credible when heavy occlusion occurs. To handle these issues, we analysis the effectiveness of various feature combinations under different occlusion conditions, and introduce a cascade of features for occluded object tracking. During the frame-to-frame tracking, we estimate the degree of occlusion and determine whether to enable or disable each feature, so the cascade is dynamic. Here, three features are used: appearance, velocity and position. In the case of slight occlusion, position and appearance is used to track the objects, in the case of partial occlusion, velocity is added to make the tracking more accurate; in the case of heavy occlusion, only position is used to predicate the correspondence.

In recent years, many efforts have been made to solve the occlusion tracking problem, but as far as we know, rare attention was paid to the analysis of the trackability of the scenes. Tao $et\ al.$[1] proposed motion layers which is a scene

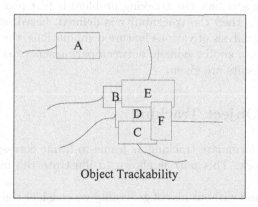

Fig. 1. During multiple objects tracking, one object is affected by other objects. In some situations, some objects are easy to be tracked well. While some objects are not. In this figure, the objects are represented as rectangle. The green lines represent object trajectories. Object A is easy to be tracked well. Because this object is not affected by other objects. Object B, C, D, E, F move together and occlude each other. The objects in this situation are not tracked well. Trackability is proposed to explain how well the object can be tracked at given situation.

representation with ordering information that contains complete information for inferring the foreground object and the background occlusion. Each moving object is modeled as a motion layer. Some background object may occluded foreground motion layers. Zhao and Nevatia[2] proposed an approach to detect and track multiple humans, especially occlusion humans in the scene. The human model is used to detect and track humans which overlap each other. Khan and Shah [3] proposed a framework to deal implicitly with occlusion and is able to correctly label people during occlusion. A person is segmented into classes of similar color using the Experctation Maximinzation algorithm first. This similar color classes has been tracked using maximum posterior probability approach. Elgammal *et al.* [4] proposed a method which tracks people using segmentation people under occlusion. Each people is initialized with an appearance model first. During occlusion, the overlapped people have been segmentation by using maximum likelihood estimation to estimate the segmentation. Then the relative depth is recovered. Perera *et al.* [5][6] proposed a trace linking method to solve the totally occlusion during long gaps. The two segments of the moving object trajectory is linked when the object has been occluded by the background for a long time.

Inferential uncertainty is an important problem in computer vision, which is related to the scene, the features we use and the information we want to obtain. Wu *et al.* [7] argued that entropy of the posterior probability is a good measure of the imperceptbility of the image understanding problem, which equals to the entropy difference of the distributions of the scene and the image. We want to measure the uncertainty of the correspondence of the objects in the consecutive frames, so we define the trackability using the entropy of the correspondence probability.

In the following sections, the tracking problem is first posed as occluded object matching task, then the trackability is defined, based on the effectiveness and performance analysis of various feature combinations, the feature cascade is proposed to obtain a good reconcile between performance and accuracy, and at last experiment results are given.

2   Multiple Object Tracking

In this paper, we formulate tracking as frame-to-frame correspondence just like much previous work. This part is shown to illustrate the method of tracking occlusion objects.

For simplicity and without loss of generality we consider the static background case. A Gaussian-mixture based adaptive background modeling algorithm[8][9] is used to detect foreground in the scene. Then morphological operators are used to remove the noise spots and fill the holes in the foreground part. Finally the method of connected component is used to label different parts of foreground mask as different moving objects. Ideally, one object will be labeled as one patch. When occlusion happens, more than one objects could be labeled together. Each labeled patch has an bounding box. These boxes have one (isolated object) or more objects (occlusion happens). Histogram match is used to track the isolated objects. Small motion is assumed in this paper. So in the adjacent frames the position and the size of the moving objects will not change a lot. The size of the isolated object is determined by the bounding boxes which are obtained from the labeled foreground patches. When the isolated objects moving across scales, the size of the objects changes gradually. The bounding boxes of the objects are also changed to adapt the size of the objects changing. The occluded and occluding objects are matched using the feature cascade proposed in this paper. The cascade is used to match each object in previous frame with sliding patch in the compounded bounding box. For example, in Fig. 3, the object 1 in frame t is tracked using histogram match and object 2 and 3 are tracked using our feature cascade.

Degree of occlusion. For one object is the measurement of the degree that the object is overlapped with other objects. In this paper, each object is represented as a bounding box which is the outer rectangle of the object region. During object occlusion, the bounding boxes of related objects are overlapped.In Fig. 2, each object is denoted as a bounding box and overlapped region is shaded. Degree of occlusion for object A and B is defined as $O_A = \frac{Area(A \cap B)}{Area(A)}$ and $O_B = \frac{Area(A \cap B)}{Area(B)}$ respectively.

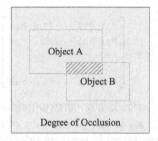

Fig. 2. Two objects move in the scene. Each object is represented as a bounding box. When occlusion happens, the two objects overlap each other. The degree of occlusion for one object is the ratio of the size of the overlapped part to the real part of the object.

3 Object Trackability

3.1 Definition of Trackability

Assume there are n objects in frame $t - 1$ and m objects in frame t. When the objects in frame $t - 1$ disappear in frame t, the death events d are introduced. We define the correspondence between frame $t - 1$ and t as the transitional **correspondence matrix Γ_t**.

$$\Gamma_t = \begin{bmatrix} \gamma_{11} & \gamma_{12} & \cdots & \gamma_{1m} & \gamma_{1d} \\ \gamma_{21} & \gamma_{22} & \cdots & \gamma_{2m} & \gamma_{2d} \\ \vdots & \vdots & \ddots & \vdots & \vdots \\ \gamma_{n1} & \gamma_{n2} & \cdots & \gamma_{nm} & \gamma_{nd} \end{bmatrix} \tag{1}$$

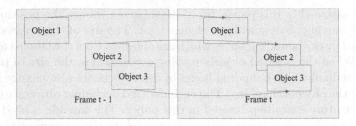

Fig. 3. After obtaining the outer rectangles in frame t, object 1 is tracked using histogram match. The occluded and occluding objects which are object 2 and 3 are tracked using the method in this paper.

where

$$\gamma_{i,j} = \begin{cases} 1 & \text{If } B_i^{t-1} \text{ and } B_j^t \text{ are correspondence to the same object} \\ 0 & \text{Otherwise} \end{cases} . \quad (2)$$

We define the **confusion matrix $\mathbf{S_t}$** to denote the probability of object correspondence.

$$\mathbf{S_t} = \begin{bmatrix} s_{11} & s_{12} & \cdots & s_{1m} & s_{1d} \\ s_{21} & s_{22} & \cdots & s_{2m} & s_{2d} \\ \vdots & \vdots & \ddots & \vdots & \vdots \\ s_{n1} & s_{n2} & \cdots & s_{nm} & s_{nd} \end{bmatrix} \quad (3)$$

where $s_{ij} = p(\gamma_{ij} = 1 | B_i^{t-1}, B_j^t)$ is the correspondence probability between the object B_i^{t-1} and object B_j^t, s_{id} is the probability of object death. If any object in frame $t-1$ does not have possible correspondence to the objects in frame t, it will be assigned to object death rather than an unlikely correspondence. The **intrackability** of the object i in frame $t-1$ can be computed by

$$\mathcal{IT}(i) = -\sum_{k \in \{1,\dots,m,d\}} s_{i,k} \log s_{i,k} \quad (4)$$

and we also introduce **trackability** to normalize the intrackability to $(0,1]$

$$\mathcal{T}(i) = \frac{1}{\mathcal{IT}(i) + 1} \quad (5)$$

Trackability is a measurement of how well the object can be tracked at given situation. The tracking features of appearance information, object velocity and object position are all used to estimate the correspondence probability. The trackability is computed using these correspondence probability.

In the case of single hypothesis tracking, we just choose the most dominant correspondence as candidate match, so **successful tracking rate** is a good indicator of trackability. Under good trackability, there are only one dominant correspondence, so the successful tracking rate is high, when the trackability is poor, many nearly equal peaks may present in the correspondence probability, and the successful tracking rate drops down (See Fig. 4). Successful tracking rate is defined as the ratio of the number of successul tracked objects in all frames and the total number of objects presented in all frames.

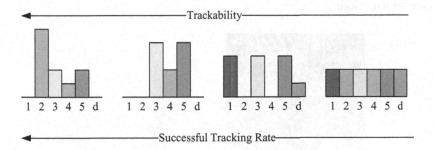

Fig. 4. In the case of single hypothesis tracking, successful tracking rate is a good indicator of trackability

3.2   Tracking Features

There are three tracking features are used in this paper: appearance, velocity and position.

Appearance. Each moving object has a block of image patch to record the appearance information. Let A_i^t represents the appearance information of object i in frame t. The more similar the appearance of the two objects is, the larger the energy is. The energy represents that the two objects are correspondence to one object. Self-correlation are used to compute the similarity of the two objects.

$$T(A_i^{t-1}, A_j^t) = \frac{\sum_{(x,y)\in A_i^{t-1}, (x',y')\in A_j^t} (A_i^{t-1}(x,y) * A_j^t(x',y'))}{\sqrt{(\sum_{(x,y)\in A_i^{t-1}} A_i^{t-1}(x,y)^2) * (\sum_{(x',y')\in A_j^t} A_j^t(x',y')^2)}} \quad (6)$$

During occlusion, the appearance information is occluded or disturbed. So, the appearance energy of the two objects is computed as follows:

$$E_A(A_i^{t-1}, A_j^t, O_j) = -T(A_i^{t-1}, A_j^t) * (1 - O_j) \quad (7)$$

where O_j is the degree of occlusion of object j defined in previous section.

Velocity. The velocity field in frame t are computed by Lucas-Kanade optical flow method. Let V_i^{t-1} denote the velocity of the object i in frame $t - 1$, which is estimated by the recent few frames. Let V_j^t denote the velocity of the object j in frame t, which is computed by the average velocity of the optical flow field covered by bounding box j. The average velocity is affected by occlusion. So the energy that the two objects are correspondence to one object is computed as follows:

$$E_V(V_i^{t-1}, V_j^t, O_j) = (||V_{i,x}^{t-1} - V_{j,x}^t|| + ||V_{i,y}^{t-1} - V_{j,y}^t||) * (1 - O_j) \qquad (8)$$

where $||V_{i,x}^{t-1} - V_{j,x}^t||$ and $||V_{i,y}^{t-1} - V_{j,y}^t||$ are the absolute value of the object velocity subtraction.

Fig. 5. Compute the velocity by optical flow

Position. Let P_i^t denote the position of object i in frame t. The predicted position of object i in frame t is computed by $\hat{P}_i^t = P_i^{t-1} + V_i^{t-1}$. Then the predicted position is compared with the position of bounding box j in frame t. The closer the predicted position of object i and the position of object j are, the larger the probability that the two objects are correspondence to the same object. The position engery is computed as follows:

$$E_P(P_i^{t-1}, P_j^t) = Dis(\hat{P}_i^t, P_j^t) \qquad (9)$$

where $Dis(\hat{P}_i^t, P_j^t)$ is the distance of the position $\hat{P}_i^t$ and P_j^t.

3.3   Computation of Trackability

In frame t, object state is denoted as $B_j^t = (A_j^t, V_j^t, P_j^t, O_j)$, where O_j is the degree of occlusion of object j. When occlusion happens, the object appearance information and velocity information (the velocity is estimated by velocity field) are disturbed. With the increasing of degree of occlusion, the reliability of the two tracking features decrease. So degree of occlusion is considered to compute object trackability.

$$p(\gamma_{ij} = 1 | B_i^{t-1}, B_j^t) \qquad (10)$$

$$= p(\gamma_{ij} = 1 | A_i^{t-1}, A_j^t, V_i^{t-1}, V_j^t, P_i^{t-1}, P_j^t, O_j) \qquad (11)$$

$$= p(\gamma_{ij} = 1 | A_i^{t-1}, A_j^t, V_i^{t-1}, V_j^t, \hat{P}_i^t, P_j^t, O_j) \qquad (12)$$

$$\propto \exp\left[-\lambda_A E_A(A_i^{t-1}, A_j^t, O_j) - \lambda_V E_V(V_i^{t-1}, V_j^t, O_j) - \lambda_P E_P(P_i^{t-1}, P_j^t)\right] \quad (13)$$

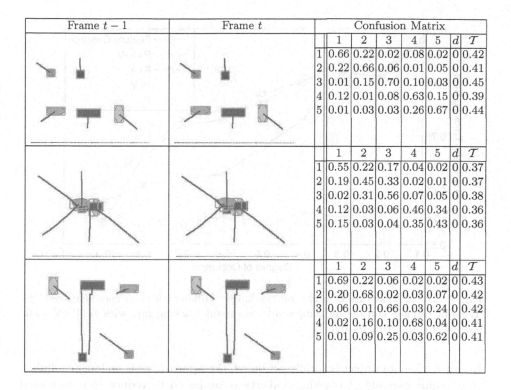

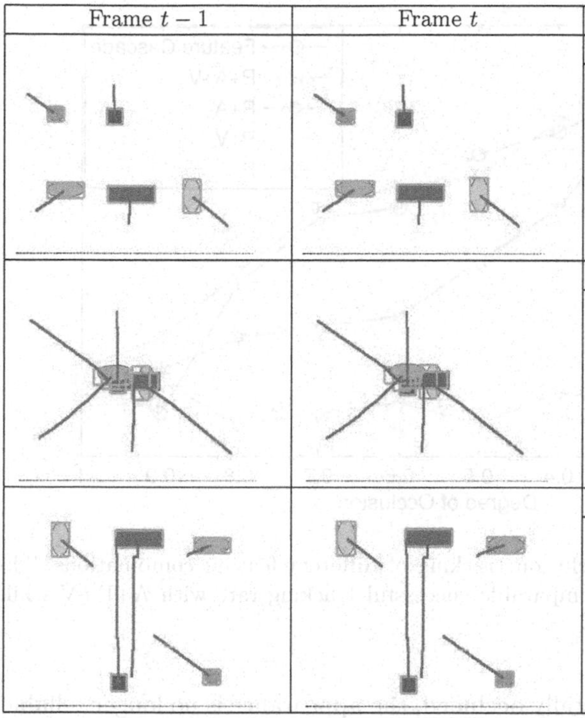
Frame $t-1$
Frame t

Confusion Matrix

	1	2	3	4	5	d	T
1	0.66	0.22	0.02	0.08	0.02	0	0.42
2	0.22	0.66	0.06	0.01	0.05	0	0.41
3	0.01	0.15	0.70	0.10	0.03	0	0.45
4	0.12	0.01	0.08	0.63	0.15	0	0.39
5	0.01	0.03	0.03	0.26	0.67	0	0.44

	1	2	3	4	5	d	T
1	0.55	0.22	0.17	0.04	0.02	0	0.37
2	0.19	0.45	0.33	0.02	0.01	0	0.37
3	0.02	0.31	0.56	0.07	0.05	0	0.38
4	0.12	0.03	0.06	0.46	0.34	0	0.36
5	0.15	0.03	0.04	0.35	0.43	0	0.36

	1	2	3	4	5	d	T
1	0.69	0.22	0.06	0.02	0.02	0	0.43
2	0.20	0.68	0.02	0.03	0.07	0	0.42
3	0.06	0.01	0.66	0.03	0.24	0	0.42
4	0.02	0.16	0.10	0.68	0.04	0	0.41
5	0.01	0.09	0.25	0.03	0.62	0	0.41

Fig. 6. The toy example with 5 objects. These objects move together and then apart. The images from top to bottom show this process. Before and after occlusion, the trackability of each object is higher than that of the object during occlusion.

A toy exmaple with 5 objects is shown in Fig. 6. They move together then depart. In the top and bottom image, thc distance of each objects is farther and do not occlude each other, so the objects have little effect of multiple object confusion and occlusion. The distributions of the correspondence probability of the objects are converged and one correspondence is dominant. So their trackabilities are good. While in the middle image, the object move together. They occlude each other and confusion arise. No dominant correspondence for one object in this situation. So their trackabilities are is poor. The objects in such situations are hard to be tracked well.

4 Dynamic Feature Cascade

During occlusion objects tracking, the more tracking features are chosen, the better tracking results can be obtained. But the performance will drop with more tracking features used. And the reliability of the tracking features such as appearance and velocity decreases with the degree of occlusion increasing.

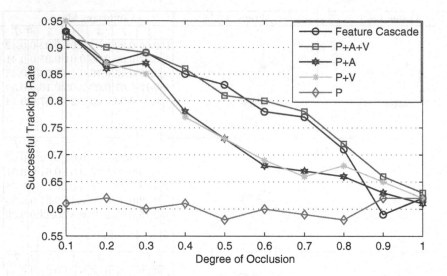

Fig. 7. The success rate of occlusion tracking of different feature combinations. The dynamic cascade can obtain comparable successful tracking rate with A+P+V while has lower computing burden.

Because when an object is totally occluded, the appearance is no longer reliable. A dynamic cascade of tracking features is proposed to reduce computational complexity when possible and to increase trackability when occlusion happens. At different degree of occlusion, each feature is determined to be enable or disable by trackability and computational complexity analysis.

When occlusion happens, the appearance information will be occluded and be disturbed by the nearby object. So the reliability of appearance and velocity are not as good as those without occlusion.

The computational complexities of the three features are sorted as follows $O(V) > O(A) > O(P)$, where $O(V)$ is the computational complexity of pixel optical flow, $O(A)$ is the computational complexity of appearance match and $O(P)$ is the computational complexity of position prediction. $O(P)$ is much smaller than $O(A)$ and $O(V)$. Position prediction cost little time during tracking and is not affected by the degree of occlusion. So position prediction should be used for all degree of occlusion. The feature of appearance and velocity should be determined whether enable or disable.

Because successful tracking rate is a good indicator of trackability, we investigated the successul tracking rate of different feature combinations at different occlusion cases. One occlusion case is one object occlusion tracking for one frame. 27062 frame occlusion objects have been test for different feature combinations. From Fig. 7, we can see that the successful tracking rate of the feature combination of position + appearance and position + velocity are similar. But $O(V)$ is greater than $O(A)$. So velocity is not singly combined with position. Then we have three feature combinations which are shown as follows:

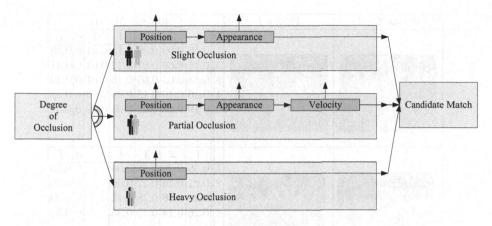

Fig. 8. The dynamic feature cascade

1. Position *(P)*
2. Position + Appearance *(P+A)*
3. Position + Appearance + Velocity *(P+A+V)*

When degree of occlusion is small, The successful tracking rates of combination *(P+A)* and *(P+A+V)* are similar and that of *(P)* is much smaller than those of *(P+A)* and *(P+A+V)*. So when degree of occlusion is small, the reliability of velocity is small. The tracking feature velocity is not used in this range of degree of occlusion.

As the degree of occlusion grows, the successful tracking rate of *(P+A+V)* is much larger than those of *(P)* and *(P+A)*, so all the tracking features should be used to obtain enough accuracy.

When the objects are almost totally occluded, the successful tracking rate of the three combinations are similar. But the computation complexity of *(P)* is much smaller than those of the other two, so only the position is used now.

During occluded objects tracking, the degree of occlusion of each object is increasing first, then begin to decrease. So in the tracking process, a step of estimating degree of occlusion will be done first. Then the tracking features are selected to be used. The tracking feature selection process is summarize in Fig. 8.

Two thresholds θ_1, θ_2 are used to divided the degree of occlusion into three levels. The degree of occlusion which is in the range $[0, \theta_1)$ is slight occlusion, while in the range $[\theta_1, \theta_2)$ is partial occlusion. The degree of occlusion which is in the range $[\theta_2, 1]$ is heavy occlusion. Thresholds θ_1, θ_2 are experientially decided considering both the successful tracking rate and computation complexity. In our experiments, the following values are chosen $\theta_1 = 0.3, \theta_2 = 0.8$.

5   Experiments

The algorithm is implemented in C++ program language and the experiments are conducted on computer with 2 Intel Pentium 2.8G CPUs and 1G memory.

Frame $t-1$	Frame t	Confusion Matrix							
			1	2	3	4	5	d	$\mathcal{T}$
		1	0.66	0.09	0.09	0.15	0.01	0	0.39
		2	0.11	0.69	0.07	0.12	0.01	0	0.41
		3	0.18	0.03	0.65	0.16	0.01	0	0.40
		4	0.16	0.10	0.15	0.69	0.03	0	0.38
		5	0.10	0.03	0.03	0.05	0.79	0	0.47
			1	2	3	4		d	$\mathcal{T}$
		1	0.39	0.20	0.33	0.08		0	0.35
		2	0.21	0.46	0.27	0.06		0	0.36
		3	0.33	0.16	0.38	0.13		0	0.35
		4	0.26	0.14	0.25	0.35		0	0.34

Fig. 9. Three humans move together in the scene. In the top images, there is no occlusion happens. So the trackability of each object is higher. While in the bottom images, the three humans walk together and occlude each other. So their trackability is lower.

Nearly 100 minutes of videos have been used to test the algorithm. These videos are taken from the street and are from different views. An avergage performance of 10 frames/second(320×288 pixels, color) can be reached.

There are 27062 frame occlusion objects which have different occlusion degree. The occlusion cases have been classified into ten categories according to degree of occlusion. And each category has ten percent degree of occlusion. Each method combination has been used to track occlusion objects. Then the successful tracking rate of each category for each feature combination can be obtained. This can be seen in Fig. 7.

Some typical frames are shown in Fig. 9, and the confusion matrix are at the right. Three humans move together in the scene. In the top images, there is no occlusion happens. So the trackability of each object is higher. While in the bottom images, the three humans walk together and occlude each other. So their trackability is lower.

6 Conclusion

Object trackability is proposed to measure how well the object can be tracked using given features. Three tracking features (appearance, velocity and position) are used to form feature combinations in this paper. Trackability is defined as the entropy of probability of the object in current frame corresponding to each object in the next frame. An example is shown in this paper to explain the trackability. A dynamic cascade structure for features is used to raise computation velocity. The degree of occlusion is estimated and the features are determined to be enable or disable. 27062 frame occlusion objects are used to get the successful tracking rate of different feature combinations. Several tracking results are shown in this paper.

Obviously, more tracking features can be easily integrated into our framework, such as shape, color, kinematic and so on, this will remain as our future work.

Acknowledgement

This work is done when the authors are at the Lotus Hill Institute for Computer Vision and Information Science. This work was supported by the National Natural Science Foundation of China under Contract 60575017, the Program for New Century Excellent Talents in University (NNCET-05-0641), and the Outstanding Youth Foundation of Hubei Province of China (2003ABB002).

References

1. Tao, H., Sawhney, H.S., Kumar, R.: Object tracking with bayesian estimation of dynamic layer representations. IEEE Transactions On Pattern Analysis And Machine Intelligence 1 (January 2002)
2. Zhao, T., Nevatia, R.: Tracking multiple humans in complex situations. IEEE Transactions On Pattern Analysis And Machine Intelligence 9 (September 2004)
3. Khan, S., Shah, M.: Tracking people in presence of occlusion. In: Asian Conference on Computer Vison (2000)
4. Elgammal, A.M., Davis, L.S.: Probabilistic framework for segmenting people under occlusion. In: Proc. ICCV (2001)
5. Kaucic, R., Perera, A.G.A., Brooksby, G., Kaufhold, J., Hoogs, A.: A unified framework for tracking through occlusions and across sensor gaps. In: Proc. CVPR (2005)
6. Perera, A.G.A., Srinicas, C., Hoogs, A., Glen Brooksby, W.H.: Multi-object tracking through simultaneous long occlusions and split-merge. In: Proc. CVPR (2006)
7. Wu, Y., Zhu, S., Guo, C.: From information scaling of natural images to regimes of statistical models. Quarterly of Applied Mathematics (2007) (to appear)
8. Stauffer, C., Grimson, E.: Adaptive background mixture models for real-time tracking. In: Proceedings of the IEEE Workshop On the Event Mining in Video (2003)
9. Stauffer, C., Grimson, E.: Learning patterns of activity using real-time tracking. IEEE Transactions On Pattern Analysis And Machine Intelligence 8 (August 2000)

Discrete Skeleton Evolution

Xiang Bai[1] and Longin Jan Latecki[2]

[1] Dept. of Electronics and Information Engineering, HuaZhong University of Science and
Technology, Wuhan, Hubei 430074, China
xiang.bai@gmail.com
[2] Dept. of CIS, Temple University, Philadelphia, PA 19122, USA
latecki@temple.edu

Abstract. Skeleton can be viewed as a compact shape representation in that the shape can be completely reconstructed form the skeleton. We present a novel method for skeleton pruning that is based on this fundamental skeleton property. We iteratively remove skeleton end braches with smallest relevance for shape reconstruction. The relevance of branches is measured as their contribution to shape reconstruction. The proposed pruning method allows us to overcome the instability of skeleton representation: a small boundary deformation leads to large changes in skeleton topology. Consequently, we are able to obtain very stable skeleton representation of planar shapes.

Keywords: Skeleton, skeleton pruning, shape reconstruction from skeleton.

1 Introduction

Skeleton, or *Medial Axis*, has been widely used for shape analysis and object recognition, such as image retrieval and computer graphics, character recognition, image processing, and analysis of biomedical images [1]. Although a lot of efforts were made to analysis the shape based on the skeletal trees/graphs [16][17][23][24][25][28], these approaches have only demonstrated applicability to objects with simple and distinctive shapes, and therefore, cannot be applied to more complex shapes like shapes in the MPEG-7 data set [20]. There are two main factors that constraint the performance of skeleton-based shape matching: 1) skeleton's sensitivity to object's boundary deformation: little noise or variation of boundary often generates redundant skeleton branches that may disturb the topology of skeleton's graph seriously; 2) the time cost for extraction of skeleton and matching skeleton trees/graphs cannot satisfy the requirement of fast shape retrieval. The performance of skeleton matching depends directly on the property of shape representation. Therefore, to prune the grassy skeletons into the visual skeletons is usually inevitable [11]. The goal of this paper is to introduce a novel method for skeleton pruning, called Discrete Skeleton Evolution (DSE). The motivation of DSE is that removing the end braches of skeletons iteratively will not change the topology of the original shape, which benefits from the recent work of Bai et al. [19]. We obtained in a natural way a hierarchical structure of simplified skeletons as illustrated in Fig. 1.

A.L. Yuille et al. (Eds.): EMMCVPR 2007, LNCS 4679, pp. 362–374, 2007.
© Springer-Verlag Berlin Heidelberg 2007

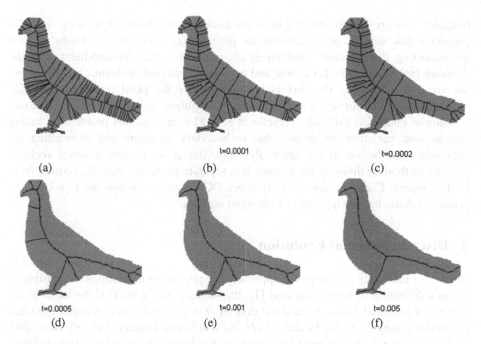

Fig. 1. The skeleton evolution process results in iterative pruning of the skeleton of a bird, (a) is the original skeleton and (b)~(f) are the pruned skeletons with the different thresholds

1.1 Related Work

Skeletonization approaches can be broadly classified into four types: thinning algorithms [5][6], discrete domain algorithms based on the Voronoi diagram [2][8][15], algorithms based on distance transform [3][4][7][9], and algorithms based on mathematical morphology [12][13][18].

All the obtained skeletons are subjected to the skeleton's sensitivity and many of them also include pruning methods along with the skeletonization. As an essential part of skeletonization algorithms, skeleton pruning algorithms usually appear in a variety of application-dependent formulations [11]. There are mainly two ways of pruning methods: (1) based on significance measures assigned to skeleton points [2] [3], [11], [27], and (2) based on boundary smoothing before extracting the skeletons [11], [21]. In particular, curvature flow smoothing still have some significant problems that make the position of skeletons shift and have difficulty in distinguishing noise from low frequency shape information on the boundary [11]. A different kind of smoothing is proposed in [10]. A great progress have been made in the type (1) of pruning approaches that define a significance measure for skeleton points and remove points whose significance is low. Shaked and Bruckstein [11] give a complete analysis and compare such pruning methods. To the common significance measures of skeleton points belong to propagation velocity, maximal thickness, radius function, axis arc length, the length of the boundary unfolded. Ogniewicz et al. [2] present a few significance measures for pruning hairy Voronoi skeletons without disconnecting the skeletons. Siddiqi et al. combine a flux measurement with the thinning process to extract a robust and accurate connected skeleton [22]. However, the error in calculating the flux is both limited by the pixel resolution and also proportional to the curvature of the boundary evolution front. This makes the exact location of endpoints difficult. Torsello et al. [18] overcome this problem by taking into account variations of density due to boundary curvature and eliminating the curvature contribution to the error. Recently, Bai et al. present a novel skeleton pruning method by dividing the contour into separate segments with the vertices from DCE (Discrete Curve Evolution) [19]. Since DCE does not change the topology, the pruned skeleton has the topology of the input skeleton.

2   Discrete Skeleton Evolution

Before we introduce the proposed approach, we give some definitions. According to Blum's definition of the medial axis [1], the skeleton S of a set D is the locus of the centers of maximal disks. A maximal disk of D is a closed disk contained in D that is interiorly tangent to the boundary ∂D and that is not contained in any other disk in D. Each maximal disc must be tangent to the boundary in at least two different points. With every skeleton point $s \in S$ we also store the radius $r(s)$ of its maximal disk.

By Theorem 8.2 in [26], the skeleton S is a geometric graph, which means that S can be decomposed into a finite number of connected arcs, called skeleton branches, composed of points of degree two, and the branches meet at skeleton joints (or bifurcation points) that are points of degree three or higher.

Definition 1. The skeleton point having only one adjacent point is an endpoint (the skeleton endpoint); the skeleton point having more than two adjacent points is a junction point. If a skeleton point is not an endpoint or a junction point, it is called a connection point. (Here we assume the curves of the skeleton is one-pixel wide)

Definition 2. A skeleton *end branch* is part of the skeleton between a skeleton endpoint and the closest junction point. Let l_i ($i = 1, 2, ..., N$) be the endpoints of a skeleton S. For each endpoint l_i, $f(l_i)$ denotes the nearest junction point. Formally, an *end branch* $P(l_i, f(l_i))$ is the shortest skeleton path between l_i and $f(l_i)$.

For example, in Fig. 2, arc from 1 to a is a skeleton end branch: $P(1, f(1)) = P(1,a)$. The arc from a to b is not an end branch; it is a skeleton (inner) branch. Observe that point a is the nearest junction point of two endpoints (1 and 7).

Based on Blum's definition of a skeleton, a skeleton point s must be the center of a maximal disk/ball contained in the shape D.

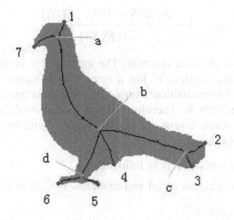

Fig. 2. The endpoints (red) and junction points (green) on the skeleton in Fig. 1(e)

Definition 3. Let $r(s)$ denotes the radius of the maximal disk $B(s, r(s))$ centered at a skeleton point s. The reconstruction of a skeleton S is denoted $R(S)$ and given by

$$R(S) = \bigcup_{s \in S} B(s, r(s)) \qquad (1)$$

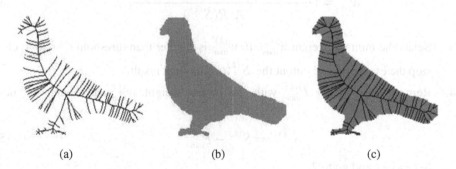

(a)   (b)   (c)

Fig. 3. The reconstruction (b) of the original skeleton (a) is very close to the original shape in (c)

As illustrated in Fig.3, we can reconstruct the original shape from its skeleton. Skeleton pruning can be seen as a simplification for skeleton, meanwhile, we hope the pruned skeleton can contain enough information of the shape.

In this paper, we introduce an iterative algorithm to prune the skeleton. In every step, we remove one end branch with the lowest weight. There are two motivations: 1) removing an end branch will not change the skeleton's topology; 2) the end branch with low contribution to the reconstruction is removed first.

We define the **weight** w_i for each end branch $P(l_i, f(l_i))$ as:

$$w_i = 1 - \frac{A(R(S - P(l_i, f(l_i))))}{A(R(S))} \tag{2}$$

where function $A(\ )$ is the area function. The intuition for skeleton pruning is that an end branch with a small weight w_i has a negligible influence on the reconstruction, since the area of the reconstruction without this branch is nearly the same as the area of the reconstruction with it. Therefore, it can be removed. The proposed skeleton pruning is based on iterative removal of end branches with the smallest weights until the desirable threshold is met.

The skeleton pruning algorithm is as follows:

1. We initialize the weights of all end branches $w_i^{(0)}$ ($i = 1, 2, \ldots, N^{(0)}$) based on the original skeleton $S^{(0)}$:

$$w_i^{(0)} = 1 - \frac{A(R(S^{(0)} - P(l_i^{(0)}, f(l_i^{(0)}))))}{A(R(S^{(0)}))} \tag{3}$$

2. In the kth iteration step, for $i = 1, 2, \ldots, N^{(k)}$ compute the weight for each end branch in the skeleton $S^{(k)}$:

$$w_i^{(k)} = 1 - \frac{A(R(S^{(k)} - P(l_i^{(k)}, f(l_i^{(k)}))))}{A(R(S^{(k)}))} \tag{4}$$

3. Select the minimal weight $w_{\min}^{(k)}$. If $w_{\min}^{(k)}$ is smaller than threshold t, go to 4; else, stop the evolution and output the $S^{(k)}$ as the final result.

4. Remove end branch $P_{\min}^{(k)}$ with the lowest weight $w_{\min}^{(k)}$ and obtain the new skeleton:

$$S^{(k+1)} = S^{(k)} - P_{\min}^{(k)} \tag{5}$$

5. Set $k=k+1$ and go to 2.

It is easy to see that this algorithm preserves the skeleton topology, since only end braches are removed. This fact is proven in Theorem 1, which states that $S^{(k+1)}$ is topologically equivalent to the original skeleton S.

Theorem 1. For every k, $S^{(k+1)} = S^{(k)} - P_{\min}^{(k)}$ is a strong deformation retract of S.

Proof: We will show that $S^{(k+1)}$, is a strong deformation retract of $S^{(k)}$. Since composition of strong deformation retractions is a strong deformation retraction, it follows that $S^{(k+1)}$, is a strong deformation retract of S.

We obtain $S^{(k+1)}$ by removing end branch $P(l_{min}, f(l_{min}))$ from $S^{(k)}$. Therefore, mapping $\pi\colon S^{(k)} \to S^{(k+1)}$ defined as identity on $S^{(k+1)}$ and $\pi(P(l_{min}, f(l_{min})))= f(l_{min})$ is a strong deformation retraction. This proves the theorem.

The robustness to noise and other boundary deformations of the proposed method follows form the fact that we compare the area. As it is well known, even significant contour noise has very small effect on the object area. We will demonstrate this fact in Section 3.1. Moreover, the area of articulated parts remains nearly constant, which makes the proposed skeleton pruning robust for articulated objects. For example, the obtained skeletons of the classes of Glas and Elephant in Fig.5 are with clear structures and insensitive to the boundary protrusions. Observe also that our approach does not shorten the remaining skeleton branches, since we only completely remove end branches. This is a desirable feature for object recognition, which we illustrate in Section 3.2.

3   Experiments and Discussions

In this section, we evaluate the performance of the proposed method in three parts. In Section 3.1, we demonstrate the stability to shape deformations and contour noise; In Section 3.2, we provide a comparison to other methods. In Section 3.3, we show that the proposed method is independent from the topology of the skeleton. This is in accord with Theorem 1, which guarantees topology preservation even in the presence of multiple loops.

3.1   Test on Kimia's Dataset

In this Section, we want to demonstrate the robustness to shape deformations and noise. We apply the proposed method to a well-known dataset for shape analysis

Fig. 4. Sample shapes in Kimia's dataset with original skeletons

(Kimia's 216 shapes), which consists of 18 classes with 12 shapes in each class [16]. The shapes in the same class are similar but slightly different, and the original skeletons are very grassy due to the many small noises or changes on their boundary as shown in Fig.4. We use the same threshold t for the shapes in the same class, and the values of t for all the classes are listed in Table.1. We put all the results in Fig.5. The obtained pruned skeletons in Fig.5 seem to be in accord with human perception and stable to the threshold t, which can be used for skeleton-based shape analysis efficiently.

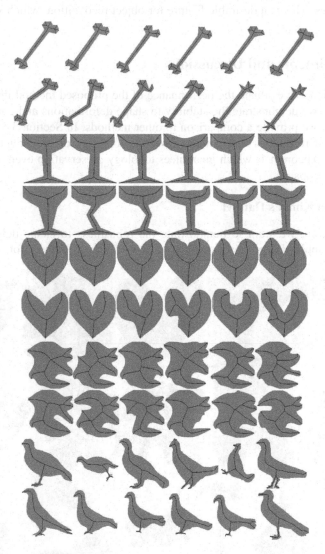

Fig. 5. The pruned skeletons of Kimia's dataset

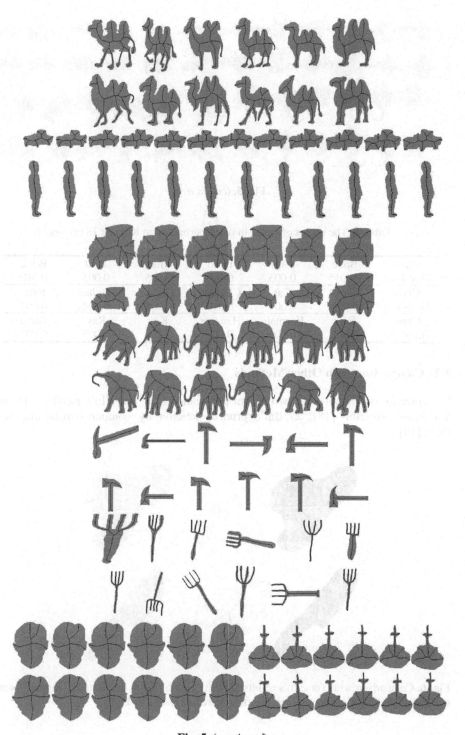

Fig. 5. (*continued*)

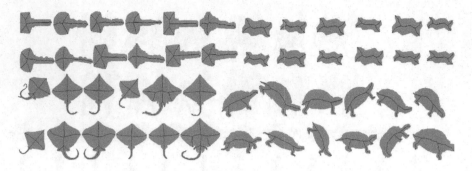

Fig. 5. (*continued*)

Table 1. The values of *t* used in our experiments on Kimia's 18 classes

Class	Bone	Glas	Heart	Misk	Bird	Brick
Threshold	0.0015	0.005	0.005	0.005	0.005	0.010
Class	Camel	Car	Children	Classic	Elephant	Face
Threshold	0.005	0.004	0.010	0.004	0.004	0.008
Class	Fork	Fountain	Hammer	Key	Ray	Turtle
Threshold	0.005	0.006	0.005	0.005	0.005	0.007

3.2   Comparison with Other Methods

We compare our result with two recent publications: 1) Torsello's modified Hamilton Jacob Skeleton [18]; 2) Bai's pruned skeleton by contour partitioning with DCE [19].

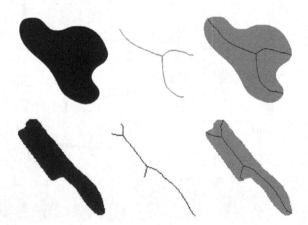

Fig. 6. Comparison with Torsello's results in [18]: column (a) are the example original shapes, column (b) are Tosello's results, and column (c) are our results by DSE

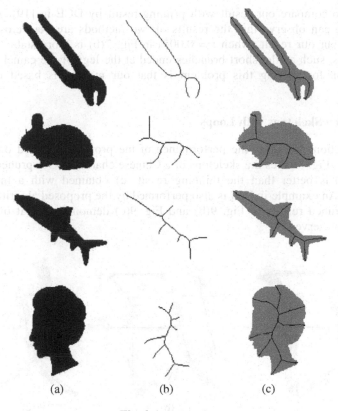

(a)   (b)   (c)

Fig. 6. (*continued*)

We test our algorithm on several representative shapes in Torsello's paper [18]. In Fig.6, it's easy to observe that our results are comparable to Torsello's results, and all the results are computed with the same t ($t = 0.005$). In addition, our method never shortens the skeleton branches. For example, in Torsello's results, some skeleton branches of women's head are much shorter than our result.

(a)   (b)

Fig. 7. Comparison with Bai's result in [19]

We also compare our result with pruning result by DCE in [19]. As shown in Fig. 7, we can observe that the results of two methods are very close with clear structure, but our result (when $t = 0.005$) in Fig. 7(b) is more stable to the small protrusions, such as the short branches ended at the legs of the camel in Fig. 7(a). The reason for solving this problem is that our method is based on skeletons directly.

3.3   Test on Skeletons with Loops

In this section, we show the performance of the proposed method on the images with holes. Fig. 8 gives the skeletons of a Chinese character. Our pruned skeleton in Fig. 8 (b) is better than the thinning result (c) obtained with a morphological thinning. An example in [19] is also performed by the proposed algorithm in Fig. 9, and the pruned results in Fig. 9(b) and Fig. 9(c) demonstrate that our method is topology preserved.

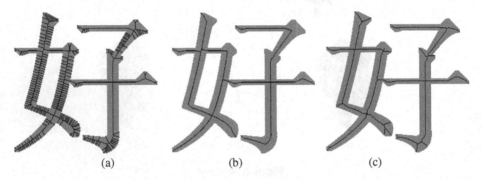

(a)   (b)   (c)

Fig. 8. The Skeletons of a Chinese character. (a) and (b) is the original skeleton and the pruned skeleton separately, and (c) is the result of morphological thinning.

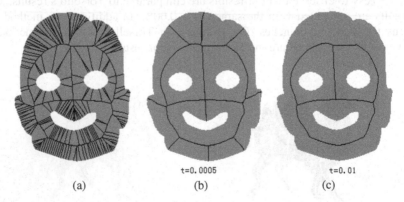

t=0.0005   t=0.01

(a)   (b)   (c)

Fig. 9. The skeletons of a face in [19]. (a) is the original skeleton, (b) and (c) are the pruned results with different thresholds.

3.4 Some Experimental Details

All the original skeletons in this paper were generated based on the distance transform by the algorithm in [4]. Our pruning algorithm is with high time cost due to the reconstruction in every iterative step, and the average time for pruning a noisy skeleton is about 4 minutes on the PC with 1.5 GHZ CPU and 512M RAM. Therefore, for increasing speed, another way is encouraged: First, prune the skeletons with the pruning algorithm in [19] in the coarse level; then, prune the pruned skeletons from algorithm [19] with the proposed algorithm. In this way, the average time cost for one pruning has been reduced to about 15 seconds.

4 Conclusions and Future Work

We present an iterative algorithm for pruning skeleton that is based on removing the end branch with the lowest weight for reconstruction in each step. The experiments prove that DSE is an efficient tool for skeleton-based representation and recognition. Even for different shapes of the same class in Kimia's dataset, we can use the same threshold, which is very important for automatic recognition with skeletons. Moreover, we have proved that DSE is topology preserved both in theorem and experiment. In the future, our work will focus on pruning 3D curve skeletons in analogous way.

Acknowledgments. We want to thank Zhuowen Tu for his useful comments and suggestions on this work. This work was supported by the Cultivating Fond for Momentous Scientific Innovation Project of Higher Education in China (Grant no. 705038) and in part by the NSF Grant IIS-0534929 and by DOE Grant DE-FG52-06NA27508.

References

1. Blum, H.: Biological Shape and Visual Science (Part I). J. Theoretical Biology 38, 205–287 (1973)
2. Ogniewicz, R.L., Kübler, O.: Hierarchic Voronoi skeletons. Pattern Recognition 28(3), 343–359 (1995)
3. Malandain, G., Fernandez-Vidal, S.: Euclidean skeletons. Image and Vision Computing 16, 317–327 (1998)
4. Choi, W.-P., Lam, K.-M., Siu, W.-C.: Extraction of the Euclidean skeleton based on a connectivity criterion. Pattern Recognition 36, 721–729 (2003)
5. Pudney, C.: Distance-Ordered Homotopic Thinning: A Skeletonization Algorithm for 3D Digital Images. Computer Vision and Image Understanding 72(3), 404–413 (1998)
6. Leymarie, F., Levine, M.: Simulating the grassfire transaction form using an active Contour model. IEEE Trans. Pattern Analysis and Machine Intell. 14(1), 56–75 (1992)
7. Golland, P., Grimson, E.: Fixed topology skeletons. In: CVPR, vol. 1, pp. 10–17 (2000)
8. Mayya, N., Rajan, V.T.: Voronoi Diagrams of polygons: A framework for Shape Representation. In: Proceedings of the IEEE Conference on Computer Vision and Pattern Recognition, pp. 638–643 (1994)

9. Ge, Y., Fitzpatrick, J.M.: On the Generation of Skeletons from Discrete Euclidean Distance Maps. IEEE Trans. Pattern Analysis and Machine Intell. 18(11), 1055–1066 (1996)
10. Gold, C.M., Thibault, D., Liu, Z.: Map Generalization by Skeleton Retraction. In: ICA Workshop on Map Generalization, Ottawa (August 1999)
11. Shaked, D., Bruckstein, A.M.: Pruning Medial Axes. Computer Vision and Image Understanding 69(2), 156–169 (1998)
12. Dimitrov, P., Damon, J.N., Siddiqi, K.: Flux Invariants for Shape. In: Int. Conf. Computer Vision and Pattern Recognition (2003)
13. Siddiqi, K., Bouix, S., Tannenbaum, A.R., Zucker, S.W.: Hamilton-Jacobi Skeletons. International Journal of Computer Vision 48(3), 215–231 (2002)
14. August, J., Siddiqi, K., Zucker, S.W.: Ligature Instabilities and the Perceptual Organization of Shape. Computer Vision and Image Understanding 76(3), 231–243 (1999)
15. Brandt, J.W., Algazi, V.R.: Continuous skeleton computation by Voronoi diagram. Comput. Vision, Graphics, Image Process 55, 329–338 (1992)
16. Sebastian, T.B., Klein, P.N., Kimia, B.B.: Recognition of shapes by editing their shock graphs. IEEE Trans. Pattern Anal. Mach. Intell. 26(5), 550–571 (2004)
17. Aslan, C., Tari, S.: An Axis-Based Representation for Recognition. ICCV, pp. 1339–1346 (2005)
18. Torsello, A., Hancock, E.R.: Correcting Curvature-Density Effects in the Hamilton-Jacobi Skeleton. IEEE Transaction on Image Processing 15(4), 877–891 (2006)
19. Bai, X., Latecki, L.J., Liu, W.Y.: Skeleton Pruning by Contour Partitioning with Discrete Curve Evolution. IEEE Trans. Pattern Anal. Mach. Intell. 29(3), 449–462 (2007)
20. Latecki, L.J., Lakämper, R., Eckhardt, U.: Shape Descriptors for Non-rigid Shapes with a Single Closed Contour. In: Proc. of IEEE Conf. on Computer Vision and Pattern Recognition (CVPR), pp. 424–429 (2000)
21. Mokhtarian, F., Mackworth, A.K.: A Theory of Multiscale, Curvature-Based Shape Representation for Planar Curves. IEEE Trans. Pattern Analysis and Machine Intelligence 14, 789–805 (1992)
22. Dimitrov, P., Phillips, C., Siddiqi, K.: Robust and EfficientSkeletal Graphs. In: Proc. IEEE Conf. Computer Vision and Pattern Recognition, pp. 1417–1423 (2000)
23. Demirci, F., Shokoufandeh, A., Keselman, Y., Bretzner, L., Dickinson, S.: Object Recognition as Many-to-Many Feature Matching. International Journal of Computer Vision 69(2), 203–222 (2006)
24. Siddiqi, K., Shokoufandeh, A., Dickinson, S.J., Zucker, S.W.: Shock Graphs and Shape Matching. International Journal of Computer Vision 35(1), 13–32 (1999)
25. Zhu, S.C., Yuille, A.L.: FORMS: A flexible object recognition and modeling system. Int. J. Comput. Vis. (IJCV) 20(3), 187–212 (1996)
26. Choi, H.I., Choi, S.W., Moon, H.P.: Mathematical Theory of Medial Axis Transform. Pacific J. Math. 181(1), 57–88 (1997)
27. van Eede, M., Macrini, D., Telea, A., Sminchisescu, C., Dickinson, S.: Canonical Skeletons for Shape Matching. ICPR, 64–69 (2006)
28. Tu, Z., Yuille, A.L.: Shape Matching and Recognition Using Generative Models and Informative Features. In: Pajdla, T., Matas, J(G.) (eds.) ECCV 2004. LNCS, vol. 3021, pp. 195–209. Springer, Heidelberg (2004)

Shape Classification Based on Skeleton Path Similarity

Xingwei Yang[1], Xiang Bai[2], Deguang Yu[1], and Longin Jan Latecki[1]

[1] Dept. of Computer and Information Sciences, Temple University, 1805 North Broad Street,
Philadelphia, PA 19122, USA
`{xingwei.yang,deguang,latecki}@temple.edu`
[2] Dept. of Electronics and Information Engineering, HuaZhong University of Science and
Technology, Wuhan, Hubei 430074, China
`xiang.bai@gmail.com`

Abstract. Most of the traditional methods for shape classification are based on contour. They often encounter difficulties when dealing with classes that have large nonlinear variability, especially when the variability is structural or due to articulation. It is well-known that shape representation based on skeletons is superior to contour based representation in such situations. However, approaches to shape similarity based on skeletons suffer from the instability of skeletons and matching of skeleton graphs is still an open problem. Using a skeleton pruning method, we are able to obtain stable pruned skeletons even in the presence of significant contour distortions. In contrast to most existing methods, it does not require converting of skeleton graphs to trees and it does not require any graph editing. We represent each shape as set of shortest paths in the skeleton between pairs of skeleton endpoints. Shape classification is done with Bayesian classifier. We present excellent classification results for complete shape.

Keywords: Skeleton pruning, skeleton path, Bayesian Classification.

1   Introduction

An important goal in image analysis is to classify and recognize objects. They can be characterized in several ways, using color, texture, shape, movement, and location. Shape, as a significant factor of objects, is an important research direction in image classification and recognition. Shape of planar objects can be described based on their contours or on skeletons.

When utilizing contours in classification and recognition, shape classes that have a large nonlinear variability of global shape, due to structural variation, articulation, or other factors, present a challenge for several existing shape recognition approaches. Approaches that match the target shape to stored example shapes require a large number of stored examples to capture the range of variability [1]. Furthermore, existing example-based and model-based approaches cannot handle object classes that have different parts or numbers of parts without splitting the class into separate subclasses. This type of structural variation can be handled by approaches that

A.L. Yuille et al. (Eds.): EMMCVPR 2007, LNCS 4679, pp. 375–386, 2007.
© Springer-Verlag Berlin Heidelberg 2007

represent part relationships explicitly and match shapes syntactically; however, these structural approaches are computationally expensive [2].

On the other hand, skeleton (or medial axis), which integrates geometrical and topological features of the object, is an important shape descriptor for object recognition [4]. Shape similarity based on skeleton matching usually performs better than contour or other shape descriptors in the presence of partial occlusion and articulation of parts [5][6][7][1]. However, it is a challenging task to automatically recognize the objects using their skeletons due to skeleton sensitivity to boundary deformation [8]. Usually the skeleton branches have to be pruned for recognition [8][9][10][11]. Moreover, another major restriction of recognition methods based on skeleton is a complex structure of obtained tree or graph representations of the skeletons. Graph edit operations are applied to the tree or graph structures, such as merge and cut operations [12][13][14][15][16], in the course of the matching process. Probably the most important challenge for skeleton similarity is the fact that the topological structure of skeleton trees or graphs of similar object may be completely different. Besides, some methods [21] have focused on utilizing geometry measures to gauge the similarity of 2D shapes by comparing their skeletons. This fact is illustrated in Fig. 1. Although the skeletons of the two horses (a) and (b) are similar, their skeleton graphs (c) and (d) are very different. This example illustrates the difficulties faced by approaches based on graph edit operations in the context of skeleton matching. To match skeleton graphs or skeleton trees like the ones shown in Fig. 1, some nontrivial edit operations (cut, merge, et al.) are inevitable.

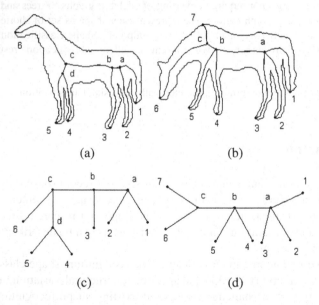

Fig. 1. Visually similar shapes in (a) and (b) have very different skeleton graphs in (c) and (d)

On the other hand, skeleton graphs of different objects may have the same topology as shown in Fig. 2. The skeletons of the brush in Fig. 2(a) and the pliers in Fig. 2(b) have the same topology as shown in Fig. 2(c).

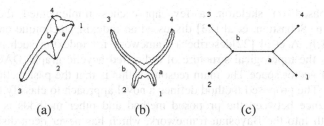

Fig. 2. Dissimilar shapes in (a) and (b) can have the same skeleton graphs (c)

The proposed method combines Bayesian classifier and a novel skeleton representation that overcomes the above limitations. This paper utilizes a three-level statistical framework including distinct models for dataset, class, and part. Bayesian inference is used to perform classification within this framework. Based on Bayes rule, the posterior probabilities of classes can be computed by the difference between skeletons of query shape and the shape in dataset. In the proposed framework, it can work well to classify complete shapes.

In section 2, the background of the related method will be discussed. The way to obtain and represent skeletons is introduced in section 3 and 4. The Bayesian framework is given in the section 5. In section 6, experimental results and analysis on two different datasets have been given. At last, conclusion and future work are drawn out.

2 Background

This section briefly introduces some recent methods developed for shape matching, including classification, detection, and retrieval. A number of approaches are based on the contour. Belongie et al. [1] proposed the concept of 'shape context', which are log-polar histograms among different points on the shape. Through finding the correspondence between points on different shapes, this approach can get the similarity between the shapes. Some methods used boosting to classify objects. Bar-Hillel, et al. [17] designed a classifier based on a part-based, generative object model. The approach given by Opelt, et al. [18] developed a novel learning algorithm which uses Adaboost to learn the shape features. Besides the learning algorithm, Gorelick et al. [19] used the Poisson Equation to extract various shape properties for shape classification. Tu and Yuille presented an algorithm for shape matching and recognition based on a generative model for how one shape can be generated by the other [26]. Sun and Super [3] used distribution of contour parts in known object classes to classify shapes with Bayesian classifier. Their classification works only for complete query shapes.

In contrast to the methods based on contour, many researchers have worked on the approaches based on skeleton. Zhu et al. matched the skeleton graphs of objects using a branch-bounding method that was limited to motionless objects [12]. Shock graph was a kind of ARG proposed by Siddiqi and Dickinson et al [24][25]., which was based on the shock Grammar. The distance between subgraphs was measured by comparing the eigenvalues of their adjacency matrices. Besides the methods for shape similarity based on skeleton, a few approaches implemented the skeleton in classification. Sebastian. et al [23] discussed an indexing technique on shock graph, Shokoufandeh, A et al [22]describe a framework for indexing such representations that embeds the topological structure of a directed acyclic graph (DAG) into a low-dimensional vector space The main reason for this is that the past methods have high complexity. The proposed method defines a novel approach to classify the shape. The main difference between the proposed method and other methods is it utilizes the skeleton path into the Bayesian framework, which has never been discussed before. The results are very promising and the complexity of the proposed method is much lower than current methods, such as shock graph [2].

3   Obtain Skeleton by Skeleton Pruning

Any topology preserving method can be used to compute skeletons. We used the method by Choi et al. [9].

The limitation of the skeleton is that it is sensitive to the boundary deformation and the noise. Therefore, it is difficult to obtain the ideal skeletons to recognize the objects. In order to solve this problem, this method utilizes skeleton pruning introduced in [10] to improve the skeleton. First, Discrete Curve Evolution (DCE) simplifies the polygon. Then the skeleton is pruned so that only branches ending at the DCE vertices remain. For example, in Fig. 3 (a) the skeleton contains a lot of noise. In Fig. 3 (b) all the endpoints (denoted by 1, 2, ..., 14) of the elephant's skeleton are vertices of the DCE simplified polygon (in red). The pruned skeleton is guaranteed to preserve the topology of the shape and it is robust to noise and boundary deformation [10]. Moreover, the skeleton endpoints are guaranteed to lie on the object contour.

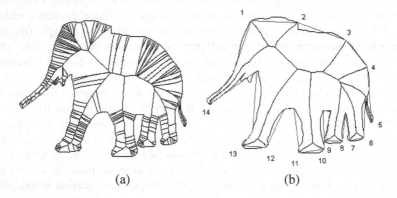

(a)   (b)

Fig. 3. (a) The original input skeleton. (b) The skeleton pruned with contour partitioning [10].

4   Shape Representation with Skeleton Paths

The endpoint in the skeleton graph is called an end node, and the junction point in the skeleton graph is called a junction node. The shortest path between a pair of end nodes on a skeleton graph is called a skeleton path. We show a few example skeleton paths in Fig 4.

Fig. 4. The elephant's skeleton and the shortest paths (in red) between the pairs of endpoints

The shortest paths between every pair of skeleton endpoints are represented as sequences of radii of the maximal disks at corresponding skeleton points.

Suppose there are N end nodes in the skeleton graph G to be matched, and let v_i (i = 1, 2, ..., N) denote the ith end node along the shape contour in the clockwise direction. Let sp(m, n) denote the skeleton path from v_m to v_n. We sample sp(m, n) with M equidistant points, which are all skeleton points. Let $R_{m,n}(t)$ denote the radius of the maximal disk at the skeleton point with index t of in sp(m, n). Let $R_{m,n}$ denote a vector of the radii of the maximal disks centered at the M sample skeleton points on sp(m, n):

$$R_{m,n}=(R_{m,n}(t))_{t=1,...,M}=(r_1,r_2,...,r_M) \qquad (1)$$

In this paper, the radius $R_{m,n}(s)$ is approximated with the values of the distance transform DT(s) at each skeleton point s. Suppose there are N_0 pixels in the original shape S. To make the proposed method invariant to the scale, we normalize $R_{m,n}(s)$ in the following way:

$$R_{m,n}(s) = \frac{DT(s)}{\dfrac{1}{N_0}\displaystyle\sum_{i=1}^{N_0} DT(s_i)} \qquad (2)$$

where s_i (i=1, 2, ..., N_0) varies over all N_0 pixels in the shape.

The shape dissimilarity between two skeleton paths is called a path distance. If R and R' denote the vectors of radii of two skeleton paths sp and sp' respectively, the path distance pd between sp and sp' is:

$$pd(R,R') = \sum_{i=1}^{M} \frac{(r_i - r_i')^2}{|r_i + r_i'|} \qquad (3)$$

5 Bayesian Classification

Compared to the method in [3], which uses contour segments and Bayesian classification to perform a recognition task, our method uses paths instead of contour segments. The basic idea is very simple, similar shape should have similar paths. Therefore, the difference of paths between similar shapes should be small. This Bayesian framework can obtain the classification by summing the difference of query shape's skeleton paths to the all of the shapes' skeleton paths in the same class. The smaller the difference is, the more possible the query shape belongs to the class.

Given a shape ω' that should be classified by Bayesian Classifier, we build the skeleton graph $G(\omega')$ of ω' and input $G(\omega')$ as the query. For a skeleton graph $G(\omega')$, if the number of end nodes is n, the corresponding number of paths is n(n-1) compared to the number of parts n! in [3]. Then, the Bayesian Classifier computes the posterior probability of all classes for each path $sp' \in G(\omega')$. By accumulating the posterior probability of all of the paths of $G(\omega')$, the system automatically yields the ranking of class hypothesis.

If two different paths have small pd value, the value of probability should be large. Otherwise, it should be small. Therefore, we use Gaussian distribution to compute the probability p:

$$p(sp' \mid sp) = \frac{1}{\sqrt{2\pi\alpha}} \exp(-\frac{pd(sp',sp)^2}{2\alpha}) \tag{4}$$

For different datasets, the α should be different. In our experiments, for the dataset of Aslan and Tari [20], $\alpha=0.15$ and $\alpha=0.05$ for Kimia dataset [2].

The class-conditional probability for observing sp' given that ω' belongs to class c_i is:

$$p(sp' \mid c_i) = \sum_{sp \in G(c_i)} p(sp' \mid sp) p(sp \mid c_i) \tag{5}$$

We assume that all paths within a class path set are equiprobable, therefore

$$p(sp \mid c_i) = 1/|G(c_i)| \tag{6}$$

c_i is one of the M classes.

The posterior probability of a class given that path $sp' \in G(\omega')$ is determined by Bayes rule:

$$p(c_i \mid sp') = \frac{p(sp' \mid c_i) p(c_i)}{p(sp')} \tag{7}$$

Similar to the above assumption, $p(c_i)=1/M$. The probability of sp' is equal to

$$p(sp') = \sum_{i=1}^{M} p(sp' \mid c_i) p(c_i) \tag{8}$$

Through the above formulas, we can get the posterior probability of all paths of $G(\omega')$. By summing the posterior probabilities of a class over the set of paths in the input shape, we obtain the probability that the input shape belongs to a given class. Obviously, the biggest one, Cm, is the class that input shape belongs to.

$$C_m = \arg\max_{i=1,...,M} \sum_{sp' \in G(\omega')} p(c_i \mid sp') \qquad (9)$$

6 Experiments

In this section, we evaluate the performance of the proposed method based on the dataset of Aslan and Tari [20]. We selected this dataset due to large variations of shapes in the same classes. As shown in Fig. 5, Aslan and Tari dataset includes 14 classes of articulated shapes with 4 shapes in each class. We use each shape in this dataset as a query, and show the classification result of our system in Fig. 6. We used leave one out classification, i.e., the query shape was excluded from its class.

Fig. 5. Aslan and Tari dataset [20] with 56 shapes

The table in Fig. 6 is composed of 14 rows and 9 columns. The first column of the table represents the class of each row. For each row, there are four experimental results which belong to the same class. Each experimental result has two elements. The first one is the query shape and the second one is the classification result of our system. If the result is correct, it should be the equal to the first column of the row. The red numbers mark the wrong classes assigned to query objects. Since there is only one error in 56 classification results, the classification accuracy in percentage by this measure is 98.2%. In fact, the only error is reasonable. Even a human can misclassify it. The query shape is very similar to star, the class 8. Therefore, in some sense, we can conclude that all of our results are correct.

class	query	result	query	result	query	result	query	result
1		1		1		11		9
2		2		2		2		2
3		3		3		6		3
4		4		4		4		4
5		5		5		5		5
6		6		6		6		6
7		7		7		7		7
8		8		8		8		8
9		9		9		9		9
10		10		10		10		10
11		11		11		11		11
12		12		12		12		12
13		13		8		13		13
14		14		14		14		14

Fig. 6. Results of the proposed method on Aslan and Tari dataset [20]. Since each class is composed of 4 shapes, the class of query and the result should be the same. Red numbers mark the results where this is not the case.

class	query	result	query	result	query	result	query	result
1		1		1		11		9
2		2		2		2		2
3		3		3		6		3
4		4		4		4		4
5		5		5		5		5
6		6		6		6		6
7		7		7		7		7
8		8		8		8		8
9		9		9		9		9
10		10		10		10		10
11		11		11		11		11
12		12		12		12		12
13		13		8		13		13
14		14		14		14		14

Fig. 7. Results of the Sun and Super's method on Aslan and Tari dataset [20]. Since each class is composed of 4 shapes, the class of query and the result should be the same. Red numbers mark the results where this is not the case.

We compared our method to the method presented by Sun and Super in [3], their method uses the same Bayesian classifier but is based on contour parts. As shown in Fig.7, their method yields 4 wrong results for 56 query shapes, so the accuracy is only 92.8%.

Moreover, the classification time for all 56 shapes with the proposed method takes only 5 minutes on the PC with 1.5 GHZ CPU and 512M RAM. However, Sun and Super's method takes 13 minutes on the same computer.

We also apply the proposed method to Kimia dataset [23] as shown in Fig. 8, which includes 18 classes, and each class is consisted of 12 shapes. In each experiment, we remove the query shape from the dataset Therefore there are 215 shapes in dataset and one query shape. Since there are only 12 errors in 216 classification results, the classification accuracy in percentage is 94.4%, which is comparable to Sun and Super's result [3]. Though the accuracy of Sebastian et al [23] on the dataset is 100 percent which is better than the proposed method, the proposed method is still promising. The classification time for all 216 shapes with the proposed method takes only nearly 25 minutes on the PC with 1.5 GHZ CPU and 512M RAM.

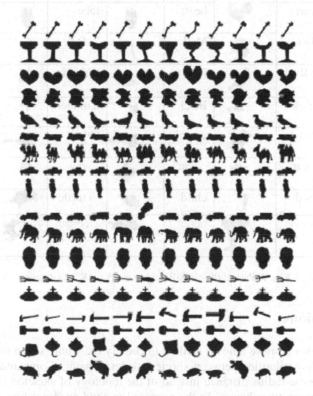

Fig. 8. Eighteen classes in Kimia dataset [23]

In Fig.9, we just give out 2 correct experimental results for each class and the last four images are chosen from the 12 error classifications, the wrong classification results are in red. The reason for the first wrong classification is the skeleton of head of the bird is similar to the same part of camel. For the glass, the skeleton is similar to the end of the bone. Moreover, the turtle is misclassified to elephant, as the tail part is like the same part of elephant. The Misk is like the brick in some sense, therefore the misclassification is reasonable.

Moreover, based on the classification results, the proposed method is rotation and scale invariant. In the experiment of the first dataset, Aslan and Tari dataset [21], the shapes have been rotated but the results are still correct. For the dataset of kimia [23], the size of the shape in the same class is different from each other. The proposed method can still obtain over 94 percent accuracy.

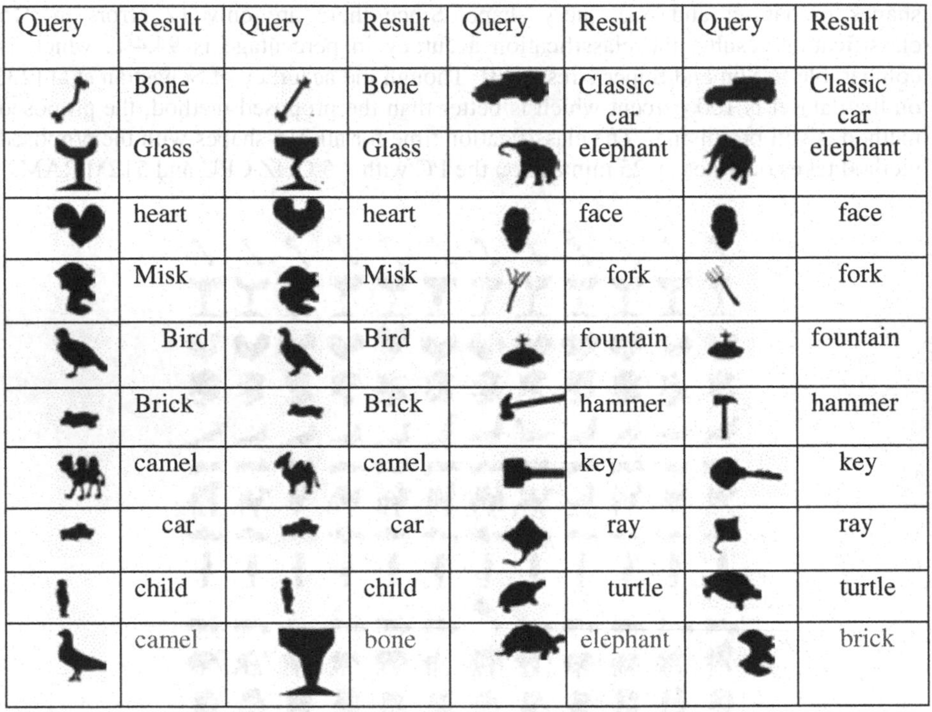

Query	Result	Query	Result	Query	Result	Query	Result
	Bone		Bone		Classic car		Classic car
	Glass		Glass		elephant		elephant
	heart		heart		face		face
	Misk		Misk		fork		fork
	Bird		Bird		fountain		fountain
	Brick		Brick		hammer		hammer
	camel		camel		key		key
	car		car		ray		ray
	child		child		turtle		turtle
	camel		bone		elephant		brick

Fig. 9. Part of the classification results on Kimia dataset

7 Conclusions

In this paper, we propose a novel method to classify the whole shape that is based on statistics of dissimilarities between shortest skeleton paths. Compared to the shock graph, we use the radius distance instead of the topology of skeleton to measure the similarity between two shapes. As the proposed method need not find corresponding between different skeleton paths, It avoids complex discussion on finding corresponding between two skeletons. Moreover, the result of two different datasets demonstrated that skeleton paths are very efficient shape representation for classification. However, as the probability of two paths is calculated based on the radius difference between two paths, if one shape's radiuses are totally different from other shapes in its class, the system may misclassify it to other classes. It is the drawback of the proposed method compared to the shock graph. In the future, our work will focus on solving this kind of problems and implementing the classification method in the part classification.

References

1. Belongie, S., Malik, J., Puzicha, J.: Shape matching and object recognition using shape contexts. IEEE Trans. PAMI 24(4), 509–522 (2002)
2. Sebastian, T., Klein, P., Kimia, B.: Shock-Based Indexing into Large Shape Databases. In: Tistarelli, M., Bigun, J., Jain, A.K. (eds.) ECCV 2002. LNCS, vol. 2359, pp. 731–746. Springer, Heidelberg (2002)
3. Sun, K.B., Super, B.J.: Classification of Contour Shapes Using Class Segment Sets. In: CVPR, pp. 727–733 (2005)
4. Blum, H.: Biological Shape and Visual Science. J. Theoretical Biology 38, 205–287 (1973)
5. Sebastian, T.B., Kimia, B.B.: Curves vs skeletons in object recognition. Signal Processing 85, 247–263 (2005)
6. Basri, R., Costa, L., Geiger, D., Jacobs, D.: Determining the Similarity of Deformable Shapes. Vision Research 38, 2365–2385 (1998)
7. Huttenlocher, D.P., Klanderman, G.A., Rucklidge, W.J.: Comparing Images Using the Hausdorff Distance. IEEE Trans. Pattern Analysis and Machine Intelligence 15(9), 850–863 (1993)
8. Shaked, D., Bruckstein, A.M.: Pruning Medial Axes. Computer Vision and Image Understanding 69(2), 156–169 (1998)
9. Choi, W.-P., Lam, K.-M., Siu, W.-C.: Extraction of the Euclidean skeleton based on a connectivity criterion. Pattern Recognition 36(3), 721–729 (2003)
10. Bai, X., Latecki, L.J., Liu, W.-Y.: Skeleton Pruning by Contour Partitioning with Discrete Curve Evolution. IEEE Trans. Pattern Analysis and Machine Intelligence 29(3), 449–462 (2007)
11. Ogniewicz, R.L., Kubler, O.: Hierarchic voronoi skeletons. Pattern Recognition 28, 343–359 (1995)
12. Zhu, S.C., Yuille, A.L.: FORMS: A flexible object recognition and modeling system. Int. J. Computer Vision (IJCV) 20(3), 187–212 (1996)
13. Liu, T., Geiger, D.: Approximate Tree Matching and Shape Similarity. In: Proc. Int. Conf. Computer Vision (ICCV), pp. 456–462 (1999)
14. Geiger, D., Liu, T., Kohn, R.V.: Representation and self-similarity of shapes. IEEE Trans. Pattern Anal. Mach. Intell. 25(1) (2003)
15. Di Ruberto, C.: Recognition of shapes by attributed skeletal graphs. Pattern Recognition 37(1), 21–31 (2004)
16. Pelillo, M., Siddiqi, K., Zucker, S.W.: Matching hierarchical structures using association graphs. IEEE Trans. Pattern Anal. Mach. Intell. 21(11), 1105–1120 (1999)
17. BarHill, A., Hertz, T., Weinshall, D.: Object class recognition by boosting a part based model. CVPR (2005)
18. Fussenegger, M., Opelt, A., Pinz, A., Auer, P.: Object Recognition Using Segmentation for Feature Detection. In: Proc. Int. Conf. Pattern Recognition (ICPR), pp. 41–44 (2004)

19. Gorelick, L., Galun, M., Sharon, E., Basri, R., Brandt, A.: Shape Representation and Classification Using the Poisson Equation. IEEE Trans. Pattern Analysis and Machine Intelligence 28(12), 1991–2005 (2006)
20. Aslan, C., Tari, S.: An Axis Based Representation for Recognition. In: ICCV, pp. 1339–1346 (2005)
21. Torsello, A., Hancock, E.R.: A skeletal measure of 2D shape similarity. Computer Vision and Image Understanding 95(1), 1–29 (2004)
22. Shokoufandeh, A., Macrini, D., Dickinson, S., Siddiqi, K., Zucker, S.W.: Indexing hierarchical structures using graph spectra. IEEE Trans. Pattern Analysis and Machine Intelligence 27(7), 1125–1140 (2005)
23. Sebastian, T.B., Klein, P.N., Kimia, B.B.: Recognition of shapes by editing their shock graphs. IEEE Trans. Pattern Analysis and Machine Intelligence 26(5), 550–571 (2004)
24. Siddiqi, K., Shokoufandeh, A., Dickenson, S.J., Zucker, S.W.: Shock graphs and shape matching. ICCV, 222–229 (1998)
25. Siddiqi, K., Shokoufandeh, A., Dickinson, S., Zucker, S.: Shock graphs and shape matching. International Journal of Computer Vision 30, 1–24 (1999)
26. Tu, Z., Yuille, A.L.: Shape Matching and Recognition Using Generative Models and Informative Features. ECCV, 195–209 (2004)

Removing Shape-Preserving Transformations in Square-Root Elastic (SRE) Framework for Shape Analysis of Curves

Shantanu H. Joshi[1], Eric Klassen[2], Anuj Srivastava[3], and Ian Jermyn[4]

[1] Dept. of Electrical Engineering, Florida State University, Tallahassee, FL 32310, USA
[2] Dept. of Mathematics, Florida State University, Tallahassee, FL 32306, USA
[3] Dept. of Statistics, Florida State University, Tallahassee, FL 32306, USA
[4] INRIA Sophia Antipolis, B.P. 93, 06902, Cedex, France

Abstract. This paper illustrates and extends an efficient framework, called the square-root-elastic (SRE) framework, for studying shapes of closed curves, that was first introduced in [2]. This framework combines the strengths of two important ideas - elastic shape metric and path-straightening methods - for finding geodesics in shape spaces of curves. The elastic metric allows for optimal matching of features between curves while path-straightening ensures that the algorithm results in geodesic paths. This paper extends this framework by removing two important shape preserving transformations: rotations and re-parameterizations, by forming quotient spaces and constructing geodesics on these quotient spaces. These ideas are demonstrated using experiments involving 2D and 3D curves.

1 Introduction

Shape analysis of closed curves, in two, three, or higher dimensions, has become an important topic of study. In particular, a large number of mathematical representations and metrics have been proposed to analyze shapes of such curves, albeit mostly in two-dimensional situations. Despite the large variety in metrics proposed, there is an emerging consensus on the suitability of the elastic metric for curve-shape analysis. This metric uses a combination of bending and stretching/compression to find optimal deformations from one shape to another. Additionally, it is invariant to re-parameterizations of curves. On pre-defined shape spaces of curves, these deformations are computed as shortest paths, or geodesics, under this chosen metric. This metric was suggested by Younes [13] and Mio et al. [6,7]; the latter developed a shooting method to compute geodesic paths between arbitrary shapes. Several other authors, including Michor and Mumford [5] and Shah [11], have also highlighted the advantages of this metric.

Although there is continuing research on various shape representations and metrics, we point out that the computational evaluations of different approaches are yet to be performed. For instance, we can ask the question: Amongst the different representations, such as coordinate functions, angle functions, curvature functions, log-speed functions, and deformation vector fields, introduced for shape analysis of parameterized curves, which is the most efficient one for elastic shape analysis? We elaborate this question further. Consider the representation of a planar curve β by its velocity vector $\dot{\beta}(s)$, seen

A.L. Yuille et al. (Eds.): EMMCVPR 2007, LNCS 4679, pp. 387–398, 2007.
© Springer-Verlag Berlin Heidelberg 2007

as a complex scalar $r(s)e^{i\theta(s)}$. Here $r(s)$ is the instantaneous speed and $\theta(s)$ is the angle made by $\dot{\beta}(s)$ with the positive X axis. Mio et al. [7] use the pair (ϕ, θ), with $\phi = \log(r)$, to represent and analyze the shape of β. Other researchers have used r directly, or its integral form $\int r(s)ds$, as representatives of speeds of curves. In this case, the elastic metric assumes a complicated form due to the requirement of invariance to parameterizations. Secondly, it may not be computationally efficient. As an example, in case of Mio et al., the elastic metric translates into the form,

$$\langle (h_1, g_1), (h_2, g_2)\rangle_{(\phi,\theta)} = \int h_1(s)h_2(s)e^{\phi(s)}ds + \int g_1(s)g_2(s)e^{\phi(s)}ds. \quad (1)$$

This metric given by Eqn. 1, under the (ϕ, θ) representation varies from point to point on the shape manifold, and is thereby complicated to implement.

Recently, Joshi et al. [2] proposed a new framework that uses the square-root of the speed of the curve and greatly simplifies the computation of geodesics. Additionally, it applies a more stable as well as an efficient, path-straightening approach for finding geodesics. This framework has been called the Square-Root Elastic (SRE) framework, since it uses the square-root representation to obtain an elastic analysis of curves. The SRE framework has the following advantages. Under this representation, the elastic metric reduces to a simple $\mathbb{L}^2$ metric. Not only is the metric the same at all points, but it is also much simpler to implement and study. The SRE framework combines the strengths of the elastic metric and the path straightening method for finding geodesics. Furthermore, there are convenient, isometric mappings from the proposed representation to other forms used previously. Finally, this approach is applicable to study of curves in $\mathbb{R}^n$ for all n and not just $n = 2$.

In this paper, our focus is on the shapes of curves, rather than the curves themselves. The paper [2] constructed a space $\mathcal{C}$ of closed curves in $\mathbb{R}^n$, and presented algorithms for computing geodesic paths between curves in $\mathcal{C}$. However, there are infinite number of elements in $\mathcal{C}$ that represent the same shape. The reason for this multiplicity is that rigid rotations and re-parameterizations of a curve can result in different elements of $\mathcal{C}$, but they all have the same shape. The sets of such representations are called orbits, and they form equivalent classes of shapes. For the purpose of shape analysis of curves, one needs to remove these shape-preserving transformations from the representation, and to compute geodesic paths in the resulting shape (quotient) space, called $\mathcal{S}$. The shape space is viewed as the quotient of $\mathcal{C}$ under the rotation $SO(n)$ and re-parametrization $\mathcal{D}$ groups, i.e. $\mathcal{S} = \mathcal{C}/(SO(n) \times \mathcal{D})$. To find geodesics in $\mathcal{S}$, one needs to find the shortest geodesic path(s) between elements of orbits in $\mathcal{C}$. Previous papers have used different techniques for removing these transformations. Since most of the past papers have studied curves in $\mathbb{R}^2$, where the rotation space is simply one-dimensional, one can do so using the exhaustive search [4]. The group of all re-parameterizations is often removed using the dynamic programming algorithm to match points across curves [7,10]. In this paper, we present a gradient approach that uses the differential geometry of these transformation groups, $SO(n)$ and $\mathcal{D}$, to match any two curves. The basic idea is to initialize a geodesic path in $\mathcal{C}$ between arbitrary elements of the two given orbits, and to use the gradient directions on these spaces to iteratively reduce the geodesic length until one reaches a geodesic path in $\mathcal{S}$.

The remainder of the paper is organized as follows. Section 2 summarizes the SRE framework for analyzing curves; it presents the square-root representation of curves and the path-straightening approach to finding geodesics in $\mathcal{C}$. Section 3 presents the main results of this paper: defining shape space $\mathcal{S}$ and computation of geodesic paths in $\mathcal{S}$. This is followed by some experimental results in Section 4 and a short summary in Section 5.

2   Square-Root Elastic (SRE) Framework

Next, we summarize ideas presented in [2] for elastic matching of closed curves in $\mathbb{R}^n$. There are two main ideas here: (i) the use of square-root representation of curves, and (ii) the use of path-straightening flows for computing geodesic paths. We summarize each of these ideas here, and refer the reader to [2] for further details.

2.1   Square-Root Representations of Curves

For the interval $I \equiv [0, 2\pi]$, let $\beta : I \to \mathbb{R}^n$ be a curve with a non-vanishing derivative everywhere. Denote its shape by the function $q : I \to \mathbb{R}^n$ as follows,

$$q(s) = \frac{\dot{\beta}(s)}{\sqrt{||\dot{\beta}(s)||}} \in \mathbb{R}^n . \tag{2}$$

Here, $s \in I$, $|| \cdot || \equiv \sqrt{(\cdot, \cdot)_{\mathbb{R}^n}}$, and $(\cdot, \cdot)_{\mathbb{R}^n}$ is taken to be the standard Euclidean inner product in $\mathbb{R}^n$. The quantity $||q(s)||$ represents the square-root of the instantaneous speed of the curve β, whereas the ratio $\frac{q(s)}{||q(s)||}$ is the unit tangent vector for each $s \in [0, 2\pi)$ along the curve. Indeed, the curve β can be recovered from q using $\beta(s) = \beta(0) + \int_0^s ||q(t)|| \, q(t) \, dt$. Let $\mathcal{Q} \equiv \{q = (q_1, \ldots, q_n) : I \mapsto \mathbb{R}^n | \forall i, \, q_i \in \mathbb{L}^2 \text{ and } \forall s, q(s) \neq 0\}$ be the space of all vector valued functions representing the curves as described above. This is an open subset of the infinite-dimensional vector space of all functions on I. Each element of this set represents an elastic curve on $\mathbb{R}^n$. Denote $\mathcal{B} \equiv \{q \in \mathcal{Q} | \int_0^{2\pi} (q(s), q(s))_{\mathbb{R}^n} ds = 1\}$ as the space of all unit-length, elastic curves. The closure condition for a curve β implies that $\int_0^{2\pi} \dot{\beta}(s) ds = 0$. For the square-root representation, this translates to $\int_0^{2\pi} ||q(s)||q(s) \, ds = 0$. Define a mapping $\mathcal{G} : \mathcal{Q} \mapsto \mathbb{R}^n$ according to $\mathcal{G}_i = \int_0^{2\pi} q_i(s) \, ||q(s)|| ds$. The space obtained by the inverse image $\mathcal{A} = \mathcal{G}^{-1}(0)$ is the space of all closed, elastic curves. Then, the subset $\mathcal{C} = \mathcal{A} \cap \mathcal{B} \subset \mathcal{Q}$ is the space of all unit-length, closed and elastic curves that are invariant to translation and scaling.

The length of a geodesic or the "shortest path" between two points on a manifold depends on the Riemannian metric, which is an inner product defined on each tangent space of the manifold. Thus we would like to construct a tangent space $T_q(\mathcal{C})$ at each point q. We observe that the tangent space of $\mathcal{Q}$ is the space of all vector valued functions with $\mathbb{L}^2$ components. We define an inner-product on $\mathcal{Q}$ as follows.

Definition 1. *Given a curve $q \in \mathcal{Q}$, and the tangent vectors $u, v \in T_q(\mathcal{Q})$ respectively, the inner product between u, v is defined as,*

$$\langle u, v \rangle = \int_0^{2\pi} (u(s), v(s))_{\mathbb{R}^n} \, ds. \tag{3}$$

In order to define the space of tangent vectors to $\mathcal{C}$, we derive the normal space of $\mathcal{C}$ at q first. As shown in [2], the normal space of $\mathcal{C}$ at q is

$$N_q(\mathcal{Q}) = \text{span} \left\{ q(s), \left(\frac{q_i(s)}{\|q(s)\|} q(s) + \|q(s)\| \mathbf{e}^i \right), i = 1, \ldots, n \right\},$$

where $\mathbf{e}^i$ is a unit vector in $\mathbb{R}^n$ along the i coordinate axis. Given a curve $q \in \mathcal{Q}$, and the tangent vector w to $\mathcal{Q}$ at q, the tangent space of $\mathcal{C}$ at q is defined as $T_q(\mathcal{C}) = \{w : I \to \mathbb{R}^n | w \perp N_q(\mathcal{Q})\}$.

2.2 Path Straightening Flows for Computing Geodesics

Given two curves q_0 and q_1, our goal is to find a geodesic between them in $\mathcal{C}$ under the Riemannian metric specified in Eqn. 3. In the path straightening approach, first introduced in [3], the given shapes are connected by an initial arbitrary path that is iteratively "straightened" so as to minimize its length. This iteration is performed using the gradient of an energy E, until one reaches a critical point of E. Let $\alpha : [0, 1] \to \mathcal{C}$ be any path in $\mathcal{C}$. Then, the critical points of the energy

$$E[\alpha] = \frac{1}{2} \int_0^1 \langle \dot{\alpha}(t), \dot{\alpha}(t) \rangle \, dt \tag{4}$$

are geodesics in $\mathcal{C}$ (see [12]). In order to minimize the integral in Eqn. 4, we need to find the gradient of the energy $E[\alpha]$ in the space of all paths on $\mathcal{C}$. For this purpose, we define $\mathcal{F}$ as the collection of all paths in $\mathcal{C}$, and $\mathcal{F}_0 \subset \mathcal{F}$ as the collection of all paths going from q_0 to q_1. Since each element along the path α is a curve in $\mathcal{C}$, the tangent space $T_\alpha(\mathcal{F})$ is written as $T_\alpha(\mathcal{F}) = \{w | w(t) \in T_{\alpha(t)}(\mathcal{C}) \, \forall t \in [0, 1]\}$. Following [3], we impose the Palais metric [8] on $T_\alpha(\mathcal{F})$ to result in a Riemannian structure on the space of all paths $\mathcal{F}$. For $u_1, u_2 \in T_\alpha(\mathcal{F})$, the Palais metric is given by the inner product,

$$\langle\langle u_1, u_2 \rangle\rangle = \langle u_1(0), u_2(0) \rangle + \int_0^1 \left\langle \frac{Du_1}{dt}(t), \frac{Du_2}{dt}(t) \right\rangle \, dt, \tag{5}$$

where $\frac{D}{dt}$ denotes a covariant derivative. The gradient of $E[\alpha]$ is a vector field in the tangent space $T_\alpha(\mathcal{F}_0)$, where $T_\alpha(\mathcal{F}_0) = \{w \in T_\alpha(\mathcal{F}) | w(0) = w(1) = 0\}$.

To derive the gradient vector field of $E[\alpha]$ on $T_\alpha(\mathcal{F})$, we state the following theorem from [3] without proof.

Theorem 1. *The gradient vector field of E is given by a v in $T_\alpha(\mathcal{F})$ such that $\frac{Dv}{dt} = \dot{\alpha}$, and $v(0) = 0$.*

Theorem 1 implies that the gradient of E in $T_\alpha(\mathcal{F})$ is given by covariantly integrating the velocity vector field $\frac{d\alpha(t)}{dt}$ along the curve α. Once the gradient in $T_\alpha(\mathcal{F})$ is obtained, it can be orthogonally projected into $T_\alpha(\mathcal{F}_0)$ to obtain the required gradient for iteratively updating α. This projection is specified using the following lemma.

Lemma 1. *The orthogonal complement of the tangent space $T_\alpha(\mathcal{F}_0)$ in $T_\alpha(\mathcal{F})$ is given by $T_\alpha^\perp(\mathcal{F}_0) \equiv \{w \in T_\alpha(\mathcal{F}) \mid \frac{D}{dt}\left(\frac{Dw}{dt}\right) = 0\}$.*

Using this Lemma, a tangent vector field $v \in T_\alpha(\mathcal{F})$ can be projected onto $T_\alpha(\mathcal{F}_0)$ by subtracting a component w of v that satisfies $\frac{D}{dt}\left(\frac{Dw}{dt}\right) = 0$; this property makes w a covariantly linear vector field. In our case, w is given by $t\tilde{v}(t)$, where $\tilde{v}(t)$ is a backward parallel transport of the vector field $v(1)$ along α. After obtaining the gradient of the energy $E[\alpha]$ in $\mathcal{F}_0$, we can update the path α in the direction of the gradient field v.

Now we can combine all the previous steps to compute geodesics in the space $\mathcal{C}$. The first step is the initialization of a path α on $\mathcal{C}$. Using the initialized path α, Algorithm 1 summarizes various steps using the path-straightening approach in computing the geodesic. The resulting geodesic distance between the two curves is then given by

---

Algorithm 1. Given $q_0, q_1 \in \mathcal{C}$, compute a geodesic between them

1: Initialize a path α between q_0 and q_1.
2: **repeat**
3:    Compute the path velocity $\alpha_t \equiv \frac{d\alpha}{dt}$ along α.
4:    Calculate the covariant integral (v) of $\frac{d\alpha}{dt}$.
5:    Parallel translate (backward) $v(1)$ along α as $\tilde{w}$.
6:    Form the gradient vector of E in $T_\alpha(\mathcal{F}_0)$ as $v = v - t\tilde{w}$.
7:    Update the path α in the direction v.
8:    Compute path energy $E = \frac{1}{2k}\sum_0^k \langle \alpha_t(\tau), \alpha_t(\tau) \rangle$.
9: **until** $\|\nabla E\| > \epsilon$

$\int_0^1 \sqrt{\langle \dot{\hat{\alpha}}(t), \dot{\hat{\alpha}}(t) \rangle}\, dt$, where $\hat{\alpha}$ is the resulting geodesic path.

3   Removing Shape Preserving Transformations

So far, we have constructed geodesics between a pair of curves in the $\mathcal{C}$. In doing so, we implicitly assumed that the starting points of both the curves were fixed, and the rotational alignment remained unchanged. However the "shape" of a curve remains invariant under rotations as well as the choice of starting point along the curve. Furthermore, the appearance of a curve, including its pose (scale, location, and orientation) is also invariant to the speed of traversal along the curve. In the following subsections, we define the space of elastic shapes, and outline an optimization algorithm that measures the "elastic" distance between curves under certain well-defined shape-preserving transformations.

3.1   Elastic Shape Space ($\mathcal{S}$)

The idea of matching shapes of objects by studying their deformations under appropriate group actions is well known [1]. Following a slightly different approach, we are interested in constructing the shape space as a quotient space of $\mathcal{C}$, modulo shape preserving transformations such as rigid rotations and re-parameterizations. In addition to translation and scaling, we identify the following re-parameterizations and group actions on the curve that preserve its shape.

1. **Placement of origin (seed):** A change in the starting point of the curve $q \in C$ is represented by the action of a unit circle $\mathbb{S}^1$ on q, according to $r \cdot q(s) = q((s - r)_{\mathrm{mod}\ 2\pi})$ for $r \in [0, 2\pi]$ with 0 and 2π identified.
2. **Rigid rotation:** A rigid rotation of a curve is considered as a group action by a n-by-n rotation matrix $O_n \in SO(n)$ on q, and is defined as $O_n \cdot q(s) = O_n q(s)$, $\forall s \in [0, 2\pi)$.
3. **Re-parametrization by speed:** A curve traveled at arbitrary speeds is said to be re-parameterized by a non-linear differentiable map γ (with a differentiable inverse) also referred to as a diffeomorphism. We define $\mathcal{D} = \{\gamma : \mathbb{S}^1 \to \mathbb{S}^1\}$ as the space of all orientation-preserving and origin-preserving diffeomorphisms. Then, the resulting variable speed parameterizations of the curve can be thought of as diffeomorphic group actions of $\gamma \in \mathcal{D}$ on the curve q. This group action is derived as follows. Let q be the representation of a curve β. Let $\alpha = \beta(\gamma)$ be a re-parametrization of β by γ. Then the respective velocity vectors can be written as $\dot{\alpha} = \dot{\gamma}\dot{\beta}(\gamma) = \dot{\gamma}q(\gamma)\|q(\gamma)\| = \|\sqrt{\dot{\gamma}}q(\gamma)\|\sqrt{\dot{\gamma}}q(\gamma)$. The re-parametrization of q by γ is defined as a right action of the group $\mathcal{D}$ on the set C and written as $q \cdot \gamma = \sqrt{\dot{\gamma}}\,(q \circ \gamma)$. Figure 1 shows an example of a re-parameterized curve by an arbitrary γ. The change in the speed due to re-parametrization is observed by discrete points plotted along the curve.

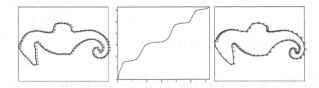

Fig. 1. From left, an unit-speed 2-D curve, γ acting on the curve, and the re-parameterized curve

Altogether, the set of curves affected by the group actions above, partition the space C into equivalent classes. We now define the elastic shape space as the quotient space $S = C/(\mathbb{S}^1 \times SO(n) \times \mathcal{D})$. The problem of finding a geodesic between two shapes in S is same as finding the shortest path between the equivalent classes of the given pair of shapes. Since the actions of the re-parametrization groups on C constitute actions by isometries, this problem also amounts to minimizing the length of the geodesic path, such that

$$d(q_0, q_1) = \min_{r \in \mathbb{S}^1, O_n \in SO(n), \gamma \in \mathcal{D}} d(q_0, (r \cdot O_n q_1) \cdot \gamma) \tag{6}$$

Conceptually, this involves finding an optimal rotational alignment ($\hat{O}$), seed ($\hat{r}$), and an optimal speed parameterization ($\hat{\gamma}$) that minimizes the distance given by Eqn. 6. As a result of this optimality, the geodesic path becomes orthogonal to the respective orbits of $SO(n)$ and $\mathbb{S}^1$. We propose an iterative solution using a gradient descent approach that successively causes the projection of the tangent vector ($\alpha_t(1)$) on each of these orbits, to be zero. In practice, we divide the problem in two steps. At every iteration, we will first find a geodesic in the quotient space under all rotations, $C/(\mathbb{S}^1 \times SO(n))$ and use it as an initial condition for computing the geodesic in the elastic shape space $S = (C/(\mathbb{S}^1 \times SO(n)))/\mathcal{D}$ until the algorithm converges.

3.2 Geodesics in $C/(\mathbb{S}^1 \times SO(n))$

We briefly discuss the computation of geodesics in the space C after removing all rotations and seed placements. We recall that the orbit of any element $q \in C$ under a group action $g \in G$ is the set $G_q = \{q \cdot g : g \in G\}$. Given a pair of shapes q_0 and q_1, the idea is to construct a tangent space of the orbit O_{q_1} of q_1 under the group action by $O_n \in SO(n)$ and iteratively make the projection of the tangent vector $\alpha_t(1)$ on $T_{q_1}(O_{q_1})$ to be zero. We will adopt the approach similar to Klassen et al. and refer the reader to [3] for details. Furthermore, the optimal seed is given by $\hat{r} = \mathrm{arginf}_r \langle q_0, r \cdot q_1 \rangle$. In practice, for a discrete representation (T samples), the optimal seed is given as $\hat{r} = \mathrm{argmin}_{r=1,\dots,T} \langle q_0, r \cdot q_1 \rangle$. Together, the above minimization approach yields a locally optimal alignment in terms of the rotation and placement of origin. Although this approach works for n-dimensional curves, one can adopt other efficient methods for low-dimensional curves. Particularly in the case of 2D curves, we can take advantage of the fact that the group of rotations $SO(2)$ is a 1-dimensional manifold. In this case, we can discretize the angle $\theta \in \mathbb{S}^1$ and using the map $\theta \mapsto \begin{bmatrix} \cos(\theta) & -\sin(\theta) \\ \sin(\theta) & \cos(\theta) \end{bmatrix}$, search in the orbit of O_{q_1} to achieve an optimal rotational alignment. As an example, Fig. 2 shows the variation of the geodesic path energy in C against the rotational alignment. The path corresponding to $O = O_{\mathrm{opt}}$ is the geodesic in $C/(\mathbb{S}^1 \times SO(n))$.

Fig. 2. Left: Path energy vs. the angle of rotation. Right: Geodesic Path corresponding to the optimal alignment.

3.3 Geodesics in S

In order to compute geodesics in the quotient space of elastic shapes, S, we follow a similar idea as above. This time, we define an orbit $\mathcal{D}_{q_1}$ of q_1 under the group action by $\gamma \in \mathcal{D}$. An optimal elastic alignment between any two shapes q_0 and q_1 is obtained, when the projection of the tangent vector $\alpha_t(1)$ on the subspace $T_{q_1}(\mathcal{D}_{q_1})$ is zero.

Instead of posing the optimization problem on the subspace $T_{q1}(\mathcal{D}_{q1})$, we consider the tangent space of $\mathcal{D}$ at identity $T_{id}(\mathcal{D})$, which is an $\mathbb{L}^2(\mathbb{S}^1)$ space and construct 1-parameter flows on $\mathcal{D}$ as follows. Let $\psi_t : T_{id}(\mathcal{D}) \to \mathcal{D}$ represent a 1-parameter flow at identity on $\mathcal{D}$ such that $\psi_0(id, g) = s$ for any tangent vector $g \in T_{id}(\mathcal{D})$. Next, we define the diffeomorphic group action $\phi : \mathcal{C} \times \mathcal{D} \to \mathcal{C}$ as $\phi_\gamma(q) = \sqrt{\dot\gamma}q(\gamma)$. Then the differential ϕ_* maps the tangent vector g to $\phi_*(g) \in T_q(\mathcal{D}_q)$ and is given by,

$$\phi_*(g) = q'(s)g(s) + \frac{1}{2}q(s)g'(s), \quad s \in [0, 2\pi) \tag{7}$$

Let $V \equiv \{v_i\}, i = 1, \ldots, d$ denote the Fourier basis for the tangent space of $\mathcal{D}$. Using the differential map $\phi_*(V)$, we can construct the basis for the tangent space $T_{q1}(\mathcal{D}_{q1})$. Following the previously outlined approach, we compute a geodesic between shapes represented by q_0 and q_1 in $\mathcal{C}/(\mathbb{S}^1 \times SO(n))$. In the process of computing this geodesic, the shape q_1 gets rotated and shifted as $\tilde{q}_1 = \hat{r} \cdot \hat{O}q_1$, where $\hat{r} \in \mathbb{S}^1$ and $\hat{O} \in SO(n)$. We then project the tangent vector $\alpha_t(1)$ on $T_{\tilde{q}_1}(\mathcal{D}_{\tilde{q}_1})$. For a tangent vector $\alpha_t(1) \in T_\alpha(\mathcal{F})$, its projection on $T_{\tilde{q}_1}(\mathcal{D}_{\tilde{q}_1})$ is defined as

$$\pi(\alpha_t(1)) = \sum_{i=1}^{d} \langle \alpha_t(1), \phi_*(v_i) \rangle \, \phi_*(v_i) \tag{8}$$

Using the inverse of ϕ_*, we can construct a tangent vector $g \in T_{id}(\mathcal{D})$ and compute 1-parameter flows on $\mathcal{D}$. The above procedure is repeated until the quantity $\langle \pi(\alpha_t(1)), \pi(\alpha_t) \rangle$ becomes zero. Shown in Fig. 3 is a cartoon diagram illustrating this process. The complete procedure of finding a geodesic in $\mathcal{S} \equiv \mathcal{C}/\mathcal{D}$ is described in Algorithm 2. As an example, Fig. 4 shows the comparisons between non-elastic

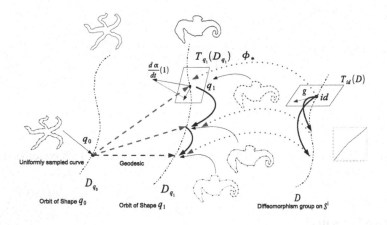

Fig. 3. Illustration of the process of finding geodesics in $\mathcal{S}$

geodesics computed using the method in [4] and elastic geodesics computed using algorithm 2.

Algorithm 2. Given two curves q_0 and q_1, compute a geodesic in $\mathcal{S}$

1: Find the geodesic between q_0 and q_1 in $\mathcal{C}/(\mathbb{S}^1 \times SO(n))$ using the approach outlined in Sec. 3.2. This also yields the tangent vector $\alpha_t(1)$ at q_1.
2: Let $\{v_i\}, i = 1, \ldots, d$ be the Fourier basis for $T_{id}(\mathcal{D})$.
3: Project the vector $\alpha_t(1)$ on $T_{q_1}(\mathcal{D}_{q_1})$ using Eqn. 8.
4: **if** $\|\pi(\mathbf{u})\|^2 < \epsilon$ **then**
5:     Stop.
6: **end if**
7: Form the tangent vector $\mathbf{g} \in T_{id}(\mathcal{D})$ as, $\mathbf{g} = \sum_{i=1}^{d} \langle \alpha_t(1), \phi_*(v_i) \rangle v_i$.
8: Compute the flow on $\mathcal{D}$ at id, such that $\tilde{\gamma} = \Psi_\epsilon(id, \mathbf{g}) = id - \epsilon \mathbf{g}$.
9: Set $q_1 = q_1 \cdot \tilde{\gamma} = \sqrt{\gamma'} q \circ \tilde{\gamma}$.
10: Go to Step 1.

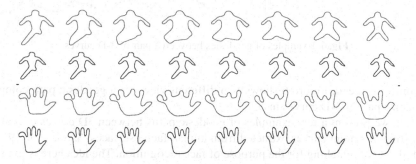

Fig. 4. Odd rows show non-elastic geodesic paths [4]. Even rows show elastic geodesics.

4   Experimental Results

In this section, we present some experimental results for computing elastic geodesics by implementing the above algorithms in MATLAB®. Figure 5 shows pairwise geodesics between 2-D curves in the shape space $\mathcal{S}$. Intermediate shapes along the geodesics have tick-marks placed around the curve, that help identify parts of the curve traversed by

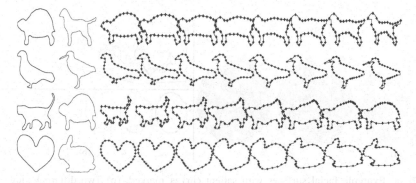

Fig. 5. Row-wise geodesic paths in $\mathcal{S}$ between the pair of curves shown to the left

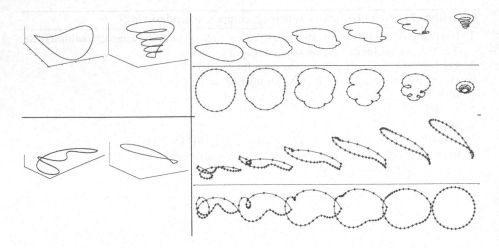

Fig. 6. Examples of geodesics between a pair of 3-D curves

non-uniform speed. Figure 6 shows two different views of a geodesic path computed between a pair of 3-D curves in S.

Next, we present a few examples of geodesic paths between 3D curves of real data consisting of salient curves extracted from human facial surfaces. Samir et al. [9] have used 3D curve matching for the purpose of face recognition. The idea here is to extract important curves from 3D scans of facial surfaces across subjects, and use pairwise geodesic distances to match them. Figure 7 (a) shows 2D views of a facial surface over-laid with 3D curves resulting from a specific depth function. Figure 7 (b) shows different views of a geodesic in S between two arbitrarily selected curves on this surface.

Finally, we present a clustering result of a sample of 25 shapes of gestures from the ASL alphabet. We compute the pairwise elastic geodesic distances between the set of shapes shown in Fig. 8 and use a simple $k-$means algorithm to automatically group them into 5 clusters. It is observed that the elastic distance captures local variabilities effectively, an important requirement in clustering of shapes.

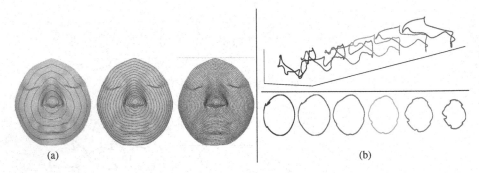

(a)   (b)

Fig. 7. (a) Example facial surfaces with salient curves marked. (b) Two different views of a geodesic between two facial curves.

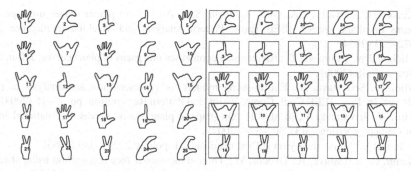

Fig. 8. Left panel shows 25 gestures from the ASL alphabet. Right panel shows row-wise clusters of gestures.

5 Summary

This paper illustrates and extends the square-root elastic (SRE) framework introduced recently in [2] for analyzing shapes of closed curves in $\mathbb{R}^n$. The novelty in this framework is that the representation of elastic curves by a single vector valued function that incorporates both stretching and bending along the curve. The elastic (Riemannian) metric reduces to a simple $\mathbb{L}^2$ form that greatly simplifies analysis and understanding of shapes of curves. This paper extends this idea by computing geodesics on the quotient spaces formed by the action of rotation and re-parametrization groups on the spaces of closed curves. The main idea is to use a gradient iteration to find a geodesic path between any two orbits. Experimental results, obtained on 2D and 3D curves, underline the utility of using elastic metrics and emphasize the generality of these ideas to higher dimensions.

Acknowledgments

We thank Chafik Samir ENIC, Lille, France, for the use of 3D facial curves. This research was partially supported by ARO W911NF-04-01-0268 and AFOSR FA9550-06-1-0324. Dr. Eric Klassen gratefully acknowledges support from the NSF grant 0430954. This research was also supported by INRIA/Florida State University Associated Team "SHAPES grant and a Visiting Professorship from INRIA to Anuj Srivastava during the summer of 2006.

References

1. Grenander, U.: General Pattern Theory. Oxford University Press, Oxford (1993)
2. Joshi, S., Klassen, E., Srivastava, A., Jermyn, I.H.: An efficient representation for computing geodesics between n-dimensional elastic shapes. In: Proc. IEEE Computer Vision and Pattern Recognition (CVPR), Minneapolis, USA (June 2007)
3. Klassen, E., Srivastava, A.: Geodesics between 3D closed curves using path straightening. In: Leonardis, A., Bischof, H., Pinz, A. (eds.) ECCV 2006. LNCS, vol. 3951, pp. 95–106. Springer, Heidelberg (2006)

4. Klassen, E., Srivastava, A., Mio, W., Joshi, S.H.: Analysis of planar shapes using geodesic paths on shape spaces. IEEE Trans. Pattern Analysis and Machine Intelligence 26(3), 372–383 (2004)
5. Michor, P.W., Mumford, D.: Riemannian geometries on spaces of plane curves. J. Eur. Math. Soc. 8, 1–48 (2006)
6. Mio, W., Srivastava, A.: Elastic-string models for representation and analysis of planar shapes. In: Proc. IEEE Conf. Comp. Vision and Pattern Recognition, pp. 10–15 (2004)
7. Mio, W., Srivastava, A., Joshi, S.H.: On shape of plane elastic curves. International Journal of Computer Vision 73(3), 307–324 (2007)
8. Palais, R.S.: Morse theory on Hilbert manifolds. Topology 2, 299–349 (1963)
9. Samir, C., Srivastava, A., Daoudi, M.: Three-dimensional face recognition using shapes of facial curves. IEEE Trans. Pattern Anal. Mach. Intell. 28(11), 1858–1863 (2006)
10. Sebastian, T.B., Klein, P.N., Kimia, B.B.: On aligning curves. IEEE Transactions on Pattern Analysis and Machine Intelligence 25(1), 116–125 (2003)
11. Shah, J.: An H^2 type riemannian metric on the space of planar curves. In: Larsen, R., Nielsen, M., Sporring, J. (eds.) MICCAI 2006. LNCS, vol. 4190, Springer, Heidelberg (2006)
12. Spivak, M.: A Comprehensive Introduction to Differential Geometry, Vol I & II. Publish or Perish, Inc. Berkeley (1979)
13. Younes, L.: Computable elastic distance between shapes. SIAM Journal of Applied Mathematics 58(2), 565–586 (1998)

Shape Analysis of Open Curves in $\mathbb{R}^3$ with Applications to Study of Fiber Tracts in DT-MRI Data

Nikolay Balov[1], Anuj Srivastava[1], Chunming Li[2], and Zhaohua Ding[2]

[1] Department of Statistics, Florida State University, Tallahassee, FL 32306
Tel.:850-222-4157; Fax.:850-644-5271
balov@stat.fsu.edu, anuj@stat.fsu.edu
[2] Institute of Imaging Sciences, Vanderbilt University, Nashville, TN 37232
chunming.li@vanderbilt.edu, zhaohua.ding@vanderbilt.edu

Abstract. Motivated by the problem of analyzing shapes of fiber tracts in DT-MRI data, we present a geometric framework for studying shapes of open curves in $\mathbb{R}^3$. We start with a space of unit-length curves and define the shape space to be its quotient space modulo rotation and reparametrization groups. Thus, the resulting shape analysis is invariant to parameterizations of curves. Furthermore, a Riemannian structure on this quotient shape space allows us to compute geodesic paths between given curves and helps develop algorithms for: (i) computing statistical summaries of a collection of curves using means and covariances, and (ii) clustering a given set of curves into clusters of similar shapes. Examples using fiber tracts, extracted as parameterized curves from DT-MRI images, are presented to demonstrate this framework.

1 Introduction

One of the main tools in medical diagnosis that relies on image data is shape analysis. Very often the health and the functionality of anatomical parts are related to their physical shapes, and can be assessed using the shape information. To study shapes of anatomical parts, it is convenient to view them as simple geometrical objects, such as curves, surfaces, patches, or volumes. Such abstractions allow researchers to impose coordinate systems on anatomies and to compare them across subjects, species, or populations. In this paper we consider the problem of analyzing shapes of fiber tracts that have been estimated previously using DT-MRI data.

Although the analysis of DT-MRI data is increasingly popular, the neuroscience/image analysis community has mainly focused on the connectivity between cortical regions, without due regard to the shape of the connecting fiber tracts. To date a plethora of methods have been proposed for estimating fiber tracts, given the DT-MRI data, with the final output being mostly a display of fibers overlaid on the reference anatomical images. There have been studies on certain coarse geometrical descriptors of fiber tracts, such as smoothness, curvature, or length, but the goal was only to validate the estimated fibers. The shape

A.L. Yuille et al. (Eds.): EMMCVPR 2007, LNCS 4679, pp. 399–413, 2007.
© Springer-Verlag Berlin Heidelberg 2007

of fiber tracts, in fact, has significant physiological and pathological relevance, and its analysis can provide important insights. For instance, when there is a brain tumor, the neuronal fiber tracts may penetrate through or detour around it, depending on the nature of the tumor. Analyzing the shape of relevant fiber tracts thus may help differentiate the tumor. Another situation which involves the shape of a large number of fiber tracts is the brain development. In the earliest stage, the brain surface is flat; the cortical surface evolves with time but the fiber tracts that connect different cortical regions pull together the connected regions so as to give rise to the gyri and sulci seen in later stages. In this situation, shape analysis of fiber tracts may allow us to peer into, tract-specifically, the development dynamics, relative rate of maturation, or help assess anomalies. An equally important area of applications is the analysis of skeletal or cardiac muscle fiber shape, which are closely related to force production. Analyzing the shape of muscle fibers may provide measurements of the mechanical performance, fatigue prediction, and so on.

While a shape analysis of fibers is a promising notion, not much attention has been paid to this area, with a few exceptions. Batchelor et al. [7] have studied the shapes of fibers as curves in $\mathbb{R}^3$ using some commonly used shape descriptors, such as the curvature scale space, Fourier descriptors, PCA of registered coordinates, and links. Sherbondy [8] describes a principal component analysis approach for analyzing shapes of the fiber tracts. Corouge et al. [12] have studied shapes of fiber tracts for clustering using average and Hausdorff distances considering the curves as point sets. A modification of Hausdorff distance is also used in [16]. An outstanding need in this area is a formal definition of shape spaces of fiber tracts and development of tools for statistical analysis on those spaces. This has been achieved in recent years on analyzing shapes of *closed* curves [4,2,10,5,9,3]. Since the fiber tracts are mostly open curves, they do not have the nonlinear closure constraint that makes the shape analysis of closed curves more complicated. So what is the main issue in shape analysis of fiber tracts? It is important to have a framework in which shape analysis in invariant to rigid motions and re-parameterizations of curves representing tracts. In other words, shape analysis should be performed in the quotient spaces under the action of these transformations.

Our approach is define a fundamental framework for representing and analyzing shapes of open curves in $\mathbb{R}^3$. We do this by defining a shape space S such that the curves of interest are elements of S. To develop the framework we consider the fiber tracts as continuous (parameterized) curves although for implementation purposes we will utilize their sampled coordinates. Another important aspect here is that shapes analysis will be invariant to rotation, translation, and re-parametrization of curves, since these transformations do not change the shape of a curve. In order to compare shapes of any two given curves, we define a Riemannian metric on S and compute a geodesic path connecting the corresponding two points in S. Geodesic paths are useful for several reasons: (i) their lengths help quantify differences between shapes of curves, (ii) they help define and compute statistical summaries, i.e. means and covariances, of a collection of curves, and (iii) using parallel transport, they can use to compare not only individual fibers but also ensemble of fibers. (The use of geodesic here is considerably different from the geodesics used in diffusion flows, e.g. in extracting fiber tracts from DT-MRI data [13]). In this paper, the discussion and the experimental results are restricted to the first two items only.

The rest of this paper is organized as follows. In Section 2, we introduce a geometric representation of curves in $\mathbb{R}^3$ using their velocity functions, and define a quotient space of comparing curves invariant to their re-parametrizations. Section 3 presents some computer experiments to demonstrate these ideas. An important tool that is developed here is the computation of geodesic paths between curves in the quotient space of shapes. We use this tool to compute sample statistics of fibers in Section 4 and to compute clustering of fiber tracts in Section 5. Finally, we outline a method to compare shapes of bundles of fiber tracts in Section 6.

2   Shape Spaces of Open Curves in $\mathbb{R}^3$

We start by identifying the space of open curves and consider its differential geometry for performing shape analysis.

2.1   Representations of Open Curves

Let Γ be the set of C^1 (parameterized) curves in $\mathbb{R}^3$. For the time being we will focus on curves of unit length by scaling all curves to make them unit length. Denote the velocity of a curve c at point $s \in I = [0,1]$ by $\gamma(s) = c'(s)$. Every curve in Γ will be represented by its velocity function γ and, hence, we can write $\Gamma = \{\gamma : [0,1] \mapsto \mathbb{R}^3\}$. Note that γ is already invariant to the translation of c in $\mathbb{R}^3$; however, it is dependent on the orientation and the choice of parametrization. Γ is an infinite-dimensional differentiable manifold. For any $\gamma \in \Gamma$, the tangent space is given by: $T_\gamma \Gamma = \{w | w : [0,1] \mapsto \mathbb{R}^3\} = \Gamma$. To impose a Riemannian structure on Γ, we assume the inner product: for $w_1, w_2 \in T_\gamma \Gamma$,

$$\langle w_1, w_2 \rangle = \int_0^1 (w_1(s), w_2(s)) ds \ ,$$

where $(\cdot, \cdot)$ is the Euclidean inner product in $\mathbb{R}^3$. With this metric, Γ becomes a Hilbert space and distances between points in Γ can be computed using the Euclidean distances:

$$d_\Gamma(\gamma_1, \gamma_2) = \sqrt{\int_0^1 \|\gamma_1(s) - \gamma_2(s)\|^2 ds} \ , \tag{1}$$

where $\| \cdot \|$ is simply the Euclidean norm in $\mathbb{R}^3$. For any two curves $\gamma_1, \gamma_2 \in \Gamma$, the geodesic connecting them in Γ is simply the "straight line": for $t \in [0, 1]$, $\Psi_t(\gamma_1, \gamma_2) = t\gamma_2 + (1 - t)\gamma_1$. So far, Γ is a vector space and geodesic paths between any two points in Γ are straight lines. In particular, the differences between any two curves are computed point wise along the curves with the chosen parameterizations of the two curves dictating the matching. This is the limitation of this idea. Different parameterizations of curves will lead to different distances between them. Ideally, one would like to treat the re-parameterizations of a curve as a shape-preserving transformation, similar to rigid rotations, translations, and scaling. This idea is developed next.

2.2   Re-parametrization and Orientation Orbits of Curves

The curves of interest can occur at arbitrary parameterizations so we have account for this variability. Our goal is to compare shapes of curves independent of their parametrization and orientation. First, we study the notion of re-parameterization of curves in Γ and then account for the rotations.

1. **Re-parametrization Orbits:** For a curve $\gamma \in \Gamma$, let $\gamma(\phi(s))$ denote a re-parametrization where ϕ is a diffeomorphism from I to itself. We define diffeomorphisms to be smooth, invertible maps, with smooth inverses. Let Φ be the set of all diffeomorphisms of I. Superposition or composition ($\circ$) is a natural associative operation in Φ. If $\phi_1, \phi_2 \in \Phi$ then $\phi_1 \circ \phi_2 \in \Phi$, and any map $\phi \in \Phi$ has inverse $\phi^{-1} \in \Phi$. Therefore, $(\Phi, \circ)$ is a Lie group with the identity map id as unit element e. Consider the mapping $\psi : \Phi \times \Gamma \to \Gamma$, given by

$$\psi(\phi, \gamma(.)) = \gamma(\phi(.))\sqrt{\dot{\phi}}.$$

We will write $\phi\gamma$ for $\psi(\phi, \gamma)$ to simplify later notation. For each $\phi \in \Phi$, $\psi_\phi : \Gamma \to \Gamma$ is diffeomorphism, $e\gamma = \gamma$ and $(\phi_1\phi_2)\gamma = \phi_1(\phi_2\gamma)$. Therefore, ψ is a group action on Γ. Furthermore, ψ_ϕ is an isometry since $d_\Gamma(\phi\gamma_1, \phi\gamma_2) = d_\Gamma(\gamma_1, \gamma_2)$ for any $\phi \in \Phi$. As described later, this has a useful implication in that

$$d_\Gamma(\phi_1\gamma_1, \phi_2\gamma_2) = d_\Gamma(\gamma_1, \phi_1^{-1}\phi_2\gamma_2).$$

We define any two elements of Γ as equivalent, if we can go from one to another using an element of Φ. In other words, we define $\gamma_1 \sim \gamma_2$ if $\gamma_2 = \phi\gamma_1$ for some $\phi \in \Phi$. This relation partitions Γ into disjoint equivalence classes of the type:

$$[\gamma]_p = \{\phi\gamma, \phi \in \Phi\} \quad \subset \Gamma,$$

which are also called *orbits*. The underscore p denotes that the orbit is under the action of Φ. Since elements of an orbit are considered equivalent, from a shape analysis perspective, the space of interest is the quotient space $\Gamma/\sim$ or Γ/Φ. The projection $\pi : \Gamma \to \Gamma/\Phi$ is given by $\pi(\gamma) = [\gamma]_p$. Since the quotient space Γ/Φ inherits a Riemannian structure from Γ, the projection π is a local diffeomorphism.

2. **Rotation Groups:** Let $o \in SO(3)$ be a rotation matrix and $o\gamma = \{o\gamma(s)|s \in I\}$ be the rotated curve. $SO(3)$ is a three-dimensional group that acts iso-metrically on Γ. Since a rotation does not change the shape of a curve, we want the shape analysis to be invariant to this group action. Similar to the re-parametrization group, we can define the orbits under $SO(3)$ as:

$$[\gamma]_o = \{o\gamma|o \in SO(3)\} \subset \Gamma,$$

and define all elements of Γ that fall in the same orbit as equivalent. Under-score o denotes that the orbit is under the action of $SO(3)$. Combining both the group actions, we can define a larger orbit as:

$$[\gamma] = \{o(\phi\gamma)|\phi \in \Phi, o \in SO(3)\} .$$

For shape analysis we consider these orbits as equivalence classes and study the quotient space $\Gamma/(\Phi \times SO(3))$ as our shape space S. Elements of S will form shapes of our interest and we are interested in statistical analysis on S.

2.3   Geodesics and Distances in Shape Space S

One of our goals is to compare shapes of curves irrespective of their parameter-izations and rotations. Towards this goal, we utilize the distance function in S that is inherited from Γ. Accordingly, the distance between any two orbits $[\gamma_1]$ and $[\gamma_2]$ is the pairwise shortest distance between elements of those two orbits. Mathematically,

$$\begin{aligned} d_s([\gamma_1], [\gamma_2]) &= \min_{\phi_1, \phi_2 \in \Phi, o_1, o_2 \in SO(3)} d_\Gamma(o_1(\phi_1\gamma_1), o_2(\phi_2\gamma_2)) \\ &= \min_{\phi \in \Phi, o \in SO(3)} d_\Gamma(\gamma_1, o(\phi\gamma_2)) \end{aligned} \tag{2}$$

The equality from the first equation to the second comes from the fact that the group actions by Φ and $SO(3)$ are isometries.

The resulting $d_s(\cdot, \cdot)$ is a well-defined distance function in S. By definition, it is invariant to the action of a $\phi \in \Phi$ and $o \in SO(3)$ on one or both the curves. Given any two curves γ_1, γ_2, this distance is based on finding an optimal ϕ and an optimal o that minimizes the right side of Eqn. 2. This joint optimization problem can be solved using repeated iterations of the following two individual problems:

1. For a fixed $o \in SO(3)$, the problem of finding an optimal ϕ is that of optimal matching or registration between different points along γ_1 and γ_1. Let $\phi^* = \text{argmin}_{\phi \in \Phi} d_\Gamma(\gamma_1, o(\phi\gamma_2))$ be the optimal re-parametrization of γ_2. This can be found using the dynamic programming algorithm as described in several earlier papers, including [1,3].
2. Similarly, for a fixed ϕ, the optimal value of $o \in SO(3)$ is computed as follows. Let

$$a = \int_0^1 \gamma_1(s)\gamma_2(\phi(s))^T ds \quad \in \mathbb{R}^{3\times 3} ,$$

and let $a = u\sigma v^T$ be the singular value decomposition of a. Then, $o^* = uv^T$ is the optimal rotation for aligning the two curves. In case the determinant of a is negative, the optimal rotation is given by $o^* = u\tilde{v}^T$ where $\tilde{v}$ is same as v except its last column is multiplied by -1.

To obtain the joint solution for (ϕ^*, o^*) we use repeated optimization, i.e. optimize ϕ while fixing o and optimize o while fixing ϕ, and repeat until convergence. The resulting geodesic path between $[\gamma_1]$ and $[\gamma_2]$ in $\mathcal{S}$ is given by:

$$\Psi_t([\gamma_1], [\gamma_2]) = to^*(\phi^*\gamma_2) + (1 - t)\gamma_1 . \tag{3}$$

2.4   Penalty Function on Optimal Matching ϕ^*

So far this matching problem has been stated without any constraints on ϕ and can result in a severe shrinkage or expansion of γ_2 by having a ϕ^* that is quite drastic. To avoid this possibility, one often adds another term that penalizes a large deformation of a curve during matching. For two curves γ_1 and γ_2 in Γ we define d to be

$$d_s([\gamma_1], [\gamma_2]) = \min_{\phi \in \Phi, o \in SO(3)} \left(d_\Gamma(\gamma_1, o(\phi\gamma_2)) + \lambda d_\phi(\phi, id)^2 \right), \tag{4}$$

where d_ϕ is a distance function on Φ and λ is a positive constant that controls the *compression* and *stretching* of the re-parametrization. There are several choices of d_ϕ:

- $\mathbb{L}^2$ distance function: $d_\phi(\phi_1, \phi_2) = \int_0^1 (\phi_1(s) - \phi_2(s))^2 ds$. The main advantage of this choice is its simplicity and low computational cost. The disadvantage is that this distance is not invariant to the action of Φ on itself, i.e. $d_\phi(\phi_1, \phi_2) \neq d_\phi(\phi \circ \phi_1, \phi \circ \phi_2)$ in general. A consequence is that the resulting d_s, as defined in Eqn. 4, is not invariant to the action of Φ on Γ and is not a proper distance on $\mathcal{S}$.
- Fisher-Rao distance function: As described in [14], the Fisher-Rao distance on Φ is the only invariant distance, and can be defined as: $d_\phi(\phi_1, \phi_2) = \cos^{-1}(\int_0^1 \sqrt{\dot{\phi}_1(s)}\sqrt{\dot{\phi}_2(s)}ds)$. If this d_ϕ is used in Eqn. 4, the resulting d_s remains invariant to the action of Φ on Γ and, thus, is a proper distance on $\mathcal{S}$.

In this paper we have used the $\mathbb{L}^2$ distance on Φ and the use of Fisher-Rao distance is left for future work. Once the optimal ϕ^* and o^* are chosen according to Eqn. 4, the geodesic path is given by Eqn. 3.

2.5   Discrete Implementation

For computer experiments, the curves have to be discretized and represented by a collection of points. Let a curve be sampled by n points selected arbitrarily. To represent this mathematically, we start with a uniform partitioning of $[0, 1]$, given by $U_n = \{0, \frac{1}{n}, \frac{2}{n}, ..., 1\}$. Then, for any $\phi \in \Phi$, the set $\phi(U_n)$ denotes another partitioning on $[0, 1]$ and $\gamma(\phi(U_n))$ represents an arbitrary sampling of the curve represented by γ using n points; we will denote it by $\bar{\gamma}(\phi) \equiv \gamma(\phi(U_n)) \in \mathbb{R}^{n \times 3}$. The orbit of a sample curve is given by:

$$[\bar{\gamma}] = \{o\gamma(\phi(U_n)) | \phi \in \Phi, o \in SO(3)\} \ .$$

With this discrete representation, the distance in Eqn. 4 becomes,

$$d_s([\bar{\gamma}_1], [\bar{\gamma}_2]) = \min_{\phi \in \Phi, o \in SO(3)} \left(d_\Gamma(\bar{\gamma}_1, o(\bar{\gamma}_2(\phi))) + \lambda d_\phi(\phi, id)^2 \right), \tag{5}$$

where $d_\Gamma(\bar{\gamma}_1(\phi_1), \bar{\gamma}_2(\phi_2)) = \sqrt{(\sum_{i=1}^{n} \|\gamma_1(\phi_1(i/n)) - \gamma_2(\phi_2(i/n))\|^2)}$. As mentioned earlier, the search for the optimal ϕ is performed using the dynamic programming technique and for optimal $o \in SO(3)$ is performed using SVD of elements of γ_1 and γ_2.

Remark: One major difference in discrete implementation and continuous theory is that we allow ϕ to be many-to-one and one-to-many in the computer implementations. In other words, ϕ may not not strictly be a diffeomorphism. This allows for the possibility of **partial matching**, that is, of one curve being matched to only a part of the other curve.

Shown in Figure 1 are two examples of optimal re-parameterizations (ϕ^*) according to Eqn. 5. The left panel is for a small value of λ while the right panel is for a larger value of λ. A larger value of λ keeps the optimal matching ϕ^* to be closed to the identity element. From the application point of view, the first result is more desirable as it preserves interesting features in going from one curve to the other. Shown in Figure 2 are two examples of geodesic paths between curves γ_1 and γ_2. These curves are shown at the top and at the bottom, respectively, in each panel. Drawn in between are curves that form equally spaced points along the geodesic that connects $[\gamma_1]$ and $[\gamma_2]$ in $\mathcal{S}$.

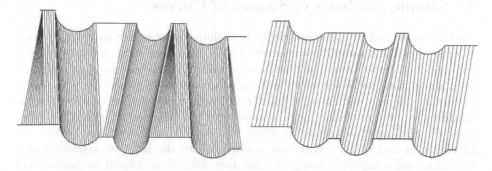

Fig. 1. Shown are optimal re-parametrisations of the bottom curve in respect to the top one, according to Eqn. 5). The value of λ in the left panel is 0.01 and 1.0 in the right one. In the later case, the optimal re-parametrization is closer to the identity. Using smaller values for λ facilitates establishing correspondence between features (arcs in this example).

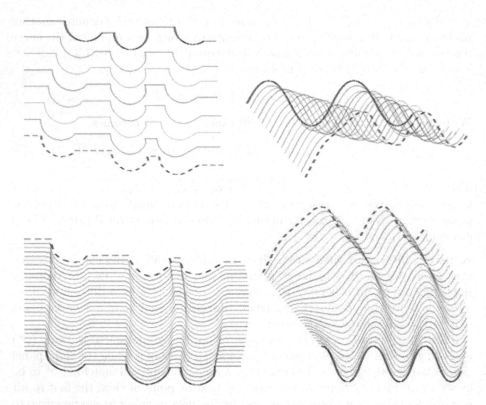

Fig. 2. Two examples of geodesic paths in $\mathcal{S}$ between two curves (γ_1 and γ_2) shown in bold at top and bottom of each column. Top row depicts the evolution in the orbit of the first curves (Φ-action: $\gamma_1 \rightarrow \phi\gamma_1 \in [\gamma_1]$). Second row shows the evolution between re-parametrized start curves and the end ones ($\phi\gamma_1 \rightarrow \gamma_2$).

3   Sample Statistics of Shapes of Curves

An important tool in statistical analysis of shapes of curves is to compute their sample means and sample covariances. The Riemannian structure defined on $\mathcal{S}$ enables us to perform such statistical analysis. There are at least two ways of defining a mean value for a random variable that takes values on a nonlinear manifold. The first definition, called the extrinsic mean, involves embedding the manifold in a larger vector space, computing the Euclidean mean in that space (e.g. Γ), and then projecting it down to $\mathcal{S}$. The other definition, called the **intrinsic mean** or the   **Karcher mean** utilizes the intrinsic geometry of $\mathcal{S}$ to define and compute a mean on that manifold. It is defined as follows: Let $d_s(\gamma_i, \gamma_j)$ denote the length of the geodesic from γ_i to γ_j in $\mathcal{S}$. To calculate the Karcher mean of given curves $\{\gamma_1, ..., \gamma_n\}$ in $\mathcal{S}$, define the variance function:

$$\mathcal{V} : \mathcal{S} \rightarrow \mathbb{R}, \mathcal{V}(\gamma) = \sum_{i=1}^{n} d(\gamma, \gamma_i)^2 \qquad (6)$$

The Karcher mean is then defined by:

$$\mu = \arg\min_{\gamma \in \mathcal{S}}(\mathcal{V}(\gamma)) \tag{7}$$

Several past papers have utilized a gradient-based approach for finding μ. We do not repeat this method here, but refer the reader to [15] or [2]. This gradient approach uses the notion of simple updates to change configurations while minimizing the cost. Simple moves consist of: (i) change velocity function (γ) of the candidate mean, and (ii) re-parameterize sample curves γ_is with respect to the candidate mean. In this implementation, the action of $SO(3)$ is not considered to reduce computational cost. It must be noted that the method does not guarantee a globally optimal solution.

Shown in Figures 3-4 are some examples of computing the Karcher means of observed curves. Figure 3 shows examples of computing mean curves in $\mathcal{S}$ when the observed samples are fiber tracts estimated from a DT-MRI data-set. Figure 4 highlights a limitation with the gradient-based search for the mean curve. As the picture shows, there are several local minimizers of $V(\gamma)$ possible, if the the sample curves are of significantly different shapes.

Fig. 3. Sample means of fiber tracts estimated using DT-MRI data

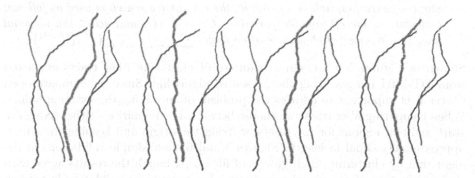

Fig. 4. Means of a bundle of four curves, quite different in shape and not smooth. Local minima vary when choose different initial curves to feed the gradient search. This behaviour is expected when large variability in the group of curves is present.

4   Clustering Bundles of Curves

We want to cluster a bundle of m curves into k classes C_i, $|C_i| = m_i, i = 1, ..., k$. One possibility it to use a k-means clustering algorithm as follows. For each cluster, we can define a sample mean and let $\mu_i = \mu(C_i)$ be the sample mean curve for class C_i. Denote variance within i-th class by V_i, $V_i = \sum_{\gamma \in C_i} d^2(\gamma, \mu_i)$. Then, k-means algorithm finds a partitioning that minimizes the total intra-cluster variance $V = \sum_{i=1}^{k} V_i$. However, this method requires updating the means of clusters at every step of the iteration and, thus, this method becomes computationally very expensive. Instead, we seek a more direct method for clustering curves that uses the distance function derived on S without using cluster means. We propose to use *estimation* of within-class variances,

$$\hat{V}_i = \min_{j=1}^{m_i} \frac{1}{m_i} \sum_{\gamma_i, \gamma_j \in C_i} d(\gamma_j, \gamma_i) \tag{8}$$

and apply a *brut-force* approach to do the clustering based on minimization of $\hat{V} = \sum_{i=1}^{k} \hat{V}_i$.

Algorithm 1 (Clustering Algorithm). *Let $C \subset \Gamma_n$, $|C| = m$ and $k \geq 2$ is a fixed number (expected number of clusters).*

1. *Calculate all pairwise distances d between curves from C. Computational cost for this step is high, the usual limitation of exhaustive algorithms.*
2. *Find a local minima of $\hat{V}$ using simple moves [10] that consist of: (i) moving a curve from one cluster to another, and (ii) inter-changing two curves from two configurations. These moves are chosen randomly and the moves are performed on the basis of costs associated with the configurations before and after the moves. For details, please refer to [10].*
3. *It must be noted that this method is not guaranteed to provide a globally optimal configuration, unless, of course, the exhaustive search is used as follows: Calculate $\hat{S}$ for all possible partitions $C = \cup_{i=1}^{k} C_i$ and select the minimal one. The number of choices is often large, $\sum_{m_1+...+m_k=m} \binom{m}{m_1}\binom{m-m_1}{m_2}...1.$*

Shown in Figures 5-9 are some examples of clustering fiber tracts extracted from DT-MRI images using the presented algorithm. Since we compare open curves it is important to address the problem of specifying the extreme points. When comparing fiber tracts we choose between two scenarios - either fixing the start and end regions for the tracts or fixing the origin and keeping the curves approximately equal in length. Figures 5 and 6 show simple validations of the algorithm by clustering small groups of fibers. Although the results agree with the anticipation of our visual perception a more rigorous validation is needed. And this is one of the directions we will pursue in the future.

A potential application of our technique is in forming consistent shape bundles from large groups of tracts, of the type shown in Figure 7. Consistency of a bundle is measured by the shape variability of the tracts within the bundle. Example 8 is particularly interesting because it shows how an anatomically correct bundle

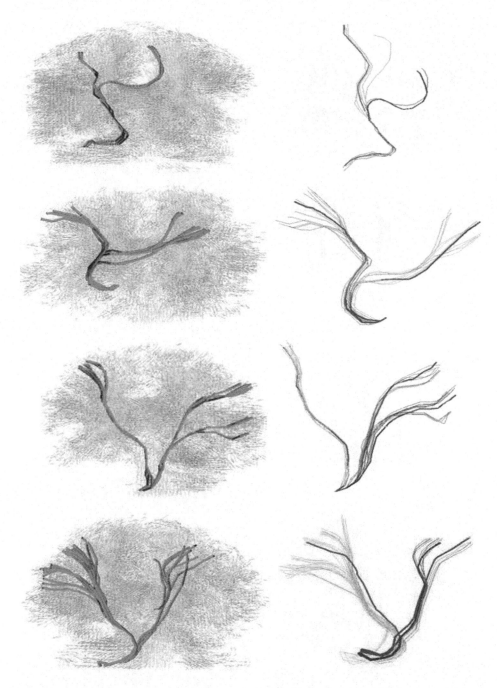

Fig. 5. Clustering of paths in Diffusion-Tensor MRI field. On the right, clusters are shown with corresonding means (ticker lines). Expected maximum number of clusters is fixed (k=3) and sometimes is larger than needed(second example). The number of curves varies from 32 to 96.

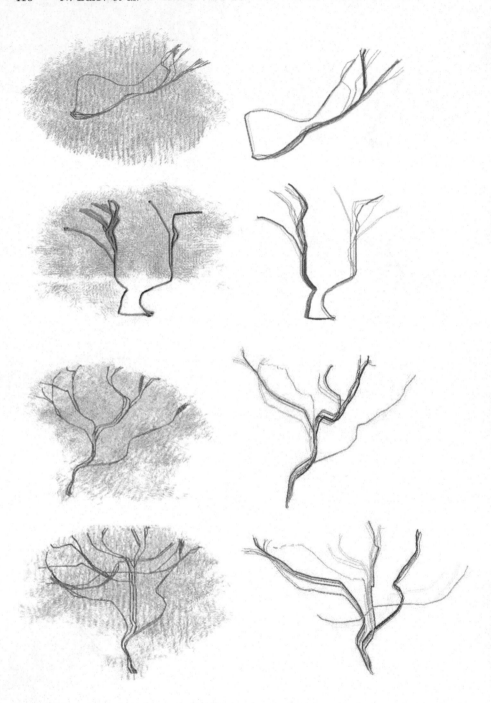

Fig. 6. Clustering of paths in Diffusion-Tensor MRI field. Expected number of clusters is set to 3 in the first two examples and to 5 for the second two. The different appearance between left and right views is due to the different projections used.

Fig. 7. Example of how the proposed clustering method extracts more consistent in shape bundle of fibers (middle) from a larger group of fibers (left), filtering out a remaining fibers (right). This automatic separation is not so easy for a human eye.

Fig. 8. Clustering of paths connecting Broka and Wernicke areas in DT-MRI. In the second example, the main bundles form two classes (red and green), while erroneous paths are separated in different class (yellow). The bundle in green is anatomically correct one between the two regions.

Fig. 9. Clustering of fiber tracts into bundles from Right Uncinate brain region for three different subjects - the right and left one are schizophrenics, the middle one is normal(control). Clear difference in number and shape of the bundles is observed. The algorithm separates 3 bundles for the left subject, 2 for the middle and 5 for the right one.

(shown in green) connecting Broka and Wernicke areas is separated from the erroneous tracts (shown in yellow).

More exciting research direction is toward relating geometric differences with anatomic functionalities, especially those manifested during a disease. A possible approach is to come up with a statistical model specifying the number and shape of the fiber bundles in carefully chosen anatomical regions. This can be done by specifying sample mean and variance for each bundle. In Figure 9 we try to illustrate this idea with samples of fibers obtained from three subjects(two schizophrenics and one control) and belonging to one and the same region (Right Uncinate). Picture like this give us an impression that there is a difference between schizophrenics and controls, but more comprehensive studies remain to be performed.

5   Summary

In this paper we have developed a framework for the geometric analysis of open curves, with the goal of using this framework for clustering and analysis of fiber tracts in DT-MRI data. Each curve is represented by its velocity function and the curves are compared in a manner that is invariant to rotation, translation, and re-parametrization. Choosing a Riemannian structure on the quotient space of curves, we describe a numerical approach for forming geodesic paths between individual curves in the shape space. Geodesic distances help provide a definition for statistical mean shapes for curves and help setup a clustering algorithm.

References

1. Sebastian, T.B., Klein, P.N., Kimia, B.B.: On aligning curves. IEEE Transactions on Pattern Analysis and Machine Intelligence 25(1), 116–125 (2003)
2. Klassen, E., Srivastava, A., Mio, W., Joshi, S.: Analysis of planar shapes using geodesic paths on shape spaces. IEEE Patt. Analysis and Machine Intell. 26(3), 372–383 (2004)
3. Mio, W., Srivastava, A., Joshi, S.H.: On Shape of Plane Elastic Curves. International Journal of Computer Vision (2007)
4. Michor, P.W., Mumford, D.: Riemannian geometries on spaces of plane curves. Journal of the European Mathematical Society 8, 1–48 (2006)
5. Younes, L.: Computable elastic distance between shapes. SIAM Journal of Applied Mathematics 58, 565–586 (1998)
6. Miller, M.I., Younes, L.: Group Actions, Homeomorphisms, and Matching: A General Framework. International Journal of Computer Vision 8, 1–48 (2002)
7. Batchelor, P.G., Calamante, F., Tournier, J.-D., Atkinson, D., Hill, D.L.G., Connelly, A.: Quantification of the shape of fiber tracts. Magnetic Resonance in Medicine 55, 894–903 (2006)
8. Sherbondy, A.: Shape analysis of fiber tractography in the human brain (2006), Website http://www.stanford.edu/sherbond/pres.pdf
9. Davies, R.H., Twining, C.J., Allen, P.D., Cootes, T.F., Taylor, C.J.: Building optimal 2D statistical shape models. Image and Vision Computing 21, 82–1171 (2003)

10. Srivastava, A., Joshi, S., Mio, W., Liu, X.: Statistical Shape Analysis: Clustering, Learning and Testing. IEEE Transactions on Pattern Analysis and Machine Intelligence 27(4), 590–602 (2005)
11. Bakircioglu, M., Grenander, U., Khaneja, N., Miller, M.: Curve matching on brain surfaces using induced fr'enet distances matrices. Special issue of Human brain mapping (2000)
12. Corouge, I., Gouttard, S., Gerig, G.: Towards a Shape Model Of White Matter Fiber Bundles Using Diffusion Tensor MRI. In: International Symposium on Biomedical Imaging (2004)
13. Fillard, P., Gilmore, J., Piven, J., Lin, W., Gerig, G.: Quantitative Analysis of White Matter Fiber Properties along Geodesic Paths. Medical Image Computing and Computer-Assisted Intervention (2003)
14. Srivastava, A., Jermyn, I., Joshi, S.: Riemannian Analysis of Probability Density Functions with Applications in Vision. In: IEEE Conference on Computer Vision and Pattern Recognition (CVPR) (2007)
15. Le, H.L., Kendall, D.G.: The Riemannian Structure of Euclidean shape spaces: a novel environment for Statistics. Annals of Statistics (1993)
16. O'Donnell, L., Kubicki, M., Shenton, M.E., Dreusicke, M.E., Grimson, W.E.L., Westin, C.-F.: A Method for Clustering White Matter Fiber Tracts. American Journal of Neuroradiology (AJNR) (2006)

Energy-Based Reconstruction of 3D Curves for Quality Control

H. Martinsson[1], F. Gaspard[1], A. Bartoli[2], and J.-M. Lavest[2]

[1] CEA, LIST, Boîte Courrier 94, F-91 191 Gif sur Yvette, France
Tel.: +33(0)1 69 08 82 98, Fax: +33(0)1 69 08 83 95
hanna.martinsson@cea.fr
[2] LASMEA (CNRS/UBP), 24 avenue des Landais, F-63 177 Aubière, France
Tel.: +33(0)4 73 40 76 61, Fax: +33(0)4 73 40 72 62
adrien.bartoli@gmail.com

Abstract. In the area of quality control by vision, the reconstruction of
3D curves is a convenient tool to detect and quantify possible anomalies.
Whereas other methods exist that allow us to describe surface elements,
the contour approach will prove to be useful to reconstruct the object
close to discontinuities, such as holes or edges.

We present an algorithm for the reconstruction of 3D parametric
curves, based on a fixed complexity model, embedded in an iterative
framework of control point insertion. The successive increase of degrees of
freedom provides for a good precision while avoiding to over-parameterize
the model. The curve is reconstructed by adapting the projections of a
3D NURBS snake to the observed curves in a multi-view setting. The
optimization of the curve is performed with respect to the control points
using an gradient-based energy minimization method, whereas the insertion procedure relies on the computation of the distance from the curve
to the image edges.

1 Introduction

The use of optical sensors in metrology applications is a complicated task when
dealing with complex or irregular structures. More precisely, projection of structured light allows for an accurate reconstruction of surface points but does not
allow for a precise localization of the discontinuities of the object. This paper
deals with the problem of reconstruction of 3D curves, given the CAD model, for
the purpose of a control of conformity with respect to this model. We dispose of
a set of images with given perspective projection matrices. The reconstruction
will be accomplished by means of the observed contours and their matching,
both across the images and to the model. We proposed a previous version of
our algorithm, based on edge distances, in [12]. The contributions of this paper
with respect to the former one resides in the energy formulation, giving a new
structure to the problem. We have also completed the experimental evaluation.

Algorithms based on active contours [11] allows for a local adjustment of the
model and a precise reconstruction of primitives. More precisely, the method

A.L. Yuille et al. (Eds.): EMMCVPR 2007, LNCS 4679, pp. 414–428, 2007.
© Springer-Verlag Berlin Heidelberg 2007

allows for an evolution of the reprojected model curves toward the image edges, thus to minimize the distance in the images between the predicted curves and the observed edges.

The parameterization of the curves as well as the optimization algorithms we use must yield an estimate that meets the requirements of accuracy and robustness necessary to perform a control of conformity. We have chosen to use NURBS curves [14], a powerful mathematical tool that is also widely used in industrial applications.

In order to ensure stability, any method used ought to be robust to erroneous data, namely the primitives extracted from the images, since images of metallic objects incorporate numerous false edges due to reflections.

Although initially defined for ordered point clouds, active contours have been adapted to parametric curves. Cham and Cipolla propose a method based on affine epipolar geometry [4] that reconstructs a parametric curve in a canonical frame using stereo vision. The result is two coupled snakes, but without directly expressing the 3D points. In [19], Xiao and Li deal with the problem of reconstruction of 3D curves from two images. However, the NURBS curves are approximated by B-splines, which makes the problem linear, at the expense of loosing projective invariance. The reconstruction is based on a matching process using epipolar geometry followed by triangulation. The estimation of the curves is performed independently in the two images, that is, there is no interactivity between the 2D observations and the 3D curve in the optimization. Kahl and August introduce in [10] a coupling between matching and reconstruction, based on an a priori known distribution of the curves and on an image formation model. The curves are expressed as B-splines and the optimization is done using gradient descent.

Other problems related to the estimation of parametric structures have come up in the area of surfaces. In [18], Siddiqui and Sclaroff present a method to reconstruct rational B-spline surfaces. Point correspondences are supposed given. In a first step, B-spline surface patches are estimated in each view, then the surface in 3D, together with the projection matrices, are computed using factorization. Finally, the surface and the projection matrices are refined iteratively by minimizing the 2D residual error. So as to avoid problems due to over-parameterization, the number of control points is limited initially, to be increased later on in a hierarchical process by control point insertion.

In the field of medical imaging, energy minimization methods have been developed to reconstruct 3D curves in a stereo setting. Sbert and Solé reconstruct in [16] a 3D curve using an energy based evolution method. The associated PDE of the energy functional, derived by the Euler-Lagrange formulation, is solved using a level-set approach. In [3], Canero et al. define in a force field by reprojecting external image forces, given by the distance to the edges. A 3D curve is then reconstructed via the evolution of an active contour, guided by the force field.

In the case of 2D curve estimation, other aspects of the problem are addressed. Cham and Cipolla adjust a spline curve to fit an image contour [5]. Control points are inserted iteratively using a new method called PERM (potential for energy-reduction maximization). An MDL (minimal description length [9]) strategy is used to define a stopping criterion. In order to update the curve, the actual curve is sampled and a line-search is performed in the image to localize the target shape. The optimization is performed by gradient descent. Brigger et al. present in [2] a B-spline snake method without internal energy, due to the intrinsic regularity of B-spline curves. The optimization is done on the knot points rather than on the control points, which allows the formulation of a system of equations that can be solved by digital filtering. So as to increase numerical stability, the method is embedded in a multi-resolution framework. In [8], Figueiredo et al. address the problem from a statistical point of view, proposing a completely automatic contour estimator, in the sense that no parameter need to be adjusted by the user. Supposing a uniform distribution of the knot points, the B-spline curve that approximates a given set of contour points at best, in the least squares sense, is given by a linear system depending only on the number of control points. This number is fixed in advance using an MDL criterion. Meegama and Rajapakse introduce in [13] an adaptive procedure for control point insertion and deletion, based on the euclidean distance between consecutive control points and on the curvature of the NURBS curve. Local control is ensured by adjustment of the weights. The control points evolve in each iteration in a small neighborhood (3 × 3 pixels). Yang et al. use a distance field computed a priori with the fast marching method in order to adjust a B-spline snake [20]. Control points are added in the segment presenting a large estimation error, due to a degree of freedom insufficient for a good fit of the curve. The procedure is repeated until the error is lower than a fixed threshold. Redundant control points are then removed, as long as the error remains lower than the threshold.

2   Problem Formulation

Given a set of images of an object, together with its CAD model, our goal is to reconstruct in 3D the curves observed in the images. The reconstruction is performed by minimizing an energy functional. In order to obtain a 3D curve that meets our requirements regarding regularity, rather than reconstructing a point cloud, we estimate a NURBS curve. Since the regularity aspects are thereby taken care of, the energy functional is defined solely based on image data. The minimization problem is formulated for a set of M images and N sample points by

$$\mathbf{C}(\mathcal{P}) = \arg\min_{\mathcal{P}} \sum_{i=0}^{M-1} \sum_{j=0}^{N-1} E(T_i(\mathbf{C}(\mathcal{P}, t_j))), \tag{1}$$

where E is the external energy functional, T_i is the projective operator for image i and $\mathcal{P}$ is the set of control points.

Our choice to use NURBS curves is justified by several reasons. First, NURBS curves have interesting geometrical properties, namely concerning regularity and continuity. An important geometrical property that will be of particular interest is the invariance under projective transformations.

3 Properties of NURBS Curves

Let $U = \{u_0, \cdots, u_m\}$ be an increasing vector, called the knot vector. A NURBS curve is a vector valued, piecewise rational polynomial over U, defined by

$$\mathbf{C}(t) = \sum_{i=0}^{n} \mathbf{P}_i R_{i,k}(t) \quad \text{with} \quad R_{i,k}(t) = \frac{w_i B_{i,k}(t)}{\sum_{j=0}^{n} w_j B_{j,k}(t)}, \qquad (2)$$

where $\mathbf{P}_i$ are the control points, $B_{i,k}(t)$ the B-spline basis functions defined over U, w_i the associated weights and k the degree.

It is a common choice to take $k = 3$, which has proved to be a good compromise between required smoothness and the problem of oscillation, inherent to high degree polynomials. For our purposes, the parameterization of closed curves, we consider periodic knot vectors, that is, verifying $u_{j+m} = u_j$. Given all these parameters, the set of NURBS defined on U forms, together with the operations of point-wise addition and multiplication with a scalar, a vector space.

For details on NURBS curves and their properties, refer to [14].

3.1 Projective Invariance

According to the pinhole camera model, the perspective projection $T(\cdot)$ that transforms a world point into an image point is expressed in homogeneous coordinates by means of the transformation matrix $\mathbf{T}_{3\times 4}$. Using weights associated with the control points, NURBS curves have the important property of being invariant under projective transformations. Indeed, the projection of (2) remains a NURBS, defined by its projected control points and their modified weights. The curve is written

$$\mathbf{c}(t) = T(\mathbf{C})(t) = \frac{\sum_{i=0}^{n} w_i' T(\mathbf{P}_i) B_{i,k}(t)}{\sum_{i=0}^{n} w_i' B_{i,k}(t)} = \sum_{i=0}^{n} T(\mathbf{P}_i) R_{i,k}'(t), \qquad (3)$$

where the $R_{i,k}'$ are the basis functions of the projected NURBS. The new weights, w_i', are given by

$$w_i' = (T_{3,1}X_i + T_{3,2}Y_i + T_{3,3}Z_i + T_{3,4})\, w_i = \mathbf{n} \cdot (\mathbf{C}_O - \mathbf{P}_i)\, w_i, \qquad (4)$$

where $\mathbf{n}$ is a unit vector along the optical axis and $\mathbf{C}_O$ the optical center of the camera.

3.2 Control Point Insertion

One of the fundamental geometric algorithms available for NURBS curves is the control point insertion. The key is the knot insertion, which is equivalent to adding one dimension to the vector space, consequently adapting the basis. Since the original vector space is included in the new one, there is a set of control points such that the curve remains unchanged.

Let $\bar{u} \in [u_j, u_{j+1})$. We insert $\bar{u}$ in U, forming the new knot vector $\bar{U} = \{\bar{u}_0 = u_0, \cdots, \bar{u}_j = u_j, \bar{u}_{j+1} = \bar{u}, \bar{u}_{j+2} = u_{j+1}, \cdots, \bar{u}_{m+1} = u_m\}$. The new control points $\bar{\mathbf{P}}_i$ are given by the linear system

$$\sum_{i=0}^{n} \mathbf{P}_i R_{i,k}(t) = \sum_{i=0}^{n+1} \bar{\mathbf{P}}_i \bar{R}_{i,k}(t). \tag{5}$$

We present the solution without proof. The new control points are written [14]

$$\bar{\mathbf{P}}_i = \alpha_i \mathbf{P}_i + (1 - \alpha_i) \mathbf{P}_{i-1}, \tag{6}$$

with

$$\alpha_i = \begin{cases} 1 & i \leq j - k \\ \dfrac{\bar{u} - u_i}{u_{i+k} - u_i} & \text{if } j - k + 1 \leq i \leq j \\ 0 & i \geq j + 1 \end{cases}. \tag{7}$$

Note that only k new control points need to be computed, due to the local influence of splines.

4 Optimization

When treating NURBS curves, the regularity aspects are taken care of implicitly by the parameterization and the energy functional can be reduced to its external energy part. We will consider two forms of energy functionals, one based on the distance from the curve to the image contours and another one based on the gradient intensity. The optimization will in both cases operate on the control points of the 3D NURBS curve.

4.1 Distance Minimization

Using a distance formulation and the properties of NURBS curves, the minimization problem (1) is written

$$\min_{\{\hat{\mathbf{P}}_l\}} \sum_{i=0}^{M-1} \sum_{j=0}^{N-1} \left(\mathbf{q}_{ij} - \sum_{l=0}^{n} T_i(\hat{\mathbf{P}}_l) R_{l,k}^{(i)}(t_j) \right)^2, \tag{8}$$

where $\mathbf{q}_{ij}$ is a contour point associated with the curve point of parameter t_j in image i and T_i is the projective operator for image i. The search for candidate contour points is carried out independently in the images using a method inspired by the one used by Drummond and Cipolla in [7].

Search for Image Contours. We sample the NURBS curve projected in the image, to use as starting points in the search for matching contour points. A line-search is performed in order to find the new position of the curve, ideally corresponding to an edge. Our approach is based solely on the contours. Due to the aperture problem, the component of motion of an edge, tangent to itself, is not detectable locally and we therefore restrict the search for the new edge position to the edge normal at each sample point. As we expect the motion to be small, we define a search range (typically in the order of 20 pixels) so as to limit computational cost. In order to find the new position of a sample point, for each point belonging to the normal within the range, we evaluate the gradient and compute a weight based on the intensity and the orientation of the gradient and the distance from the sample point. The weight function v_j for a sample point p_j and the candidate point p_ξ will be of the form

$$v_j(p_\xi) = \varphi_1(|\nabla I_\xi|) \cdot \varphi_2 \left(\frac{\hat{n}_j \cdot \nabla I_\xi}{|\nabla I_\xi|} \right) \cdot \varphi_3(|p_j - p_\xi|),$$

where $\hat{n}_j$ is the normal of the projected curve at sample point j, ∇I_ξ is the gradient at the candidate point and the φ_k are functions to define. The weight function will be evaluated for each candidate p_ξ and the point p'_j with the highest weight, identified by its distance from the original point $d_j = |p_j - p'_j|$, will be retained as the candidate for the new position of the point.

The bounded search range and the weighting of the point based on their distance from the curve yield a robust behavior, close to that of an M-estimator.

4.2 Gradient Energy Minimization

Using the classical energy formulation and the properties of NURBS curves, the minimization problem (1) is written

$$\min_{\{\hat{\mathbf{P}}_l\}} \sum_{i=0}^{M-1} \sum_{j=0}^{N-1} E \left(\sum_{l=0}^{n} T_i(\hat{\mathbf{P}}_l) R_{l,k}^{(i)}(t_j) \right), \tag{9}$$

where $R_{l,k}^{(i)}$ are the basis functions for the projected NURBS curve in image i. The energy functional E can, as already mentioned, be restricted to its external part, due to the use of NURBS. A common choice is to use the gradient intensity. We will however include local information on the curve, namely its normal direction, using the intensity of the gradient projected onto the curve normal.

4.3 Distance Versus Gradient Energy

For comparison, we have implemented the two methods in the iterative setting that will be introduced in the following section. Both methods yielded similar results and converge after a number of iterations to an asymptotic lower limit. The 3D error with respect to the true curve is however somewhat lower for the gradient-based method. The results are given in Fig. 2. The difference is partly explained by the noise and the parallel structures perturbing the edge tracking algorithm. An example of candidate points located on a parallel image contour, due to specularities, is given in Fig. 1. Although the gradient intensity method outperforms the distance method, the distance-based cost function will prove to be useful in the iterative framework that will embed the curve optimization.

Fig. 1. Problems related to specularities and to the search for candidate points. Starting at the projection of the initial curve (in blue), some candidate points (in magenta) belong to a parasite edge.

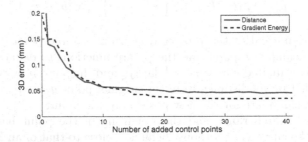

Fig. 2. The evolution of the error, with respect to the true 3D curve, for an optimization using cost functions based on the distance to the image contours and on the gradient intensity respectively. Whereas the distance method seems to yield good results in the start, its asymptotic limit is somewhat higher than that of the gradient intensity method.

5   Curve Estimation

The problem has two parts. First, the optimization of the 3D NURBS curve by energy minimization on a fixed number of control points, then the control point insertion procedure. For the fixed size optimization problem, we use the non-linear Levenberg-Marquardt minimization method. This step allows the control points to move in 3D, but does not change their number. In order to obtain an optimal reconstruction of the observed curve, we iteratively perform control point insertion. So as to avoid over-parameterization for stability reasons, the first optimization is carried out on a limited number of control points. Their number is then increased by iterative insertion, so that the estimated 3D curve fits correctly also in high curvature regions. As mentioned earlier, the insertion of a control point is done without influence on the curve and a second optimization is thus necessary in order to take advantage from the increased number of degrees of freedom.

5.1   Optimization on the Control Points

The first step of the optimization consists in projecting the curve in the images. Since the surface model is known, we can identify the visible parts of the curve in each image and retain only the sample points corresponding to visible parts.

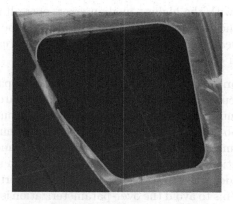

Fig. 3. Due to the use of 3D objects, auto-occlusions cause parts of the curve to be invisible from some viewpoints

During the iterations, to keep the same cost function, the residual error must be evaluated in the same points in each iteration. Supposing small displacements, we can consider that visible pieces will remain visible throughout the optimization. See Fig. 3 for an example of occlusions due to the 3D structure of the objects.

The optimization of (9) is done on the 3D control point coordinates, leaving the remaining parameters of the NURBS curve constant. The weights associated with the control points are modified by the projection giving 2D weights varying with the depth of each control point, according to the formula (4), but they are not subject to the optimization.

5.2 Control Point Insertion

Due to the use of NURBS, we have a method to insert control points. What remains is to decide where to place them. We also need a criterion to decide when to stop the control point insertion procedure.

Position of the New Control Point. Several strategies have been used. Cham and Cipolla consider in [5] the dual problem of knot insertion. They define an error energy reduction potential and propose to place the knot point so as to maximize this potential. The control point is placed using the method described earlier. In our algorithm, since every insertion is followed by an optimization that adjusts the control points, we settle for choosing the interval where to place the point. Since the exact location within the interval is not critical, the point is placed at its midpoint. Dierckx suggests in [6] to place the new point at the interval that presents the highest error. This is consistent with an interpretation of the error as the result of a lack of degrees of freedom that inhibits a good description of the curve. If, however, the error derives from other sources, this solution is not always optimal.

In our case, a significant mean error could also indicate the presence of parasite edges or that of a parallel structure close to the target curve. We will therefore choose the interval with the highest median error, over all images. The error is defined as the distance from a sample point to its corresponding contour point in the image. The search for candidate contour points is carried out using the method described in 4.1.

Stopping Criterion. One of the motives for introducing parametric curves was to avoid treating all curve points, as only the control points are modified during the optimization. If the number of control points is close to the number of samples, the benefit is limited. Too many control points could also cause numerical instabilities, due to an over-parameterization of the curve on the one hand and the size of the non-linear minimization problem on the other hand. It is thus necessary to define a criterion that decides when to stop the control point insertion.

A strategy that aims to avoid the over-parameterization is the use of statistical methods inspired by the information theory. Based in a Maximum Likelihood environment, these methods combine a term equivalent, in the case of a normal distributed errors, to the sum of squares of the residual errors with a term penalizing the model complexity. Given two estimated models, in our case differentiated by their number of control points, the one with the lowest criterion will be retained. The first criterion of this type, called AIC (Akaike Information Criterion), was introduced by Akaike in [1] and is written, in the case of normally distributed errors,

$$AIC = 2k + n \ln \frac{RSS}{n}, \tag{10}$$

where k is the number of control points, n is the number of observations and RSS is the sum of the squared residual errors. Another criterion, based on a bayesian formalism, is the BIC (Bayesian Information Criterion) presented by Schwarz [17]. It stresses the number of data points n, so as to ensure an asymptotic consistency and is written, also in the case of normally distributed errors,

$$BIC = 2k \ln n + n \ln \frac{RSS}{n}. \tag{11}$$

Another family of methods uses the MDL [15] formulation, which consists in associating a cost with the quantity of information necessary to describe the curve. Different criteria follow, depending on the formulation of the estimation problem. In the iterative control point insertion procedure of Cham et Cipolla [5], the stopping criterion is defined by means of MDL. The criterion depends, on the one hand on the number of control points and on the residual errors, on the other hand on the number of samples and on the covariance.

Yet another way of choosing an appropriate model complexity is the classical method of cross-validation. The models are evaluated based on their capacity to describe the data. A subset of the data is used to define a fixed complexity model, while the rest serve to validate it. The process is repeated and a model is retained if its performance is considered good enough.

We have chosen to use the BIC, computed using the contour points found with the method presented in 4.1, for this first version of our algorithm. A more thorough study of the influence of the stopping criterion in our setting will be performed at a later stage.

5.3   Algorithm

The algorithm we implemented has two layers. The optimization of a curve using a fixed complexity model is embedded in an iterative structure that aims

Table 1. Reconstruction algorithm presented

OBJECTIVE

Given a 3D NURBS curve extracted from the CAD model, (partially) seen in M images, sampled in N points. We want to reconstruct the 3D curve observed in the images in order to compare it to the model.

ALGORITHM

- **Visibility Check** Identification of the visible parts

$$\chi_{ij} = \begin{cases} 1 \text{ if } \sum_{l=0}^{n-1} T_i(\mathbf{P}_l)\mathbf{R}_{l,k}^{(i)}(t_j) \text{ visible} \\ 0 \text{ sinon} \end{cases}$$

- **Optimization** on the control points

$$\min_{\{\hat{\mathbf{P}}_l\}} \sum_{i=0}^{M-1} \sum_{j=0}^{N-1} \chi_{ij} E\left(\sum_{l=0}^{n-1} T_i(\hat{\mathbf{P}}_l)\mathbf{R}_{l,k}^{(i)}(t_j) \right)$$

- **Line-search** for contour points $\mathbf{q}_{ij}$ matching $\mathbf{p}_{ij} = \sum_{l=0}^{n-1} T_i(\mathbf{P}_l)\mathbf{R}_{l,k}^{(i)}(t_j)$

$$\mathbf{q}_{ij} = \arg_{\mathbf{p}_{ij}^m} \max_{-d \leq m \leq d} v_j(\mathbf{p}_{ij}^m,) \text{ where } \mathbf{p}_{ij}^m = \mathbf{p}_{ij} + m \cdot \hat{\mathbf{n}}_{ij}$$

- **Computation of the** BIC

$$BIC_0 = k \ln(N \cdot M) + N \cdot M \cdot \ln\left(\sum_{i=0}^{M-1} \sum_{j=0}^{N-1} \chi_{ij} \left(\mathbf{q}_{ij} - \sum_{l=0}^{n-1} T_i(\mathbf{P}'_l)\mathbf{R}_{l,k}^{(i)}(t_j) \right)^2 / (N \cdot M) \right)$$

- **do** (control point insertion)
    - **Line-search** for contour points $\mathbf{q}_{ij}$
    - **Computation of the median error** for each interval I_K.

$$m_K = \operatorname*{med}_{E_K} |\mathbf{q}_{ij} - \sum_{l=0}^{n-1} T_i(\mathbf{P}'_l)\mathbf{R}_{l,k}^{(i)}(t_j)|$$

    where $E_K = \{(i,j) \mid 0 \leq i < M, t_j \in I_K, \chi_{ij} \neq 0\}$.
    - **Knot point insertion** at the midpoint of interval $I = \arg\min_K m_K$.
    - **Visibility Check** Identification of the visible parts
    - **Optimization** on the control points
    - **Computation of the** BIC_J
- **while** $(BIC_J < BIC_{J-1})$

---

to increase the number of control points. The non-linear optimization of the 3D curve is performed by the Levenberg-Marquardt algorithm, using a cost function based on an energy formulation. The control point insertion procedure uses a search for contour points in the images in order to compute the median as well as the RSS error of the projected curve. The mechanism of our method is outlined in Table 1.

6   Experimental Evaluation

6.1   Virtual Images

In order to validate our algorithms for image data extraction and for curve reconstruction, we have performed a number of tests on virtual images. The virtual setting also allows us to simulate deformations of the target object.

We construct a simplified model of an object, based on a single target curve. We then apply our 3D reconstruction algorithm, starting at a modified "model curve", on a set of virtual views, see Fig. 4. The image size is 1284×1002 pixels. The starting curve has 10 control points, to which 45 new points are added. The sampling used for the computations is of 200 points. To fix the scale, note that at the mean distance from the object curve, one pixel corresponds roughly to 0.22 mm. The evaluation of the results is done by measuring the distance from a set of sampled points from the estimated curve to the target model curve. The distances from the target curve are shown in Fig. 4. We obtain the following results:

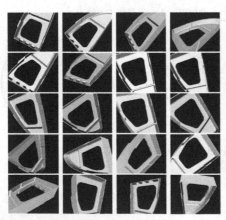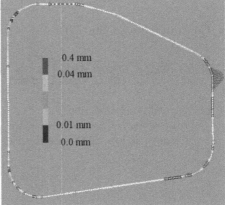

Fig. 4. *Left:* Some of the 32 virtual images used for the reconstruction of the central curve. *Right:* The distances from the sampled points from the reconstructed 3D curve to the model curve. The cloud of sample points from the estimated curve is shown together with the target curve. The starting curve is shown in black. The differences are represented by lines with length proportional to the distance between the curve and the target, using a scale factor of 20.

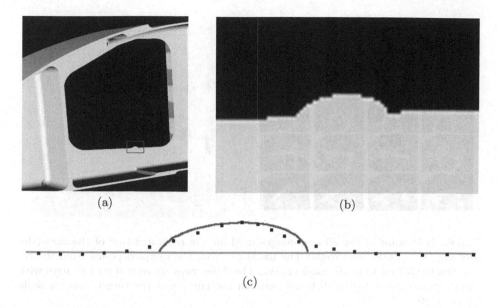

(a) (b)

(c)

Fig. 5. Reconstruction of a nonconformity based on a series of virtual images of an object with an anomaly. The object is shown in (a) with the anomaly marked in red, with a close-up in (b). The result of the reconstruction around the anomaly is shown in (c), with the original curve in green, the anomaly in red and the reconstructed points in black.

Mean error	0.0336 mm
Median error	0.0228 mm
Standard deviation	0.0485 mm

We note that the error corresponds to less than a pixel in the images, which indicates a sub-pixel image precision.

Using a series of virtual images of an object presenting a minor anomaly, we have also tested the capacity of our system to detect nonconformities, see Fig. 5. Based on 27 images and starting at the model curve, our algorithm manages to reconstruct the curve and its anomaly with a mean error of 0.0765 mm. Although the reconstruction is good, the error is concentrated around the anomaly, which is somewhat smoothed out.

6.2 Real Images

We also consider a set of real images, see Fig. 6, with the same target curve, using the same starting "model curve" as in the virtual case. We now need to face the problem of noisy image data, multiple parallel structures and imprecision in the localization and the calibration of the views. The image size is 1392×1040 pixels. The starting curve has 10 control points, to which 48 new points are added. The sampling used for the computations is of 200 points. At the mean distance from

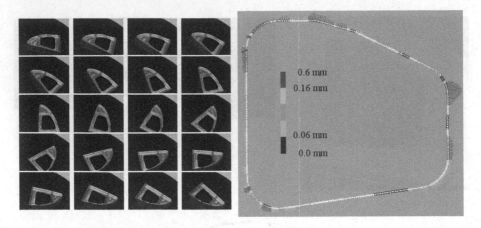

Fig. 6. *Left:* Some of the 36 real images used for the reconstruction of the curve describing the central hole. *Right:* The distances from the sampled points from the reconstructed 3D curve to the model curve. The differences are represented by lines with length proportional to the distance between the curve and the target, using a scale factor of 20.

the object curve, one pixel corresponds roughly to 0.28 mm. The distances from the target curve are shown in Fig. 6. We obtain the following results:

Mean error	0.137 mm
Median error	0.125 mm
Standard deviation	0.074 mm

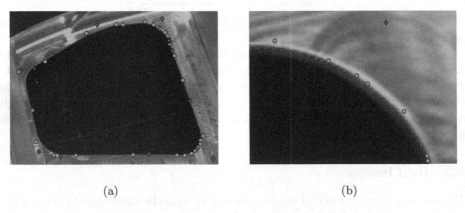

(a)   (b)

Fig. 7. (a) Evolution of the control points. In red the 10 initial control points, in green the 30 control points after optimization and in blue the final curve. (b) A detail, showing the curve between two similar contours, the parallel model curve to the lower left and a false edge caused by specularities to the upper right. Thanks to the multi-view setting, the reconstructed curve corresponds to the true 3D curve.

Even if the errors are higher than in the case of virtual images, we note that they still correspond to less than a pixel in the images. The difference is explained by the noise and to some extent by specularities, causing parallel structures perturbing the minimization algorithm, see Fig. 7(b).

The evolution of the control points is demonstrated in Fig. 7(a), where the set of initial control points is shown, together with the final curve and its control points. As expected, the control points inserted are concentrated in the regions of high curvature, such as the corners.

7 Conclusions

We have presented an adaptive 3D reconstruction method using parametric curves, limiting the degrees of freedom of the problem. An algorithm for 3D reconstruction of curves using a fixed complexity model is embedded in an iterative framework, allowing an enhanced approximation by control point insertion. The optimization of the curve with respect to the control points is performed by means of a minimization of an gradient-based energy functional, whereas the insertion procedure is based on the distance from the curve to the observed image contours. An experimental evaluation of the method, using virtual as well as real images, has let us validate its performance in some simple, nevertheless realistic, cases with specular objects subject to occlusions and noise.

Future work will be devoted to the integration of knowledge of the CAD model in the image based edge tracking. Considering the expected neighborhood of a sample point, the problem of parasite contours should be controlled and has limited impact on the obtained precision.

We also plan to do a deeper study around the stopping criterion used in the control point insertion process, using cross-validation.

References

1. Akaike, H.: A new look at the statistical model identification. IEEE Transactions on Automated Control 19(6), 716–723 (1974)
2. Brigger, P., Hoeg, J., Unser, M.: B-spline snakes: A flexible tool for parametric contour detection. IEEE Trans. on Image Processing 9(9), 1484–1496 (2000)
3. Canero, C., Radeva, P., Toledo, R., Villanueva, J.J., Mauri, J.: 3D curve reconstruction by biplane snakes. In: 15th International Conference on Pattern Recognition (ICPR'00), vol. 4, pp. 563–566 (2000)
4. Cham, T.-J., Cipolla, R.: Stereo coupled active contours. In: Conference on Computer Vision and Pattern Recognition, pp. 1094–1099. IEEE Computer Society, Los Alamitos (1997)
5. Cham, T.-J., Cipolla, R.: Automated B-spline curve representation incorporating MDL and error-minimizing control point insertion strategies. IEEE Transactions on Pattern Analysis and Machine Intelligence 21(1), 49–53 (1999)
6. Dierckx, P.: Curve and Surface Fitting with Splines. Oxford University Press, Inc, New York (1993)

7. Drummond, T., Cipolla, R.: Real-time visual tracking of complex structures. IEEE Transactions on Pattern Analysis and Machine Intelligence 7, 932–946 (2002)
8. Figueiredo, M., Leitao, J., Jain, A.K.: Unsupervised contour representation and estimation using B-splines and a minimum description length criterion. IEEE Transactions on Image Processing 9(6), 1075–1087 (2000)
9. Hansen, M.H., Yu, B.: Model selection and the principle of minimum description length. Journal of the American Statistical Association 96(454), 746–774 (2001)
10. Kahl, F., August, J.: Multiview reconstruction of space curves. In: 9th International Conference on Computer Vision, vol. 2, pp. 1017–1024 (2003)
11. Kass, M., Witkin, A., Terzopoulos, D.: Snakes: Active contour models. International Journal of Computer Vision 4(1), 321–331 (1987)
12. Martinsson, H., Gaspard, F., Bartoli, A., Lavest, J-M.: Reconstruction of 3d curves for quality control. In: 15th Scandinavian Conference on Image Analysis (to appear, 2007)
13. Meegama, R.G.N., Rajapakse, J.C.: NURBS snakes. Image and Vision Computing 21, 551–562 (2003)
14. Piegl, L., Tiller, W.: The NURBS book. In: Monographs in visual communication, 2nd edn. Springer, Heidelberg (1997)
15. Rissanen, J.: Modeling by shortest data description. Automatica 14, 465–471 (1978)
16. Sbert, C., Solé, A.F.: Stereo reconstruction of 3d curves. In: 15th International Conference on Pattern Recognition (ICPR'00), vol. 1 (2000)
17. Schwarz, G.: Estimating the dimension of a model. Ann. of Stat. 6, 461–464 (1978)
18. Siddiqui, M., Sclaroff, S.: Surface reconstruction from multiple views using rational B-splines and knot insertion. In: First International Symposium on 3D Data Processing Visualization and Transmission, pp. 372–378 (2002)
19. Xiao, Y.J., Li, Y.F.: Stereo vision based on perspective invariance of NURBS curves. In: IEEE International Conference on Mechatronics and Machine Vision in Practice, vol. 2, pp. 51–56 (2001)
20. Yang, H., Wang, W., Sun, J.: Control point adjustment for B-spline curve approximation. Computer-Aided Design 36, 639–652 (2004)

3D Computation of Gray Level Co-occurrence in Hyperspectral Image Cubes

Fuan Tsai, Chun-Kai Chang, Jian-Yeo Rau, Tang-Huang Lin, and Gin-Ron Liu

Center for Space and Remote Sensing Research
National Central University
300 Zhong-Da Road
Zhongli, Taoyuan 320 Taiwan
Tel.: +886-3-4227151 ext. 57619,
Fax: +886-3-4254908
ftsai@csrsr.ncu.edu.tw

Abstract. This study extended the computation of GLCM (gray level co-occurrence matrix) to a three-dimensional form. The objective was to treat hyperspectral image cubes as volumetric data sets and use the developed 3D GLCM computation algorithm to extract discriminant volumetric texture features for classification. As the kernel size of the moving box is the most important factor for the computation of GLCM-based texture descriptors, a three-dimensional semi-variance analysis algorithm was also developed to determine appropriate moving box sizes for 3D computation of GLCM from different data sets. The developed algorithms were applied to a series of classifications of two remote sensing hyperspectral image cubes and comparing their performance with conventional GLCM textural classifications. Evaluations of the classification results indicated that the developed semi-variance analysis was effective in determining the best kernel size for computing GLCM. It was also demonstrated that textures derived from 3D computation of GLCM produced better classification results than 2D textures.

1 Introduction

Texture is one of the most important features used in various computer vision and image applications. In visual interpretation as well as digital processing and analysis of remote sensing images, texture is regarded as an essential spatial characteristics and commonly used as an index for feature extraction and image classification, especially when working on high resolution airborne and satellite imagery. Computerized texture analysis focuses on structural and statistical properties of spatial patterns on digital images. These methods have been applied successfully to solve sophisticated problems, such as image segmentation [1], content-based image retrieval [2] and detecting invasive plant species [3]. Previous studies [4, 5, 6] indicated that statistics-based texture approaches are very suitable for analyzing images of natural scenes and perform well in image classification. Among the various texture computing methods, gray level co-occurrence

A.L. Yuille et al. (Eds.): EMMCVPR 2007, LNCS 4679, pp. 429–440, 2007.
© Springer-Verlag Berlin Heidelberg 2007

matrix (GLCM) originally presented by Haralick et al. [7] is probably the most commonly adopted algorithm, especially for textural feature extraction and classification of remote sensing images.

Conventional texture analysis algorithms compute texture properties in a two-dimensional (2D) image space. This may work well in panchromatic (single-band) images and multispectral imagery with limited and discrete spectral bands. However, as imaging technologies evolve, new types of image data with volumetric characteristics have emerged, for example, magnetic resonance imaging (MRI) in medical imaging and hyperspectral images in remote sensing. Directly applying traditional 2D texture analysis algorithms to these new types of imaging data will not able to fully explore three-dimensional (3D) texture features in the volumetric data sets. To address this issue, this study undertook the development of extending conventional 2D GLCM texture computation into a 3D form for better texture feature extraction and classification of hyperpectral remote sensing images.

2   Hyperspectral Volumetric Texture Analysis

Hyperspectral imaging is an emerging technology in remote sensing. With tens to hundreds of contiguous spectral bands covering visible to short-wavelength infrared spectral regions, hyperspectral remote sensing data provide rich information about ground coverage. Because of the high resolution and abundant details in the spectral domain, most existing hyperspectral analysis algorithms focused on extracting spectral features from the data sets. For example, the minimum noise fraction (MNF) transformation, spectrally segmented principal component analysis [8] and derivative spectral analysis [9,10], all aimed at extracting useful spectral features from complex hyperspectral data sets. For texture analysis of hyperspectral imagery, most researchers applied conventional 2D texture algorithms to a single band at a time and collected these 2D textures for subsequent analysis. However, with the contiguous spectral sampling, a hyperspectral data set can be considered as an image cube with volumetric characteristics as illustrated in Fig. 1. Consequently, it should be possible to treat hyperspectral imagery as volumetric data and investigate texture features in a 3D manner.

Currently, related works and applications in volumetric texture analysis are still limited. A voxel co-occurrence matrix similar to GLCM was introduced by Gao [11] to visualize and interpret 3D seismic data. A similar approach was also used in analyzing MRI data [12]. Bhalerao and Reyes-Aldasoro [13] also demonstrated a volumetric texture description for MRI based on a sub-band filtering technique similar to the Gabor decomposition [14]. Another texture description for medical imagery based on gray level run-length and class distance was proposed and achieved promising results [1, 15]. Suzuki et al. [16] also extended HLAC (higher order local autocorrelation) shape descriptors into 3D mask patterns for the classification of solid textures. These methods had one thing in common, i.e. they all dealt with isolating specific objects (body parts, organ tissues etc.) from volumetric data sets. Although they worked well in identifying target boundaries (shapes), they might not be suitable for extracting general

Fig. 1. Hyperspectral imagery as an image cube

texture features in hyperspectral imagery. Other types of texture description, such as models derived from Markov Random Field [17] and fractal geometry [18], might be able to extend to 3D forms, but the complexity and expense in computation could seriously limit their usability in analyzing hyperspectral image cubes. For hyperspectral remote sensing data containing natural scenes, a general gray level statistics based texture descriptor might be more appropriate and likely to achieve satisfactory feature extraction and classification results.

3 Methods and Materials

Texture features derived from GLCM are so-called second order texture calculations because they are based on the joint co-occurrence of gray values for pairs of pixels at a given distance and direction.

3.1 3D GLCM Computation

For a hyperspectral image cube with n levels of gray values, the co-occurrence matrix, M, is a n by n matrix. Values of the matrix elements within a moving box, W, at a given displacement, $d = (dx, dy, dz)$, are defined as

$$M(i,j) = \sum_{z=1}^{W_z - d_z} \sum_{x=1}^{W_x - d_x} \sum_{y=1}^{W_y - d_y} CONDITION$$

$$CONDITION = (G(x,y,z) = i \wedge G(x + d_x, y + d_y, z + d_z) = j)?1:0$$

(1)

where x, y, z are denoted as the position in the moving box. In other words, the value of a 3D GLCM element, $M(i,j)$, reflects that within a moving box, how often the gray levels of two pixels, $G(x,y,z)$ and $G(x+d_x, y+d_y, z+d_z)$, with the spatial relationship of d, are equal to i and j, respectively. Theoretically, there can be numerous combinations of the spatial relationship or the displacement vector, d. However, for the simplification of computation, it is usually set as one pixel in distance and 13 combinations in horizontal and vertical directions.

The original GLCM reference [7] suggested 14 statistical measures to evaluate the properties of GLCM. However, some of them are highly correlated and only a few are recommended for use with remote sensing imagery because they are more suitable for describing features in natural scenes [19,20,21]. Four statistical measures were used in this study, including contrast (CON), entropy (ENT), homogeneity (HOM) and angular second moment (ASM) as listed from Eq. (2) to Eq. (5).

$$CON = \sum\sum \left[(i-j)^2 M_{ij} \right] \tag{2}$$

$$ENT = \sum\sum (M_{ij} \cdot \log M_{ij}) \tag{3}$$

$$HOM = \sum\sum \left\{ \frac{1}{1+(i-j)^2} M_{ij} \right\} \tag{4}$$

$$ASM = \sum\sum M_{ij}^2 \tag{5}$$

3.2   Semi-variance Analysis

Among the parameters affecting GLCM-based texture analysis, the size of the moving box (kernel) has the most significant impact. A previous study demonstrated that kernel size accounted for 90% of the variability in textural classification [22]. During the evaluation, it usually requires a large kernel size to obtain meaningful descriptions of the entire data set. However, for texture segmentation, a small moving box size is preferred in order to accurately locate boundaries between different texture regions. Therefore, it is critical to determine the most appropriate moving box size for GLCM calculations. In this regard, semi-variance analysis has been proved an effect method to find the best moving box size for GLCM computation [23,3].

Let $Z(x_i)$ and $Z(x_i + d)$ be two pixels with a lag of d (a vector of specific direction and distance) in three dimension. For all pixel pairs in a volumetric data set, the semi-variance is defined as

$$\gamma(d) = \frac{1}{2N(d)} \sum [Z(x_i) - Z(x_i + d)]^2 \tag{6}$$

where $N(d)$ is the number of pixel pairs in the data set. A typical semi-variance curve is shown in Fig. 2. In practice, training regions of interested targets were selected from the data set to produce variance curves of different targets. The purpose was to find the range where the semi-variance would reach its maximum (sill).

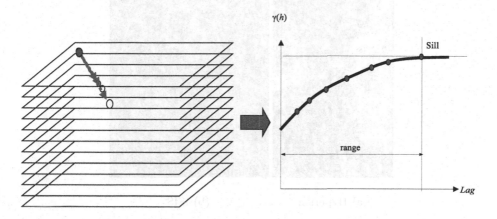

Fig. 2. Typical semi-variance curve of a 3D image cube

3.3   Test Data

Two hyperspectral data sets as displayed in Fig. 3 were used to test the performance of the developed algorithms of 3D GLCM computation and semi-variance analysis. The first data set was an EO-1 Hyperion image acquired in Jan. 2004, which covers the Heng Chun peninsula of southern Taiwan. Hyperion is a spaceborne hyperspectral imaging spectrometer (http://eo1.usgs.gov/hyperion.php). It has 220 spectral bands covering 400-2500 nm in wavelength at a spectral sampling interval of 10 nm and a nominal 30 meter spatial resolution. Because of the low signal-to-noise ratio in the longer wavelength region, only forty five continuous bands (band-11 to band-55) in the visible to near infrared (up to the first water absorption region) were extracted from the original scene and resulting in a 481x255x45 image cube for testing.

The second image cube used was acquired with an experimental high resolution airborne hyperspectral imager called Intelligent Spectral Imaging System (ISIS) (http://www.itrc.org.tw/Publication/Newsletter/no75/p08.php) in Sep. 2006. ISIS is a pushbroom instrument with 218 spectral bands (430-945 nm at 3.5-5 nm spectral resolution). The ISIS scene has a 1.5 meter spatial resolution and covers a mountainous area with rich natural and planted forests in central Taiwan. Spectral bands (band-20 to band-210) of the same wavelength region used in the Hyperion data set were extracted from the original ISIS imagery. An 800 pixels by 800 pixels sub-image centered with nadir track was selected as the test area to minimize variations caused by the spectral "smile" effect [24] commonly seen in pushbroom sensors. Therefore, the testing ISIS data set was a 800x800x190 image cube.

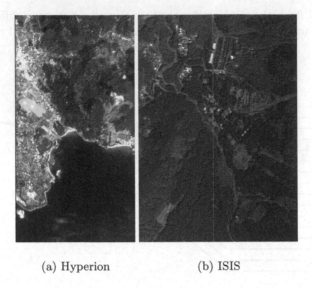

(a) Hyperion   (b) ISIS

Fig. 3. False color hyperspectral imagery

Other supplementary data included photo-maps, high resolution aerial photographs and landcover maps of the study areas. These data were primarily used for geo-referencing (registering) the original images, selection of training regions for semi-variance analysis and supervised classification as well as evaluating classification results.

4   Results and Discussions

Several tests were conducted on the two image cubes to evaluate performance of the developed algorithms for 3D computation of GLCM. First, a series of 3D semi-variance analysis were applied to the two image cubes. Fig. 4 shows examples of the semi-variance curves of four different targets to classify in the Hyperion data. Different colored curves in Fig. 4 represent semi-variances at different directions (azimuth, zenith) as labeled in the bottom of each plot. One thing to note in the plots of Fig. 4 is the divergency of the red curve for each target. The red semi-variance curves were computed along direction $(0, 0)$. Unlike MRI or other solid data sets, the third (Z) axis of a hyperspectral image cube is the spectrum instead of a geometric axis. Therefore, direction $(0, 0)$ tried to calculate variance of the same pixel at two wavelengths without any spatial consideration, thus diverging as the lag increased.

Semi-variance analysis in Fig. 4 indicated that 5 was the best kernel size for the textural analysis of the Hyperion data. To test this hypothesis, three GLCMs were generated with 3x3x3, 5x5x5 and 7x7x7 moving boxes. Supervised classifications were conducted on aforementioned four statistical measures with exactly the same training and verification data randomly selected from ground truth

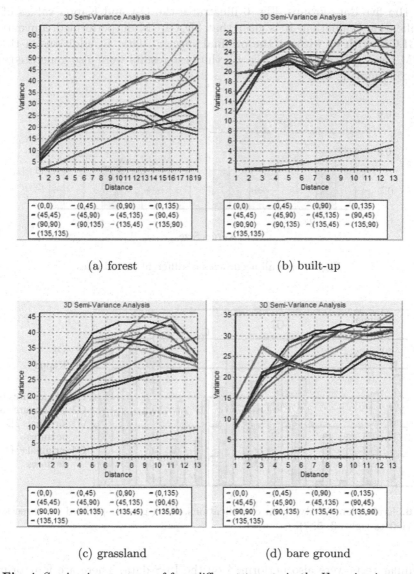

(a) forest

(b) built-up

(c) grassland

(d) bare ground

Fig. 4. Semivariance curves of four different targets in the Hyperion imagery

landcover maps. The overall accuracies (OA) of the classifications are plotted in Fig. 5. In this test, moving box of 5x5x5 produced the best results except the CON. This has validated the effectiveness of the developed 3D semi-variance analysis. In addition, to compare 3D GLCM computation with 2D GLCM, more thorough classifications were conducted on features extracted from five feature collections, including original spectral data, textures from 2D GLCM, textures from 3D GLCM, original plus 2D textures and original plus 3D textures with the three moving box sizes. Principal component analysis was used to select features from the five data groups. Fig. 6 illustrates the OA and Kappa values of the classification evaluation. It is clear that 3D GLCM outperformed conventional 2D GLCM with or without the original spectral data. In addition, in the 3D GLCM cases (G2 and G4 in Fig. 6), the best results were also generated from the 5x5x5 moving box.

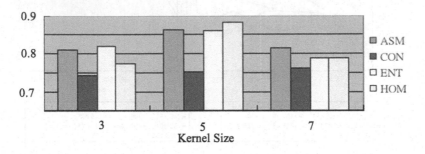

Fig. 5. Overall accuracies of different kernel sizes

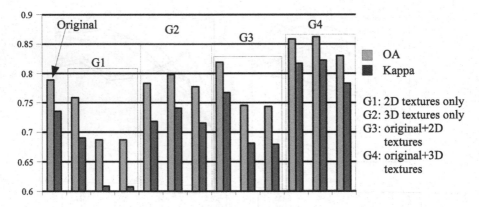

Fig. 6. Evaluations of Hyperion classifications. Each group operated on three kernel sizes (left to right: 3, 5, 7).

Similar tests were also performed on the ISIS data. There are four primary vegetation ground coverages in the ISIS scene, including Taiwania fir, Japanese cedar, maple, and bamboo. Fig. 7 displays the training regions selected for semi-variance analysis and the classification results based on 2D and 3D textures calculated with a kernel size of 5. The four vegetation types are color coded as red, dark red, green and blue, respectively in Fig. 7. The training regions were selected according to landcover maps provided by a local forestry administration agency. A visual inspection on Fig. 7 reveals that 2D textural classification had completely misclassified the fir and cedar classes as maple or bamboo, while 3D textures identified most of the two classes (as well as the other two) correctly.

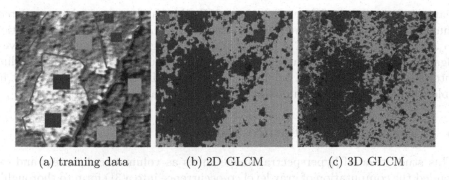

(a) training data   (b) 2D GLCM   (c) 3D GLCM

Fig. 7. Classification results of the ISIS data with 2D and 3D GLCM features

Semi-variance analysis on the ISIS data set suggested that 5x5x5 and 7x7x7 moving boxes were the best for 3D computation of GLCM. A series of classifications similar to the ones applied to the Hyperion data were also carried out on the ISIS data for a quantitative evaluation. The evaluation results are displayed in Fig. 8. In general, the best classification was resulted from features extracted from original spectral data plus 3D textures computed with a 7x7x7 moving box. However, it was noted that the OA differences between 3D and 2D textural classifications were insignificant. Part of the reason is because OA is an overly optimistic evaluation for classification accuracy since it does not account for omission errors. This can be contended by the observation that OA values in Fig. 8 do not reflect the high omission errors in the 2D textural classification result of Taiwania fir (red) and Japanese cedar (dark red) categories as displayed in Fig. 7. The relatively lower Kappa of 2D textural classification results is another indication of the uncertainty. Another possible reason might have to do with the characteristics of ISIS data. Because of the fine spectral resolution, texture

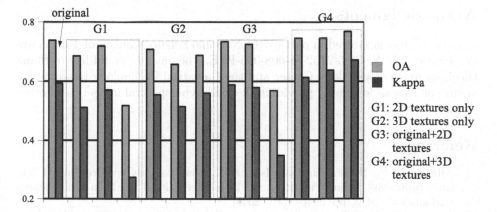

Fig. 8. Evaluations of ISIS classifications. Each group operated on three kernel sizes (left to right: 3, 5, 7).

features derived from 3D computation of GLCM may become highly correlated, thus degrading the classification performance. The impact of spectral resolution to 3D computation of GLCM in hyperspectral image cubes is still under investigation. Nonetheless, resampling the spectral resolution to a broader sampling interval (for example, from 3.5-5 nm to 10 nm as the Hyperion data) might be able to enhance the discriminability of hyperspectral 3D textures.

5   Conclusion and Future Work

This study treated hyperspectral image cubes as volumetric data sets and extended the computation of gray level co-occurrence into a 3D form to thoroughly explore volumetric texture features of hyperspectral remote sensing data. A 3D semi-variance analysis algorithm was also developed to obtain appropriate moving box (kernel) sizes for computing gray level co-occurrence in 3D image cubes. Results of tests conducted on two hyperspectral image cubes validated that the developed semi-variance algorithm was effective in determining the best moving box sizes for 3D texture description. The experiments also demonstrated that texture features derived from 3D computation of GLCM provided better classification results than features collected from conventional 2D GLCM calculations.

The results of this study suggest that 3D computation of gray level co-occurrence should be a viable approach to extract volumetric texture features from hyperspectral image cubes for classification. It is also possible to apply these techniques to other remote sensing data with volumetric characteristics, such as LiDAR data with multiple returns or electro-magnetic scans. However, there are still issues for improvement. For example, the impact of spectral resolution to the correlations of generated texture features will need to be studied in detail. Another interested research topic derived from this study will be to further develop a third-order texture descriptor to truly represent three-dimensional texture features of complicated hyperspectral and other volumetric data.

Acknowledgments

This study was supported in part by the National Science Council of Taiwan under Project No. NSC-95-2752-M-008-005-PAE. The authors would like to thank the Instrument Technology Center and the Industrial Technology Research Institute of Taiwan for kindly providing the ISIS hyperspectral imagery and other data.

References

1. Albregtsen, F., Nielsen, B., Danielsen, H.E.: Adaptive gray level run length features from class distance matrices. In: 15th International Conference on Pattern Recognition. vol. 3, pp. 3746–3749 (2000)
2. Jhanwar, N., Chaudhuri, S., Seetharaman, G., Zavidovique, B.: Content based image retrieval using motif cooccurrence matrix. Image and Vision Computing 22(12), 1211–1220 (2004)

3. Tsai, F., Chou, M.J.: Texture augmented analysis of high resolution satellite imagery in detecting invasive plant species. Journal of the Chinese Institute of Engineers 29(4), 581–592 (2006)
4. du Buf, J.M.H., Kardan, M., Spann, M.: Texture feature performance for image segmentation. Pattern Recognition 23(4), 291–309 (1990)
5. Ohanian, P.P., Dubes, R.C.: Performance evaluation of four classes of texture features. Pattern Recognition 25(8), 819–833 (1992)
6. Reed, T.R., du Buf, J.M.H.: A review of recent texture segmentation and feature extraction techniques. CVGIP: Image Understanding 57(3), 359–372 (1993)
7. Haralick, R.M., Shanmugan, K., Dinstein, I.: Texture features for image classification. IEEE Trans. Systems, Man Cybernetics 3(6), 610–621 (1973)
8. Tsai, F., Lin, E.K., Yoshino, K.: Spectrally segmented principal component analysis of hyperspectral imagery for mapping invasive plant species. International Journal of Remote Sensing 28(5-6), 1023–1039 (2007)
9. Tsai, F., Philpot, W.: A derivative-aided image analysis system for land-cover classification. IEEE Transactions on Geoscience and Remote Sensing 40(2), 416–425 (2002)
10. Tsai, F., Philpot, W.D.: Derivative analysis of hyperspectral data. Remote Sensing of Environment 66(1), 41–51 (1998)
11. Gao, D.: Volume texture extraction for 3D seismic visualization and interpretation. Geophsics 68(4), 1294–1302 (2003)
12. Mahmoud-Ghoneim, D., Toussaint, D., Constans, J.M., de Certaines, J.D.: Three dimensional texture analysis in MRI: a preliminary evaluation in glioma. Magnetic Resonance Imaging 21(9), 983–987 (2003)
13. Bhalerao, A., Reyes-Aldasoro, C.C.: Volumetric texture description and discriminant feature selection for MRI. In: Moreno-Díaz Jr., R., Pichler, F. (eds.) EUROCAST 2003. LNCS, vol. 2809, pp. 573 584. Springer, Heidelberg (2003)
14. Unser, M.: Texture classification and segmentation uisng wavelet frames. IEEE Trans. on Image Processing 4(11), 1549–1560 (1995)
15. Nielsen, B., Albregtsen, F., Danielsen, H.E.: Low dimensional adaptive texture feature vectors from class distance and class difference matrices. IEEE Trans. Medical Imaging 23(1), 73–84 (2004)
16. Suzuki, M.T., Yoshino, Y., Osawa, N., Sugimoto, Y.Y.: Classification of solid texture using 3D mask patterns. In: 2004 IEEE International Conference on Systems, Man and Cybernetics, pp. 6342–6347 (2004)
17. Cross, G.R., Jain, A.K.: Markov Random Field texture models. IEEE Trans. on PAMI 5(1), 25–39 (1983)
18. Keller, J.M., Chen, S.: Texture description and segmentation through fractal geometry. Computer Vision, Graphics and Image Processing 45, 150–160 (1989)
19. Baraldi, A., Parmiggiani, F.: An investigation of the textural characteristics associated with gray level cooccurrence matrix statistical parameters. IEEE Transactions on Geoscience and Remote Sensing 33, 293–303 (1995)
20. Clausi, D.A.: An analysis of co-occurrence statistics as a function of grey level quantization. Canadian J. Remote Sensing 28, 45–62 (2002)
21. Jobanputra, R., Clausi, D.A.: Preserving boundaries for image texture segmentation using grey level co-occurring probabilities. Pattern Recognition 39, 234–245 (2006)

22. de Martino, M., Causa, F., Serpico, S.B.: Classification of optical high resolution images in urban environment using spectral and textural information. In: IEEE International Geoscience and Remote Sensing Symposium (IGARSS'03). vol. 1, pp. 467–469 (2003)
23. Kourgli, A., Belhadj-Aissa, A.: Texture primitives description and segmentation using variography and mathematical morphology. In: IEEE International Conference on Systems, Man and Cybernetics. vol. 7, pp. 6360–6365 (2004)
24. Datt, B., McVicar, T.R., van Niel, T.G., Jupp, D.L.B., Pearlman, J.S.: Preprocessing EO-1 Hyperion hyperspectral data to support the application of agricultural indexes. IEEE Transactions on Geoscience and Remote Sensing 41, 1246–1259 (2003)

Continuous Global Optimization in Multiview 3D Reconstruction

Kalin Kolev[1], Maria Klodt[1], Thomas Brox[1],
Selim Esedoglu[2], and Daniel Cremers[1]

[1] Department of Computer Science, University of Bonn, Germany
[2] Department of Mathematics, University of Michigan, USA

Abstract. In this work, we introduce a robust energy model for multiview 3D reconstruction that fuses silhouette- and stereo-based image information. It allows to cope with significant amounts of noise without manual pre-segmentation of the input images. Moreover, we suggest a method that can globally optimize this energy up to the visibility constraint. While similar global optimization has been presented in the discrete context in form of the maxflow-mincut framework, we suggest the use of a continuous counterpart. In contrast to graph cut methods, discretizations of the continuous optimization technique are consistent and independent of the choice of the grid connectivity. Our experiments demonstrate that this leads to visible improvements. Moreover, memory requirements are reduced, allowing for global reconstructions at higher resolutions.

1 Introduction

We consider the classical problem of inferring a dense 3D structure reconstruction of an object from a collection of views calibrated to a common world coordinate system. Among the multitude of existing methods one can distinguish between two major classes of techniques according to the exploited image information: shape from silhouettes and stereo.

In case of sparsely textured objects, silhouette-based methods exhibit favorable performance. Most of them aim at approximating the *visual hull* [18] of the imaged object. The visual hull is an outer approximation of the observed solid, constructed as the intersection of the visual cones associated with all image silhouettes. The earliest attempts use a volumetric representation of the scene, where each voxel is labeled as opaque or transparent according to each projection onto the images [20]. Latter developments led to the use of surface-based representations, which allow to impose regularization in an energy minimization framework. These methods are able to reconstruct a smooth version of the visual hull from the raw input images without the immediate need for manually outlined silhouettes [24,28]. This is because the segmentation of each image is obtained through the evolution of a single surface in 3D rather than separate contours in 2D. As a result, such methods exhibit considerable robustness to outliers and erroneous camera calibration. In [16] the robustness to noise and

A.L. Yuille et al. (Eds.): EMMCVPR 2007, LNCS 4679, pp. 441–452, 2007.
© Springer-Verlag Berlin Heidelberg 2007

initialization is further increased by incorporating all available observations into a probabilistic framework.

The main drawback of silhouette-based approaches is their inability to reconstruct concavities, since these do not affect the silhouettes. Stereo-based methods capture such indentations by measuring photoconsistency of surface patches in space. The fundamental idea is that only points on the object's surface have a consistent appearance in the input images, while all other points project to incompatible image patches. The earliest algorithms use carving techniques to obtain a volumetric representation of the scene by repeatedly eroding inconsistent voxels [17]. They do not enforce the smoothness of the surface, which often results in rather noisy reconstructions. Later, energy minimization techniques based on the integration of the data fidelity criterion on the unknown surface, have become more popular [9,11,19]. In these works, one seeks the surface with the smallest weighted area, where the weights reflect the local photoconsistency.

Some recent approaches use a fusion of silhouette constraints and stereo information in order to achieve consistency in terms of silhouettes as well as image patches. Generally, there are two types of techniques to combine silhouette information and photoconsistency. The first strategy integrates silhouette constraints into stereo-based optimization [10,23,26]. The alternative is to use the visual hull merely as initialization for a stereo-based technique [27].

In this paper, we present an energy model which generalizes [16] by imposing photoconsistency constraints. Since for computing photoconsistency one needs the visibility of surface points, the photoconsistency term is collapsed at the beginning. With the resulting approximate visibility information, we can globally optimize the energy that includes both constraints. Our approach is related to the one introduced in [27]. However, the sought surface is not restricted to lie within some predefined band around the visual hull, which imposes different weighting of silhouette and stereo term. Another closely related work is the one of [19]. However, in this approach the silhouette constraint is replaced by a constant ballooning term that persistently prefers larger surfaces. To this end, visibility estimation is based on local graph edge orientations.

All previous methods use either local optimization, which is prone to instabilities and getting stuck in local minima, or *discrete* global optimization based on graph cuts. However, graph cuts can only minimize a certain class of discrete energies that are inconsistent to a corresponding continuous formulation, i.e., the solution does not converge to the continuous solution for finer grids. Thus, graph cuts are not rotationally invariant and favor polyhedral structures. In the scope of multiview reconstruction, an additional practical limitation is the relatively large memory consumption of graph cut methods, which can be decisive when computing reconstructions at a high resolution. The main contribution of the present work is the development of a novel technique for continuous global optimization for multiview reconstruction, which allows to avoid previously mentioned limitations. Similar techniques were recently proposed in the context of image segmentation [8]. In [1] another method for global optimization has been proposed, which has been extended in [2] to 3D segmentation. However, this

Table 1. Optimization techniques used in image segmentation and multiview reconstruction

	continuous local optimization	discrete global optimization	continuous global optimization
image segmentation	Snakes [13] Level Sets [5]	Graph Cuts [3]	TV-L1 [8] CSPs [1]
multiview reconstruction	Mesh-based [10] Level Sets [11]	Graph Cuts [19,23,26,27]	this work

approach does not allow to incorporate regional information, which makes it inappropriate for our model. Both techniques were inspired by the original works of [12], [25]. Table 1 provides a number of representative works on local optimization, discrete and continuous global optimization in the context of image segmentation and multiview reconstruction, respectively.

The paper is laid out as follows. The next section contains a brief reviewing of related continuous global optimization techniques in the context of image segmentation. In Section 3 we present and discuss the energy model. Section 4 is devoted to the optimization technique including implementation details. We show experimental results in Section 5 and conclude the paper with a brief summary in Section 6.

2 Convex Formulations of Image Segmentation

In a series of works [8,4,7] image segmentation functionals, namely the two-phase piecewise constant Mumford-Shah model [21] and the snakes [14] were addressed by means of convex formulations. The key idea is to represent region-integrals by means of a binary variable $u : \Omega \subset \mathbb{R}^2 \to \{0,1\}$ indicating foreground and background. The weighted length term proposed in the snakes and the geodesic active contours [6,15] can then be expressed by means of a weighted total variation (TV) norm [22,4]:

$$TV_g(u) = \int_{\Omega} g(|\nabla I|) \, |\nabla u| \, dx, \tag{1}$$

with an edge indicator function $g(|\nabla I|)$ that provides the local metric.

Since the space of binary functions is a non-convex space, also the respective optimization problems are non-convex. However, in [8] it was found that when minimizing the total variation norm over all real-valued functions $u : \Omega \to \mathbb{R}$, the values of $u(x)$ converge to $\pm\infty$ almost everywhere. Therefore the segmentation can be cast as a convex problem on the convex space of functions $u : \Omega \to [0,1]$ by enforcing $0 \leq u(x) \leq 1$ via a convex penalizer [8]

$$\theta(u) := \max \left\{ 0, 2 \left| u - \frac{1}{2} \right| - 1 \right\}. \tag{2}$$

Minimization over the space of real-valued functions and subsequent threshold will then lead to a global minimizer of the respective segmentation problems.

In this work, we will revisit these ideas and show that under appropriate assumptions the multiview reconstruction problem can be cast as a spatially continuous convex optimization problem. Moreover we will introduce an efficient numerical solution by means of Successive Overrelaxation (SOR).

3   A Continuous Energy Model for Multiview Reconstruction

Let $V \subset \mathbb{R}^3$ be a volume, which contains the scene of interest, and $I_1, \ldots, I_n :$ $\Omega \to \mathbb{R}^3$ a collection of calibrated color images with perspective projections $\pi_1, \ldots, \pi_n$. We are looking for some surface $\hat{S} \subset V$ that gives rise to these images. This can be formulated as the energy minimization problem

$$E(S) = -\int_{R_{obj}^S} \log P_{obj}(x) \, dx - \int_{R_{bck}^S} \log P_{bck}(x) \, dx + \nu \int_S \rho(x) \, dA \to \min. \quad (3)$$

The energy consists of two parts. The first two terms impose the silhouette constraint via a probabilistic segmentation of the volume into object and background. The third term acts as a constraint both for smoothness and photoconsistency by seeking the minimal surface with respect to a Riemannian metric. The parameter ν controls the weighting of both parts of the energy.

The definition of the probability terms follows [16]. Regarding the silhouette constraint, according to a certain surface estimate S, all points in V can be divided into two classes: lying inside S or belonging to the background, i.e. $V = R_{obj}^S \cup R_{bck}^S$, where R_{obj}^S denotes the interior and R_{bck}^S the exterior. Considering the given image content we can assign each point $x \in V$ two probabilities $P_{obj}(x)$ and $P_{bck}(x)$ associated with R_{obj}^S and R_{bck}^S, respectively. More precisely

$$\begin{aligned} P_{obj}(x) &:= P(\{I_l(\pi_l(x))\}_{l=1,\ldots,n} \mid x \in R_{obj}^S) \\ P_{bck}(x) &:= P(\{I_l(\pi_l(x))\}_{l=1,\ldots,n} \mid x \in R_{bck}^S). \end{aligned} \quad (4)$$

Note that in this formulation $P_{obj}(x)$ and $P_{bck}(x)$ will generally not sum to 1. Considering dependence of the image observations we can write

$$P_{obj}(x) = \sqrt[n]{\prod_{i=1}^n P(I_i(\pi_i(x)) \mid x \in R_{obj}^S)}$$

$$P_{bck}(x) = 1 - \sqrt[n]{\prod_{i=1}^n [1 - P(I_i(\pi_i(x)) \mid x \in R_{bck}^S)]}. \quad (5)$$

The probability of a voxel being part of the foreground is equal to the probability that *all* cameras observe this voxel as foreground, whereas the probability of background membership describes the probability of *at least one* camera seeing

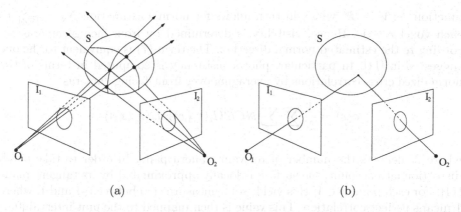

Fig. 1. Orthogonal image features used for multiview 3D reconstruction: shape from silhouettes vs. shape from stereo. (a) Region-based color information. Surface concavities are not presented in the reconstruction. (b) Stereo-based point matching. Surface indentations can also be captured.

background. The root is for normalization with respect to the number of camera views, since both products will converge to 0 for $n \rightarrow \infty$. Thus, dependency between single image observations is expressed in terms of their geometric mean. Note that the fusion of all available image observations allows for quite robust silhouette-based surface estimation.

The foreground/background probabilities for the single image observations

$$P(I_i(\pi_i(x)) \mid x \subset R^S_{obj}) \sim \mathcal{N}(\mu_{obj}, \Sigma_{obj})$$
$$P(I_i(\pi_i(x)) \mid x \in R^S_{bck}) \sim \mathcal{N}(\mu_{bck}, \Sigma_{bck}). \tag{6}$$

are modeled to be Gaussian distributed. The parameters of both models, i.e. mean vectors and covariance matrices, can be updated during optimization by projecting the current surface estimate onto the images in order to collect pixels, which belong to the respective regions. However, in our implementation we replace this iterative scheme by estimating the parameters interactively marking a small object and background region in one of the images. This is a requirement for the energy to be globally minimizable. Minimization of the first two terms in (3) results in the most probable surface with respect to the probability distributions P_{obj} and P_{bck}.

The last term in (3)

$$E_{stereo}(S) = \int_S \rho(x) \, dA \tag{7}$$

accounts for photoconsistency and smoothness of the sought surface. It is particularly important in order to reconstruct concavities that are not visible from the silhouettes; see Figure 1 for an illustration of the conceptual difference between the silhouette- and stereo-based constraints. Computation of ρ requires visibility estimation. To this end, we minimize the energy with Euclidean regularizer $\rho(x) = 1$. From the resulting surface, one can compute a signed distance function $\phi : V \to \mathbb{R}$, which in turn allows for normal estimation $N_x = \frac{\nabla \phi}{|\nabla \phi|}$ to each voxel $x \in V$. Hence, visibility is determined by front-facing cameras according to the estimated normal direction. The term (7) is equivalent to the one suggested in [11]. In particular, photoconsistency is computed in terms of the normalized cross-correlations by averaging over front-facing cameras

$$c(x) = \frac{1}{N} \sum_i \sum_j NCC(I_i(\pi_i(x)), I_j(\pi_j(x))), \tag{8}$$

where N denotes the number of relevant camera pairs. In order to take patch distortion into account, the surface is locally approximated by its tangent plane [11]. For each point $x \in V$ this yields some measure $c(x)$ between -1 and 1, where 1 means perfect correlation. This value is then mapped to the unit interval $[0, 1]$ using the following function proposed in [27]:

$$\rho(x) = 1 - \exp\left(-\tan\left(\frac{\pi}{4}(c(x) - 1)\right)^2 / \sigma^2 \right). \tag{9}$$

Smoothness is implicitly enforced since minimizing (7) corresponds to finding the minimal surface with respect to a Riemannian metric [5]. Note that global optimization of this energy alone yields the empty surface. In our energy (3), the silhouette-based terms naturally prevent the empty surface without requiring additional knowledge about the scene.

4   Continuous Global Optimization

4.1   An Equivalent Convex Formulation

Energy (3) can be globally optimized, provided the object and background parameters of the Gaussian distribution and the visibility of points are given. In this paper we build upon the optimization technique described in Section 2 by formulating (3) as a continuous convex optimization problem.

To this end, the surface S is represented implicitly by the characteristic function $u : V \to \{0, 1\}$ of R_{bck}^S, i. e. $u = \mathbf{1}_{R_{bck}^S}$ and $1 - u = \mathbf{1}_{R_{obj}^S}$. Hence, changes in the topology of S are handled automatically without reparametrization. With the implicit surface representation we have the following constrained, non-convex energy minimization problem corresponding to (3):

$$E(u) = \int_V (\log P_{obj}(x) - \log P_{bck}(x)) u(x)\, dx + \nu \int_V \rho(x) |\nabla u|\, dx \to \min, \tag{10}$$

$$\text{s. t. } u \in \{0, 1\}.$$

The minimization problem stated in (10) is non-convex, since the optimization is carried out over a non-convex set of binary functions. However, relaxing the binary condition and extending the optimization to all functions $u : V \to \mathbb{R}$, where also intermediate values can be taken, will cause the values of $u(x)$ to converge to $\pm\infty$ almost everywhere. In order to circumvent this difficulty, one can restrict the domain by enforcing $0 \le u(x) \le 1$ via a convex penalizer $\theta(u)$:

$$E(u) = \int_V \Big(\log P_{obj}(x) - \log P_{bck}(x) \Big) u(x) + \nu\rho(x)|\nabla u| + \alpha\theta(u(x)) \, dx, \quad (11)$$

where α has to be chosen sufficiently large in order to ensure that u does not leave the interval $[0,1]$. This leads to a convex formulation, which allows for global optimization by using standard techniques like gradient descent. Finally, we come up with a global minimizer of the original non-convex functional (10) by thresholding the result at any $\mu \in (0,1)$. In our experiments, we chose $\mu = 0.5$, but we obtained virtually the same results with $\mu \in [0.1, 0.9]$.

In summary, the optimization can be split into two steps:

1. Find a minimizer u of (11).
2. Threshold the result: $R_{obj}^S = \{x \in V \mid u(x) < \mu \text{ for some } \mu \in (0,1)\}$.

A necessary condition for a minimum of (11) is stated by the associated Euler-Lagrange equation

$$0 = (\log P_{obj} - \log P_{bck}) - \nu\rho \operatorname{div}\left(\frac{\nabla u}{|\nabla u|}\right) - \langle \nabla\rho, \frac{\nabla u}{|\nabla u|}\rangle + \alpha\theta'_\epsilon(u)$$

$$= (\log P_{obj} - \log P_{bck}) - \nu \operatorname{div}\left(\rho\frac{\nabla u}{|\nabla u|}\right) + \alpha\theta'_\epsilon(u), \quad (12)$$

where θ_ϵ is a regularized version of the derivative of θ with respect to its argument.

4.2   Fast Minimization by Successive Overrelaxation

Discretization of the Euler-Lagrange equation (12) leads to a sparse nonlinear system of equations, which can be solved via gradient descent. However, gradient descent converges very slowly. Thus, we suggest to use a fixed point iteration scheme that transforms the nonlinear system into a sequence of linear systems. These can be efficiently solved with iterative solvers, such as Gauss-Seidel, successive over-relaxation (SOR), or even multi-grid methods.

Neglecting the term $\alpha\theta'_\epsilon(u)$, which can in practice be replaced by simply clipping values of u that fall out of the interval $[0,1]$, the only source of nonlinearity in (12) is the diffusivity $g := \frac{1}{|\nabla u|}$. Starting with an initialization $u^0 = 0.5$, we can compute g and keep it constant. For constant g, (12) yields a linear system of equations, which we solve with the SOR method. This means, we iteratively compute an update of u at voxel i by

$$u_i^{l,k+1} = (1-\omega)u_i^{l,k} + \omega \frac{\nu \sum\limits_{j\in\mathcal{N}(i),j<i} \rho_j g_{i\sim j}^l u_j^{l,k+1} + \nu \sum\limits_{j\in\mathcal{N}(i),j>i} \rho_j g_{i\sim j}^l u_j^{l,k} - b_i}{\nu \sum\limits_{j\in\mathcal{N}(i)} \rho_j g_{i\sim j}^l} \quad (13)$$

Fig. 2. Some of the input images used for 3D reconstruction. The image sequence is pretty challenging because of the presence of reflections and illumination artefacts.

where $\mathcal{N}(i)$ denotes the neighborhood of i, $g_{i \sim j}$ denotes the diffusivity between voxel i and its neighbor j, and the vector b_i contains the constant part of (12) that does not depend on u, i.e. the fidelity term $b_i = \log P_{obj,i} - \log P_{bck,i}$. The over-relaxation parameter ω has to be chosen in the interval $(0, 2)$ for the method to converge. The optimal value depends on the linear system to be solved. Empirically we obtained the fastest convergence rate for $\omega = 1.85$. After being sufficiently close to a fixed point u^l (we iterated for $k = 1, ..., 10$), one can update the diffusivities and solve the next linear system. Iterations are stopped as soon as the energy decay in one iteration is in the area of number precision.

5   Experiments

Figure 2 depicts 3 of 33 input images of resolution 640×480 used for reconstruction. The input images are pretty challenging because of the presence of illumination artefacts and specular reflections. Note that automatic color-based segmentation of the single images is infeasible due to the similarity in color of the bunny figure and the illumination effects in the background.

Figure 3 shows reconstructions from the above image sequence by using discrete graph cuts and the proposed continuous optimization technique applied on the model described in Section 3. Both reconstructions look accurate. However, the graph cut reconstruction looks generally slightly oversmoothed because of the discrete approximation of the smoothness term. In addition, the proposed minimization exhibits considerable reductions in memory compared to graph cuts (in our implementation about a factor of 20), which allows to perform global optimization at higher volume resolutions. We ran the proposed optimization on an architecture with 2 GB of main memory and volume of more than 20 million voxels (see Figure 4). The corresponding graph cut computation is infeasible for this resolution.

The evolution of an initial surface towards the final result is depicted in Figure 5. Note that the final reconstruction does not depend on the initialization, since the used cost function is minimized globally. A closer look at the evolution process reveals the difference to local optimization techniques like level sets, where the surface evolves coherently, i.e. there are no unnecessary topological changes.

Fig. 3. Multiview reconstruction of the sequence in Figure 2. **First row:** reconstruction with the proposed method. **second row:** reconstruction obtained by minimizing the same energy functional (3) via graph cuts. Volume resolution was set to $108 \times 144 \times 162$. At this resolution both reconstructions look quite similar.

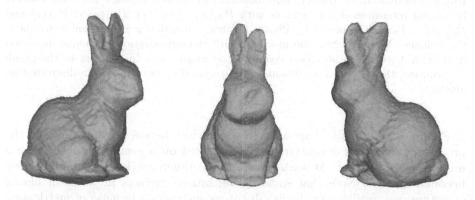

Fig. 4. Reconstruction obtained by the proposed approach at a volume resolution of $216 \times 288 \times 324$. Increasing the resolution by a factor of 2 in each dimension allows for the emergence of fine-scale details (compare to Figure 3). Graph cut reconstruction at such a resolution was infeasible on our machines due to memory overflow.

Fig. 5. Surface evolution towards the final result. Intermediate surfaces were generated by thresholding the evolving function u at 0.5 (see Section 4). In contrast to level set schemes the evolution process is not coherent.

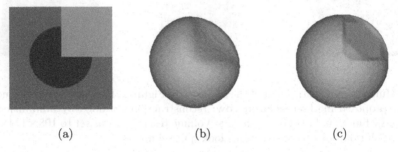

       (a)   (b)   (c)

Fig. 6. Continuous vs. discrete optimization. (a) A slice through the data volume. Increasing intensities denote regions with $P_{obj}(x) > P_{bck}(x)$, $P_{obj}(x) < P_{bck}(x)$ and $P_{obj}(x) = P_{bck}(x)$ respectively. Photoconsistency function ρ is constant throughout the volume. (b) Reconstruction obtained with the optimization technique described in Section 4. (c) Reconstruction computed by graph cuts. In contrast to the graph cut solution, the proposed continuous optimization does not suffer from discretization artefacts.

Figure 6 additionally emphasizes a comparison between graph cuts and the proposed continuous optimization when applied on a synthetic sphere with a missing piece of data. At such locations the difference between both models becomes obvious. Note that some discretization artefacts in terms of blocky structures are available in the graph cut reconstruction because of metrication errors, even with 26-neighborhood system. In addition, sharp corners occur, since the discrete model does not take the curvature of the surface into account. In contrast, the continuous optimization achieves nice and smooth continuation of the missing part of the surface.

6   Summary

In this paper an energy model for multiview 3D reconstruction allowing global optimization is proposed. To the best of our knowledge this is the first work to cast multiview 3D reconstruction as a continuous convex optimization problem (up to visibility). As for graph cuts this allows to compute globally optimal shapes. However, in contrast to discrete techniques, the proposed continuous formulation does not suffer from metrication errors. Moreover, it requires considerably less memory, thereby allowing for optimal reconstructions at higher resolutions. All these properties are demonstrated experimentally.

Acknowledgments

We thank Reinhard Klein and his group for helping us with the data acquisition.

References

1. Appleton, B., Talbot, H.: Globally optimal geodesic active contours. J. Math. Imaging Vis. 23(1), 67–86 (2005)
2. Appleton, B., Talbot, H.: Globally minimal surfaces by continuous maximal flows. IEEE Trans. Pattern Anal. Mach. Intell. 28(1), 106–118 (2006)
3. Boykov, Y., Veksler, O., Zabih, R.: Fast approximate energy minimization via graph cuts. IEEE Transactions on Pattern Analysis and Machine Intelligence 23(11), 1222–1239 (2001)
4. Bresson, X., Esedoglu, S., Vandergheynst, P., Thiran, J.P., Osher, S.: Global minimizers of the active contour/snake model. Technical Report CAM-05-04, Department of Mathematics, University of California at Los Angeles, CA (January 2005)
5. Caselles, V., Catté, F., Coll, T., Dibos, F.: A geometric model for active contours in image processing. Numerische Mathematik 66, 1–31 (1993)
6. Caselles, V., Kimmel, R., Sapiro, G.: Geodesic active contours. In: Proc. Fifth International Conference on Computer Vision, pp. 694–699. IEEE Computer Society Press, Cambridge, MA (1995)
7. Chambolle, A.: Total variation minimization and a class of binary MRF models. In: Rangarajan, A., Vemuri, B., Yuille, A.L. (eds.) EMMCVPR 2005. LNCS, vol. 3757, pp. 136–152. Springer, Heidelberg (2005)
8. Chan, T., Esedoglu, S., Nikolova, M.: Algorithms for finding global minimizers of image segmentation and denoising models. SIAM Journal on Applied Mathematics 66(5), 1632–1648 (2006)
9. Duan, Y., Yang, L., Qin, H., Samaras, D.: Shape reconstruction from 3D and 2D data using PDE-based deformable surfaces. In: Proc. European Conference on Computer Vision, pp. 238–251 (2004)
10. Esteban, C.H., Schmitt, F.: Silhouette and stereo fusion for 3D object modeling. Computer Vision and Image Understanding 96(3), 367–392 (2004)
11. Faugeras, O., Keriven, R.: Variational principles, surface evolution, PDE's, level set methods, and the stereo problem. IEEE Transactions on Image Processing 7(3), 336–344 (1998)
12. Hu, T.C.: Integer Programming and Network Flows. Addison-Wesley, Reading, MA (1969)

13. Kass, M., Witkin, A.: Analyzing oriented patterns. Computer Vision, Graphics and Image Processing 37, 362–385 (1987)
14. Kass, M., Witkin, A., Terzopoulos, D.: Snakes: Active contour models. International Journal of Computer Vision 1, 321–331 (1988)
15. Kichenassamy, S., Kumar, A., Olver, P., Tannenbaum, A., Yezzi, A.: Gradient flows and geometric active contour models. In: Proc. Fifth International Conference on Computer Vision, pp. 810–815. IEEE Computer Society Press, Cambridge, MA (1995)
16. Kolev, K., Brox, T., Cremers, D.: Robust variational segmentation of 3D objects from multiple views. In: Franke, K., Müller, K.-R., Nickolay, B., Schäfer, R. (eds.) Pattern Recognition. LNCS, vol. 4174, pp. 688–697. Springer, Heidelberg (2006)
17. Kutulakos, K.N., Seitz, S.M.: A theory of shape by space carving. International Journal of Computer Vision 38(3), 199–218 (2000)
18. Laurentini, A.: The visual hull concept for visual-based image understanding. IEEE Transactions on Pattern Analysis and Machine Intelligence 16(2), 150–162 (1994)
19. Lempitsky, V., Boykov, Y., Ivanov, D.: Oriented visibility for multiview reconstruction. In: Leonardis, A., Bischof, H., Pinz, A. (eds.) ECCV 2006. LNCS, vol. 3953, pp. 226–238. Springer, Heidelberg (2006)
20. Martin, W.N., Aggarwal, J.K.: Volumetric descriptions of objects from multiple views. IEEE Transactions on Pattern Analysis and Machine Intelligence 5(2), 150–158 (1983)
21. Mumford, D., Shah, J.: Optimal approximations by piecewise smooth functions and associated variational problems. Communications on Pure and Applied Mathematics 42, 577–685 (1989)
22. Rudin, L.I., Osher, S., Fatemi, E.: Nonlinear total variation based noise removal algorithms. Physica D 60, 259–268 (1992)
23. Sinha, S., Pollefeys, M.: Multi-view reconstruction using photo-consistency and exact silhouette constraints: A maximum-flow formulation. In: Proc. International Conference on Computer Vision, pp. 349–356. IEEE Computer Society, Washington, DC (2005)
24. Snow, D., Viola, P., Zabih, R.: Exact voxel occupancy with graph cuts. In: Proc. International Conference on Computer Vision and Pattern Recognition, vol. 1, pp. 345–353 (2000)
25. Strang, G.: Maximal flow through a domain. Mathematical Programming 26, 123–243 (1983)
26. Tran, S., Davis, L.: 3D surface reconstruction using graph cuts with surface constraints. In: Leonardis, A., Bischof, H., Pinz, A. (eds.) ECCV 2006. LNCS, vol. 3952, pp. 219–231. Springer, Heidelberg (2006)
27. Vogiatzis, G., Torr, P., Cippola, R.: Multi-view stereo via volumetric graph-cuts. In: Proc. International Conference on Computer Vision and Pattern Recognition, pp. 391–399 (2005)
28. Yezzi, A., Soatto, S.: Stereoscopic segmentation. International Journal of Computer Vision 53(1), 31–43 (2003)

A New Bayesian Method for Range Image Segmentation

Smaine Mazouzi[1] and Mohamed Batouche[2]

[1] LERI-CReSTIC, Université de Reims, B.P. 1035, 51687, Reims, France
mazouzi@leri.univ-reims.fr
[2] Département d'informatique, Université de Constantine, 25000, Algérie
batouche@wissal.dz

Abstract. In this paper we present and evaluate a new Bayesian method for range image segmentation. The method proceeds in two stages. First, an initial segmentation is produced by a randomized region growing technique. The produced segmentation is considered as a degraded version of the ideal segmentation, which should be then refined. In the second stage, pixels not labeled in the first stage are labeled by using a Bayesian estimation based on some prior assumptions on the regions of the image. The image priors are modeled by a new Markov Random Field (MRF) model. Contrary to most of the authors in range image segmentation, who use only surface smoothness MRF models, our MRF takes into account also the smoothness of region boundaries. Tests performed with real images from the ABW database show a good potential of the proposed method for significantly improving the segmentation results.

Keywords: Image Segmentation, Range Image, Randomized Region Growing, Bayesian Estimation, Markov Random Field.

1 Introduction

The segmentation of an image is often necessary to provide a compact and convenient description of its content, suitable for high level analysis and understanding. It consists in assigning pixels to homogenous and disjoint sets called image regions. Pixels that belong to the same region share a common feature called "region homogeneity criterion". In range images, segmentation methods can be divided into two distinct categories: edge-based segmentation methods, and region-based segmentation methods. In the first category, pixels that correspond to discontinuities in depth or in surface normals are selected and chained in order to delimit the regions in the image [10,6,13]. Edge-based methods are well known for their low computational cost; however they are very sensitive to noise. On the other hand, region-based methods use geometrical surface proprieties to gather pixels with the same proprieties in disjoint regions [21,14,5,1]. The region growing technique is widely used. First, region seeds are selected; then regions are enlarged by recursively including homogenous surrounding pixels. Compared to edge-based methods, region-based methods are more stable and less sensitive to noise. However, their efficiency depends strongly on the selection of the region

A.L. Yuille et al. (Eds.): EMMCVPR 2007, LNCS 4679, pp. 453–466, 2007.
© Springer-Verlag Berlin Heidelberg 2007

seeds. Some authors have used hybrid approaches. For these approaches, often a region-based method and an edge-based one are combined so that the detected edges are used to initialize and steer a region-based segmentation [19].

Few authors have integrated Bayesian inference in range image segmentation. Lavalle and Hutchinson [16] have used a Bayesian test to merge regions in both range and textured images. Region merging is based on some observation vectors, and some image priors. The merging of two regions depends on the probability that the resulting region is homogenous. Jain and Nadabar [11] have proposed a Bayesian method for edge detection in range images. Considering the smoothness of image surfaces as a prior, they use the Line Process (LP) Markov random field (MRF) model [8] to label image pixels as EDGE or NON-EDGE pixels. Wang and Wang [20] have presented a hybrid scheme for range image segmentation. First, they proposed a joint Bayesian estimation of both pixel labels, and surface patches. Next, the solution is improved by combining the Scan Line algorithm for edge detection [13], and the Multi-Level Logistic (MLL) MRF model [4]. They aim at reducing the model complexity by estimating the number and the parameters of the image regions. Li proposes in [17] a Markov random field model for range image smoothing with discontinuity preserving. The utilization of the MAP-MRF (maximum a posterior - Markov random field) framework has allowed region smoothing with preserving of both step and roof edges.

In spite of various contributions of the works previously cited, some aspects inherent to range image segmentation were omitted. Indeed, most of the works use Markovian models that are based exclusively on the surface smoothness prior. Moreover, the proposed methods proceed by labeling pixels without ensuring the continuity of the resulting regions. Typically, in the approach proposed by Wang and Wang [20], pixels belonging to coplanar regions may be assigned equally to any of these regions. The spatial continuity constraint of resulting regions seems that it was not taken into account.

The method proposed in this paper aims first at providing an initial degraded segmentation version, using an improved region growing technique, and then to refine this version by a Bayesian-MRF labeling. The refinement of the initial segmentation consists in a Bayesian regularization of unlabeled pixels. These latter are mostly close to region boundaries. A new Markov random field model is used to model the prior information on image regions, by considering both surface and edge smoothness. In the first stage, the image regions are extracted using a randomized region growing technique. This latter is based on random sampling of region seeds. In the second stage, unlabeled pixels are labeled using Bayesian estimation, based on two distinct priors. The first one consists of the surface smoothness prior, which is modeled by the MLL model [4]. The second one which is introduced in this work, consists of the edge smoothness prior. The new MRF model uses a high-order neighborhood system, and is based on the assumption that edge pixels are situated on straight lines which represent region boundaries. The use of the ICM algorithm (Iterated Conditional Modes) [3] to search for the optimal solution has allowed us to formulate region continuity by defining a constraint on the possible labels of a given pixel. Indeed, the label of a given pixel is selected among the set of labels corresponding to the regions to which the pixel is close. The experimentations performed with real images from the ABW database [9] show the good potential of the proposed method to provide an accurate segmentation of range images.

The remainder of the paper is organized as follows: In Section 2, we introduce the image segmentation by randomized region growing. Section 3 is devoted to the proposed Bayesian approach for segmentation refinement. We present in this section the new Markov Random Field (MRF) model, as well as the adaptation of the ICM algorithm for the optimal solution search. The experimental results are shown in Section 4, in which we present respectively the evaluation framework, parameter selection and the comparative results. Finally, a conclusion summarizes our contribution and presents some possible future investigations.

2 Image Segmentation by Randomized Region Growing

2.1 Surface Modeling

A range image is a discretized two-dimensional array where at each pixel (x, y) is recorded the distance $d(x, y)$ between the range finder plane and the corresponding point of the scene. Regions in such an image are the visible patches of object surfaces. Let d^* a new representation of the row image, where $d^*(x, y)$ represents the tangent plane to the surface at (x, y). The best tangent plane at (x, y) is obtained by the multiple regression method using the set of neighboring pixels $\chi(x, y)$. The neighborhood $\chi(x, y)$ is made up of pixels (x', y') situated within a 3×3 window centred at (x, y), and whose depths $d(x', y')$ are close, according to a given threshold (Tr_h). The plane equation in a 3-D coordinate system may be expressed as follows:

$$z = ax + by + c \tag{1}$$

where $(a, b, -1)^T$ is a normal vector to the plane, and $|c|/\sqrt{a^2 + b^2 + 1}$ is the orthogonal distance between the plane and the coordinate origin. The plane parameters a, b and c at (x_0, y_0) are obtained by the minimization of the function Φ, defined as follows:

$$\Phi(a, b, c) = \sum_{(x', y') \in \chi(x_0, y_0)} (ax' + by' + c - d(x', y'))^2 \tag{2}$$

with
$\chi(x, y) = \{(x+i, y+j); (i, j) \in \{-1, 0, +1\}^2 \text{ and } |d(x+i, y+j) - d(x, y)| < Tr_h\}$
The quality of estimation according to the regression model is also computed:

$$q(x, y) = \sum_{(x', y') \in \chi(x, y)} \frac{(ax + by + c - d(x', y'))^2}{(d(x', y') - \overline{d(x, y)})^2} \tag{3}$$

Operations performed on the new image are based on the comparison of two planes. Indeed, we consider that two planes : $z = ax + by + c$ and $z = a'x + b'y + c'$

are equal if they have, according to some thresholds, the same orientation and the same distance to the coordinate origin. Let $v = (a, b, -1)^T$ and $v' = (a', b', -1)^T$, and let θ be the angle between v and v' and h the distance between the two planes: $\sin(\theta) = \|v \otimes v'\|/\|v\|\|v'\|$ and $h = |c/\|v\| - c'/\|v'\||$. So, the two planes are considered equal if $\sin(\theta) \leq Tr_\theta$ and $h \leq Tr_h$, where Tr_θ and Tr_h are respectively the angle and the distance thresholds. Plane comparison is first used to test if a given pixel belongs to a planar region, given its plane equation. It is also used to test if the pixel is, or is not, a pixel of interest (edge or noise pixel). In this case, the pixel in question is considered as a pixel of interest if at least one of its neighbors has a different plane equation, according the previous thresholds.

2.2   Region Growing by Randomized Region Seed Sampling

Inspired from the RANSAC algorithm [7], our region growing technique is based on random sampling of region seeds. A given seed centred at (x_t, y_t) is formed by the pixels in a $W \times W$ window, and belong all to the same plane. A generated seed is accepted if only the surface estimation quality q at this seed is greater than a given threshold Q. The estimation quality of a seed is represented by the minimum of estimation qualities of pixels that form the seed. For every accepted seed, a region growing is performed by recursively including homogenous pixels situated on the borders of the region in growth. Selection-growing process is repeated until no new region can be created.

Random sampling of region seeds permits to select the best seeds for region growing. The selected seeds are characterized by a good quality which allows to include in a given region the largest possible set of homogenous pixels. Indeed, several seeds within the same region can be generated; however none of these seeds is accepted. The first generated seed for which the quality q is greater than the threshold Q will be accepted and considered for region growing. The randomized region growing algorithm is described as follows:

```
t=0
Repeat
    Generate a random position (xt,yt)
    If seed quality q(xt,yt)>Q then
        Perform a region growing starting from (xt,yt)
    EndIf
    t=t+1
Until none new region was generated since t-DT
// DT a given interval: DT>>1
```

For each generated region R_l (labeled l), the residual variance σ_l^2 is calculated as follows:

$$\sigma_l^2 = \sum_{(x,y) \in R_l} (a_l x + b_l y + c_l - d(x, y))^2 \qquad (4)$$

where (a_l, b_l, c_l) are the plane equation coefficients of the region R_l. This parameter will be used in Bayesian edge regularization.

Note that in 2-D images, the variance σ^2 depends only on the noise, and consequently it is considered constant for all the regions of the image. However, in range images σ^2 depends on both noise and surface orientation regarding the plane of the range finder. Its value is proportional to the angle of the surface inclination.

Region growing by randomized region seed sampling has provided better results compared to deterministic region growing (see Fig. 1b,c). However, the resulting segmentation often remains unsatisfactory. In Fig. 1c, we can note that most of the unlabeled pixels are those close to region boundaries. We present in the next section a new Bayesian method which allows refining the resulting segmentation by a reliable labeling of unlabeled pixels.

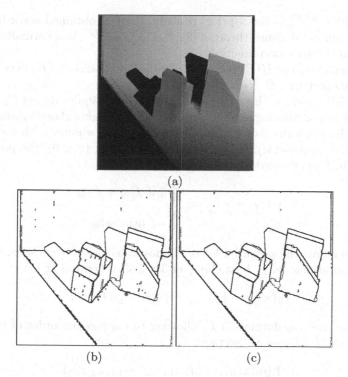

(a)

(b)　　　　　　　　　(c)

Fig. 1. Region growing. (a) Range image (abw.test.6); (b) segmentation result by deterministic region growing; (c) segmentation result by randomized region growing

3 Edge Regularization by Bayesian Inference

3.1 MAP-MRF Pixel Labeling

We have used the piecewise smoothness of image surfaces as well as the piecewise smoothness of region boundaries as priors to model distributions of pixel labels (MRF) in range images. Let S denote the image lattice. At each site $(x, y) \in S$,

$d(x, y)$ is the depth at the site, and $d^*(x, y)$ represents the corresponding plane equation parameters: $d^*(x, y) = (a_{x,y}, b_{x,y}, c_{x,y})$. Let M be the number of regions in the image. So, each site (x, y) can take a label $f_{x,y}$ from the set of labels $L = \{l_1, \ldots, l_M\}$. The labeling set $F = \{f_{x,y}, (x, y) \in S, f_{x,y} \in L\}$, represents a segmentation of the image. If we assume that F is Markovian, segmenting S according to the MAP-MRF framework [18] is equivalent to calculate the maximum a posteriori (MAP) of the distribution of the set F: $P(F/d)$, by considering F as a Markov Random Field (MRF). According to Bayes' rule, the maximum a posteriori $P(F/d)$ is expressed as follows:

$$P(F/d) = \frac{p(d/F)P(F)}{p(d)} \tag{5}$$

$P(F) = Z^{-1}e^{-U(F)}$ is the a priori probability of F obtained according to the Markov-Gibbs equivalence theorem [2]. $Z = \sum_F e^{-U(F)}$ is a normalization constant called the partition function.

The a priori energy $U(F)$ is a sum of clique potentials $V_c(F)$ over the set of all possible cliques C: $U(F) = \sum_{c \in C} V_c(F)$.

In our MRF model we have considered two sets of cliques: the set C_1 of cliques, formed of two neighboring sites according to the 4-neighborhood system, and the set C_2 of cliques, formed of 9 sites located in a 3×3 window with the center as an edge pixel (unlabeled). By using the parameter ζ, $(\zeta < 0)$, the potential V^1 of cliques in C_1 is defined as follows:

$$V^1(f_{x,y}, f_{x',y'}) = \begin{cases} \zeta \text{ if } f_{x,y} = f_{x',y'} \\ \\ -\zeta \text{ otherwise} \end{cases} \tag{6}$$

In order to define the potential V^2 of cliques in blocks of 3×3 sites, we use the following notations: let c_9 be a clique of 3×3 sites centred at (x, y):

$$c_9(x, y) = \{f_{x+i,y+j}, (i, j) \in \{-1, 0, 1\}^2\} \tag{7}$$

Let's define the transformation Γ, allowing to express the order of the sites in cliques of C_2; $\Gamma : C_2 \to F^9$, so that:

$$\Gamma(c_9(x, y)) = (f_{x-1,y-1}, \cdots, f_{x+1,y+1}) \tag{8}$$

By using the parameter κ, $(\kappa < 0)$, and considering possible configurations of cliques in C_2 (see Fig. 2) the potential V^2 can thus be expressed as follows:

$$V^2(\Gamma(c_9(x, y))) = \begin{cases} \kappa \text{ if } \exists(x', y'), (x'', y'') \mid f_{x,y} = f_{x',y'} = f_{x'',y''} \\ \text{and } \phi((x', y'), (x, y), (x'', y'')) = \pi \\ \\ 0 \text{ if } \exists(x', y'), (x'', y'') \mid f_{x,y} = f_{x',y'} = f_{x'',y''} \\ \text{and } \phi((x', y'), (x, y), (x'', y'')) = 2\pi/3 \\ \\ -\kappa \text{ otherwise} \end{cases} \tag{9}$$

where $\phi((x', y'), (x, y), (x'', y''))$ is the angle between the two vectors $(x' - x, y' - y)^T$ and $(x'' - x, y'' - y)^T$.

The potential V^1 models the surface smoothness, whereas V^2 models the edge smoothness. Configurations used to define V^2 depend on the surface type. For images containing polyhedral objects, considered in this work, V^2 is defined on the basis that the boundary between two adjacent regions is formed by pixels belonging to the same straight line (Fig. 2). So, configurations that correspond to locally unsmooth edges are penalized by using a positive clique potential $(-\kappa)$.

The likelihood distribution $p(d/F)$, is obtained by assuming that the observations $\{d(x, y), (x, y) \in S\}$ are degraded by an independent Gaussian noise:

$$d(x, y) = a_{f_{x,y}} x + b_{f_{x,y}} y + c_{f_{x,y}} + e(x, y) \tag{10}$$

with $e(x, y) \sim N(0, \sigma^2_{f_{x,y}})$. So, the likelihood distribution is expressed as follows:

$$p(d/F) = \frac{1}{\prod_{(x,y)\in S} \sqrt{2\pi\sigma^2_{f_{x,y}}}} e^{-U(d/F)} \tag{11}$$

with the likelihood energy $U(d/F)$ defined by:

$$U(d/F) = \sum_{(x,y)\in S} (a_{f_{x,y}} x + b_{f_{x,y}} y + c_{f_{x,y}} - d(x, y))^2 / 2\sigma^2_{f_{x,y}} \tag{12}$$

Since $p(d)$ is constant for a fixed d, the optimal solution F^* is obtained by maximizing the a posteriori probability $P(F/d) \propto p(d/F)P(F)$, which is equivalent to minimizing the a posteriori energy $U(F/d) = U(d/F) + U(F)$:

$$F^* = argmin\{U(d/F) + U(F)\} \tag{13}$$

3.2 Computation of the Optimal Solution

By assuming that F is Markovian, and the observations $\{d(x, y); (x, y) \in S\}$ are conditionally independent, we have used the ICM algorithm to minimize the a posteriori energy $U(F/d)$. By considering $U(F/d)$ as a sum of energies over all the sites: $U(F/d) = \sum_{(x,y)\in S} U(f_{x,y}/d(x, y))$, we can separate it in two terms:

$$U(F/d) = \sum_{(x,y)\in S'} U(f_{x,y}/d(x, y)) + \sum_{(x,y)\in S-S'} U(f_{x,y}/d(x, y)) \tag{14}$$

where S' is the set of sites that have not been labeled in the first stage (by region growing): $S' = \{(x, y) \in S \mid f_{x,y} \text{ is undefined }\}$.

Assuming the correctness of the labeling of the set $S - S'$ (performed by region growing), the term $\sum_{(x,y)\in S-S'} U(f_{x,y}/d(x, y))$ is thus constant. Minimizing the energy $U(F/d)$ is equivalent to minimizing the energy $U'(F/d)$ which corresponds to the sites in S':

$$U'(F/d) = \sum_{(x,y)\in S'} U(f_{x,y}/d(x, y)) \tag{15}$$

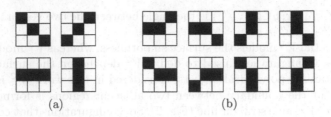

(a)   (b)

Fig. 2. Clique potential $V^2(c_9)$ defined according to the edge smoothness prior. (a) Full locally smooth edge : $V^2(c_9) = \kappa$; (b) partial locally smooth edge : $V^2(c_9) = 0$; otherwise, the edge is not locally smooth : $V^2(c_9) = -\kappa$.

The assumption of the correctness of the labeling of $S - S'$ also allows to define a constraint on the set of possible values that a site in S' can have during the execution of the ICM algorithm. Indeed, the label $f^k_{(x,y)}$ at the iteration k, of a site (x, y) is chosen among the set $L'(x, y) \subset L$ containing the labels of the sites, labeled in the first stage, and located in a $W \times W$ window centred at (x, y). Formally, $L'(x, y)$ is defined as follows:

$$L'(x,y) = \{l | \exists (x', y') \in S - S', (x' - x, y' - y) \in [-W/2, W/2]^2 \wedge f_{x',y'} = l\} \quad (16)$$

The two previous heuristics allow to speed up the calculation of the minimum of the a posteriori energy $U'(F/d)$. They allow also to satisfy the region continuity constraint. For the latter problem, if we assume that the distance between two coplanar regions R and R' is greater than W, labels l_R and $l_{R'}$ corresponding respectively to R and R', cannot belong to the same set $L'(x, y)$. For example, if the site (x, y) belongs to R, it can not be labeled $l_{R'}$, although energies $U'(l_R/d(x, y))$ and $U'(l_{R'}/d(x, y))$ are equal.

4   Experimentation and Discussion

4.1   Evaluation Framework

Hoover et al. have proposed a dedicated framework for the evaluation of range image segmentation algorithms [9], which has been used in several related works [13,12,5,1]. The framework consists of a set of real range images, and a set of objective performance metrics. It allows to compare a machine-generated segmentation (MS) with a manually-generated segmentation, supposed ideal and representing the ground truth (GT). The most important performance metrics are the numbers of instances respectively of correctly detected regions, over-segmented regions, under-segmented regions, missed regions, and noise regions. Region classification is performed according to a compare tool tolerance T; $50\% < T \leq 100\%$ which reflects the strictness of the classification. The 40 real images of the ABW database are divided into two subsets: 10 training images, and 30 test images. The training images are used to estimate the parameters of a given segmentation method. Using the obtained parameters, the method is applied to the test images. The performance metrics are computed and stored in order to be used to compare the involved methods. In our case, four methods, namely USF, WSU, UB and UE, cited in [9] are involved in the comparison.

4.2 Parameter Selection

Since the evaluation framework provides a set of training images with ground truth segmentation (GT), we have opted to a supervised approach for the estimation of the used parameters.

For the proposed method, named BRIS for Bayesian Range Image Segmentation, six parameters should be fixed: Tr_θ, Tr_h, W, Q, ζ and κ. The performance criterion used in parameter selection is the average number of correctly detected regions with the compare tool tolerance T set to 80%. The parameters are divided into two subsets: 1) Tr_θ, Tr_h, W and Q which represent respectively the angle threshold, the depth threshold, the window size, and the seed quality threshold. These parameters are used by the randomized region growing algorithm. 2) ζ and κ which express respectively the clique potentials V^1 and V^2, and they are used for edge regularization.

For the first subset of parameters, 256 combinations namely $(Tr_\theta, Tr_h, W, Q) \in \{15°, 18°, 21°, 24°\} \times \{12, 16, 20, 24\} \times \{5, 7, 9, 11\} \times \{0.90, 0.95, 0.97, 0.99\}$, were run on the training images. The threshold Tr_θ was set to 21°. Note that higher values of this parameter under-differentiate regions regarding their orientations, and lead to an under-segmentation of the image. However, lower values over-differentiate pixels and lead to an over-segmentation. It results in a high number of false and small regions, which should be merged in the true neighboring regions. The threshold Tr_h is set to 16. Note that values significantly greater than 16 can lead to wrongly merge some parallel overlapped regions. However, if Tr_h is significantly less than 16, highly sloped regions cannot be detected as planar regions [13]. This results in a high rate of missed regions. Parameters W and Q were set respectively to 7 and 0.97. The selected value of W permits to estimate the plane equation by considering a wide neighborhood (W^2 pixels), whereas Q ensure that the plane parameters are reliable, and the window $W \times W$ is not located between two different regions.

We have used the Coding method [2] to estimate the parameters ζ and κ. For each image in the training set, a pair of values of these parameters is calculated. The average values are then computed and used as the parameter values. We have used a single coding of the set $S - S'$, corresponding to cliques of 3×3 sites. Indeed, we assume that the regularization of the region boundaries is more convenient for the labeling of unlabeled pixels, because these pixels are mostly close to the region boundaries. The optimal values for each training image is calculated by the simulated annealing algorithm [15], using a Gibbs sampler [8]. The average values of ζ and κ obtained with the training set were respectively -0.37×10^{-4} and -0.21×10^{-4}.

4.3   Performance Evaluation and Comparison

Fig. 3 illustrates the impact of the Bayesian edge regularization on the segmentation results, with all the test images. The gap between the two curves shows that the segmentation results are significantly improved for the high values of the compare tool tolerance T. Indeed, edge regularization in range images allows to improve segmentation accuracy, by optimal labeling of pixels close to region boundaries. It was reported [9,12] that segmentation methods provide better results when they pay particular attention to process region boundaries.

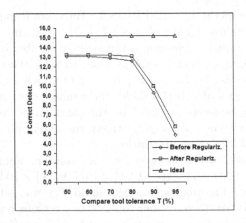

Fig. 3. Comparison of average results before and after edge regularization of the test images, according to T ; $0.5 < T \le 1.0$

Fig. 4 shows the segmentation results of the image abw.test.8, with the compare tool tolerance T set to 80%. This image was considered as a typical image to compare the involved methods [9,5]. Fig. 4a shows the range image, and Fig. 4b shows the ground truth segmentation (GT). Fig. 4c, 4d 4e and 4f represent the segmentation results obtained respectively by USF, WSU, UB and UE methods. Fig. 4g presents the segmentation result obtained by our method. Metrics in table 1 show that all image regions detected by the best-referenced segmenter (UE) were detected by our method. Due to the amount of noise, all methods

Table 1. Comparison results with abw.test.8 image for T=80%

Method	GT region	Correct detection	Over-segmentation	Under-segmentation	Missed	Noise
USF	21	17	0	0	4	3
WSU	21	12	1	1	6	4
UB	21	16	2	0	3	6
UE	21	18	1	0	2	2
BRIS	21	18	2	0	1	1

Table 2. Average results of the different involved methods for T=80%

Method	GT region	Correct detection	Over-segmentation	Under-segmentation	Missed	Noise
USF	15.2	12.7	0.2	0.1	2.1	1.2
WSU	15.2	9.7	0.5	0.2	4.5	2.2
UB	15.2	12.8	0.5	0.1	1.7	2.1
UE	15.2	13.4	0.4	0.2	1.1	0.8
BRIS	15.2	13.1	0.4	0.1	1.7	0.9

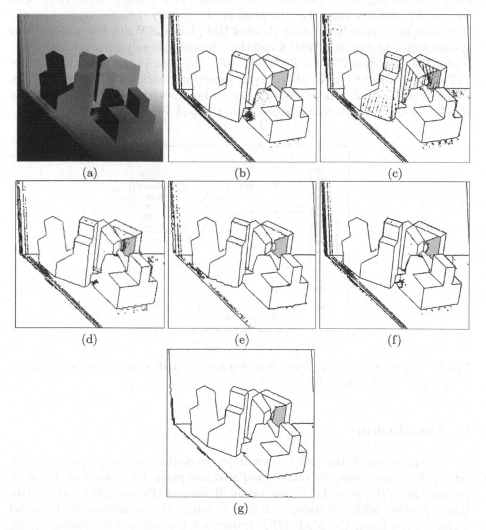

(a)   (b)   (c)

(d)   (e)   (f)

(g)

Fig. 4. Segmentation results of abw.test.8 image. (a) Range image; (b) ground truth segmentation (GT); (c) USF result; (d) WSU result; (e) UB result; (f) UE result; (g) BRIS result.

have failed to detect the shadowed region. The incorrectly detected regions are those with small sizes, and situated on the horizontal support. Compared to the involved methods, the values of the incorrect detection metrics are also good. Our method is equivalent to UE, and scores higher than the others.

Table 2 shows the average results obtained with all test images, and for all performance metrics. The compare tool tolerance was set to the typical value 80%. By considering both correct detection and incorrect detection metrics, obtained results show a good efficiency of our method, compared to the others.

Fig. 5 shows the average numbers of correctly detected regions for all test images, according to the compare tool tolerance T; $T \in \{51\%, 60\%, 70\%, 80\%, 90\%, 95\%\}$. Results show that the number of correctly detected regions by our method is in average better than those of USF, UB and WSU. For instance, our system scored higher than WSU for all the values of the compare tool tolerance T. It scored higher than USF for $T \in \{80\%, 90\%, 95\%\}$, and better than UB for $T \in \{50\%, 60\%, 70\%, 80\%\}$. For all incorrect detection metrics (Over-segmentation, Under-segmentation, Missed, Noise), our method has equivalent scores to those of UE and USF, which scored higher than UB and WSU.

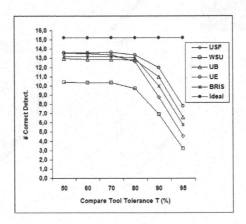

Fig. 5. Average results of correctly detected regions of all methods, according to the compare tool tolerance T ; $0.5 < T \leq 1.0$

5   Conclusion

We have presented in this paper a new Bayesian method for range image segmentation. Region growing by randomized seed sampling, introduced in this work provides an initial degraded segmentation. Results at this stage were better than those obtained with deterministic region growing. The refinement of the initial segmentation using the MAP-MRF framework has allowed improving significantly the segmentation results. We have presented a new MRF model which allows to model both surface, and edge smoothness, considered as prior assumptions on regions in range images. Extensive tests were performed on real images from the ABW database. Obtained results show a good potential of the proposed method for providing an efficient and accurate range image segmentation. In our future works we extend the proposed method to curved objects, by defining the surface proprieties specific to these objects, as well as the appropriate MRF models.

References

1. Bab Hadiashar, A., Gheissari, N.: Range image segmentation using surface selection criterion. IEEE Transactions on Image Processing 15(7), 2006–2018 (2006)
2. Besag, J.E.: Spatial interaction and statistical analysis of lattice systems. Journal of the Royal Statistical Society, Series B 36, 192–236 (1974)
3. Besag, J.E.: On the statistical analysis of dirty pictures. Journal of the Royal Statistical Society, Series B 48, 259–302 (1986)
4. Derin, H., Elliott, H.: Modeling and segmentation of noisy and textured image using gibbs random fields. IEEE Transactions on Pattern Analysis and Machine Intelligence 9(1), 39–55 (1987)
5. Ding, Y., Ping, X., Hu, M., Wang, D.: Range image segmentation based on randomized hough transform. Pattern Recognition Letters 26(13), 2033–2041 (2005)
6. Fan, T.J., Medioni, G.G., Nevatia, R.: Segmented description of 3-D surfaces. IEEE Journal of Robotics and Automation 3(6), 527–538 (1987)
7. Fischler, M.A., Bolles, R.C.: Random sample consensus: a paradigm for model fitting with applications to image analysis and automated cartography. Readings in computer vision: issues, problems, principles, and paradigms, pp. 726–740 (1987)
8. Geman, S., Geman, D.: Stochastic relaxation, Gibbs distributions, and the Bayesian restoration of images. IEEE Transactions on Pattern Analysis and Machine Intelligence 6(6), 721–741 (1984)
9. Hoover, A., Jean-Baptiste, G., Jiang, X., Flynn, P.J., Bunke, H., Goldgof, D.B., Bowyer, K.W., Eggert, D.W., Fitzgibbon, A.W., Fisher, R.B.: An experimental comparison of range image segmentation algorithms. IEEE Transactions on Pattern Analysis and Machine Intelligence 18(7), 673–689 (1996)
10. Inokuchi, S., Nita, T., Matsuda, F., Sakurai, Y.: A three dimensional edge-region operator for range pictures. In: 6th International Conference on Pattern Recognition, Munich, pp. 918–920 (1982)
11. Jain, A.K., Nadabar, S.G.: MRF model-based segmentation of range images. In: International Conference on Computer Vision, Osaka, Japan, pp. 667–671 (1990)
12. Jiang, X., Bowyer, K.W., Morioka, Y., Hiura, S., Sato, K., Inokuchi, S., Bock, M., Guerra, C., Loke, R.E., du Hans, J.M.: Buf. Some further results of experimental comparison of range image segmentation algorithms. In: 15th International Conference on Pattern Recognition, Barcelona, Spain, vol. 4, pp. 4877–4882 (2000)
13. Jiang, X., Bunke, H.: Edge detection in range images based on Scan Line approximation. Computer Vision and Image Understanding 73(2), 183–199 (1999)
14. Kang, S.B., Ikeuchi, K.: The complex EGI: A new representation for 3-D pose determination. IEEE Transactions on Pattern Analysis and Machine Intelligence 15(7), 707–721 (1993)
15. Kirkpatrick, S., Gelatt, C.D., Vecchi, M.P.: Optimization by simulated annealing. Readings in Computer Vision: Issues, Problems, Principles, and Paradigms, pp. 606–615 (1987)

16. LaValle, S.M., Hutchinson, S.A.: Bayesian region merging probability for parametric image models. In: IEEE Conference on Computer Vision and Pattern Recognition, New York, pp. 778–779 (1993)
17. Li, S.Z.: Roof-edge preserving image smoothing based on MRFs. IEEE Transactions on Image Processing 9(6), 1134–1138 (2000)
18. Li, S.Z.: Markov random field modeling in image analysis. Springer-Verlag New York, Inc. Secaucus, NJ (2001)
19. Lim, A.W.T., Teoh, E.K., Mital, D.P.: A hybrid method for range image segmentation. Journal of Mathematical Imaging and Vision 4, 69–80 (1994)
20. Wang, X., Wang, H.: Markov random field modeled range image segmentation. Pattern Recognition Letters 25(3), 367–375 (2004)
21. Yang, H.S., Kak, A.C.: Determination of the identity, position and orientation of the topmost object in a pile. Computer Vision, Graphics, and Image Processing 36(2-3), 229–255 (1986)

Marked Point Process for Vascular Tree Extraction on Angiogram

Kaiqiong Sun, Nong Sang, and Tianxu Zhang

Institute for Pattern Recognition and Artificial Intelligence
Huazhong University of Science and Technology, Wuhan 430074, P.R. China
kqsunn@gmail.com, {nsang,txzhang}@hust.edu.cn

Abstract. This paper presents a two-step algorithm to perform automatic extraction of vessel tree on angiogram. Firstly, the approximate vessel centerline is modeled as marked point process with each point denoting a line segment. A *Double Area* prior model is proposed to incorporate the geometrical and topological constraints of segments through potentials on the interaction and the type of segments. Data likelihood allows for the vesselness of the points which the segment covers, which is computed through the Hessian matrix of the image convolved with 2-D Gaussian filter at multiple scales. Optimization is realized by simulated annealing scheme using a Reversible Jump Markov Chain Monte Carlo (RJMCMC) algorithm. Secondly, the extracted approximate vessel centerline, containing global geometry shape as well as location information of vessel, is used as important guide to explore the accurate vessel edges by combination with local gradient information of angiogram. This is implemented by morphological homotopy modification and watershed transform on the original gradient image. Experimental results of clinical digitized coronary angiogram are reported.

Keywords: Angiogram, Vascular tree, Segmentation, Stochastic geometry, Marked point process, Markov Chain Monte Carlo.

1 Introduction

Vessel segmentation algorithm is critical component of circulatory blood vessel analysis systems. It provides important information of quantitative analysis about cardiac and cerebral disease. The difficulties of vessel segmentation mainly come from the weak contrast between vascular trees and the background, an advance unknown and easily deformable shapes of the vessel tree, sometimes overlapping strong shadows of bones and so on.

Many techniques about the problem of coronary vessel extraction begin with some local optimization process to character the vessel structure in the form of different description, operator and model. They have been built upon the pixel domain and different feature spaces of image. Region growing techniques mainly discriminate the vessel part from angiogram by intensity similarity and spatial proximity. Skeleton-based methods aim at extracting blood vessel centerlines, from which the whole vessel tree is reconstructed [1]. The ridge-based methods

A.L. Yuille et al. (Eds.): EMMCVPR 2007, LNCS 4679, pp. 467–478, 2007.
© Springer-Verlag Berlin Heidelberg 2007

make use of intensity ridges to approximate the skeleton of the tubular objects while the grays image is treated as 3D elevation maps [2]. Ridge points can be obtained by tracing the intensity map from arbitrary point, along the steepest ascent direction of intensity. The Hessian matrix, containing the second-order differential properties of image, are also been used to track the ridge point [3] [4]. Matching filter approach convolves the image with multiple matched filters for the extraction of vessel structure [5]. The essence of this method existing in describes or approaches local lattices of image with the convolution kernel used. This method compares to the mathematical morphology schemes, which apply structuring elements to "match" the image with morphologic operators. The combination of several operators has also been proposed to complement one or more of them. For the incompleteness of these local descriptions for vessel structure, they usually have a heuristic post-processing step.

The methods based on local process usually lack detailed description of the geometry and topology of the global vessel structure, and might be very sensitive to local minima. Thus, besides the local description of vessel structure, other methods for vessel segmentation have also been proposed to character some relation between the local feature and shape of the whole structure [6],[7]. These methods usually are expressed with some model. Deformable models are such techniques that find object contours using parametric curves that deform under the influence of internal and external forces. It can either be parametric [8] or geometric [9]. The curvature and the gradient, acting as description for shape and intensity feature of vessel structure, are used to define an energy function. The final contour fits the vessel boundaries following a differential equation whose solution corresponds to a local minimum of the energy. In these methods, the constraint for curvature can be treated as the prior information about the vessel structure.

The prior information about vessel, however, can also be embedded into some model for the region of vessel structure instead of the edges. Based on such consideration, a two-step method, which combines the global geometry and local operator, for vessel extraction is proposed to achieve both robustness and accuracy in this paper. The basic idea is to inference the global geometry information, the approximate vessel centerline, and makes use of it to regularize a local edge detection algorithm. Where the abstract geometry embodied in object acts as important carrier of detailed quantities information of object.

The stochastic geometry model is explored to capture the geometry and topology of vessel tree. Assume the vessel centerline consists of local linear segments with certainly length and orientation, which is called approximate vessel centerline. In the first step, approximate vessel centerline is modeled as marked point process [10]. A *Double Area* model is proposed to incorporate the geometric and topological knowledge of vessel structure. The connectivity of structure is characterized by distance of extreme point of segments. The orientation consistency of connected segment characters the alignment of vessel structure. The optimization is done via simulated annealing using a Reversible Jump Markov Chain Monte Carlo (RJMCMC) algorithm [11],[12]. We design well balanced

Markov chains to explore the solution space. Moreover, Data-Driven techniques are utilized to compute heuristic information in the cue space [13].

As some abstract geometric descriptor of vessel structure, the approximate centerline contains important shape and location information of vessel tree. Thus, in the second step, the centerline is used as important guide to explore the accurate vessel edges by combination with local gradient information of angiogram. This is implemented by morphological homotopy modification and watershed transform on the original gradient image.

Several contributions have been proposed for vessel extraction with point process under the stochastic geometry framework. A Gaussian intensity model developed by E.Thönnes et al. [14] is adopted and used as the observed data under a Bayesian framework. Vascular image is modeled with random tree models(RTMs). A multi-scale approach based on marked point processes is proposed in [15]. Compared with the existing object-orientated methods, the proposed method in this paper model the shape or geometry rather than the vessel object (region or edge) itself. It takes advantage of the point process in a mediate way for vessel extraction. The approximate centerline explored acts as a hidden variable [16], [17] in the sense of computing for vessel extraction. The robustness is assumed by the geometry model and the accuracy is kept for the local operator used.

This paper is organized as follows: In the next section, the Double Area model is described and used for modeling the approximate vessel centerline. Next, the RJMCMC dynamics is built to simulate the model. In section 3, experiments on simulation of the point process with real clinic x-ray angiogram are reported. Section 4 describes the process of watershed technology used to detect accuracy vessel boundaries with the extracted vessel approximate centerline as marker. Finally, some discussion is presented.

2   Model for Approximate Vessel Centerline Extraction

Among the stochastic methods widespread in image analysis, the marked point process has the advantage of combing information "globally" to identify geometrical shape and is acknowledged more appropriate prior model than discrete Markov random fields [18] to use in object recognition and some other "high-level" vision problem. It is adopted in order to solve image analysis problems using an object-oriented rather than a pixel-oriented approach [16]. In this section, we firstly turn the vessel segmentation into a shape recognition problem and make use of marked point process to model approximate vessel centerline, which contains the shape and topology information of vessel.

The approximate vessel centerline is described as a configuration of line segments set. A segment is given by $s_i = (p_i, m_i)$, with $p_i = (x_i, y_i) \in \wedge \subset R^2$, the coordinates of its center. The label of a segment $m_i = (l_i, \theta_i) \in \Omega_M$ are its length and orientation. $\wedge$ is the lattices the image covers. Ω_M is the marker space $[l_{min}, l_{max}] \times [0, \pi]$. The line segments set $S = \{s_i, i = 1, ..., n \in N\}$ is considered as a realization of a point process on $\wedge \times \Omega_M$.

The real vessel centerline is generally characterized by several strong constraints such as structure continuity and consistency of local orientation. These constrains can be considered as interactions between segments of the point process which can either penalize or favor some particular configuration through potentials in the density of process. Two components constitute the probability density of the point process. The first is the interaction model (Double Area model), which is determined by the interaction between segments: attraction, rejection, and the dimension of the segments set. The second term is the data model, which gives the location in the image of the different segment with the centerline.

Within the framework of a Gibbs point process [16], the probability density of the proposed model is:

$$f(S) \propto \beta^n \exp(-E(S)) = \beta^n exp - (E_p(S) + E_d(S)) \ , \tag{1}$$

where $E_p(S)$ is the interaction energy, and $E_d(S)$ is the data energy. The estimate of the approximate centerline is obtained by minimizing the energy function $E(S)$:

$$S^* = \arg\min\{E_p(S) + E_d(S) - n \log \beta\} \ . \tag{2}$$

The term $-n \log \beta$ is the energy term corresponding to the Poisson process to which the density of point process with respect, and may be interpreted as a penalty on the total number of segments.

The global minimum of the energy function $E(S)$ is found by a simulated annealing technique. This algorithm iteratively simulates the law:

$$f(S,T) = [f(S)]^{\frac{1}{T}} \ , \tag{3}$$

while slowly decreasing the temperature T. When $T \to 0$, the result of the simulations converges in probability to the global minimum.

2.1   Double Area Model

The Double Area model is based on the types of segments and two relations of interaction between segments, R_a (attraction) and R_r (rejection). The energy of this prior model is:

$$E_p(S) = \lambda_0 n + \lambda_1 n_f + \lambda_2 n_s + \lambda_a \sum_{<s_i,s_j>R_a} g_a(s_i, s_j) + \lambda_r \sum_{<s_i,s_j>R_r} g_r(s_i, s_j) \ . \tag{4}$$

In the above prior energy, n, n_f and n_s are respectively the number of total segments, free segments and single segments. λ_0, λ_1, and λ_2 are the penalty constant for the number of them. λ_0 is $-\log \beta$ in formula (2). We follow the definition of segment types in [19]; two segments are said to be connected if two of their extremities are closer than a constant ϵ. This relation defines three types of segments. Free segments are those which are not connected, single ones are those with only one of their endpoints connected to other segments, and double segments have their two endpoints connected.

$< s_i, s_j > R_a$ is a pair of interacting segments of attraction, and $g_a(s_i, s_j)$ is the potential function with respect to R_a. The attractive interaction, R_a, is defined to favor the connectivity of pairs of segments. A segment s has two extremities to which another segment can be connected, U_s and V_s (see Fig.1(a)). An attractive region W_a is defined for each segment. This region is represented by disks centered at two extremities of a segment: $W_a(s) = C_a(U_s, r_a = l_s/4) \cup C_a(V_s, r_a = l_s/4)$. Fig.1(a) illustrates this definition. Two segments $s_i = (p_i, m_i)$ and $s_j = (p_j, m_j)$ have attractive interaction if the attractive region of the two segments intersect, i.e. $W_a(s_i) \cap W_a(s_j) \neq \varnothing$. The potential function between $< s_i, s_j > R_a$ is intersection area of the two region to that of the minimal one:

$$g_a(s_i, s_j) = \sum_{K1=U_{s_i}, V_{s_i}} \sum_{K2=U_{s_j}, V_{s_j}} \frac{A(C_a(K1) \cap C_a(K2))}{min(A(C_a(K1)), A(C_a(K2)))} , \qquad (5)$$

where $A(.)$ denotes the area of $(.)$. This interaction describes the connectivity of segments and is assigned a negative weight λ_a to favor connectivity of segments.

Fig.1(b) shows the attractive interactions between several segments. Two segments with the minimal energy state $(g_a = 1)$ defined by above potential function may take on different orientation relation of them, or different curvature. The real vessel structure will prefer $< s_1, s_2 > R_a$ to $< s_1, s_3 > R_a$ (see Fig.1(b)). We define rejective interaction to distinguish the connected segments having different orientation consistency. A rejective region $W_r(s)$ is defined for each segment: $W_r(s) = C_r(P_s, r_r = l_s/2)$. Two segments $s_i = (p_i, m_i)$ and $s_j = (p_j, m_j)$ have rejective interaction if the rejective region of the two segments intersect, i.e. $W_r(s_i) \cap W_r(s_j) \neq \varnothing$. The potential function for rejection interaction between $< s_i, s_j > R_r$ is:

$$g_r(s_i, s_j) = \frac{A(W_r(s_i) \cap W_r(s_j))}{min(A(W_r(s_i)), A(W_r(s_j)))} , \qquad (6)$$

with $A(.)$ being the area of $(.)$ also. This definition is also applicable to the segments which are not connected. Fig.2(a) shows the rejective region of a segment and (b) shows several examples of rejection configuration.

Similar segment process is devised and used to extract line network from remotely sensed image in [20]. Except the slight difference of the property of expected object, a distinct peculiarity of the proposed model in this paper is that no threshold for orientation is used to forbid any unexpected configuration.

2.2 The Data Term

To check the fitness of a segment s to the data, we consider the set of pixels Q_s covered by s in the image. Data potential is defined with the vesselness of Q_s computed by a multiscale vessel enhancement measure, based on the work of [4] on ridge filters. Having extracted the two eigenvalues of the Hessian matrix

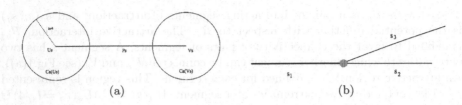

(a) (b)

Fig. 1. Attractive interaction of segments. (a) A segment s has two extremity U_s and V_s, and two disk attractive region centered at U_s and V_s. (b) Three segments s_1, s_2, s_3 share a common endpoint, and both $g_a(s_1, s_2)$ and $g_a(s_1, s_3)$ have the same attraction potential, while the two point pairs have different consistency of orientation.

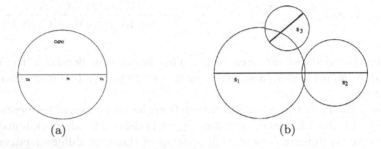

(a) (b)

Fig. 2. Rejective interaction of segments (a) A segment s has one disk rejective region centered at P_s. (b) $g_r(s_1, s_2)$ and $g_r(s_1, s_3)$ have different potential values.

computed at scale σ, ordered $|\lambda_1| \leq |\lambda_2|$, a vesselness function is defined at each pixel:

$$\nu(\sigma) = \begin{cases} 0, & \text{if} \lambda_2 \geq 0 \\ \exp\frac{-R_B^2}{2\beta^2}(1 - \exp\frac{-S^2}{2c^2}), & \text{otherwise} . \end{cases} \tag{7}$$

where $R_B = \frac{|\lambda_1|}{|\lambda_2|}$, and $S = \sqrt{\lambda_1^2 + \lambda_2^2}$. A detailed explanation of each parameter in this measure is in [4]. A vesselness image can be taken with the maximum of the response of the filter across several selected scales. We expect that the pixels in Q_s have big vesselness, thus define the data energy of the point process S as

$$E_d(S) = \lambda_d \sum_{s \in S} \psi(\nu(Q_s)) . \tag{8}$$

$\nu(Q_s)$ is the mean vesselness of all the pixels in Q_s. $\psi(x) = (\frac{1-x}{1+x})^k$ is a decreasing function. This function casts a range (depending on k) of vesselness into energy near zero.

2.3 Optimization by Data-Driven MCMC

A solution for the point process is represented by a point set

$$S = (N, \{(x_i, y_i), l_i, \Theta_i\}; i = 1, 2 \ldots N) , \tag{9}$$

where the number of point N is unknown and the solution space is a union of many subspaces Ω_N of varying dimensions:

$$\Omega = \cup_{N=0}^{\infty} \Omega_N, \Omega_N = \wedge \times \Omega_M , \tag{10}$$

where Ω_N is the subspace with exactly N point. It is further decomposed into location and marker space.

Designing Ergodic Markov Chain Dynamics. The search algorithm should make the markov chain can visit any state in the solution in finite time steps. It requires both jump dynamics which move between subspace of varying dimensions and diffusion dynamics which move within a subspace of a fixed dimension. That the Markov chain have $f(S)$ as its invariant probability at equilibrium is assured by detailed balance at every move and the reversibility of each move.

We use the Reversible Jump Markov Chain Monte Carlo (RJMCMC) algorithm with a Metroplis-Hasting-Green dynamics [11],[12] to simulate the process distribution $f(S,T)$ specified by the density $[f(S)]^{\frac{1}{T}}$. At any state of the Markov chain, a propose kernel is proposed and the transition is accepted with a probability given by the Green's ratio. This accepted rate is computed so that the detailed balance condition is verified, condition under which this algorithm converge to $f(S,T)$.

We adopt three types of dynamics in the evolution of Markov chain, which are used randomly with probabilities q_b, q_d, q_m respectively.

Dynamics 1: Birth of a point. It is a process of jump between spaces of different size. Suppose at a certain time step, we propose to birth a point, thus move the Markov chain from current state S to $S \cup \{s'\}$. By the classic Metropolis-Hastings method [21], we need two proposal probabilities $p(S \rightarrow S \cup \{s'\})$ for the move and $p(S \cup \{s'\} \rightarrow S)$ for moving back. Then the proposal move is then accepted with probability

$$\alpha(S \rightarrow S \cup \{s'\}) = min(1, \frac{p(S \cup \{s'\} \rightarrow S)f^{\frac{1}{T}}(S \cup \{s'\})}{p(S \rightarrow S \cup \{s'\})f^{\frac{1}{T}}(S)}) . \tag{11}$$

Dynamics 2: Death of a point. It is the reversible jump of Dynamics 1. In this case of death of a segment s', the probability to accept is

$$\alpha(S \rightarrow S \setminus s') = min(1, \frac{p(S \setminus s' \rightarrow S)f^{\frac{1}{T}}(S \setminus s')}{p(S \rightarrow S \setminus s')f^{\frac{1}{T}}(S)}) . \tag{12}$$

Dynamics 3: Diffusion of the length and orientation of a point. It is the modification of a randomly chosen object according to a symmetrical transformation. The transformation can be stretching of length or changing the orientation of the considered segment:

$$S \rightarrow (S \setminus s(p_s, l_s, \theta_s)) \cup \{s'(p_s, (l_s + d_l)[l_{min}, l_{max}], \theta_s + d_\theta[0, \pi])\} , \tag{13}$$

where [.] denotes the module function. $d_{(.)}$ is a small step of (.). The acceptance probability for this dynamic is

$$\alpha(S \to (S \setminus s) \cup \{s'\}) = min(1, \frac{f^{\frac{1}{T}}((S \setminus s) \cup \{s'\})}{f^{\frac{1}{T}}(S)}) \ . \tag{14}$$

Computing Important Proposal Probabilities in Cue Space. The effectiveness of MCMC depends critically on the design of the proposal probability. The important proposal probability can be computed using Data-Driven methods [13]. We use the computation of vesselness, presented in section 2.2 for data term, to define a birth kernel.

To compute $p(S \to S \cup \{s'\}$ in formula (11), the route first chooses a birth move with probability q_b, then chooses a point p_i from the pixel domain $\wedge$; this probability is denoted by $q(p_i)$. Given p_i, it chooses a length l_i and an orientation θ_i with probabilities $q(l_i|p_i)$ and $q(\theta_i|p_i)$ respectively. Thus,

$$p(S \to S \cup \{s'\}) = q_b q(p_i) q(l_i|p_i) q(\theta_i|p_i) \ . \tag{15}$$

$q(p_i)$ is often decided by the goodness of fit on a point. A point with good fit has a higher chance to birth. From the vesselness computed in section 2.2 of each point, we can define an inhomogeneous birth kernel for the midpoint coordinate of a segment

$$q(p_i) = \frac{\nu(p_i)}{\sum_{p_j \in \wedge} \nu(p_j)} \ , \tag{16}$$

where $\nu(p_i)$ is the vesselness of point p_i. Similarly, we can also define an inhomogeneous kernel for the orientation given the segment midpoint. It is computed by the mean vesselness of a segment under all the possible orientation with a fixed length. Lastly, $q(l_i|p_i)$ is the uniform distribution in the length space.

The proposal probability for a death move is

$$p(S \cup \{s'\} \to S) = q_d \frac{1}{n(S) + 1} \ , \tag{17}$$

where $n(S)$ is the number of segments in S. It means choosing a point in the current configuration randomly under uniform distribution for a death move.

3   Experiments on Simulation of the Point Process

The proposed method is implemented on the clinical coronary and some of the results are reported in this section. Fig.3(a) is a part of original angiogram. The minimal and maximal diameter of vessel tree is about 5 and 17 pixels respectively. We use three scale, σ being 2, 4, 8, respectively, to compute the vesselness image by formula (7). Fig.3(b) shows the result of this vesselness filter. The normalized vesselness image is treated as a distribution for the proposal probability for birth a segment, which is expressed by formula (16). The proposal probability of the orientation given a point also is computed from the vesselness image. Fig.3(c) is the result of simulation of the point process. The interaction parameters were fixed to $\lambda_0 = 1$, $\lambda_1 = 0.7$, $\lambda_2 = 0.4$, $\lambda_a = 1.75$, $\lambda_r = 5.5$, $\lambda_d = 0.9$.

Other results are shown in Fig.4; (a) and (d) are the original angiogram; (b) and (e) are the extracted approximate centerline by the point process; (c) and (f) are the results of segmentation by watershed, which will be presented in next section. It can be seen that some small branching of vessel is missed by the point process in (e). It shows that there is need for more consideration for the parameter to balance the prior and data term.

4   Extraction of Coronary Artery Edge with Marker Watershed

The centerline extracted above provides important clues of form as well as locations of vessel and many local operators can be used to detect the accurate vessel boundary based on it. Though the gray intensity of background varies on the angiogram, there exists gray level different between the vessel and the background everywhere. The implementation of morphological watershed transform on gradient image of original angiogram can extract the watershed lines (vessel boundary) which separate the homogeneous gray region [22]. Oversegmentation is prevented by only using the regional minima defined by the centerline.

The edge extraction process is implemented on each branching of centerline respectively. Fig.5(a) is a branching of centerline, which acts as the vessel marker. Then dilating of it produces the marker of background. See the blank region in Fig.5(b). Then obtained marker of vessel and background is lastly used to modify gradient image of original angiogram, and the edge of vessel is achieved by watershed on the modified gradient image. Fig.5(c) is the morphological gradient image of part region in original image of Fig.3(a). Fig.5(d) is the gradient image with the region of vessel marker and background marker being zero. The suppression of the minima not related to the markers is achieves by applying geodesic reconstruction techniques [23] [24]. Fig.5(e) shows the modified gradient image with such operator. Taking watershed operator on Fig.5(e) get the edge of vessel shown in Fig.5(f). Union of the result of each branching produces the last vessel segmentation result shown in Fig.3(d) imposed on original angiogram. The results of the other two angiogram are shown in Fig.4(c) and (f).

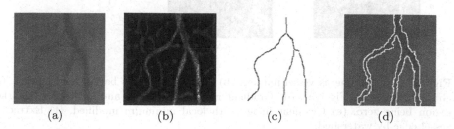

(a)   (b)   (c)   (d)

Fig. 3. Approximate centerline extraction. (a) Original angiogram. (b) Vesselness image. (c) Approximate centerline. (d) Edges imposed on original image.

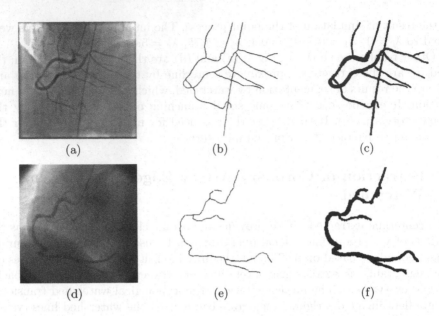

Fig. 4. (a),(d) Original angiogram. (b),(e) Extracted approximate centerline. (c),(f) Segmentation result.

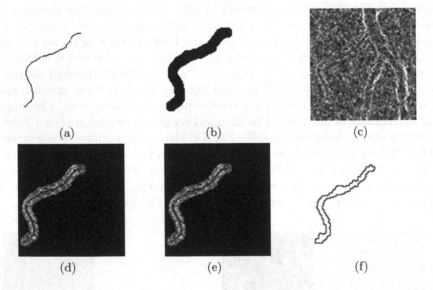

Fig. 5. (a) Centerline as vessel marker. (b) Dilation of (a) as background marker. (c) Gradient image of Fig.1(a). (d) Gradient image with vessel and background marker region being zero. (e) Gradient image with local minimum modified. (f) Extracted vessel edge by watershed.

5   Discussion

In this paper, we show a method to obtain the important information of the shape and the location of the coronary artery tree by the marked point process. The accurate edge is obtained with local operator guided by the shape and location information.

The properties of the proposed prior model need to be more deeply investigated. Relevant moves require to be defined to accelerate the convergence of the Markov chain. We use a simple measure for the data likelihood in this paper. However, many existed methods for vessel extraction may be used to compute important probabilities and data energy term. When the vessel structure takes on vessel very complex curvature relation, the ability of the proposed point process will be restricted by the length space of the segment. A possible method solving this problem is to generalize the segment process into the pure curve process, which is nearer to the real vessel structure than line segment.

Acknowledgments. This work was supported by 973 Program of China (No: 2003CB716105).

References

1. Sorantin, E., Halmai, C., Erbohelyi, B., Palagyi, K., Nyul, K., Olle, K., Geiger, B., Lindbichler, F., Friedrich, G., Kiesler, K.: Spiral-CT-based assesment of Tracheal Stenoses using 3D Skeletonization. IEEE Trans. Med. Imag. 21, 263–273 (2002)
2. Eberly, D., Gardner, R.B., Morse, B.S., Pizer, S.M., Scharlach, C.: Ridges for image analysis. JMIV 4, 351–371 (1994)
3. Wink, O., Niessen, W.J.: Max, A.: Viergever. Multiscale Vessel Tracking. IEEE Trans. Med. Imag. 23 (January 2004)
4. Frangi, A.F., Niessen, W.J., Vincken, K.L., Viergever, M.A.: Vessel enhancement filtering. In: Proc. Medical Image Computing and Computer- Assisted Intervention, pp. 130–137 (1998)
5. Sato, Y., Nakajima, S., Shiraga, N., Atsumi, H., Yoshida, S., Koller, T., Gerig, G., Kikinis, R.: 3d multi-scale line filter for segmentation and visualization of curvilinear structures in medical images. Medical Image Analysis 2, 143–168 (1998)
6. Quek, F., Kirbas, C.: Vessel extraction in medical images by wave propagation and traceback. IEEE Trans. on Med. Img. 20, 117–131 (2001)
7. Nain, D., Yezzi, A., Turk, G.: Vessel Segmentation Using a Shape Driven Flow. In: Medical Imaging Copmuting and Computer-Assisted Intervention (2004)
8. Kass, M., Witkin, A., Terzoopoulos, D.: Snakes: Active contour models. Int. J. of Comp. Vision 1, 321–331 (1988)
9. Caselles, V., Catte, F., Coll, T., Dibos, F.: A geometric model for active contours in image processing. Numerische Mathematik 66, 1–32 (1993)
10. van Lieshout, M.: Markov Point Processes and Their Applications. Imperial College Press (2000)
11. Geyer, C.J., Møler, J.: Simulation and Likelihood Inference for Spatial Point Process. Scandinavian J. Statistics, Series B 21, 359–373 (1994)

12. Green, P.: Reversible Jump Markov Chain Monte-Carlo Computation and Bayesian Model Determination. Biometrika 57, 97–109 (1995)
13. Tu, Z.W., Zhu, S.C.: Image segmentation by data-driven Markov chain Monte Carlo. IEEE Trans. Pattern Anal. Machine Intell. 24, 657–673 (2002)
14. Thönnes, E., Bhalerao, A., Kendall, W., Wilson, R.: A Bayesian Approach to inferring vascular tree structure from 2D imagery. In: Proceedings of IEEE ICIP, Rochester, New York, vol. II, 937–939 (2002)
15. Lacoste, C., Finet, G., Magnin, I.E.: Coronary Tree Extraction from x-ray Angiograms using Marked Point Process. In: Proceedings of IEEE ISBI, Arlington, Virginia, pp. 157–160 (2006)
16. Baddeley, A.J., van Lieshout, M.N.M.: Stochastic geometry models in high-level vision. In: Mardia, K.V., Kanji, G.K. (eds.) Statistics and Images, (Advances in Applied Statistics, a supplement to Journal of Applied Statistics), Nashville, TN, Abingdon, vol. 1, pp.231–256 (1993)
17. Zhu, S.C.: Statistical modeling and conceptualization of visual patterns. IEEE Trans. Pattern Anal. Machine Intell. 25, 691–712 (2003)
18. Geman, S., Geman, D.: Stochastic Relaxation, Gibbs Distributions, and the Bayesian Restoration of Images. IEEE Trans. Pattern Anal. Machine Intell. 6, 721–741 (1984)
19. Stoica, R., Descombes, X., Zerubia, J.: A Gibbs Point Process for Road Extraction in Remotely Sensed Images. Int'l J. Computer Vision 57(2), 121–136 (2004)
20. Lacoste, C., Descombes, X., Zerubia, J.: Point Processes for Unsupervised Line Network Extraction in Remote Sensing. IEEE Trans. Pattern Anal. Machine Intell. 27, 1568–1579 (2005)
21. Metropolis, N., Rosenbluth, M.N., Rosenbluth, A.W., Teller, A.H., Teller, E.: Equations of State Calculations by Fast Computing Machines. J. Chemical Physics 21, 1087–1092 (1953)
22. Vincent, L., Soille, P.: Watersheds in digital spaces: An efficient algorithm based on immersion simulations. IEEE Trans. Pattern Anal. Machine Intell. 13(6), 583–598 (1991)
23. Vincent, L.: Morphological grayscale reconstruction in image analysis: Applications and efficient algorithms. IEEE Trans. Image Processing 2(8), 176–201 (1993)
24. Haris, K., Efstratiadis, S., Maglaveras, N., Pappas, C., Gourassas, J., Louridas, G.: Model-Based Morphological Segmentation and Labeling of Coronary Angiograms. IEEE Trans. Med. Imag. 18(10), 1003–1015 (1999)

Surface Reconstruction from LiDAR Data with Extended Snake Theory

Yi-Hsing Tseng, Kai-Pei Tang, and Fu-Chen Chou

Department of Geomatics, National Cheng Kung University
No.1 University Road, Tainan, Taiwan
tseng@mail.ncku.edu.tw

Abstract. Surface reconstruction from implicit data of sub-randomly distributed 3D points is the key work of extracting explicit information from LiDAR data. This paper proposes an approach of extended snake theory to surface reconstruction from LiDAR data. The proposed algorithm approximates a surface with connected planar patches. Growing from an initial seed point, a surface is reconstructed by attaching new adjacent planar patches based on the concept of minimizing the deformable energy. A least-squares solution is sought to keep a local balance of the internal and external forces, which are inertial forces maintaining the flatness of a surface and pulls of observed LiDAR points bending the growing surface toward observations. Experiments with some test data acquired with a ground-based LiDAR demonstrate the feasibility of the proposed algorithm. The effects of parameter settings on the delivered results are also investigated.

Keywords: Snake theory, surface reconstruction, LiDAR, laser scanning.

1 Introduction

Light detection and ranging (LiDAR) systems are active sensors capable of collecting accurate 3D coordinates of scanned points densely and sub-randomly distributed on scanned object surfaces [1, 2]. The observed data are commonly called point clouds. A set of point cloud data implies abundant implicit spatial information which can be turned into explicit through some methods of data processing or segmentation. Developing a computerized algorithm to explore valuable spatial information from LiDAR data becomes an active research topic, for example extracting object features or surfaces from ground-based LiDAR and extracting digital elevation model, buildings, and trees from airborne LiDAR data.

It has long been recognized that extracting features from implicit data is the first step of deriving explicit information from data. Due to the property of LiDAR scanning, 3D surfaces are prominent features in point clouds. Surface reconstruction is, therefore, the essential processing of LiDAR data. The first goal of surface reconstruction should be reached is segmentation of point clouds into clusters that are distributed on a continuous surface. A point cluster then is able to form a TIN (Triangulated Irregular

A.L. Yuille et al. (Eds.): EMMCVPR 2007, LNCS 4679, pp. 479–492, 2007.
© Springer-Verlag Berlin Heidelberg 2007

Network) structure to represent an extracted surface. This result keeps the fidelity of the original observations. However, unwanted surface roughness caused by observation noises will be kept as well. The second goal of surface reconstruction is mathematic modelling of the reconstructed surface. In contrast to keeping fidelity, this result is a smoothed version of the observed surface.

This paper develops an algorithm for surface reconstruction from LiDAR data based on the proposed theory of extended snake, which is an extension of the snake theory commonly applied for extracting curves from digital images [3, 4]. This algorithm approximates a surface with connected planar patches. Growing from an initial seed patch, a surface is reconstructed by attaching new adjacent planar patches based on the concept of minimizing the deformable energy. A program of least-squares computation is implemented to keep a local balance of the internal and external forces, which are inertial forces maintaining the flatness of a surface and pulls of observed LiDAR points bending the growing surface toward observations.

2   Surface Reconstruction from LiDAR Data

Featured of point distribution on scanned object surfaces, LiDAR data provide 3D information of sensed objects. For many applications, it is required to construct a digital model for an explicit representation of the acquired objects. A modeling procedure is frequently composed of a serious of data processing. For example, to extract building models from LiDAR data involves data structuralization, segmentation, feature extraction, surface reconstruction and modeling etc. The principle of each data processing step is to organize the sub-randomly distributed points into higher level geometric primitives. Furthermore, a complete surface model could be constructed by extracted primitives.

The task of surface reconstruction refers to forming a suitable surface model representing a group of points which are sub-randomly distributed on a smooth surface. Two major works are involved in surface reconstruction: segmentation/organization of point clouds and representation of surface model. A segmentation algorithm of point clouds is usually based on a TIN or regular grid structure to organize huge number of points in point clouds, and furthermore to group points into clusters based on properties of point distribution. Different segmentation approaches have been proposed, including region growing, clustering analysis, tenser voting, split-and-merge based on an octree structure, etc [5-9]. To obtain an explicit representation of surface model, segmented point data sets are used to form geometric models of surfaces, such as a TIN structure of a surface model or a mathematic description of connected planar or curve patches [6, 7, 10-12]. Similar research works have been done in the field of computer vision. Adaptive or deformable models have raised much interests and show potential for shape reconstruction, especially for the reconstruction of 3D objects [4, 13, 14].

Following the idea of snake theory, i.e. the concept of minimizing the deformable energy, this paper develops an extended snake theory for surface reconstruction. In the proposed algorithm, a locally continuous surface is approximated as connected planar patches. Growing from an initial seed patch, a smooth surface is reconstructed by attaching new adjacent planar patches, which are determined by the theory of extended snake theory. During the surface reconstruction, the algorithm continuously keeps a local balance of the internal and external forces. The internal forces are inertial forces maintaining the flatness of a surface. The external forces are pulls of observed LiDAR points bending the growing surface toward observations. Figure 1 depicts the concept of the surface reconstruction.

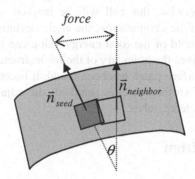

Fig. 1. The concept of surface reconstruction with extended snake theory

For the implementation, observed LiDAR point data are partitioned into point sets in the cells of a 3D gird structure, as shown in Figure 2. A surface patch is formed within a cell containing LiDAR points, so that the grid size determines the patch size. In practice, the grid size should be pre-defined based on the distribution density of points. The rule of thumb is setting the grid size to about 5 times of average point interval on a scanned object surface.

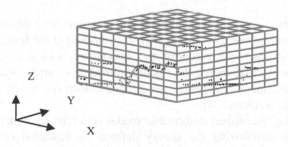

Fig. 2. A 3D gird structure

The seed patch is initiated by specifying a seed point. The grid cell contains the seed point becomes the seed cell and can be located easily by checking the coordinates of the seed point. The seed patch is then determined by the points contained in the seed cell and its neighbouring cells. It means that each surface patch is determined based on the LiDAR points contained in 27 (3*3*3) neighbouring cells. The seed patch is calculated by the least-squares fit of an unknown plane to the points in the 27 cells. The calculation actually is the same as the least-squares computation described in the next section, but there are no internal forces involved in the calculation of the seed patch.

From the seed patch, the algorithm searches possible extensions of connected patches in the neighbouring cells. If a neighbouring cell contains more than 3 LiDAR points, a new patch is determined as a candidate of connected patches through the least-squares solution described in the next section. The new patch is labelled as a connected patch, if its normal vector does not diverge too much from the normal vector of the seed patch. Otherwise, this cell will be marked as a boundary cell of the reconstructed surface. In the computation, the angle deviation is interpreted as a total energy value. The threshold of the total energy value can be a user-defined number, which indirectly determines the boundary of the reconstructed surface.

Once an extending surface patch has been found, it becomes the new seed patch to find new extensions of surface patches, until all the adjacent cells are labelled as boundary cells or connected patches.

3   Proposed Algorithm

3.1   Snake Theory

A Snake or active contour model is an energy-minimizing spline guided by external constrained forces and image forces to represent 2D image features such as lines or edges [3]. More generally, it is an example of matching a deformable model to an image by means of minimizing energy. If the deformable model on an image is represented by a vector function $v(s)$, than the deformable energy is defined as:

$$E_{snake}(v) = E_{int}(v) + E_{ext}(v) = \int_a^b (\alpha(s)|v_s(s)|^2 + \beta(s)|v_{ss}(s)|^2) ds + \int_a^b \gamma(s) P(v(s)) ds \qquad (1)$$

where $\alpha(s)$, $\beta(s)$, and $\gamma(s)$ are the weight functions with respect to the vector functions. The internal energy of spline E_{int} is composed of the first and second-order community constrains, which control the smoothness and bending of the snake respectively. The external energy E_{ext} is determined by external forces $P(v)$ derived from the extracted image features or user-defined constrains, which makes the snake trend to the image features[3, 4].

In practice, the established deformable model is solved by matching model with observations. To minimizing the energy defined by Equation (1), the necessary condition is given by the Euler-Lagrange equations. Representing Equation (1) by $F(s, v, v_s, v_{ss})$, the condition can be written as:

$$F_v - \frac{\partial}{\partial s} F_{v_s} + \frac{\partial^2}{\partial s^2} F_{v_{ss}} = 0 \qquad (2)$$

Therefore, the solution of Snakes is derived by solving the partial differential equations of Eq. (2).

3.2 Extended Snake Theory

Snake theory was originally developed for extracting linear features from a 2D image. It soon has been extended and applied for 3D surface reconstructions [4, 14, 15]. Specialized in handling LiDAR data, this paper proposes a scheme of extended Snake theory for 3D surface reconstruction from LiDAR data.

Based on the concept of minimizing the deformable energy, a continuous surface is approximated as connected planar patches. A planar patch (i) is formulated as follows:

$$S_i : A_i x + B_i y + C_i z + D_i = 0 \tag{3}$$

In practice, a 3D surface to be reconstructed is composed of limited number of planar patches, and can be expressed as:

$$S_{obj} : \{S_1, S_2, S_3, ..., S_m\} \tag{4}$$

In order to maintain the smoothness of a reconstructed surface, a constraint on the growing of surface patches is set by checking the differential of local surface. For the implementation, the angle between the normal vectors of two adjacent planar patches is minimized. Let S be the planar patch to be reconstructed and S_i be one of its adjacent planar patches have been marked as connected patches. Corresponding to S and S_i, let $\vec{n}$ and $\vec{n}_i$ denote the normal vectors of these two adjacent planar patches and θ_i denote the angle between the vectors. Therefore, the cosine of θ_i should be close to 1 and is formulated as:

$$\cos \theta_i = \frac{\vec{n} \cdot \vec{n}_i}{|\vec{n}||\vec{n}_i|} \Rightarrow 1 \tag{5}$$

When $\vec{n}$ and $\vec{n}_i$ are always scaled to be unit vectors, the internal force can be expressed as the modulus of $(\cos\theta_{ij} - 1)$, and can be formulated as:

$$f_i = |\vec{n} \cdot (\vec{n} + \Delta\vec{n}_i) - 1| = |A(A + \Delta A_i) + B(B + \Delta B_i) + C(C + \Delta C_i) - 1| \tag{6}$$

where $\vec{n} = (A, B, C)$ and $\Delta\vec{n}_i = (\Delta A_i, \Delta B_i, \Delta C_i)$. This internal force will be introduced as a constraint in surface reconstruction.

The internal force maintains the flatness during the reconstruction of a surface. However, the surface should also bend toward observed LiDAR points as if there were pulls of the points. The pulls of LiDAR points represent the external force, and can be modelled as functions of distances from the points to the corresponding planar patch. The distance from a point (x_i, y_i, z_i) to a planar patch can be expressed as:

$$d_i = \frac{|Ax_i + By_i + Cz_i + D|}{\sqrt{A^2 + B^2 + C^2}} \tag{7}$$

Because Equation (6) is established under the condition of that the normal vector is a unit vector, an additional constraint function should be introduced to the scheme. This function can be expressed as:

$$g = A^2 + B^2 + C^2 - 1 = 0 \tag{8}$$

Because d_i and g are nonlinear functions, the Newton's method is applied for the iterative least-squares scheme to achieve the minimum energy solution. Therefore, Equations (6), (7) and (8) are rewritten as series of Taylor expansion. They are formulated as follows:

$$v_{f_i} = (f_i)_0 + \left(\frac{\partial f_i}{\partial A}\right)_0 \Delta A + \left(\frac{\partial f_i}{\partial B}\right)_0 \Delta B + \left(\frac{\partial f_i}{\partial C}\right)_0 \Delta C \tag{9}$$

$$v_{d_i} = (d_i)_0 + \left(\frac{\partial d_i}{\partial A}\right)_0 \Delta A + \left(\frac{\partial d_i}{\partial B}\right)_0 \Delta B + \left(\frac{\partial d_i}{\partial C}\right)_0 \Delta C + \left(\frac{\partial d_i}{\partial D}\right)_0 \Delta D \tag{10}$$

$$v_g = g_0 + \left(\frac{\partial g}{\partial A}\right)_0 dA + \left(\frac{\partial g}{\partial B}\right)_0 dB + \left(\frac{\partial g}{\partial C}\right)_0 dC + \left(\frac{\partial g}{\partial D}\right)_0 dD \tag{11}$$

In which, $(..)_0$ represents a function value with given approximations of unknown parameters. Equation (9), (10) and (11) can be combined and expressed as a matrix form:

$$
\begin{bmatrix} v_{f1} \\ \vdots \\ v_{fm} \\ v_{d1} \\ \vdots \\ v_{dn} \\ v_g \end{bmatrix}
=
\begin{bmatrix}
\left(\frac{\partial f_1}{\partial A}\right)_0 & \left(\frac{\partial f_1}{\partial B}\right)_0 & \left(\frac{\partial f_1}{\partial C}\right)_0 & 0 \\
\vdots & \vdots & \vdots & \vdots \\
\left(\frac{\partial f_m}{\partial A}\right)_0 & \left(\frac{\partial f_m}{\partial B}\right)_0 & \left(\frac{\partial f_m}{\partial C}\right)_0 & 0 \\
\left(\frac{\partial d_1}{\partial A}\right)_0 & \left(\frac{\partial d_1}{\partial B}\right)_0 & \left(\frac{\partial d_1}{\partial C}\right)_0 & \left(\frac{\partial d_1}{\partial D}\right)_0 \\
\vdots & \vdots & \vdots & \vdots \\
\left(\frac{\partial d_n}{\partial A}\right)_0 & \left(\frac{\partial d_n}{\partial B}\right)_0 & \left(\frac{\partial d_n}{\partial C}\right)_0 & \left(\frac{\partial d_n}{\partial D}\right)_0 \\
\left(\frac{\partial g}{\partial A}\right)_0 & \left(\frac{\partial g}{\partial B}\right)_0 & \left(\frac{\partial g}{\partial C}\right)_0 & \left(\frac{\partial g}{\partial D}\right)_0
\end{bmatrix}
\begin{bmatrix} dA \\ dB \\ dC \\ dD \end{bmatrix}
+
\begin{bmatrix} (f_1)_0 \\ \vdots \\ (f_m)_0 \\ (d_1)_0 \\ \vdots \\ (d_n)_0 \\ g_0 \end{bmatrix}
\tag{12}
$$

In which, m represents the number of adjacent planar patches and n represents the number of corresponding LiDAR points. This matrix form can be symbolized as $V = GX - L$. To adjust the influences of internal and external forces, one can set some weight functions according to the data quality and desired smoothness of surface. The weight functions can be modelled as weight matrices associated with the internal and external forces. The optimal solution can be achieved by minimizing energy using the least-squares condition as:

$$V^T PV = (V_f{}^T P_f V_f + V_g{}^T P_g V_g) + V_d{}^T P_d V_d \approx w_{\text{int}} E_{\text{int}} + w_{\text{ext}} E_{\text{ext}} \Rightarrow \min \quad (13)$$

where P_f, P_g and P_d are the related weight matrices. The weight matrices can be simplified to weight numbers associated with the internal and external energy as w_{int} and w_{ext}, if all of internal forces and external forces are even. By the theory of least-squares adjustment, the increments of unknown parameters of a growing planar patch can be solved as follows:

$$X = (G^T PG)^{-1} G^T PL \quad (14)$$

Adding the solved increments to the approximations to obtain new approximations for the next iterative computation, the unknown parameters of a growing planar patch are solved iteratively. In the final run of the iteration, the calculated value of $V^T PV$ is representative of the total energy.

4   Experiments

Three data sets, acquired with a ground-based LiDAR of Optech ILRIS 3D, were processed for surface reconstruction by the proposed algorithm. The targets scanned in the data sets are a bell monument, a cylindrical building, and a sculpture. Figures 3(a), 4(a) and 5(a) show the pictures of the targets respectively. Through the process of the

(a)   (b)

Fig. 3. (a) The first experimental target: a bell monument; (b) The segmented point clusters

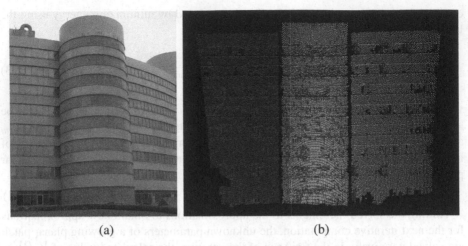

(a)   (b)

Fig. 4. (a) The second experimental target: a cylindrical building; (b) The segmented point clusters

(a)   (b)

Fig. 5. (a) The third experimental target: a sculpture; (b) The segmented point clusters

developed program of the proposed algorithm, the point clouds were segmented into clusters representing different surfaces. Each cluster of segmented points is then able to form a surface using a TIN structure. Surfaces modelled by connected planar patches can be constructed as well. Satisfied results could be delivered by given proper seed points and parameter settings based on the data conditions. Figures 3(b), 4(b) and 5(b) show segmented point clusters in colour superimposed on the point clouds in gray. One can see that an object composed of several planes or curves can be segmented into parts of smooth surfaces quite well. Figure 6 displays a reconstructed surface model of the bell monument, which is shown as a TIN structure formed by the centre points of planar patches. Similarly, Figure 7 demonstrates a reconstructed curve surface of the cylindrical building.

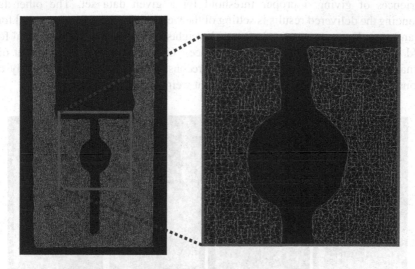

Fig. 6. A reconstructed surface model of the bell monument

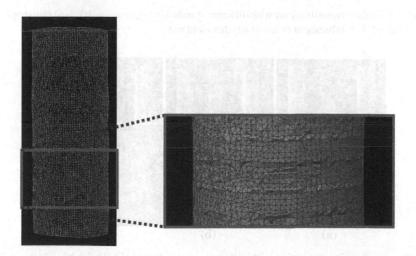

Fig. 7. A reconstructed curve surface of the cylindrical building

Selection of an initial seed point is critical in the interactive process of surface reconstruction. Given an improper seed point may lead to an incomplete reconstruction of a surface. Learned from practice, it is suggested that the seed point should be selected from a group of densely and evenly distributed points on a smooth surface portion. The completeness of a reconstructed surface will also be affected by the given threshold of total energy value. On the one hand a curve surface may not grow extending to its border if the threshold is too small, and on the other two discrete surfaces may be combined if the threshold is too large. An operator needs to learn some experiences of giving a proper threshold for a given data set. The other factor influencing the delivered results is setting of the weights of internal and external forces, w_{int} and w_{ext}. For most cases, given equal weights for the internal and external forces would be satisfied with regular curve surfaces. One may increase the weight of the external force, if a rough surface is to be reconstructed. The following study cases demonstrate the influences of giving different weights and thresholds.

Fig. 8. Planar surface reconstruction with different thresholds ($w_{int} : w_{ext} = 1{:}1$): (a) threshold =1, (b) threshold = 1.5, (c) threshold = 2, and (d) threshold = 3

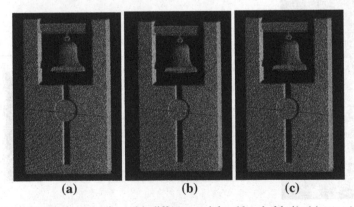

Fig. 9. Planar surface reconstruction with different weights (threshold=1): (a) $w_{int} : w_{ext} = 2{:}8$, (b) $w_{int} : w_{ext} = 1{:}1$, and (c) $w_{int} : w_{ext} = 8{:}2$

The first study case is to reconstruct a planar surface of the monument. Under the condition of setting $w_{int} : w_{ext} = 1{:}1$, the segmentation results with respect to using energy thresholds of 1, 1.5, 2 and 3 are shown in Figure 8, in which segmented points are shown as green dots. All test cases seem to deliver a correct result except the centre portion of a slightly indented circle plate. It is hard to judge which one is correct or wrong. One could tune the threshold to fit different requirements of various applications. By setting energy threshold to 1, the segmentation results with respect to using 3 different weight ratios are shown in Figure 9. Similarly, when the weight of external force is much larger than the weight of internal force, the circle plate part will be included. The delivered results make sense to setting those parameters. It would not be difficult for a user to choose a set of proper parameters.

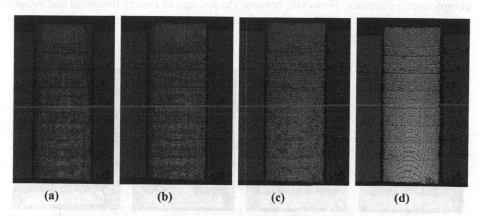

Fig. 10. Cylindrical surface reconstruction with different thresholds (w_{int} : w_{ext} = 1:1): (a) threshold =1, (b) threshold = 1.5, (c) threshold = 2, and (d) threshold = 3

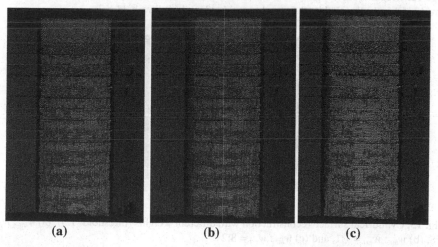

Fig. 11. Cylindrical surface reconstruction with different weights (threshold=1.5): (a) w_{int} : w_{ext} = 2:8, (b) w_{int} : w_{ext} = 1:1, and (c) w_{int} : w_{ext} = 8:2

The second study case is to reconstruct a cylindrical surface of a cylindrical building. Under the condition of setting w_{int} : w_{ext} = 1:1, the segmentation results with respect to using energy thresholds of 1, 1.5, 2 and 3 are shown in Figure 10, in which segmented points are shown as green dots. The reconstructed surface tends to more complete when we increased the given value of the energy threshold. It suggests a use of large energy threshold for reconstructing a curve surface. By setting energy threshold to 1.5, the segmentation results with respect to using 3 different weight ratios are shown in Figure 11. By observing the results, one can discover that changing the weight ratio would also alter the completeness of reconstructed surface. It means that using a small weight ratio not only tends to reconstruct a rough surface but also helps for obtaining a complete curve surface. However, because the settings of energy threshold and weight ratio are more or less application dependent, some operation experiences are required for a user to obtain a proper result.

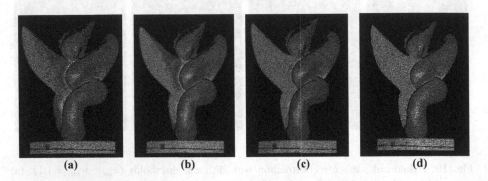

Fig. 12. Irregular curve surface reconstruction with different thresholds ($w_{int} : w_{ext} = 1{:}1$): (a) threshold =1, (b) threshold = 1.5, (c) threshold = 2, and (d) threshold = 3

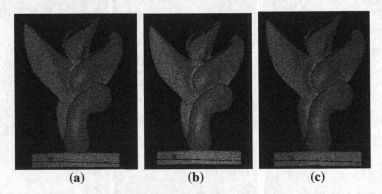

Fig. 13. Cylindrical surface reconstruction with different weights (threshold=1.5): (a) $w_{int} : w_{ext} =$ 2:8, (b) $w_{int} : w_{ext} = 1{:}1$, and (c) $w_{int} : w_{ext} = 8{:}2$

The third study case is to reconstruct an irregular curve surface of the scanned sculpture. Using the same parameter settings as the previous case, the delivered results are shown in Figure 12 and 13. In Figure 12, one can see that the reconstructed surface extends across weak boundaries when a larger value of energy threshold is used. It means that the threshold value indirectly determines boundaries of a surface can extend to. All of the results listed in Figure 12 and 13 make sense under a definition of surface boundary. Again, a proper surface boundary should also be application dependent.

5 Conclusions

A feasible algorithm of surface reconstruction from LiDAR data is proposed and tested. By giving a proper seed point and a threshold of total energy value, scanned surfaces in the scene can be extracted properly. The algorithm can deliver the results of point segmentation as well as mathematic models of extracted surfaces by the definition of surface reconstruction. However, this preliminary version of the algorithm is still sensitive to the given seed point, energy threshold and weight ratio of internal to external forces. Further research work will focus on stabilizing the computation with any given seed point and parameter settings.

The algorithm can also be tuned by improving the surface model. The current surface model is an integration of connected planar patches. Instead of modelling with planar patches, curves of spline patches would form finer surfaces than the use of planar patches.

Acknowledgments

This research work is sponsored under the project grants: NSC 94-2211-E-006-020 and NSC 95-2211-E-006-463. The authors appreciate for the support of the National Science Council, Taiwan.

References

[1] Ackermann, F.: Airborne laser scanning—present status and future expectations. ISPRS Journal of Photogrammetry & Remote Sensing 54, 64–67 (1999)

[2] Wehr, A., Lohr, U.: Airborne laser scanning—an introduction and overview. ISPRS Journal of Photogrammetry & Remote Sensing 54, 68–82 (1999)

[3] Kass, M., Witkin, A., Terzopoupos, D.: Snakes: active contour models. International Journal of Computer Vision 1, 321–331 (1988)

[4] Terzopoulous, G., Metaxas, D.: Dynamic 3D models with local and global deformations: deformable Superquadrics. IEEE Transactions on Pattern Analysis and Machine Intelligence 13, 703–714 (1991)

[5] Filin, S., Pfeiper, N.: Segmentation of airborne laser scanning data using a slope adaptive neighborhood. ISPRS Journal of Photogrammetry and Remote Sensing 60, 71–80 (2006)

[6] Hansen, W.V., Michaelsen, E., Thonnessen, U.: Cluster analysis and priority sorting in huge points clouds for building reconstruction. In: Proceedings of the 18th International Conference on Pattern Recognition, pp. 23–26 (2006)

[7] Schuster, H.-F.: Segmentation of LiDAR data using tersor voting framework. In: International Archives of Photogrammetry and Remote Sensing, Istanbul (2004)

[8] Wang, M., Tseng, Y.H.: LiDAR data segmentation and classification based on octree structure (2004)

[9] Lucieer, A., Stein, A.: Texture-based landform segmentation of LiDAR imagery. International Journal of Applied Earth Observation and Geoinformation 6, 261–270 (2005)

[10] Gruen, A., Wang, X.: CC-Modeler: a topology generator for 3-D city models. ISPRS Journal of Photogrammetry & Remote Sensing 53, 286–295 (1998)

[11] Tognola, G., Parazzini, M., Ravazzani, P., Grandori, F., Svelto, C.: Adaptive algorithm for accurate reconstruction of solid surface from unorganized data. In: IEEE International Workshop on Imaging Systems and Techniques, pp. 44–47 (2004)

[12] Filin, S.: Surface Clustering from Airborne Laser Scanning Data. International Archives of Photogrammetry and Remote Sensing, Part 3A/B 34, 119–124 (2002)

[13] Bulpitt, A.J., Efford, N.D.: An efficient 3D deformable model with a self-optimising mesh. Image and Vision Computing 14, 573–580 (1996)

[14] Montagnat, J., Delingette, H., Ayache, N.: A review of deformable surfaces: topology, geometry and deformation. Image and Vision Computing 19, 1023–1040 (2001)

[15] Brovelli, M.A., Cannata, M.: Digital terrain model reconstruction in urban areas from airborne laser scanning data: the method and an example for Pavia (northern Italy). Computers & Geosciences 30, 325–331 (2004)

Author Index

Lecture Notes in Computer Science

For information about Vols. 1–4557

please contact your bookseller or Springer